Advances in Emerging Financial Technology and Digital Money

The financial sector is witnessing rapid technological innovations, leading to the emergence of Fintech (financial technologies), revolutionizing national and international financial landscapes. Fintech is expanding and enhancing financial products and services, making them more accessible and affordable while transforming customer relationships, payment methods, financing, and transfers.

Advances in Emerging Financial Technology and Digital Money provides a platform for collective reflection, bringing together institutions, policymakers, digital and financial service providers, professionals, and academics from various disciplines. The aim is to clarify the challenges, opportunities, and socio-economic impacts of innovations in finance and technology on citizens and businesses in Morocco, Africa, and worldwide. This comprehensive collection offers valuable insights into the current state and prospects of financial technology and digital money.

This book covers all the essential topics, including:

- AI and Machine Learning in Fintech and Beyond;
- Financial Inclusion, Literacy, and Behavior;
- Fintech Ecosystems, Collaboration, and Analysis;
- Blockchain, Security, and Sustainability;
- Fintech Innovations and Applications.

In this new book, the authors share their experiences to provide a comprehensive and well-researched overview of the technologies and concepts that will transform the banking industry as we know it. It aspires to be a useful reference for executive managers, CIOs, Fintech professionals, and researchers interested in exploring and implementing an efficient Fintech strategy. The book also presents selected papers from International Fintech Congress (IFC 2022).

Advances in Emerging Financial Technology and Digital Money

Edited by
Yassine Maleh, Justin Zhang,
and Abderrahim Hansali

CRC Press
Taylor & Francis Group
Boca Raton London New York

CRC Press is an imprint of the
Taylor & Francis Group, an **informa** business

Designed cover image: ©ShutterStock Images

First edition published 2024
by CRC Press
2385 NW Executive Center Drive, Suite 320, Boca Raton FL 33431

and by CRC Press
4 Park Square, Milton Park, Abingdon, Oxon, OX14 4RN

CRC Press is an imprint of Taylor & Francis Group, LLC

© 2024 selection and editorial matter, Yassine Maleh, Justin Zhang, and Abderrahim Hansali; individual chapters, the contributors

Library of Congress Cataloging-in-Publication Data
Names: Maleh, Yassine, 1987- editor. | Zhang, Justin, editor. | Hansali,
Abderrahim, editor.
Title: Advances in emerging financial technology & digital money / edited
by Yassine Maleh, Justin Zhang, and Abderrahim Hansali.
Other titles: Advances in emerging financial technology and digital money
Description: First edition. | Boca Raton, FL : CRC Press, 2024. |
Includes index.
Identifiers: LCCN 2023044202 (print) | LCCN 2023044203 (ebook) |
ISBN 9781032667447 (hbk) | ISBN 9781032667461 (pbk) | ISBN 9781032667478 (ebk)
Subjects: LCSH: Digital currency. | Finance—Technological innovations.
Classification: LCC HG1710 .A39 2024 (print) | LCC HG1710 (ebook) |
DDC 332.4—dc23/eng/20231226
LC record available at https://lccn.loc.gov/2023044202
LC ebook record available at https://lccn.loc.gov/2023044203

ISBN: 9781032667447 (hbk)
ISBN: 9781032667461 (pbk)
ISBN: 9781032667478 (ebk)

DOI: 10.1201/9781032667478

Typeset in Times
by codeMantra

Contents

Preface ix
Editors x
Contributors xi

Chapter 1   Emerging Fintech and Digital Money: Current Trends and Future Perspectives 1

Abdelkebir Sahid and Yassine Maleh

SECTION 1   AI and Machine Learning in Fintech and Beyond

Chapter 2   E-Payment Fraud Detection in E-Commerce using Supervised Learning Algorithms 27

Manal Loukili, Fayçal Messaoudi, and Hanane Azirar

Chapter 3   Artificial Intelligence and Stock Market 36

Karim Amzile

Chapter 4   Machine Learning Clustering: Application on Aggregated and Non-Aggregated Financial Data 54

El Aidouni Salma, Benhouad Mohamed, Mestari Mohammed, and El Mansouri Adnane

Chapter 5   Candlestick Patterns Recognition in Bitcoin Price Action Graph Chart Using Deep Learning 66

Abdellah El Zaar, Nabil Benaya, Abderrahim El Allati, and Toufik Bakir

Chapter 6   Prediction of Stock Markets Using Deep Learning Architectures 77

Khalid Bentaleb, Benhouad Mohamed, and Mohammed Mestari

Chapter 7   Improving Credit Card Fraud Detection with Distributed Machine Learning and Bfloat-16 84

Bushra Yousuf, Rejwan Bin Sulaiman, and Musarrat Saberin Nipun

Chapter 8   A Proposed Semi-decentralized Approach of Federated Learning for Enhancing CCFD Using CS-SVM and K-Means Algorithms 105

Rejwan Bin Sulaiman

Chapter 9   Medical-DCGAN: Deep Convolutional GAN for Medical imaging 123

Rguibi Zakaria, Hajami abedlamjid, Dya Zitouni, and Amine ElQaraoui

SECTION 2 *Financial Inclusion, Literacy, and Behavior*

Chapter 10 Financial Inclusion and Economic Growth in the Presence of a
Cross-Sectional Dependence 137

Huynh Thi Thanh Truc

Chapter 11 The Role of Tax Policy in Stimulating and Encouraging Investment in Jordan 143

Yaser Arabyat and Hussam-eldin Daoud

Chapter 12 Financial Development and Economic Growth: An Experimental Evidence
Based on Quantile Regression 156

Van Chien Nguyen

Chapter 13 Importance of Financial Literacy Education and Financial Behavior for
Developing Nations in the World 163

Venkata Vara Prasad Janjanam and SubbaLakshmi A.V.V.S.

Chapter 14 A Literature Review on Financial Stability 181

*El Mansouri Adnane, Benhouad Mohamed, Mestari Mohammed,
and El Aidouni Salma*

Chapter 15 The Behavior of Firms Regarding the Introduction of Financing by Profit- and
Loss-Sharing Products in Morocco 189

Safa Ougoujil and Sidi Mohamed Rigar

Chapter 16 Influence of Financial Innovation on Business Performance:
Evidence from the SME Food Industry in Malaysia 203

Shreen Almas Mohamed Buhary and Hussen Nasir

SECTION 3 *Fintech Ecosystems, Collaboration, and Analysis*

Chapter 17 The Role of Collaborative Skills and Knowledge Sharing in the Emergence of
Fintech Ecosystems: Case of Casablanca Finance City 213

Smyej Oumaima and Si Mohammed Ben Massou

Chapter 18 Toward Characterization of the Fintech Ecosystem: A Systematic
Literature Review 224

Ali Mwase, Ernest Ketcha Ngassam, and Singh Shawren

Chapter 19 Analyzing the Impact of the Fear of COVID-19 on Stock Market Returns
Using Twitter Text Mining 238

Ayoub Razouk, Youness Madani, and Fatima Touhami

Chapter 20 Behavior Analysis of Lenders in P2P Lending Platforms: Identifying
Cognitive Effort by Response Time Method 250

Benhmama Asmaa, Sabiri Brahim, and Melliani Hamza

SECTION 4 *Blockchain, Security, and Sustainability*

Chapter 21 Conceiving a Blockchain-Based Upstream Supply Chain Management System
Enhancing Innovation and Sustainability 261

Ahmed El Maalmi, Kaoutar Jenoui, and Laila El Abbadi

Chapter 22 Green Finance in the Moroccan Mining Sector: Cases of Sustainable
Development and CSR 271

Insaf El Atillah and Mohamed Azeroual

Chapter 23 Cloud Data Integrity Auditing and Deduplication Using an Optimized Method
Based on Blockchain and MAS 283

*Mohamed El Ghazouani, Abdelouafi Ikidid, Charafeddine Ait Zaouiat,
Layla Aziz, Yassine El Khanboubi, Moulay Ahmed El Kiram,
and Latifa Er-Rajy*

Chapter 24 Securing Crowdfunding Platforms with Blockchain to Boost the
Real Estate Sector 293

Hibatou Allah Boulsane, Karim Afdel, and Salma El Hajjami

Chapter 25 Blockchain Technology: A Proposed Solution to Hike the Tax-to-GDP
Ratio in Bangladesh 302

Prianka Ghosh Puja and Md Sadik Adnan

Chapter 26 Approach for Strengthening the Network Security Based on Boosting
Algorithms: Performance Study..313

Sabrine Ennaji, Nabil El Akkad, and Khalid Haddouch

Chapter 27 Smart Cities Technologies in the Covid-19 Context: A Bibliometric Analysis 324

*Elafri Nedjwa, Boumali Badreldinne, Lalmi Andallah,
Sassi boudemagh Souad, Yassine Maleh, Rose Bertrand,
and Hemza Barkani*

SECTION 5 *Fintech Innovations and Applications*

Chapter 28 Fintech Innovations for Supply Chain Resilience 335

Asma Boujrouf, Sidi Mohamed Rigar, and L'houssaine Mounaim

Chapter 29 Strategic Tools for the Decision to Outsource Maintenance Activities in
Moroccan Airports 343

Ahlam Boutahar, Mohamed Ben Ali, and Said Rifai

Chapter 30 Sustainable Finance and FinTech: Facilitating a Sustainable Future
with the Utilization of Socio-Economic Financial Services 352

Zerina Bihorac, Azra Zaimovic, and Tarik Zaimovic

Chapter 31 Information Society Services in Morocco Regulatory Status
and Outlook for Development 361

Khalid Abouelouafa and Hafid Barka

Chapter 32 Cartography of Mobile Payment Technologies Used in Morocco 369

Youness Khourdifi and Abderrahim Hansali

Chapter 33 A Machine Learning-Based Recommendation System for Smart
Mobility Trip Planning in Morocco 379

El Attar Chaimae, Daoudi Najima, Abourezq Manar, El Ghali Btihal,
Hilal Imane, and Hnida Meriem

Index 389

Preface

In the rapidly evolving financial landscape, the intersection of technology and digital currencies has given rise to a new era of financial innovation. Financial technology (Fintech) has emerged over the last few decades as a disruptive force, transforming traditional financial services and reshaping how businesses and individuals interact with the financial system. From its early beginnings with the rise of the Internet and online banking to the recent developments in artificial intelligence, blockchain, and digital currencies, Fintech has profoundly impacted the global economy. *Advances in Emerging Financial Technology and Digital Money* is a comprehensive book that explores the rapidly evolving landscape of financial technology (Fintech) and digital currencies. It provides valuable insights into the latest research, trends, and innovations in the field, offering a glimpse into financial technology's current state and prospects.

This book is divided into five sections, each focusing on a specific aspect of the Fintech and digital money domains. It covers topics such as the applications of artificial intelligence and machine learning in the financial industry, the impact of Fintech on financial inclusion and literacy, the emergence of Fintech ecosystems and collaboration, the potential of blockchain for security and sustainability, and various Fintech innovations and applications.

Advances in Emerging Financial Technology and Digital Money is a valuable resource for readers interested in understanding the transformative power of Fintech and digital currencies. It aims to inspire further research and development in this rapidly evolving field while fostering a global community dedicated to tackling the challenges and leveraging the opportunities presented by the dynamic world of finance and technology.

This comprehensive volume comprises 33 chapters that delve into various aspects of financial technology, from artificial intelligence and machine learning applications to blockchain, security, and sustainability. It is divided into five sections, each focusing on a unique aspect of the rapidly evolving financial technology landscape. Section 1 delves into the applications of AI and Machine Learning in Fintech, exploring various algorithms and models that aim to improve efficiency and security in the financial industry. Section 2 concentrates on financial inclusion, literacy, and behavior, examining the role of mobile payments, financial education, and the impact of financial innovations on businesses. Section 3 investigates Fintech ecosystems, collaboration, and analysis, discussing the emergence of Fintech ecosystems, the importance of collaboration and knowledge sharing, and the effects of external factors on stock market returns. Section 4 focuses on blockchain, security, and sustainability, highlighting the potential of blockchain-based solutions for supply chain management, data integrity, and tax collection. Finally, Section 5 showcases Fintech innovations and applications, encompassing various use cases such as supply chain resilience, network intrusion detection, and mobile payment technologies. This comprehensive collection of chapters offers valuable insights into financial technology and digital money's current state and future prospects.

We extend our gratitude to all the contributors for their remarkable efforts and insights, making this book possible. We hope that readers, including scholars, researchers, policymakers, and practitioners, will find this compilation an essential resource for understanding and shaping the future of financial technology and digital money.

Yassine Maleh, Khouribga, Morocco
Justin Zhang, New York, USA
Abderrahim Hansali, Marrakech, Morocco

Editors

Yassine Maleh holds a Ph.D. in computer sciences from the University Hassan 1st, Morocco, since 2013. He holds a second Ph.D. in IT Management from the National School of Business and Management, Settat, Morocco, since 2022. He is Professor of Cybersecurity and IT Governance at the National School of Applied Sciences, Khouribga, Morocco, and was a former IT Service Manager at the National Port Agency, Morocco. He is a Senior Member of IEEE, Member of the International Association of Engineers (IAENG), and a member of the Machine Intelligence Research Labs. Dr. Maleh has contributed to information security and privacy, Internet of Things Security, and Wireless and Constrained Networks Security. His research interests include Information Security and Privacy, Internet of Things, Networks Security, Information System, and IT Governance. He has published over 140 papers (book chapters, international journals, and conferences/workshops), 30 edited books, and 5 authored books. He is the Editor-in-Chief of the *International Journal of Information Security and Privacy* (IJISP). He is an Associate Editor for IEEE Access and the *International Journal of Digital Crime and Forensics* (IJDCF).

Justin Zhang is a faculty member at the Coggin College of Business at the University of North Florida. He received his Ph.D. in Business Administration with a concentration in Management Science and Information Systems from Pennsylvania State University, University Park. His research interests include economics of information systems, knowledge management, electronic business, business process management, information security, and social networking. He is the Editor-in-Chief of the *Journal of Global Information Management*, an ABET program evaluator, and an IEEE senior member.

Abderrahim Hansali, has a PhD in Economics and Management, is a graduate from the Higher Institute of Administration in Rabat and ENA in Paris. He is currently a Research Teacher at ENCG in Marrakech and a consultant in digital strategies and change management. His career began as a founding teacher of BTS preparatory classes in Morocco. After working for the Agence Nationale de Réglementation des Télécommunications in Rabat on the implementation of a quality management system (2007) and as an auditor for the Inspectorate General of the Ministry of National Education (2008), he was appointed as Deputy Director and National Coordinator of the GENIE program within the same ministry (2009), where he helped set up the Laboratoire National des Ressources Numériques. Since 2016, he has taken up teaching and research duties in management sciences at Cadi Ayyad University in Marrakech. In 2014, he set up the FONDATION MAROC NUMERIQUE, a non-governmental organization dedicated to supporting digital strategies and promoting a citizen digital culture. In 2019, he helped set up ACISS (African Center for Innovative Sustainable Solutions) at Cadi Ayyad University in Marrakech.

Contributors

Laila El Abbadi
Engineering Sciences Laboratory, National
    School of Applied Sciences
Ibn Tofail University
Kenitra, Morocco

Hajami Abedlamjid
Laboratory Watch Laboratory for Emerging
    Technologies
Hassan First University of Settat
Settat, Morocco

Khalid Abouelouafa
EM2TI, National Institute of Posts and
    Telecommunications
Rabat, Morocco

Md Sadik Adnan
BBA, Army Institute of Business
    Administration
Bangladesh University of Professionals
Savar, Bangladesh

El Mansouri Adnane
Hassan II University, ENSET
Mohammedia, Morocco

Karim Afdel
Faculty of Sciences
Ibn Zohr University
Agadir, Morocco

Nabil El Akkad
National School of Applied Sciences
University of Sidi Mohamed Ben Abdellah
Fez, Morocco

Abderrahim El Allati
Laboratory of R&D in Engineering Sciences,
    FST Al Hoceima
Abdelmalek Essaadi University
Tetouan, Morocco

Karim Amzile
Faculty of Law, Economics and Social Sciences
    Agdal
Mohammed V University
Rabat, Morocco

Lalmi Andallah
University of Constantine 3
Constantine, Algeria

Yaser Arabyat
Department of Economics
Al-Balqa' Applied University
Salt, Jordan

Benhmama Asmaa
Hassan II University
Casablanca, Morocco

Insaf El Atillah
LEG Laboratory, Polydisciplinary Faculty
Sultan Moulay Slimane University
Beni Mellal, Morocco

Mohamed Azeroual
LEG Laboratory, Polydisciplinary Faculty
Sultan Moulay Slimane University
Beni Mellal, Morocco

Hanane Azirar
Faculty of Juridical, Economic and Social
    Sciences
Sidi Mohamed Ben Abdellah University
Fez, Morocco

Layla Aziz
Computer Science Department, Laboratory
    LISI
Cadi Ayyad University
Marrakesh, Morocco

Boumali Badreldinne
University of Constantine 3
Constantine, Algeria

Toufik Bakir
ImViA Laboratory
Bourgogne University
Dijon, France

Hafid Barka
EM2TI
National Institute of Posts and
    Telecommunications
Rabat, Morocco

Hemza Barkani
University of Oum elbouaki
Oum elbouaki, Algeria

Mohamed Ben Ali
LMPGI Laboratory, Higher School of
    Technology

Nabil Benaya
Laboratory of R&D in Engineering Sciences,
    FST Al Hoceima
Abdelmalek Essaadi University
Tetouan, Morocco

Khalid Bentaleb
Laboratory of SSDIA, ENSET
Bd Hassan II
Mohammedia, Morocco

Rose Bertrand
Strasbourg University
Strasbourg, France

Zerina Bihorac
School of Economics and Business
University of Sarajevo
Sarajevo, Bosnia and Herzegovina

Asma Boujrouf
Cadi Ayyad University
Marrakech, Morocco

Hibatou Allah Boulsane
Faculty of Sciences
Ibn Zohr University
Agadir, Morocco

Ahlam Boutahar
LMPGI Laboratory, Higher School of
    Technology (EST), Superior National School
    of Electricity and Mechanics (ENSEM)
Hassan II University of Casablanca
Casablanca, Morocco

Sabiri Brahim
Department of Economics and Business
    Management
Hassan II University
Casablanca, Morocco

El Ghali Btihal
ITQAN Team, LyRica Lab
Information Sciences School
Rabat, Morocco

Shreen Almas Mohamed Buhary
Faculty of Business and Communication
Malaysia Universiti Malaysia Perlis
Arau, Malaysia

El Attar Chaimae
ITQAN Team, LyRica Lab
Information Sciences School
Rabat, Morocco

Hussam-eldin Daoud
Department of Economics
Mutah University
AL-Karak, Jordan

Mohamed El Ghazouani
Polydisciplinary Faculty of Sidi Bennour
Chouaîb Doukkali University
El Jadida, Morocco

Salma El Hajjami
Faculty of Sciences
Ibn Zohr University
Agadir, Morocco

Yassine El Khanboubi
Faculty of Science Ben M'Sik
Hassan II University
Casablanca, Morocco

Moulay Ahmed El Kiram
Computer Science Department, Laboratory
LISI
Cadi Ayyad University
Marrakesh, Morocco

Ahmed El Maalmi
Engineering Sciences Laboratory, National
School of Applied Sciences
Ibn Tofail University
Kenitra, Morocco

Abdellah El Zaar
Laboratory of R&D in Engineering Sciences,
FST Al Hoceima
Abdelmalek Essaadi University
Tetouan, Morocco

Amine ElQaraoui
Laboratory Watch Laboratory for Emerging
Technologies
Hassan First University of Settat
Settat, Morocco

Sabrine Ennaji
National School of Applied Sciences
University of Sidi Mohamed Ben Abdellah
Fez, Morocco

Latifa Er-Rajy
Computer Science Department, Laboratory
LISI
Cadi Ayyad University
Marrakesh, Morocco

Khalid Haddouch
National School of Applied Sciences
University of Sidi Mohamed Ben Abdellah
Fez, Morocco

Melliani Hamza
Faculty of Economics and Management
Sultan Moulay Slimane University
Beni Mellal, Morocco

Abderrahim Hansali
ENCG
Cadi Ayyad University
Marrakech, Morocco

Abdelouafi Ikidid
Computer Science Department, Laboratory
LISI
Cadi Ayyad University
Marrakesh, Morocco

Hilal Imane
ITQAN Team, LyRica Lab
Information Sciences School
Rabat, Morocco

Venkata Vara Prasad Janjanam
Department of Management Studies
VIT-AP University
Amaravati, India

Kaoutar Jenoui
Laboratory Smartilab Sciences
Moroccan School of Engineering Sciences
Rabat, Morocco

Youness Khourdifi
Laboratory of Sciences of Materials,
Mathematics and Environment
Sultan Moulay Slimane University,
Polydisciplinary Faculty of Khouribga
Beni Mellal, Morocco

Manal Loukili
National School of Applied Sciences
Sidi Mohamed Ben Abdellah University
Fez, Morocco

Youness Madani
Sultan Moulay Slimane University
Beni Mellal, Morocco

Yassine Maleh
LaSTI Laboratory, ENSA Khouribga
Sultan Moulay Slimane University
Beni Mellal, Morocco

Abourezq Manar
ITQAN Team, LyRica Lab
Information Sciences School
Rabat, Morocco

Si Mohamed Ben Massou
ENCGM
Marrakech, Morocco

Hnida Meriem
ITQAN Team, LyRica Lab
Information Sciences School
Rabat, Morocco

Fayçal Messaoudi
National School of Applied Sciences
Sidi Mohamed Ben Abdellah University
Fez, Morocco

Benhouad Mohamed
Department of Economics and Business
   Management
Hassan II University, ENSET
Casablanca, Morocco

Mestari Mohammed
Hassan II University, ENSET
Mohammedia, Morocco

L'houssaine Mounaim
Cadi Ayyad University
Marrakech, Morocco

Ali Mwase
Department of Marketing and Management
Makerere University Business School
Kampala, Uganda

Daoudi Najima
ITQAN Team, LyRica Lab
Information Sciences School
Rabat, Morocco

Hussen Nasir
Faculty of Business and Communication
Malaysia Universiti Malaysia Perlis
Arau, Malaysia

Elafri Nedjwa
University of Constantine 3
Constantine, Algeria

Ernest Ketcha Ngassam
School of Computing
University of South Africa
Pretoria, South Africa

Van Chien Nguyen
Faculty of Economics (Finance and Banking)
Thu Dau Mot University
Thu Dau Mot, Vietnam

Musarrat Saberin Nipun
Brunel University
London, United Kingdom

Safa Ougoujil
Interdisciplinary Laboratory of Research and
   Studies in Management of Organizations
   and Corporate Law, Faculty of Law,
   Economic and Social Sciences
Cadi Ayyad University
Marrakech, Morocco

Smyej Oumaima
ENCGM
Cadi Ayyad University
Marrakech, Morocco

Prianka Ghosh Puja
BBA, Army Institute of Business
   Administration
Bangladesh University of Professionals
Savar, Bangladesh

Ayoub Razouk
Economics and Management Laboratory
   USMS
Sultan Moulay Slimane University
Beni Mellal, Morocco

Said Rifai
LMPGI Laboratory, Higher School of
   Technology
Hassan II University
Casablanca, Morocco

Sidi Mohamed Rigar
Interdisciplinary Laboratory of Research and
   Studies in Management of Organizations
   and Corporate Law, Faculty of Law,
   Economic and Social Sciences
Cadi Ayyad University
Marrakech, Morocco

Abdelkebir Sahid
ENCG
Hassan 1st University
Settat, Morocco

El Aidouni Salma
Laboratory of 2IACS
Hassan II University, ENSET
Mohammedia, Morocco

Singh Shawren
University of South Africa
Pretoria, South Africa

Sassi Boudemagh Souad
University of Constantine 3
Constantine, Algeria

A.V.V.S. Subba Lakshmi
Department of Management Studies
VIT University
Vellore, India

Rejwan Bin Sulaiman
Northumbria University
Newcastle Upon Tyne, United Kingdom

Huynh Thi Thanh Truc
Faculty of Economics
Thu Dau Mot University
Thu Dau Mot, Vietnam

Fatima Touhami
Sultan Moulay Slimane University
Beni Mellal, Morocco

Bushra Yousuf
Department of Computer Science
SZABIST
Dubai, U.A.E

Azra Zaimovic
School of Economics and Business
University of Sarajevo
Sarajevo, Bosnia and Herzegovina

Tarik Zaimovic
School of Economics and Business
University of Sarajevo
Sarajevo, Bosnia and Herzegovina

Rguibi Zakaria
Laboratory Watch Laboratory for Emerging
    Technologies
Hassan First University of Settat
Settat, Morocco

Charafeddine Ait Zaouiat
Polydisciplinary Faculty of Sidi Bennour
Chouaîb Doukkali University
El Jadida, Morocco

Dya Zitouni
Laboratory Watch Laboratory for Emerging
    Technologies
Hassan First University of Settat
Settat, Morocco

1 Emerging Fintech and Digital Money
Current Trends and Future Perspectives

Abdelkebir Sahid and Yassine Maleh

1.1 INTRODUCTION

The financial sector is one of the areas where technological innovations are proliferating, and there is even a term dedicated to it: Fintech (Financial Technologies). Fintech is revolutionizing the national and international financial landscape, expanding products and services, and making them better, more accessible, and less expensive. It is transforming customer relationships, payment methods, and financing and transfer methods.

Economically, according to a report by the McKinsey Institute, the potential of Fintech in emerging economies is estimated at 6% of additional GDP and 95 million new jobs by 2025. However, according to the same report, two billion people and 200 million micro, small, and medium enterprises in emerging countries do not have access to credit, and those that do still face high costs.

In addition, according to the World Bank's Global Findex, 1.7 billion adults worldwide do not have a bank account, yet 1.1 billion of them have a cell phone. In emerging and developing countries worldwide, particularly in Africa, Fintech offers immense opportunities to enhance financial inclusion and economic and social development. In just six years, 1.2 billion people have gained access to financial services through technology.

Recent advances have come from digital payments, government policies, and a new generation of mobile and internet-based financial services. In the financial sector, the risks are the most varied and important, particularly concerning money laundering, terrorist financing, cybersecurity, and consumer and personal data protection, all of which have become significant concerns locally and internationally. Due to the magnitude of these economic and social implications, public authorities are called upon to put adequate regulatory frameworks in place, ensure an environment conducive to technological innovations, and seize the opportunities offered by financial technologies. This will reduce the financial inclusion gap, especially for the benefit of small businesses, young people, rural populations, and especially women.

For all the players in the financial ecosystem, central banks and private banks in particular, the changes brought about by Fintech challenge them in several respects, particularly regarding their fundamental missions: money creation, supervision, financial regulation, etc. These efforts favoring digital financial inclusion ultimately contribute to achieving the first 17 United Nations Sustainable Development Goals to eradicate poverty.

This book, *Advances in Emerging Financial Technology and Digital Money*, presents a comprehensive compendium of cutting-edge research, showcasing the latest findings, trends, challenges,

DOI: 10.1201/9781032667478-1

and opportunities in the rapidly evolving world of Fintech. This book is divided into five sections, each focusing on a unique aspect of financial technology, providing valuable insights into the current state and future prospects of financial technology and digital money.

1.1.1 PURPOSE AND SCOPE OF THIS BOOK

The rapid advancements in technology and the increasing influence of the internet have drastically transformed the financial sector, giving rise to a new era of financial technology (Fintech) and digital money. As the Fintech landscape continues to evolve, it is crucial to understand its various aspects, implications, and potential for driving economic and social development. The primary purpose of this book, *Advances in Emerging Financial Technology and Digital Money*, is to provide a comprehensive and multidisciplinary analysis of the latest developments, trends, challenges, and opportunities in Fintech and digital money.

This book aims to serve as a valuable resource for researchers, policymakers, industry professionals, and students interested in the rapidly evolving world of Fintech and digital money. It covers various topics, including artificial intelligence (AI), machine learning (ML), blockchain, cryptocurrencies, digital payments, peer-to-peer lending, crowdfunding, and financial inclusion. By bringing together contributions from experts across diverse fields and geographies, this book offers a holistic and global perspective on the current state and prospects of Fintech and digital money.

1.1.2 STRUCTURE OF THIS BOOK

The book is divided into five sections, each focusing on a specific aspect of the Fintech and digital money landscape:

Section 1: Artificial Intelligence and Machine Learning in Finance This section explores the growing role of artificial intelligence and machine learning in various financial applications, from fraud detection to stock market prediction.

Section 2: Blockchain and Distributed Ledger Technologies This section delves into the potential of blockchain and other distributed ledger technologies to revolutionize financial services, supply chain management, and more.

Section 3: Digital Payments and Financial Inclusion This section discusses the rise of digital payments and mobile financial services, examining their potential to promote financial inclusion and foster economic development.

Section 4: Peer-to-Peer Lending, Crowdfunding, and Alternative Finance This section focuses on the emergence of alternative finance models, such as peer-to-peer lending and crowdfunding platforms, and their impact on traditional financial institutions and the overall financial ecosystem.

Section 5: Regulatory Challenges and the Future of Fintech This section addresses the regulatory and compliance issues surrounding Fintech and digital money and the challenges and opportunities that lie ahead in the ever-evolving financial landscape.

Each section consists of several chapters that provide in-depth analyses, case studies, and empirical research on the respective topics. By examining the interconnections and implications of these facets of Fintech and digital money, this book aims to better understand the complex and dynamic world of financial technology and its potential to shape the future of finance.

1.2   THE EVOLUTION OF FINANCIAL TECHNOLOGY AND DIGITAL MONEY

The emergence of financial technology (Fintech) and digital money has roots in the broader history of technological advancements and financial services. This background section will provide an overview of the key milestones that have shaped the modern Fintech landscape and the evolution of digital money, leading up to the current era of innovation and disruption (Palmié et al., 2020).

1.2.1   Early Beginnings of Financial Technology

The concept of Fintech can be traced back to the early days of banking, when financial services began leveraging technology to streamline operations and improve customer experiences. The introduction of the telegraph in the 19th century enabled the first real-time, long-distance communication, laying the groundwork for modern financial markets. In the 20th century, the advent of mainframe computers and electronic data processing revolutionized the way financial institutions managed their operations, paving the way for innovations like automated teller machines and electronic funds transfers (Derry & Williams, 1960) (Figure 1.1).

1.2.2   The Emergence of the Internet and E-Commerce

The widespread adoption of the internet in the 1990s marked a turning point for Fintech, opening up new possibilities for online banking and e-commerce. The development of secure communication protocols like SSL (Secure Sockets Layer) facilitated the growth of online financial services, allowing consumers and businesses to conduct online transactions. This period also saw the establishment of early internet-based payment systems like PayPal, which played a crucial role in the rise of e-commerce giants like eBay and Amazon (Goldstein & O'Connor, 2000).

1.2.3   Fintech 2.0: The Post-2008 Financial Crisis Landscape

The global financial crisis of 2008 created an environment ripe for Fintech innovation. Public distrust in traditional financial institutions, coupled with tight regulations and the need for greater transparency, fueled the growth of alternative financial services. Fintech startups emerged, offering innovative solutions across various sectors, such as peer-to-peer lending, robo-advisory, and mobile payments. The rise of smartphones and mobile internet further accelerated the adoption of Fintech services, enabling financial services to reach previously underserved populations (Arner et al., 2015).

1.2.4   The Advent of Blockchain and Cryptocurrencies

The introduction of Bitcoin in 2009 marked the beginning of the digital money revolution. Developed by the pseudonymous Satoshi Nakamoto, Bitcoin introduced the world to blockchain, a decentralized,

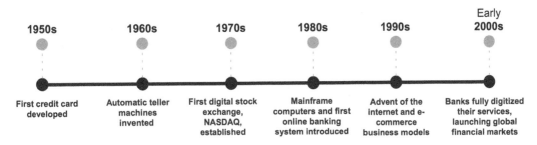

FIGURE 1.1   Evolution of modern Fintech.

distributed ledger technology (DLT) that underpins most cryptocurrencies (Nakamoto, 2008). The blockchain's ability to provide secure, transparent, and tamper-proof records of transactions made it an attractive solution for various financial applications. Since the creation of Bitcoin, thousands of other cryptocurrencies have emerged, along with numerous blockchain-based projects aimed at transforming industries beyond finance.

1.2.5 THE CURRENT STATE OF FINTECH AND DIGITAL MONEY

Today, the Fintech and digital money ecosystems are rapidly evolving, characterized by the rise of innovative startups, the adoption of new technologies by traditional financial institutions, and the growing interest from regulators and policymakers. AI, ML, and big data analytics have enabled more sophisticated financial services, such as personalized wealth management, fraud detection, and algorithmic trading. The increased focus on financial inclusion has led to innovative solutions catering to unbanked and underbanked populations, fostering economic growth and development in regions with limited access to traditional financial services (Gomber et al., 2017).

New challenges and opportunities emerge as Fintech, and digital money continues to reshape the global financial landscape. This background section has laid the foundation for understanding the historical context and key milestones that have shaped the current state of Fintech and digital money. The following chapters will delve into specific aspects of this exciting and rapidly evolving domain, exploring cutting-edge technologies, applications, and trends defining the future of finance (Knewtson & Rosenbaum, 2020).

1.3 KEY TRENDS AND INNOVATIONS IN FINTECH AND DIGITAL MONEY

Some of the institutions surveyed have already been putting into practice for several years the utilization and analysis of massive volumes of data (also known as "big data"). These institutions have created internal platforms or data lakes to centralize group data and make its use more convenient. It is now widely accepted that AI technologies cannot exist independently of big data. These technological breakthroughs can shake up established monetary systems and open new doors to opportunities for expansion, inclusivity, and increased productivity (Maleh et al., 2021).

1.3.1 APPLICATIONS OF AI AND ML IN FINTECH

AI and ML techniques have a wide range of applications in the Fintech domain, including:

- Risk assessment and credit scoring: By analyzing large volumes of data, AI and ML algorithms can help financial institutions accurately assess the creditworthiness of borrowers, reducing the risk of defaults and improving the accuracy of credit scoring models (Oualid et al., 2023).
- Fraud detection and prevention: AI and ML can detect and prevent fraudulent activities in real time by analyzing transaction data and identifying patterns or anomalies that may indicate fraud (Bao et al., 2022).
- Algorithmic trading and investment management: AI and ML can develop sophisticated trading algorithms and investment strategies that can analyze market data, identify trends, and make informed decisions to optimize returns (Brogaard et al., 2014; Treleaven et al., 2013).
- Customer service and support: AI-powered chatbots and virtual assistants can provide personalized and efficient customer support, addressing customer inquiries and resolving issues quickly (Xu et al., 2017).
- Optimization of the banking value chain: Optimization of internal processes and cost reduction through task prioritization, workflow management, and advanced document

reading. AI enables the processing of unstructured data and allows automation of previously manual processes. In parallel with the rise of self-care, AI contributes to freeing branch advisors from tasks with less added value.

Despite the numerous benefits of AI and ML in Fintech, several challenges must be addressed to ensure their successful implementation:

- Data privacy and security: The use of AI and ML relies heavily on collecting and analyzing large amounts of data, raising concerns about data privacy and security. Fintech organizations must implement robust data protection measures and comply with relevant regulations to ensure customer trust and avoid potential legal issues (Gai et al., 2017).
- Bias and fairness: AI and ML models can inadvertently perpetuate bias if the data used to train them contain discriminatory patterns. To tackle this issue, Fintech companies must develop transparent and unbiased algorithms, regularly audit their models, and strive for fairness in their decision-making processes (Cao et al., 2021).
- Explainability and transparency: The complexity of some AI and ML models can make it difficult to understand how they arrive at specific decisions. This "black box" problem can undermine trust and hinder regulatory compliance. Fintech companies must prioritize explainability and transparency in their AI and ML implementations (Deo & Sontakke, 2021).
- Talent and skill gaps: As the demand for AI and ML expertise grows, Fintech organizations must invest in talent development and training programs to ensure they have the necessary skills to develop, implement, and maintain these technologies effectively (Lukonga, 2021).

Despite these challenges, the opportunities presented by AI and ML in Fintech are immense. By embracing these technologies and addressing the associated difficulties, Fintech organizations can unlock new levels of efficiency, innovation, and customer satisfaction, paving the way for a more advanced and inclusive financial ecosystem.

1.3.2 BLOCKCHAIN AND DISTRIBUTED LEDGER TECHNOLOGIES

Blockchain and DLTs have significantly impacted the financial industry by providing secure, transparent, and efficient solutions for various applications (Maleh et al., 2022). These technologies can potentially revolutionize financial transactions by reducing costs and increasing trust among participants. Key developments in this area include the following:

- Cryptocurrencies: Digital currencies like Bitcoin and Ethereum have introduced new ways of transferring value and conducting transactions without intermediaries. They offer lower fees, faster processing times, and increased privacy compared to traditional payment methods (Melanie, 2015).
- Smart contracts: Self-executing contracts with the terms of the agreement directly written into code offer increased efficiency and automation. They enable trustless transactions and have the potential to streamline various financial processes, such as securities trading, insurance claims, and supply chain management (Wang et al., 2019).
- DeFi: DeFi platforms leverage blockchain technology to provide financial services, such as lending, borrowing, and asset management, without traditional intermediaries. DeFi can potentially increase financial inclusion and democratize access to financial products and services (Chen & Bellavitis, 2020). Figure 1.2 shows DeFi financial services.

    As a decentralized system, DeFi enables users worldwide to access their funds effortlessly. Users maintain complete ownership of their digital assets and have full autonomy in

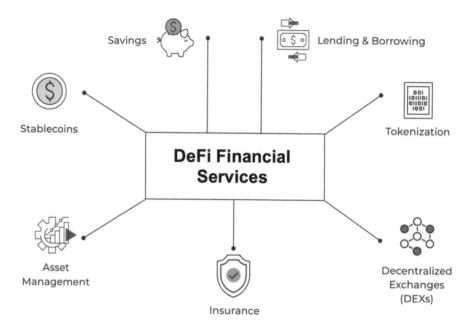

FIGURE 1.2 DeFi financial services.

managing their cryptocurrency wallets. Users can benefit from reduced transaction fees, increased returns, and complete transparency in all executed transactions by eliminating the need for intermediary parties.

DeFi enables individuals to conduct financial activities such as transferring money, obtaining loans, investing in derivative stocks, and more without intermediaries. Traditional middle agents, such as bank managers, brokers, and guarantors, are no longer involved in DeFi transactions. Utilizing smart contracts and blockchain technologies facilitates efficient, cost-effective, and peer-to-peer transactions. Figure 1.3 shows the DeFi transactions.

Cross-border payments and remittances: Blockchain-based solutions can significantly reduce the cost and time associated with international money transfers, making them more accessible and efficient for individuals and businesses (Qiu et al., 2019). Figure 1.4 presents the progress of Cross-border eCommerce transactions from purchase to confirmation.

- Digital identity and KYC/AML: DLT can provide secure, verifiable, and user-controlled digital identities, streamlining the KYC (Know Your Customer) and AML (Anti-Money Laundering) processes, and reducing the risk of identity theft and fraud (Sullivan & Burger, 2019).

1.3.3 DIGITAL PAYMENTS AND MOBILE FINANCIAL SERVICES

Digital payments and mobile financial services have transformed how consumers and businesses conduct transactions, offering greater convenience, speed, and security. The proliferation of smartphones, mobile applications, and internet connectivity has enabled the rapid growth of these services. Key developments in this area include the following:

- Mobile wallets and payment apps: Services like Apple Pay, Google Pay, and Alipay allow users to store their payment information securely on their mobile devices and make contactless payments at participating merchants. These solutions provide a seamless and convenient payment experience for both online and offline transactions (Kapoor et al., 2020).

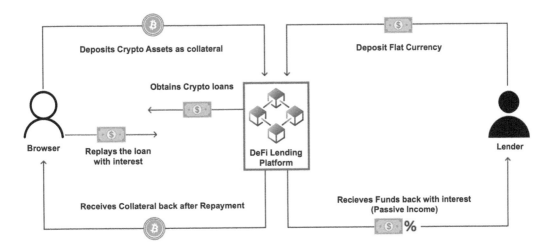

FIGURE 1.3 DeFi transactions.

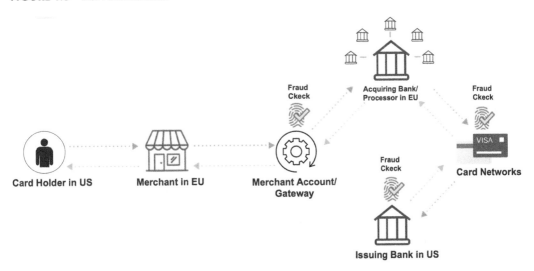

FIGURE 1.4 Cross-border eCommerce transactions progress from purchase to confirmation.

- QR code payments: QR (Quick Response) code-based payments allow users to make transactions by scanning a unique code with their smartphones, simplifying the payment process and reducing the need for physical cards or cash. This technology has been widely adopted in countries like China and India, driving the growth of digital payments (Yan et al., 2021). Digital wallets employ various technologies, such as near-field communication, secure magnetic transmission, and QR codes, to ensure secure payment experiences. Typically, the customer's device uses near-field communication to interact with the merchant's payment terminal during an in-person purchase. Once the customer verifies the transaction using a password, fingerprint, or facial recognition technology, payment tokenization securely transmits the payment information, completing the transaction, as shown in Figure 1.5.
- Digital-only banks and neo-banks: These financial institutions operate online without physical branches, offering banking services such as checking and savings accounts, loans, and investment products. Digital and neo-banks leverage technology to provide user-friendly interfaces, lower fees, and personalized customer experiences (Bradford, 2020).

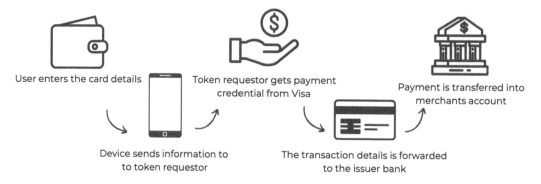

FIGURE 1.5   How does a digital wallet work?

- Mobile money services: In developing countries with limited access to traditional banking infrastructure, services like M-Pesa have been crucial in promoting financial inclusion (Jack & Suri, 2011). These services allow users to send and receive money, pay bills, and access other financial services using their mobile phones, even without a bank account.

1.3.4   PEER-TO-PEER LENDING AND CROWDFUNDING

Peer-to-peer (P2P) lending and crowdfunding are innovative financing models that have disrupted traditional banking and investment systems. These platforms leverage technology to connect borrowers and investors directly, facilitating access to capital and creating new investment opportunities. Key aspects of P2P lending and crowdfunding include the following:

- P2P lending platforms: These online platforms facilitate loans between individual borrowers and investors, often without the intermediation of a traditional financial institution. P2P lending platforms like LendingClub, Prosper, and Funding Circle leverage algorithms and data analytics to assess borrowers' credit risk, set interest rates, and match them with suitable investors. These platforms can offer lower interest rates for borrowers and higher returns for investors than traditional bank loans (Havrylchyk & Verdier, 2018).
- Equity crowdfunding: Equity crowdfunding platforms like Kickstarter and Indiegogo allow entrepreneurs to raise capital by offering equity stakes in their businesses to individual investors. This model enables startups and small businesses to access funding that may not be available through traditional channels while allowing investors to invest in early-stage companies with the potential for high returns (Mochkabadi & Volkmann, 2020).
- Reward-based crowdfunding: This allows individuals to contribute to projects or businesses in exchange for non-financial rewards, such as products, services, or experiences. Platforms like Kickstarter and Indiegogo enable entrepreneurs to raise funds for their projects by offering tiered rewards to backers based on their contribution levels (Frydrych et al., 2014).
- Debt crowdfunding: Also known as crowdlending or debt-based securities, this model allows investors to fund loans or bonds issued by businesses, earning interest on their investments. Debt crowdfunding platforms like Bondora and LendInvest cater to individual and institutional investors, offering a range of investment opportunities with varying risk and return profiles (Everett, 2019).
- Real estate crowdfunding: Platforms like Fundrise and RealtyMogul enable investors to pool their resources to invest in real estate projects, such as residential properties, commercial buildings, or land development. These platforms provide investors access to real estate investments that may have been previously inaccessible due to high entry costs or lack of expertise (Garcia-Teruel, 2019).

These innovative financing models have democratized access to capital and broadened investment opportunities for individuals and institutions alike. However, they also pose new regulatory challenges and risks, necessitating effective oversight and investor protection measures.

1.3.5   Regtech and SupTech

Regtech, or regulatory technology, and SupTech, or supervisory technology, are emerging trends in Fintech that focus on leveraging technology to improve regulatory compliance and supervisory processes within the financial industry. Regtech and SupTech aim to enhance efficiency, transparency, and security while reducing the burden of compliance and oversight (Armstrong, 2018).

- Regtech: Regtech solutions are designed to help financial institutions comply with regulatory requirements more efficiently and cost-effectively. These solutions employ advanced technologies such as AI, ML, big data analytics, and blockchain to automate and streamline compliance processes. Key areas where Regtech is making an impact include:
  - AML and KYC: Regtech solutions can automate customer due diligence processes, making it easier to detect and report suspicious activities while reducing the time and resources spent on manual checks.
  - Risk management and reporting: Regtech tools can help financial institutions identify, assess, and mitigate risks more effectively, as well as generate accurate and timely regulatory reports.
  - Data management and aggregation: By automating data collection, validation, and analysis, Regtech solutions can help institutions maintain data accuracy, integrity, and consistency across various regulatory frameworks.
- SupTech: SupTech refers to adoption of advanced technologies by regulatory and supervisory authorities to enhance their oversight capabilities. These technologies can help regulators monitor and assess financial institutions' compliance more effectively, enabling them to identify potential risks and vulnerabilities in the financial system. Key areas where SupTech is making an impact include:
  - Data collection and analysis: Supervisory authorities can use big data analytics and ML algorithms to process and analyze large volumes of structured and unstructured data, improving their ability to detect trends, anomalies, and potential risks.
  - Automated reporting and data sharing: SupTech solutions can streamline reporting processes and facilitate data sharing between regulatory authorities and financial institutions, improving transparency and collaboration.
  - Supervisory models and stress testing: Advanced analytics and simulation techniques can help supervisory authorities develop more accurate and robust supervisory models, enabling them to conduct stress tests and assess the resilience of financial institutions.

As Regtech and SupTech continue to evolve, they hold significant potential for improving the efficiency and effectiveness of regulatory compliance and supervision in the financial sector. However, their widespread adoption also raises questions about data privacy, security, and the potential for regulatory fragmentation, necessitating careful consideration and collaboration among stakeholders.

The summary presented in Table 1.1 highlights the key trends and innovations in Fintech and digital money, emphasizing the benefits, potential challenges, and real-world applications. The discussion below provides an analysis and synthesis of the information presented in the table.

AI and ML have emerged as crucial components in the Fintech landscape. These technologies enable enhanced decision-making and predictive capabilities, critical for improved risk management and fraud detection (Jakšič & Marinč, 2019). Additionally, AI- and ML-driven personalized financial services and engagement redefine the customer experience. However, these technologies

TABLE 1.1

Summary of Key Trends and Innovations in Fintech and Digital Money

Key Trends	Innovation/Trend	Key Benefits	Potential Challenges	Real-World Applications
Artificial intelligence and machine learning	Artificial intelligence and machine learning	Improved decision-making and predictive capabilities, enhanced risk management and fraud detection, personalized financial services, and customer engagement	- Ensuring data privacy, - Addressing algorithmic biases, - Ensuring explainability and transparency	Algorithmic trading, credit scoring models, Chatbots, and virtual assistants
Blockchain and distributed ledger technologies	Blockchain and distributed ledger technologies	Increased transparency, security, and trust, streamlined and cost-efficient processes, enabling decentralized financial applications and services	- Scalability concerns, interoperability issues, - Regulatory compliance	Digital identity solutions, smart contracts, cross-border payments and remittances
Digital payments and mobile financial services	Digital payments and mobile financial services	- Increased convenience and ease of transactions, - Financial inclusion for the unbanked and underbanked, - Real-time and low-cost money transfers	Security and fraud concerns, infrastructure and technological challenges, regulatory compliance	Mobile wallets, contactless payments, international remittances
Peer-to-peer lending and crowdfunding	Peer-to-peer lending and crowdfunding	Alternative financing options, lower interest rates, and fees, enhanced accessibility for SMEs and entrepreneurs	Credit risk assessment, investor protection, regulatory compliance	P2P lending platforms, equity crowdfunding, invoice financing
Regtech and SupTech	Regulatory technology and supervisory technology	Streamlined compliance processes Enhanced monitoring, and reporting Improved risk management	Data privacy and security, adoption and integration challenges, ensuring regulatory alignment	Automated compliance reporting, AML/KYC solutions, real-time fraud detection

also present challenges related to data privacy, algorithmic biases, and ensuring explainability and transparency. These concerns must be addressed to reap the full benefits of AI and ML in the financial sector.

Blockchain and DLTs are transforming the financial industry by providing increased transparency, security, and trust. These technologies enable streamlined and cost-efficient processes, and facilitate the development of decentralized financial applications and services. However, scalability, interoperability, and regulatory compliance remain significant challenges that must be overcome to fully harness the potential of blockchain and DLT.

Digital payments and mobile financial services are rapidly gaining popularity, driven by increased convenience and ease of transactions. These technologies promote financial inclusion for the unbanked and underbanked populations and facilitate real-time, low-cost money transfers. Nevertheless, security and fraud concerns, infrastructure and technological challenges, and regulatory compliance must be addressed to ensure the continued growth and success of digital payments and mobile financial services.

Peer-to-peer (P2P) lending and crowdfunding platforms are emerging as viable alternatives to traditional financing options. These platforms offer lower interest rates and fees while providing enhanced accessibility for SMEs and entrepreneurs. However, credit risk assessment, investor protection, and regulatory compliance are significant challenges that need to be addressed to sustain the growth and success of P2P lending and crowdfunding platforms.

Regtech and SupTech are transforming the way financial institutions and regulators approach compliance and supervision. These technologies streamline compliance processes, enhance monitoring and reporting, and improve risk management. Despite the potential benefits, data privacy and security, adoption and integration challenges, and ensuring regulatory alignment remain significant obstacles that must be addressed to fully leverage the potential of Regtech and SupTech solutions.

The Fintech landscape is characterized by continuous innovation and rapid development. However, the benefits and potential of these innovations can only be fully realized by addressing the inherent challenges and ensuring that stakeholders, including financial institutions, regulators, and customers, work together collaboratively.

1.4 CHALLENGES AND OPPORTUNITIES IN THE FINTECH ECOSYSTEM

1.4.1 Financial Inclusion and Empowering the Unbanked

Financial inclusion is a critical aspect of building a more equitable financial system. It seeks to provide affordable, convenient, and secure financial services to all individuals, particularly those who are unbanked or underbanked. Fintech innovations have played a significant role in advancing financial inclusion and empowering the unbanked population by addressing major barriers that hinder their access to traditional financial services. This section explores how Fintech contributes to financial inclusion and empowers the unbanked (Bhyer & Lee, 2019).

1.4.1.1 Digital and Mobile Payments

One of the most significant Fintech innovations contributing to financial inclusion is the development of digital and mobile payment solutions. These technologies enable users to make transactions, pay bills, and access other financial services through smartphones or digital devices. By leveraging mobile technology, Fintech companies have reached remote and underserved communities, providing them access to basic financial services without needing physical bank branches.

1.4.1.2 Microfinance and Peer-to-Peer Lending

Fintech has also contributed to financial inclusion by facilitating the growth of microfinance and peer-to-peer (P2P) lending platforms. These platforms provide small loans and other financial products to low-income individuals and small businesses, enabling them to access credit and financial services that are otherwise unavailable through traditional financial institutions. By leveraging technology and innovative credit scoring models, Fintech companies can assess the creditworthiness of borrowers more accurately, allowing them to extend credit to underserved segments of the population (Yum et al., 2012).

1.4.1.3 Digital Savings and Investment Platforms

Digital savings and investment platforms have emerged as critical tools for promoting financial inclusion, enabling individuals to save and invest in various financial instruments easily. By offering low-cost investment options and user-friendly interfaces, these platforms have made it easier for individuals, particularly those in low-income segments, to build wealth and secure their financial futures. Fintech companies have also developed innovative savings products, such as goal-based savings accounts and automated savings tools, to encourage users to adopt positive financial habits (Koomson et al., 2022).

1.4.1.4 Financial Literacy and Education

Fintech solutions have also played a vital role in promoting financial literacy and education, which are essential to financial inclusion. Many Fintech companies have developed digital platforms and tools to educate users on personal finance, investment strategies, and responsible money management. By leveraging gamification, interactive content, and personalized learning experiences, Fintech firms can engage users and help them improve their financial knowledge and decision-making abilities (Hastings et al., 2013).

1.4.1.5 Insurtech and Access to Insurance Services

The growth of insurtech, or the application of technology in the insurance sector, has also contributed to financial inclusion by making insurance products more accessible and affordable (Yan et al., 2018). Fintech companies have developed innovative insurance products, such as microinsurance and parametric insurance, which cater to the specific needs of low-income individuals and those living in remote areas. Additionally, insurtech platforms have simplified purchasing and managing insurance policies, enabling users to access insurance services more easily and conveniently.

1.4.2 CYBERSECURITY AND DATA PRIVACY

As Fintech continues to gain traction and reshape the financial services landscape, cybersecurity and data privacy have become critical concerns for consumers and industry stakeholders. The increasing reliance on digital platforms and the vast amounts of sensitive financial data being transmitted and stored online have made the Fintech ecosystem particularly vulnerable to cyber threats and privacy breaches.

Fintech companies face diverse cyber threats, including phishing attacks, malware, ransomware, distributed denial-of-service attacks, and data breaches. These attacks can result in significant financial losses, reputational damage, and loss of customer trust. Furthermore, Fintech companies often rely on third-party service providers and cloud-based infrastructure, which can introduce additional vulnerabilities and complexities to their cybersecurity posture (Kaur, Habibi Lashkari, et al., 2021). The cyber risk covers several risks:

- The possibility of data being lost or the service being interrupted as a result of an attack (in particular, a "denial of service" or ransomware assault);
- Risk of unauthorized exposure of sensitive and strategic information;
- Risk of sensitive and strategic information being leaked;
- Risk of fraud.

At the same time, and even though it is not technically a cyber risk in and of itself, the regulatory environment is also becoming more demanding in terms of cyber security (European Banking Authority (EBA) Guidelines, Digital Operational Resilience Act (DORA), General Data Protection Regulation (GPDR), and Operators of Essential Services (OSE)). As a result, financial institutions must take the steps necessary to protect themselves from these emerging dangers.

Banks have identified many tiers of action to take in order to lower their cyber risk. To begin, financial institutions are participating in a mapping effort. In a purely technical sense, improving information system cybersecurity impacts the entirety of the system's life cycle. The "Security by Design" method can be utilized when creating services to incorporate these concerns during development (DevSecOps). After the system has been put into production, the integration of changes has to occur much more quickly. To lessen a company's vulnerability to cyberattacks, it is necessary to establish audits, penetration testing, and various other internal security procedures to secure information systems. This focus extends to the risks posed by partners: inspections are carried out in the pre-contractual phase via an evaluation of partners and subcontractors.

In terms of resilience, financial institutions are attempting to establish ex-ante reaction plans for cyber security disasters. These plans will include risk categorization criteria, tools for assessing the reasons for the interruption (forensic analysis), and remedies to assure business continuity (doubling up databases, tools integrating the possibility of operating in "degraded mode").

Other tangible actions may be taken to decrease the risks associated with cyberspace, such as improving the availability of technical teams, deploying new tools, accelerating the updates (patching) offered by the primary IT suppliers in the event of a vulnerability, etc. In this regard, technologies based on AI are becoming more acknowledged for their capacity to identify assaults more effectively or to filter questionable information more rapidly (or even automatically). The strengthening of information technology (IT) security, which goes beyond the employment of cybersecurity professionals, drives banks to educate business teams on cybersecurity risks better in order to prevent phishing attempts. This is true from the perspective of human resources (Maleh et al., 2020).

1.4.2.1 Cyber Threats and Vulnerabilities in Fintech

All institutions surveyed noted that digitization is accompanied by increased exposure to cyber risk. Several factors can explain this growing vulnerability: the multiplication of partnerships (which involve the exchange of data and the expansion of points of vulnerability), the growing complexity of information systems with multiple "layers", and the increasing dependence on tools or service providers that can be affected by cyberattacks, which are themselves becoming more frequent and sophisticated.

1.4.2.2 Data Privacy and Regulatory Compliance

Data privacy is another critical concern for Fintech companies, as they handle vast amounts of sensitive customer information, such as personal identification details, financial transaction data, and credit histories. Ensuring the privacy and security of this data is essential for maintaining customer trust and complying with increasingly stringent data protection regulations, such as the General Data Protection Regulation and the California Consumer Privacy Act. Failure to comply with these regulations can result in significant fines and reputational damage (Hussain et al., 2021).

1.4.2.3 Building a Robust Cybersecurity Framework

To address cybersecurity challenges, Fintech companies must adopt a comprehensive and proactive approach to safeguarding their digital assets and customer data. This includes implementing robust security measures, such as encryption, multi-factor authentication, and intrusion detection systems, and continuously monitoring and updating their security infrastructure to stay ahead of emerging threats. Fintech companies should also adopt a risk-based approach to cybersecurity, identifying and prioritizing the most critical assets and vulnerabilities to allocate resources effectively (Kaur, Lashkari, et al., 2021).

1.4.2.4   Collaboration and Information Sharing

Collaboration and information sharing are essential for enhancing cybersecurity in the Fintech sector. Fintech companies should actively engage with industry peers, government agencies, and cybersecurity experts to share threat intelligence, best practices, and other relevant information. Participation in industry associations, such as the Financial Services Information Sharing and Analysis Center, can facilitate this collaboration and help companies stay informed about the latest cyber threats and trends (Liu et al., 2014).

1.4.2.5   Cyber Training and Awareness

Cybersecurity awareness and employee training programs are critical components of a Fintech company's security strategy. Employees should be educated on their potential cyber risks, the importance of adhering to security policies and procedures, and how to identify and report suspicious activities. Regular training and assessments can help ensure that employees remain vigilant and contribute to building a strong security culture within the organization (Kaur, Lashkari, et al., 2021).

1.4.3   REGULATORY CHALLENGES AND THE NEED FOR COLLABORATION

The rapid growth of Fintech has introduced a new set of regulatory challenges for the industry and the authorities responsible for overseeing it. As these innovative technologies and services disrupt traditional financial models, regulators must balance fostering innovation, protecting consumers, and ensuring financial stability. This section explores the regulatory challenges the Fintech ecosystem faces and the importance of collaboration among stakeholders to address these issues effectively (Ahern, 2021).

1.4.3.1   Navigating a Complex Regulatory Landscape

Fintech companies often face a complex and fragmented regulatory landscape, with different jurisdictions adopting varying approaches to oversight of new technologies and business models. This can create significant compliance challenges for Fintech firms, particularly those operating across multiple markets. Moreover, regulatory frameworks may not always keep pace with the rapid rate of innovation in the industry, resulting in a lack of clarity or outdated rules that may stifle growth and hinder the adoption of new services.

1.4.3.2   Balancing Innovation and Regulation

Regulators must balance supporting Fintech innovation with ensuring consumer protection, financial stability, and the financial system's integrity. This requires a flexible and adaptive regulatory approach to accommodate new technologies and business models without imposing undue burdens on the industry. Some jurisdictions have adopted regulatory "sandbox" initiatives, which allow Fintech companies to test their products and services in a controlled environment with relaxed regulatory requirements. These programs can help regulators better understand the risks and benefits of innovations while providing Fintech firms with valuable feedback and guidance on regulatory compliance (Bromberg et al., 2017).

1.4.3.3   Cross-Border Collaboration and Harmonization

As Fintech continues to drive the globalization of financial services, the need for cross-border collaboration and harmonization of regulatory frameworks becomes increasingly important. Regulatory authorities should work together to develop common standards, share best practices, and facilitate information exchange to support the growth of Fintech on a global scale. Such collaboration can help reduce regulatory arbitrage, promote market efficiency, and ensure a level playing field for all market participants (Ahmed, 2019).

1.4.3.4 Public–Private Partnerships and Stakeholder Engagement

Effective Fintech regulation requires close collaboration between public and private sector stakeholders, including regulators, industry participants, and consumer advocacy groups. Public–private partnerships can help bridge the gap between the fast-paced world of Fintech innovation and the more deliberate pace of regulatory development. By engaging with the industry and other stakeholders, regulators can better understand emerging technologies, identify potential risks, and develop targeted, risk-based regulatory approaches that support innovation while safeguarding the public interest (Yang & Tsang, 2018).

1.4.3.5 Building Regulatory Capacity and Expertise

As Fintech continues to evolve and grow in complexity, regulators must also invest in building their capacity and expertise in this area. This may involve hiring specialized staff, providing training and development opportunities for existing personnel, and leveraging external resources and expertise. Developing a robust understanding of Fintech and its implications for financial services will enable regulators to make more (Yang & Tsang, 2018).

1.4.4 Talent and Human Capital Development

Digital transformation is not just a technical process, but also involves major transformations in the various business lines of banking establishments: the tools to be mastered are changing, modes of communication with customers are evolving, and employees' expectations are evolving. This transformation of business lines and needs must be anticipated at the recruitment stage, but it also requires change management support for existing employees. As well as developing the technical skills available, banks are seeking to promote a culture of innovation internally.

1.4.4.1 Need for New Jobs and Skills

Several organizations have established forward-looking policies as part of the forward-looking management of jobs and skills (GPEC). The goal of these policies is to identify jobs and skills, as well as managerial and organizational models, in order to guarantee that requirements and human resources are matched. In their responses, all of the financial institutions have recognized an increased demand for expert profiles on technical subjects such as data, AI, IT, and the cloud, with specializations in data science, data analysis, data engineering, user experience design (UX Design), architecture, or software development. However, this demand also encompasses hybrid profiles that combine corporate expertise (such as legal or financial knowledge) with an understanding of emerging technology. The organizational charts are being updated to reflect the addition of new jobs and titles.

In a labor market that is both scarce and extremely competitive, banks are reporting possible challenges in hiring and keeping people with these attributes (the number of candidates is much lower than the demand). On the other hand, for the banking industry, these challenges are more of a focus of emphasis than a hurdle that cannot be overcome: the great majority of banks can meet their recruiting objectives.

In response to these challenges, efforts are being focused on identifying potential candidates as early on in the process as possible (for instance, through partnerships with engineering schools, training institutes, etc.), expanding potential employment pools (opening additional IT sites, setting up operations abroad, etc.), and taking better account of the expectations of future employees. Specifically, these partnerships can be formed with engineering schools, training institutes, etc. (adapting pay scales, working hours, hierarchical organization, management methods, etc.) (Doherty & Stephens, 2023).

1.4.4.2 Education and Training

In addition to hiring personnel with the appropriate levels of competence, businesses need to launch significant training initiatives to familiarize workers with the latest digital technologies, associated tools, and operational procedures. The purpose of this effort is to guarantee that the human resources made available in some domains as a result of new technical solutions are reallocated to jobs with a higher value added to them. The development of skills focuses primarily on the acquisition of new skills, whether this acquisition is a response to the transformation of the needs of the company (reskilling) or whether it aims to increase the skills of employees (upskilling) so that they can integrate the digital transformation into their everyday lives. But developing talents also involves "soft skills" and ways of company management; managers are the primary focus of attention inside financial institutions. Several of the programs mentioned by the latter are explicitly directed toward them. These include seminar cycles for senior or middle managers, targeted training on integrating the impacts of digital technology into their business line, meetings with entrepreneurs who have created Fintech or researchers working on new technologies, etc. Other organized participatory efforts, like hackathons or collaborative platforms for exchanging best practices, are also being established within banking organizations (Fahy, 2022).

1.4.4.3 Fostering a Culture of Innovation

To stay competitive in the rapidly evolving Fintech landscape, organizations must foster a culture of innovation that encourages employees to think creatively and take risks. This can be achieved by promoting a collaborative work environment, rewarding innovative ideas, and providing employees with the tools and resources they need to experiment and develop new solutions. By fostering a culture of innovation, organizations can attract and retain top talent, drive technological advancements, and maintain a competitive edge in the market (Fahy, 2022).

To advance this culture of innovation, banks must insist on:

- The role that each of these topics plays in the overall governance structure and strategy of the organization;
- A proactive recruiting approach, in addition to a training and ongoing acculturation program for workers;
- Encouragement of internal innovation, particularly via the use of incubators and other programs geared toward fostering intrapreneurship;
- Increasing interactions with creative participants in the ecosystem, including product alliances, investments, and conversation.

Organizations believe that the fragmentation of activities into "siloed" business divisions, the burdensome nature of decision-making procedures, and the rigidity of project management structures are the primary obstacles to adopting a culture of innovation. On the other hand, financial institutions frequently highlight the "agile" management method advancements they have made and continue to make as a point of contrast. When applied to an innovative project, this collaborative development logic translates into a shared vision of the stages of innovation deployment. These stages begin with an exploration and ideation phase, and continue with an incubation phase that should lead to a prototype (Proof of Concept) and then, after successive iterations, to a "minimum viable product" – before the launch of new products. These stages are outlined below.

1.4.4.4 Promoting Diversity and Inclusion

A diverse and inclusive Fintech workforce is essential for driving innovation and ensuring the development of products and services that cater to a broad range of customers. Organizations should prioritize diversity and inclusion in their recruitment and talent management strategies by actively seeking candidates from underrepresented groups, promoting equal opportunities, and creating an inclusive work environment that embraces different perspectives and ideas. By fostering a diverse and inclusive workforce, Fintech companies can tap into a wider talent pool, drive innovation, and ensure the industry's long-term success (Bähre et al., 2020).

1.4.4.5   Collaboration and Partnerships

Collaboration and partnerships between Fintech organizations, academic institutions, and industry associations can help accelerate talent and human capital development. These collaborations can facilitate knowledge sharing, provide access to specialized expertise, and create joint research and development opportunities. By working together, stakeholders can pool resources and expertise, drive innovation, and develop the skilled workforce needed to support the growth and competitiveness of the Fintech ecosystem (Ruhland & Wiese, 2022).

1.4.5   ETHICAL CONSIDERATIONS AND RESPONSIBLE INNOVATION

As Fintech continues to revolutionize the financial services industry, addressing ethical considerations and promoting responsible innovation are crucial. This ensures that the Fintech ecosystem remains sustainable and inclusive while protecting consumers and maintaining market integrity. The following are key aspects of ethical considerations and responsible innovation in Fintech.

1.4.5.1   Data Privacy and Security

With the increasing reliance on data-driven technologies, Fintech companies must prioritize data privacy and security to protect sensitive customer information. Ensuring robust data protection measures and adhering to privacy regulations, such as General Data Protection Regulation and CCPA, are essential to maintaining customer trust and preventing unauthorized access to sensitive information. Fintech companies should adopt transparent data management practices and clearly communicate their data usage policies to users.

1.4.5.2   Fairness and Non-discrimination

Fintech companies should strive for fairness and non-discrimination in their products and services. This includes addressing potential biases in the algorithms used for decision-making, credit scoring, and risk assessment. By incorporating principles of fairness and transparency into their AI and ML models, Fintech firms can ensure that their services do not unfairly disadvantage certain customer segments and contribute to a more inclusive financial ecosystem.

1.4.5.3   Financial Inclusion and Accessibility

Responsible innovation in Fintech involves promoting financial inclusion and making financial services more accessible to underserved populations. Fintech companies should design their products and services with accessibility, catering to the unique needs of different customer segments, including those with low income, limited financial literacy, or physical disabilities. By leveraging technology to bridge the financial services gap, Fintech firms can reduce inequalities and promote economic growth.

1.4.5.4   Environmental, Social, and Governance (ESG) Factors

Fintech companies should incorporate environmental, social, and governance (ESG) factors into their business strategies and decision-making processes. By focusing on sustainable and responsible practices, Fintech firms can create long-term value for their stakeholders and contribute to a more sustainable financial system. This includes adopting green finance initiatives, supporting social impact investments, and ensuring good corporate governance (Nicoletti & Nicoletti, 2021).

1.4.5.5   Collaboration with Regulators and Industry Stakeholders

To promote ethical considerations and responsible innovation in the Fintech ecosystem, Fintech companies need to engage in dialogue and collaborate with regulators and industry stakeholders. This can help develop appropriate regulatory frameworks, share best practices, and foster a culture of responsible innovation. Fintech firms should actively participate in industry forums, engage in public–private partnerships, and contribute to developing industry-wide standards and guidelines (Anagnostopoulos, 2018).

Ethical considerations and responsible innovation are vital in ensuring the sustainable growth of the Fintech industry. By focusing on data privacy and security, fairness and non-discrimination, financial inclusion and accessibility, ESG factors, and collaboration with regulators and industry stakeholders, Fintech companies can contribute to building a more equitable and responsible financial ecosystem.

1.4.6   CROSS-BORDER COLLABORATION AND MARKET EXPANSION

Cross-border collaboration and market expansion are crucial aspects of the Fintech ecosystem, enabling businesses to extend their reach, explore new markets, and create growth opportunities. The increasing globalization of the financial services sector has paved the way for Fintech companies to capitalize on this trend by forming strategic partnerships, adopting innovative business models, and leveraging technology to break geographical barriers. The following are some key factors driving cross-border collaboration and market expansion in the Fintech industry.

1.4.6.1   Regulatory Harmonization

Regulatory harmonization is critical to cross-border collaboration and market expansion for Fintech companies. By aligning regulatory frameworks and supervisory practices across jurisdictions, regulators can create a more conducive environment for Fintech companies to operate in multiple markets. Collaborative efforts, such as regulatory sandboxes and passporting arrangements, can facilitate smoother market entry for Fintech firms and help them navigate complex regulatory landscapes (Bromberg et al., 2018).

1.4.6.2   Strategic Partnerships and Alliances

Fintech companies can leverage strategic partnerships and alliances to expand their presence in foreign markets and tap into new customer segments. By forming partnerships with local financial institutions, Fintech firms can access local market knowledge, distribution networks, and customer bases, accelerating their growth and market penetration. In addition, collaboration between Fintech companies and traditional financial institutions can foster innovation and enhance the overall customer experience (Thomas & Hedrick-Wong, 2019).

1.4.6.3   Technology-Driven Solutions

Advancements in technology, particularly in mobile connectivity, cloud computing, and blockchain, have made it easier for Fintech companies to scale their operations across borders. These technologies enable Fintech firms to deliver financial services more efficiently and cost-effectively, making it more feasible for them to enter new markets and serve customers in remote or underserved areas.

1.4.6.4   Market Demand and Competitive Landscape

The growing demand for digital financial services and the increasingly competitive landscape drive Fintech companies to explore cross-border market expansion opportunities. As the adoption of digital financial services continues to rise, Fintech firms are seeking to capitalize on this trend by expanding their reach and capturing market share in new regions. Moreover, the increasing competition within the Fintech industry is encouraging companies to differentiate themselves by exploring untapped markets and establishing a strong presence in emerging economies.

1.4.6.5   Talent and Knowledge Sharing

Cross-border collaboration and market expansion also facilitate the exchange of talent, ideas, and knowledge within the Fintech ecosystem. By expanding into new markets, Fintech companies can tap into diverse pools of talent and access innovative ideas, helping them improve their products and services and stay ahead in the rapidly evolving Fintech landscape. Furthermore, knowledge sharing and collaboration between Fintech companies and local stakeholders can contribute to the overall development of the Fintech ecosystem in different regions.

1.4.7   FINTECH AND SUSTAINABLE DEVELOPMENT

Fintech holds great promise for promoting sustainable development and addressing ESG challenges. By leveraging cutting-edge technologies and innovative business models, Fintech can contribute to a more inclusive, responsible, and sustainable financial system (Zhang et al., 2021). The following are some key areas in which Fintech can drive sustainable development:

1.4.7.1   Green Finance and Impact Investing

Fintech can be pivotal in channeling investments toward projects and companies prioritizing sustainability and ESG factors. Digital platforms like robo-advisors, crowdfunding, and peer-to-peer lending can facilitate access to green finance and impact investing opportunities. By providing investors with more accessible and transparent information about sustainable investments, Fintech can promote responsible investing practices and help drive capital toward projects that contribute to sustainable development goals (Berrou et al., 2019).

1.4.7.2   ESG Data and Analytics

Fintech can enable better ESG data collection, analysis, and reporting by harnessing the power of AI, ML, and big data analytics. Enhanced ESG data and analytics can help investors make more informed decisions, identify potential risks and opportunities, and drive demand for sustainable products and services. Furthermore, ESG data and analytics advancements can promote greater transparency and accountability in corporate sustainability reporting, fostering a culture of continuous improvement and responsible business practices (In et al., 2019).

1.4.7.3   Climate Risk Assessment and Resilience

Fintech can contribute to building resilience against climate-related risks by providing tools and solutions for better understanding and managing such risks. Advanced data analytics, ML, and AI can help financial institutions assess climate-related risks, enabling them to better price insurance products, manage portfolios, and allocate capital. Additionally, Fintech can help businesses and governments develop and implement more effective strategies for adapting to and mitigating the impacts of climate change (Ng & Kwok, 2017).

1.4.7.4   Circular Economy and Resource Efficiency

Fintech can support the transition toward a circular economy by enabling more efficient use of resources and promoting waste reduction. For example, blockchain technology can facilitate more transparent and traceable supply chains, allowing businesses and consumers to make more sustainable choices. Similarly, innovative financing models, such as pay-per-use or performance-based financing, can incentivize resource efficiency and waste reduction by aligning financial incentives with sustainable outcomes (Pizzi et al., 2021).

1.4.7.5   Financial Inclusion and Social Impact

Fintech solutions can promote financial inclusion and address social challenges by providing affordable and accessible financial services to underserved populations. Mobile money, digital wallets, and alternative lending platforms can enable access to essential financial services, such as savings, credit, and insurance, for those traditionally excluded from the formal financial system. By fostering financial inclusion, Fintech can contribute to poverty alleviation, economic growth, and social development (Milana & Ashta, 2020). Table 1.2 presents an overview of the challenges and opportunities in the Fintech ecosystem.

TABLE 1.2

Challenges and Opportunities in the Fintech Ecosystem

Aspect	Opportunities	Challenges
Financial inclusion and empowering the unbanked	- Expand access to financial services for unbanked and underbanked populations - Encourage innovation in products tailored to customer needs	- Ensuring digital and financial literacy among users - Addressing infrastructure limitations, e.g., internet access and connectivity - Designing effective regulatory frameworks for emerging markets
Cybersecurity and data privacy	- Foster advanced security technologies to protect user data - Implement robust authentication mechanisms - Promote collaboration between Fintech firms and regulatory authorities	- Managing cyber threats and data breaches - Balancing data privacy with the need for innovation - Ensuring compliance with evolving data protection regulations
Regulatory challenges and the need for collaboration	- Adopt agile and flexible regulatory frameworks to encourage innovation - Facilitate public–private partnerships to shape policy and industry standards	- Keeping pace with rapid technological advancements - Addressing potential regulatory arbitrage - Coordinating international regulatory efforts and harmonizing rules
Talent and human capital development	- Attract top talent to drive Fintech innovation - Develop interdisciplinary training programs combining finance, technology, and entrepreneurship	- Addressing skills gaps and shortages in the Fintech workforce - Balancing the need for traditional financial expertise with emerging technology skills - Fostering a culture of continuous learning and development
Ethical considerations and responsible innovation	- Promote responsible innovation through transparent and ethical practices - Develop guidelines for AI and data usage in Fintech - Encourage the integration of ESG factors in Fintech offerings	- Ensuring algorithmic fairness and avoiding biases in AI-driven Fintech applications - Addressing potential social and economic inequalities stemming from Fintech adoption - Maintaining trust in the financial system as technology advances

Fintech has the potential to greatly expand financial inclusion and empower the unbanked population. However, it faces challenges in ensuring digital and financial literacy, addressing infrastructure limitations, and designing effective regulatory frameworks. Cybersecurity and data privacy are crucial aspects of Fintech, with opportunities in advanced security technologies and collaboration between firms and regulatory authorities. The challenges lie in managing cyber threats, balancing data privacy, and ensuring compliance with evolving regulations.

Regulatory challenges require agile and flexible frameworks that encourage innovation while maintaining stability in the financial sector. Collaboration between public and private stakeholders is essential to shaping policies and industry standards. Talent and human capital development are crucial for driving Fintech innovation, but addressing skills gaps and fostering a culture of continuous learning remain key challenges.

Ethical considerations and responsible innovation are essential for maintaining trust in the financial system. Opportunities lie in promoting responsible practices, developing AI and data usage guidelines, and integrating ESG factors into Fintech offerings. Ensuring algorithmic fairness, addressing potential inequalities stemming from Fintech adoption, and maintaining trust in the financial system must be addressed as the Fintech landscape evolves.

1.5 CONCLUSION

In conclusion, this introductory chapter presents a detailed and holistic overview of the Fintech landscape by examining key trends, innovations, and their transformative impact on the financial industry. This chapter offers valuable insights into the evolving financial ecosystem by analyzing the influence of technologies such as AI, ML, blockchain, digital payments, and mobile financial services. Furthermore, it sheds light on the challenges and opportunities associated with these developments, addressing essential aspects such as financial inclusion, cybersecurity, data privacy, regulatory compliance, talent development, and ethical considerations.

This comprehensive analysis significantly contributes to the scholarly understanding of the dynamic and rapidly evolving Fintech domain. By exploring the potential advantages and drawbacks of these innovations, this chapter provides a balanced perspective on this critical area of contemporary finance. As a result, readers are equipped with a thorough understanding of the current state of Fintech and are better prepared to navigate and adapt to the ongoing changes within the financial landscape.

REFERENCES

Ahern, D. (2021). Regulatory lag, regulatory friction and regulatory transition as FinTech disenablers: Calibrating an EU response to the regulatory sandbox phenomenon. *European Business Organization Law Review*, 22(3), 395–432.

Ahmed, U. (2019). The Importance of cross-border regulatory cooperation in an era of digital trade. *World Trade Review*, 18(S1), S99–S120.

Anagnostopoulos, I. (2018). Fintech and regtech: Impact on regulators and banks. *Journal of Economics and Business*, 100, 7–25.

Armstrong, P. (2018). Developments in RegTech and SupTech. *Paris: Paris Dauphine University. Available Online at: https://www.esma.europa.eu/sites/default/files/library/esma71-99-1070_speech_on_regtech.pdf.*

Arner, D. W., Barberis, J., & Buckley, R. P. (2015). The evolution of Fintech: A new post-crisis paradigm. *Georgetown Journal of International Law*, 47, 1271.

Bähre, H., Buono, G., & Elss, V. I. (2020). Fintech as a mean for digital and financial inclusion. *LUMEN Proceedings*, 14, 205–211.

Bao, Y., Hilary, G., & Ke, B. (2022). Artificial intelligence and fraud detection. *Innovative Technology at the Interface of Finance and Operations*: Volume I, 223–247. doi: 10.2139/ssrn.3738618.

Berrou, R., Dessertine, P., & Migliorelli, M. (2019). An overview of green finance. *The Rise of Green Finance in Europe*: Opportunities and Challenges for Issuers, Investors and Marketplaces, 3–29. doi: 10.1007/978-3-030-22510-0_1.

Bhyer, B. S., & Lee, S. (2019). Banking the Unbanked and Underbanked: RegTech as an Enabler for Financial Inclusion. *The RegTech Book*. doi: 10.1002/9781119362197.ch61.

Bradford, T. (2020). Neobanks: Banks by any other name. *Federal Reserve Bank of Kansas City, Payments System Research Briefing*, 12, 1–6.

Brogaard, J., Hendershott, T., & Riordan, R. (2014). High-frequency trading and price discovery. *The Review of Financial Studies*, 27(8), 2267–2306. doi: 10.1093/rfs/hhu032.

Bromberg, L., Godwin, A., & Ramsay, I. (2017). Fintech sandboxes: Achieving a balance between regulation and innovation. *Journal of Banking and Finance Law and Practice*, 28(4), 314–336.

Bromberg, L., Godwin, A., & Ramsay, I. (2018). Cross-border cooperation in financial regulation: Crossing the Fintech bridge. *Capital Markets Law Journal*, 13(1), 59–84.

Cao, L., Yang, Q., & Yu, P. S. (2021). Data science and AI in FinTech: An overview. *International Journal of Data Science and Analytics*, 12, 81–99.

Chen, Y., & Bellavitis, C. (2020). Blockchain disruption and decentralized finance: The rise of decentralized business models. *Journal of Business Venturing Insights*, 13, e00151.

Deo, S., & Sontakke, N. (2021). User-centric explainability in fintech applications. *HCI International 2021-Posters: 23rd HCI International Conference, HCII 2021, Virtual Event, July 24–29, 2021, Proceedings, Part II 23*, 481–488.

Derry, T. K., & Williams, T. I. (1960). *A Short History of Technology from the Earliest Times to AD 1900* (Vol. 231). Courier Corporation, Chelmsford, MA.

Doherty, O., & Stephens, S. (2023). Hard and soft skill needs: Higher education and the Fintech sector. *Journal of Education and Work*, 36, 1–16.

Everett, C. R. (2019). Origins and development of credit-based crowdfunding. *Available at SSRN 2442897.*

Fahy, L. A. (2022). Fostering regulator-innovator collaboration at the frontline: A case study of the UK's regulatory sandbox for fintech. *Law & Policy*, 44(2), 162–184.

Frydrych, D., Bock, A. J., Kinder, T., & Koeck, B. (2014). Exploring entrepreneurial legitimacy in reward-based crowdfunding. *Venture Capital*, 16(3), 247–269.

Gai, K., Qiu, M., Sun, X., & Zhao, H. (2017). Security and privacy issues: A survey on FinTech. *Smart Computing and Communication: First International Conference, SmartCom 2016, Shenzhen, China, December 17-19, 2016, Proceedings 1*, 236–247.

Garcia-Teruel, R. M. (2019). A legal approach to real estate crowdfunding platforms. *Computer Law & Security Review*, 35(3), 281–294.

Goldstein, A., & O'Connor, D. (2000). *E-commerce for development: prospects and policy issues.* Working Paper No. 164, OECD Development Centre.

Gomber, P., Koch, J.-A., & Siering, M. (2017). Digital finance and FinTech: current research and future research directions. *Journal of Business Economics*, 87, 537–580.

Hastings, J. S., Madrian, B. C., & Skimmyhorn, W. L. (2013). Financial literacy, financial education, and economic outcomes. *Annual Review of Economics*, 5(1), 347–373.

Havrylchyk, O., & Verdier, M. (2018). The financial intermediation role of the P2P lending platforms. *Comparative Economic Studies*, 60, 115–130.

Hussain, M., Nadeem, M. W., Iqbal, S., Mehrban, S., Fatima, S. N., Hakeem, O., & Mustafa, G. (2021). Security and privacy in FinTech: A policy enforcement framework. In Information Resources Management Association (ed.) *Research Anthology on Concepts, Applications, and Challenges of FinTech* (pp. 372–384). IGI Global, Hershey, PA.

In, S. Y., Rook, D., & Monk, A. (2019). Integrating alternative data (also known as ESG data) in investment decision making. *Global Economic Review*, 48(3), 237–260.

Jack, W., & Suri, T. (2011). *Mobile Money: The Economics of M-PESA.* National Bureau of Economic Research, Cambridge, MA.

Jakšič, M., & Marinč, M. (2019). Relationship banking and information technology: The role of artificial intelligence and FinTech. *Risk Management*, 21, 1–18.

Kapoor, A., Sindwani, R., & Goel, M. (2020). Mobile wallets: Theoretical and empirical analysis. *Global Business Review*, 0972150920961254.

Kaur, G., Habibi Lashkari, Z., Habibi Lashkari, A., Kaur, G., Habibi Lashkari, Z., & Habibi Lashkari, A. (2021). Cybersecurity threats in FinTech. In G. Kaur, Z. H. Lashkari, & A. H. Lashkari (eds.), *Understanding Cybersecurity Management in FinTech: Challenges, Strategies, and Trends,* (pp. 65–87). Springer, Cham.

Kaur, G., Lashkari, Z. H., & Lashkari, A. H. (2021). *Understanding Cybersecurity Management in FinTech.* Springer, Cham.

Knewtson, H. S., & Rosenbaum, Z. A. (2020). Toward understanding FinTech and its industry. *Managerial Finance*, 46(8), 1043–1060.

Koomson, I., Martey, E., & Etwire, P. M. (2022). Mobile money and entrepreneurship in East Africa: The mediating roles of digital savings and access to digital credit. *Information Technology & People*, 36, 996–1019.

Liu, C. Z., Zafar, H., & Au, Y. A. (2014). Rethinking fs-isac: An it security information sharing network model for the financial services sector. *Communications of the Association for Information Systems*, 34(1), 2.

Lukonga, I. (2021). Fintech and the real economy: Lessons from the Middle East, North Africa, Afghanistan, and Pakistan (MENAP) region. In M. Pompella & R. Matousek (eds.), *The Palgrave Handbook of Fintech and Blockchain* (pp. 187–214). Palgrave Macmillan Cham. doi: 10.1007/978-3-030-66433-6_8.

Melanie, S. (2015). *Blockchain: Blueprint for a New Economy.* O'Reilly Media, Sebastopol, CA.

Maleh, Y., Lakkineni, S., Tawalbeh, L., & AbdEl-Latif, A. A. (2022). Blockchain for cyber-physical systems: Challenges and applications. In Y. Maleh, L. Tawalbeh, S. Motahhir, & A. S. Hafid (eds.), *Advances in Blockchain Technology for Cyber Physical Systems* (pp. 11–59). Springer International Publishing, Cham. doi: 10.1007/978-3-030-93646-4_2.

Maleh, Y., Mamoun, A., Imed, R., Youssef, B., & Loai, T. (eds.). (2021). *Artificial Intelligence and Blockchain for Future Cybersecurity Applications* (Vol. 90). Springer International Publishing, Cham. doi: 10.1007/978-3-030-74575-2.

Maleh, Y., Shojafar, M., Alazab, M., & Romdhani, I. (2020). *Blockchain for Cybersecurity and Privacy: Architectures, Challenges, and Applications.* CRC Press, London, UK. doi: 10.1201/9780429324932.

Milana, C., & Ashta, A. (2020). Microfinance and financial inclusion: Challenges and opportunities. *Strategic Change*, 29(3), 257–266.

Mochkabadi, K., & Volkmann, C. K. (2020). Equity crowdfunding: A systematic review of the literature. *Small Business Economics*, 54, 75–118.

Nakamoto, S. (2008). Bitcoin: A peer-to-peer electronic cash system. *Decentralized Business Review*, 21260.

Ng, A. W., & Kwok, B. K. B. (2017). Emergence of Fintech and cybersecurity in a global financial centre: Strategic approach by a regulator. *Journal of Financial Regulation and Compliance*, 25, 422–434.

Nicoletti, B., & Nicoletti, B. (2021). Proposition of value and Fintech organizations in Banking 5.0. In B. Nicoletti (ed.), *Banking 5.0: How Fintech Will Change Traditional Banks in the'New Normal'Post Pandemic* (pp. 91–152). Springer Nature, Cham. doi: 10.1007/978-3-030-75871-4_4.

Oualid, A., Maleh, Y., & Moumoun, L. (2023). Federated learning techniques applied to credit risk management: A systematic literature review. *EDPACS*, 68(1), 42–56. doi: 10.1080/07366981.2023.2241647.

Palmié, M., Wincent, J., Parida, V., & Caglar, U. (2020). The evolution of the financial technology ecosystem: An introduction and agenda for future research on disruptive innovations in ecosystems. *Technological Forecasting and Social Change*, 151, 119779.

Pizzi, S., Corbo, L., & Caputo, A. (2021). Fintech and SMEs sustainable business models: Reflections and considerations for a circular economy. *Journal of Cleaner Production*, 281, 125217.

Qiu, T., Zhang, R., & Gao, Y. (2019). Ripple vs. SWIFT: Transforming cross border remittance using blockchain technology. *Procedia Computer Science*, 147, 428–434.

Ruhland, P., & Wiese, F. (2022). FinTechs and the financial industry: Partnerships for success. *Journal of Business Strategy*, 44, 228–237.

Sullivan, C., & Burger, E. (2019). Blockchain, digital identity, e-government. *Business Transformation through Blockchain: Volume II*, 233–258. doi: 10.1007/978-3-319-99058-3_9.

Thomas, H., & Hedrick-Wong, Y. (2019). How digital finance and fintech can improve financial inclusion1. In *Inclusive Growth* (pp. 27–41). Emerald Publishing Limited, Bingley.

Treleaven, P., Galas, M., & Lalchand, V. (2013). Algorithmic trading review. *Communications of the ACM*, 56(11), 76–85. doi: 10.1145/2500117.

Wang, S., Huang, C., Li, J., Yuan, Y., & Wang, F.-Y. (2019). Decentralized construction of knowledge graphs for deep recommender systems based on blockchain-powered smart contracts. *IEEE Access*, 7, 136951–136961. doi: 10.1109/ACCESS.2019.2942338.

Xu, A., Liu, Z., Guo, Y., Sinha, V., & Akkiraju, R. (2017). A new chatbot for customer service on social media. In *Proceedings of the 2017 CHI Conference on Human Factors in Computing Systems*, 3506–3510. doi: 10.1145/3025453.3025496.

Yan, L.-Y., Tan, G. W.-H., Loh, X.-M., Hew, J.-J., & Ooi, K.-B. (2021). QR code and mobile payment: The disruptive forces in retail. *Journal of Retailing and Consumer Services*, 58, 102300.

Yan, T. C., Schulte, P., & Chuen, D. L. K. (2018). InsurTech and FinTech: banking and insurance enablement. *Handbook of Blockchain, Digital Finance, and Inclusion, Volume 1*, 249–281. doi: 10.1016/B978-0-12-810441-5.00011-7.

Yang, Y.-P., & Tsang, C.-Y. (2018). Regtech and the new era of financial regulators: envisaging more public-private-partnership models of financial regulators. *University of Pennsylvania Journal of Business Law*, 21, 354.

Yum, H., Lee, B., & Chae, M. (2012). From the wisdom of crowds to my own judgment in microfinance through online peer-to-peer lending platforms. *Electronic Commerce Research and Applications*, 11(5), 469–483.

Zhang, Y., Chen, J., Han, Y., Qian, M., Guo, X., Chen, R., Xu, D., & Chen, Y. (2021). The contribution of Fintech to sustainable development in the digital age: Ant forest and land restoration in China. *Land Use Policy*, 103, 105306.

Section 1

AI and Machine Learning
in Fintech and Beyond

2 E-Payment Fraud Detection in E-Commerce using Supervised Learning Algorithms

Manal Loukili, Fayçal Messaoudi, and Hanane Azirar

2.1 INTRODUCTION

E-commerce has become a part of our daily lives. It is the process of buying and selling products or services via the internet. We can buy anything from books, groceries, clothes, and even cars online. The importance of e-commerce in our daily lives is that it saves time and money. With e-commerce, we don't have to drive around looking for a store that sells what we need. We can search for it on our phones or laptops and order it immediately. This makes shopping faster and easier than ever before.

Forecasters predict that 95% of all purchases will be online by 2040 [1]. As mobile devices and the Internet of Things make it easier than ever to buy items on demand, consumers seem to be finding the ability to shop online seamless. According to Statista, this will contribute to the growing volume of e-commerce sales worldwide, forecast to peak at $4.9 billion in 2021 and nearly $7.3 billion by 2025, as shown in Figure 2.1.

In e-commerce, credit cards are the most popular and widespread payment channel used to pay for purchases [2]. The rapid growth of e-commerce has led to an increase in the use of credit cards in the electronic world, making credit cards a target for fraudsters. Credit card fraud data shows that credit card fraud has both increased and diversified with the increase in e-shopping. Traditional fraud detection systems are inadequate to detect the variety of credit card frauds. As a result, cybercrime is flamingly exasperated worldwide every day. Credit card fraud is essentially based on the unauthorized and intrusive usage of a credit card for financial gain without the cardholder's permission using different methods that can be classified into two broad categories: offline fraud and online fraud.

We can define offline fraud as a physical operation such as the theft of a handbag or portfolio containing valuable items, such as payment cards and IDs, and the use of the crucial information they have. On the other hand, online fraud refers to when the fraudster utilizes an online platform or creates a website and pretends to be authentic to collect important personal data and make illicit payments on the different accounts of the customers. Other means used by fraudsters to collect or steal personal data are phishing, hacking, ID fraud, shoulder surfing, spyware, dumping diving, etc. [3].

Owing to the notable amounts and the illegal activities of fraudsters, e-payment fraud detection in e-commerce is becoming increasingly critical and imperative. However, traditional fraud detection solutions cannot adequately detect credit card fraud in diversified e-commerce environments. As artificial intelligence methods have evolved and advanced computers with high computational capability have become widespread, machine learning and artificial intelligence techniques have been used to identify credit card fraud more rapidly and accurately. Both fraudsters and fraud detectors are constantly fighting to detect credit card scams. Machine learning methods empower fraud detection experts.

DOI: 10.1201/9781032667478-3

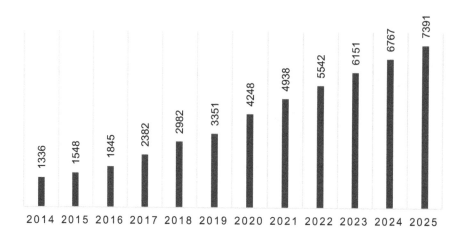

FIGURE 2.1 Global e-commerce retail sales between 2014 and 2025 in billions of U.S. dollars.

In this background, this chapter focuses on implementing a machine learning system to detect fraudulent e-payments by comparing three supervised machine learning models, CatBoost, AdaBoost, and XGBoost, based on their metrics and latency.

This chapter is structured into five sections: The next section presents some recent related work. Section 2.3 briefly defines the machine learning algorithms used. Section 2.4 describes the methodology adopted and the results obtained. Section 2.5 concludes the chapter.

2.2 LITERATURE REVIEW

The importance of machine learning in the e-business world is undeniable, as it provides effective tools to address many business challenges. Among the most common applications are customer churn prediction [4], market segmentation [5], sentiment analysis [6], and payment fraud detection. In particular, detecting payment fraud is essential for online businesses as it can help avoid significant financial losses.

Specific algorithms based on machine learning, deep learning, and data mining techniques are developed and applied to detect online fraudulent payment cases with high accuracy, precision, and a minimum of time, effort, and cost. However, the data distribution used in fraud prevention is very unbalanced. Thus, to overcome this drawback, undersampling and oversampling methods are employed to generate relatively balanced data [7].

In [8], the authors discussed the challenges and issues related to detecting fake online payments and fraud in e-commerce transactions. They presented data analysis and artificial intelligence techniques used to prevent, detect, and mitigate fraudulent online payments in e-commerce and financial transactions.

In [9], the authors highlighted the opportunities, challenges, and various threats related to electronic payments, particularly fraud, as one of the most severe risks in e-payments, resulting in considerable financial losses. They also discussed the various types of e-payments, the advantages, and the prospects of e-payments.

The authors of [10] implemented a machine learning-based credit card fraud detection system applying a genetic algorithm for feature selection. The system uses machine learning classifiers, including Random Forest, Decision Tree, Artificial Neural Network, Logistic Regression, and Naive Bayes. Experimental results showed that this credit card fraud detection system outperforms existing systems.

In [11], the author proposed an improved two-level credit card fraud detection system from imbalanced datasets using k-means semantic fusion and the artificial bee colony algorithm to increase the identification accuracy and speed up the detection convergence. This model filters features from the dataset with an embedded rule engine to assess whether the transaction is legitimate or fraudulent according to numerous customer behavior parameters, including use frequency, geographic location, and book balance. It was found that the proposed solution can significantly improve the classification accuracy of fraudulent transactions.

In [12], the author discussed the different types of credit card fraud in the age of advanced technology. He also reviewed the different deep learning methods. Finally, he suggested that organizations upgrade their security measures by moving from conventional data protection approaches to deep learning techniques such as IPS and IDS to ensure real-time fraud detection and execution of security measures to mitigate maliciousness.

The authors in [13] proposed a hybrid model based on the combination of TabNet and XGBoost, with a transactional data set that shows whether each payment is fraudulent. Finally, when comparing their proposed model with two classical models, they found that their hybrid model outperformed the other models in terms of accuracy of 89.2% and AUC-ROC score of 97.2%.

2.3 BACKGROUND

Machine learning is a branch of artificial intelligence science that employs mathematical and statistical methods to provide machines with the capacity to learn directly from data, i.e., enhance their performance in sorting out tasks without getting explicitly programmed for each task. To a large extent, it concerns the analysis, design, development, optimization, and implementation of these approaches.

Machine learning generally involves two stages [14]. The first step consists of estimating a model based on data, known as observations, which are available and in finite numbers at the system's design phase. Estimating the model entails carrying out a practical task, such as translating a speech, estimating a probability density, recognizing the presence of a horse in a picture, or participating in driving an autonomous vehicle. This "learning" or "training" process usually occurs before the model is used in practice.

The subsequent stage relates to the creation dispatch: the model is resolved, and presently data would then be able to be submitted to acquire the desired outcome. In practice, some systems may continue their learning once in production, as long as they have means of obtaining feedback on the quality of the results produced.

During the learning phase, the nature of learning is determined by the type of information available. If the data is labeled, meaning that the answer to the task is already known for this data, it is considered to be supervised learning. In supervised learning, the labels can be discrete for classification or ranking tasks or continuous for regression tasks. If the data is not labeled, it is unsupervised learning. We talk about clustering or association. We talk about classification or clustering if we have labeled and unlabeled data. It is semi-supervised learning. If the model is learned incrementally based on a reward received by the program for each of the actions undertaken, this is referred to as reinforcement learning.

This chapter focuses on supervised learning as it seeks to solve a classification problem (whether the payment is fraudulent or legitimate). Several classifiers include Decision Tree Classifier, k-Nearest Neighbors, Random Forest Classifier, Gradient Boosting Classifier, Support Vector Machine, and Naive Bayes. In this chapter, boosting algorithms are used for classification.

Boosting is an ensemble learning technique that merges a group of weak learners into one strong learner to reduce training errors [15]. A random sample of data is collected and equipped with a model, which is then trained sequentially, i.e., each model attempts to compensate for the weaknesses of its predecessor.

In this chapter, we will implement three unsupervised boosting algorithms: CatBoost, AdaBoost, and XGBoost.

- AdaBoost, also known as Adaptive Boosting, is an algorithm that takes advantage of mis-classified instances from previous models to create a new weighted data sample where those misclassified instances are given more importance [16]. This results in a new model that is more capable of resolving the errors of the original model.
- XGBoost, or Extreme Gradient Boosting, is an improved version of gradient boosting that aims to increase computational resources for the technique [17]. It includes two key features: regularization to reduce overfitting and optimization of the ordering through parallel execution to increase execution speed. The algorithm also prunes the decision tree based on the maximum depth parameter to significantly reduce execution time.
- CatBoost, or Categorical Boosting, automatically handles categorical variables without the need for conversion to numeric values [18]. This method performs well when there are multiple categorical variables to be processed.

2.4   METHODOLOGY

This chapter aims to predict fraudulent e-payment transactions, i.e., whether a transaction is fraudulent or legitimate. We used classification-supervised machine learning algorithms to simplify the fraudulent payment detection task and make it a binary classification problem. Thus, we used boosting methods because it is an approach that gathers many algorithms that rely on sets of binary classifiers, as boosting optimizes their performance. To do this, we follow the steps presented in the following:

- Data set description

    The data set used was generated by an agent-based bank payment simulator from a sample of aggregated transactional data provided by a bank. It aims to generate artificial data that can be used for research on fraud detection. This payment simulator was run for 180 steps, equivalent to six months of transactions, to obtain a distribution close enough to be reliable for testing. The system generated 594,643 transactions, of which 587,443 are legitimate payments and 7200 are fraudulent (Table 2.1).
- Data preprocessing

    The purpose of building a fraud classification model is to assign to each new incoming transaction with high certainty a probability of it being a fraud. Hence, any illegal attempt can be avoided. Figure 2.2 shows the normalization of transaction amounts.

    Large transactions are more susceptible to fraud. Therefore, comparing the values of two transactions for different clients is incorrect because the transaction amounts vary by client. Thus, the transactions for each client were compared by dividing the transaction value for each client by the maximum transaction amount that the same client holds. This method allowed the model to determine abnormal behavior for each customer transaction. Since the dataset consists of 4112 customers and each customer has a certain number of transactions, normalizing the transaction amount will range from 0 to 1, as shown in Figure 2.2.
- Feature engineering

    Feature engineering is among the most important steps in preprocessing a modeling problem. Its objective is to transform the original raw data into dimensions more accurately represented by a predictive model, thereby enhancing the evaluation metrics considerably. Ensuring that none of the features used to train the model will be available again when a new case arises is another important step in the feature engineering process.
- Hot encoding

    Hot coding was employed, a common approach for categorical preprocessing features in machine learning models [19]. This method involves creating a new binary feature for each possible category and assigning a value of 1 to the feature for each sample that corresponds to its original classification.

TABLE 2.1
Data Set

Step		Customer	Age	Gender	zipcodeOri	Merchant	zipMerchant	Category	Amount	Fraud
0	0	C1093826151	'4'	'M'	28007	"M348934600"	28007	'es_transportation'	4.55	0
1	0	'C352968107'	2	'M'	28007	"M348934600"	28007	'es_transportation'	39.68	0
2	0	C2054744914	'4	'F'	28007	"M1823072687	28007	'es_transportation'	26.89	0
3	0	C17606127907	3	'M'	28007	'M348934600'	28007	'es_transportation'	17.25	0
4	0	C757503768	5	'M'	28007	'M348934600'	28007	'es_transportation'	35.72	0
...			-							
594638	179	C1753498738	'3'	'F	28007	"M1823072687	28007	"es_transportation'"	20.53	0
594639	179	C650108285	'4'	'F	28007	'M1823072687	28007	'es_transportation'	50.73	0
594640	179	C123623130	2	'F	28007	M349281107	28007	'es_transportation'	22.44	0
594641	179	C1499363341	'5	'M	28007	'M1823072687	28007	'es_transportation'	14.46	0
594642	179	C616528518	'4'	'F	28007	'M1823072687	28007	'es_transportation'	26.93	0

594643 row × 10 columns

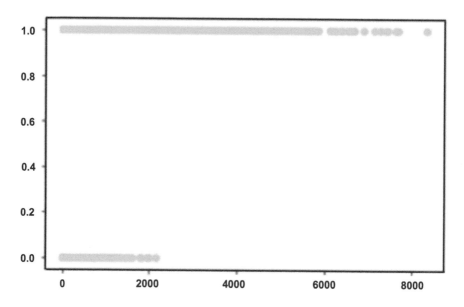

FIGURE 2.2　Normalization of transaction amounts.

- Data splitting

  Extract approximately 20% of the data set for production to simulate the actual data, which will be used to test the model. This data set is different from the testing data set, which is also used for testing the model.
- Oversampling

  As stated before, the fraudulent and legitimate cases available in the current data set have extremely imbalanced weights, with only 1.2% of the cases being fraud. Therefore, only 7,000 observations are labeled as fraud, while the rest, 580,000 observations, are labeled as clean transactions. To overcome this problem, the SMOTE (Synthetic Minority Oversampling Technique) technique is an oversampling algorithm where synthetic samples are generated for the minority class [20]. This algorithm enables surmounting the overfitting.
- Model evaluation and discussion
  - Confusion matrix

    Several parameters were used to evaluate the performance and predictive capacity of the various models implemented for predicting fraudulent transactions on the test set, including precision, recall, accuracy, and F-score [21]. These metrics are derived from the confusion matrix of each model, where true positives (TPs) and false positives (FPs) are represented as TP and FP, and false negatives and true negatives are defined as FN and TN, respectively.
  - Performance metrics
    - Recall: The recall is the TP rate and is calculated as follows:

      $$Recall = TP/(TP+FN) \qquad (1)$$
    - Precision: The precision is the ratio of predicted correct to fraudulent. Its formula is as follows:

      $$Precision = TP/(TP+FP) \qquad (2)$$
    - Accuracy: The accuracy is the ratio of the number of correct predictions and is written as:

      $$Accuracy = (TP+TN)/(TP+FP+TN+FN) \qquad (3)$$
    - F-Score: The F1 score is the harmonic mean of precision and recall and is calculated as follows:

      $$F\text{-}score = (2*Precision*Recall)/(Precision+Recall) \qquad (4)$$

Table 2.2 shows the parameters of the confusion matrices (TP, TN, FP, FN) for each of the XGBoost, CatBoost, and AdaBoost models.

We computed the performance measures from the confusion matrix of the models used (Table 2.1): precision, recall, accuracy, and f1-score. Tables 2.3–2.5 present the performance measures for each proposed model as well as their execution time.

TABLE 2.2
Confusion Matrix of the Model Used

Model	TP	TN	FP	FN
XGBoost	990	91361	2262	30
CatBoost	1020	88089	5534	0
AdaBoost	1003	91288	2335	17

TABLE 2.3
XGBoost Performance Metrics

Class	Precision	Recall	F-score
Fraudulent	0.30	0.97	0.46
Legitimate	1	0.98	0.99
Macro avg	0.65	0.97	0.73
Accuracy	0.98		
Execution time	52.34s		

TABLE 2.4
CatBoost Performance Metrics

Class	Precision	Recall	F-score
Fraudulent	0.16	1	0.27
Legitimate	1	0.94	0.97
Macro avg	0.58	0.97	0.62
Accuracy	0.94		
Execution time	0.02s		

TABLE 2.5
AdaBoost Performance Metrics

Class	Precision	Recall	F-score
Fraudulent	0.30	0.98	0.99
Legitimate	1	0.98	0.99
Macro avg	0.65	0.98	0.72
Accuracy	0.98		
Execution time	5.90s		

While the CatBoost model has the lowest execution time and predicted all fraudulent transactions, it predicted 5534 legitimate transactions out of 93623 legitimate transactions as fraudulent transactions, which reduced its accuracy.

The XGBoost model correctly predicted 990 fraudulent transactions out of 1020 fraudulent transactions, but the response time of XGBoost is the highest.

The AdaBoost model has the same accuracy as XGBoost but a better execution time than XGBoost, as well as correctly predicting more fraudulent transactions than CatBoost, which makes it at the top of the list.

2.5 CONCLUSION

Due to the significant and ongoing financial losses incurred by organizations and the complexity of detecting credit card fraud, detecting online credit card fraud is paramount to better using credit cards and e-commerce platforms.

This chapter suggests simple machine learning methods for fraudulent transaction detection. We performed an experimental comparison study between the three boosting algorithms (XGBoost, AdaBoost, and CatBoost) using several experiments with a synthetic transaction data set. The performance of the proposed methods was evaluated using performance parameters. The experimental results showed that the AdaBoost model outperformed the other algorithms and attained the best performance in terms of accuracy and latency, at 98% and 5.9s, respectively.

REFERENCES

1. Statista. https://www.statista.com/statistics/379046/worldwide-retail-e-commerce-sales/.
2. Oualid, A., Maleh, Y., Moumoun, L.: Federated learning techniques applied to credit risk management: A systematic literature review. *EDPACS*, 68, 1–15 (2023).
3. Singh, P., Singh, M.: Fraud detection by monitoring customer behavior and activities. *International Journal of Computer Applications*, 111(11), 23–32 (2015). doi: 10.5120/19584-1340.
4. Loukili, M., Messaoudi, F., El Ghazi, M.: Supervised learning algorithms for predicting customer churn with hyperparameter optimization. *International Journal of Advances in Soft Computing & Its Applications*, 14(3), 49–63 (2022). doi: 10.15849/IJASCA.221128.04.
5. Ahani, A., Nilashi, M., Ibrahim, O., Sanzogni, L., Weaven, S.: Market segmentation and travel choice prediction in Spa hotels through TripAdvisor's online reviews. *International Journal of Hospitality Management*, 80, 52–77 (2019).
6. Loukili, M., Messaoudi, F., El Ghazi, M.: Sentiment analysis of product reviews for e-commerce recommendation based on machine learning. *International Journal of Advances in Soft Computing and its Application*, 15(1), 1–13 (2023). doi: 10.15849/IJASCA.230320.01.
7. Ogwueleka, F. N.: Data mining application in credit card fraud detection system. *Journal of Engineering Science and Technology*, 6, 12 (2011).
8. Hasan, I., Rizvi, S. A. M.: AI-driven fraud detection and mitigation in e-commerce transactions. In: *Proceedings of Data Analytics and Management*, pp. 403–414. Springer, Singapore (2022). doi: 10.1007/978-981-16-6289-8_34.
9. Nasr, M. H., Farrag, M. H., Nasr, M.: e-payment systems risks, opportunities, and challenges for improved results in e-business. *International Journal of Intelligent Computing and Information Sciences*, 20(1), 16–27 (2020).
10. Shah, S., Shah, D., Shah, N.: Credit card fraud detection system using machine learning. *International Journal of Research in Engineering and Science*, 10(5), 09–14 (2022).
11. Darwish, S. M.: A bio-inspired credit card fraud detection model based on user behavior analysis suitable for business management in electronic banking. *Journal of Ambient Intelligence and Humanized Computing*, 11(11), 4873–4887 (2020).
12. Mohammed, Y. A.: Application of deep learning in fraud detection in payment systems. *International Journal of Innovative Research and Advanced Studies*, 9(4), 81–85 (2022).

13. Cai, Q., He, J.: Credit payment fraud detection model based on TabNet and Xgboot. In: *2022 2nd International Conference on Consumer Electronics and Computer Engineering (ICCECE)*, pp. 823–826. IEEE (2022, January).
14. El Orche, A., Bahaj, M.: Approach to use ontology based on electronic payment system and machine learning to prevent fraud. In: *Proceedings of the 2nd International Conference on Networking, Information Systems & Security*, pp. 1–6 (2019, March).
15. Vanhoeyveld, J., Martens, D., Peeters, B.: Customs fraud detection. *Pattern Analysis and Applications*, 23(3), 1457–1477 (2020).
16. Gedela, B., Karthikeyan, P. R.: Credit card fraud detection using adaboost algorithm in comparison with various machine learning algorithms to measure accuracy, sensitivity, specificity, precision and F-score. In: *2022 International Conference on Business Analytics for Technology and Security (ICBATS)*, pp. 1–6. IEEE (2022, February).
17. Allawala, A., Ramteke, A., Wadhwa, P.: Performance impact of minority class reweighting on XGBoost-based anomaly detection. *International Journal of Machine Learning and Computing*, 12(4), 143–148 (2022).
18. Nguyen, N., Duong, T., Chau, T., Nguyen, V. H., Trinh, T., Tran, D., Ho, T.: A proposed model for card fraud detection based on catboost and deep neural network. *IEEE Access*, 10, 96852–96861 (2022).
19. Mohana, M., Kumaran, K., Nandhini, N., Ananthi, N., Naresh, J., Madhavan, K. R.: Credit card fraud detection using neural network auto encoders. In: *2022 International Conference on Advances in Computing, Communication and Applied Informatics (ACCAI)*, pp. 1–7. IEEE (2022, January).
20. Obiedat, R., Qaddoura, R., Al-Zoubi, A. M., Al-Qaisi, L., Harfoushi, O., Alrefai, M. A., Faris, H.: Sentiment analysis of customers' reviews using a hybrid evolutionary SVM-based approach in an imbalanced data distribution. *IEEE Access*, 10, 22260–22273 (2022).
21. Aslam, F., Hunjra, A. I., Ftiti, Z., Louhichi, W., Shams, T.: Insurance fraud detection: Evidence from artificial intelligence and machine learning. *Research in International Business and Finance*, 62, 101744 (2022).

3 Artificial Intelligence and Stock Market

Karim Amzile

3.1 INTRODUCTION

Computer, robotics, machine learning, and data storage technology advancements have made it simpler to apply current technologies in business (Alsheibani et al., 2018). Artificial intelligence (AI), one of the most cutting-edge technologies, has gained more attention over the past several decades in a range of social, industrial, and business contexts, most notably the financial one.

Due to the rapid changes in stock prices and, of course, stock market index, predicting the value of stock market index becomes a difficult and challenging task (Obthong et al., 2020). As well as the recent progress of the stock markets have caused significant effects on finance, which can also be more complicated to predict the index.

Nowadays, most financial analysts are directly related to this subject, and the more the technology develops, the more they need to know and predict the index during the periods of crisis and financial instability, which leads them to be interested in the prediction of the index based on sophisticated tools, such as the techniques derived from AI (Qiu et al., 2016). This leads us to question the performance of the techniques stemming from AI in the modeling of the value of stock market index (Anggoro, 2020).

However, AI has opened up fascinating new views on the stock market, especially when it comes to estimating the worth of stock market index. Artificial neural networks (ANN), a subset of AI, have shown they are capable of deciphering intricate data and spotting minute patterns (Dwivedi et al., 2019). By using historical and training data, these ANNs can be trained to recognize patterns and predict future stock index fluctuations with some accuracy. This predictive capability offers investors and traders a considerable advantage, enabling them to make more informed decisions and adjust their strategies accordingly (Gandhmal and Kumar, 2019). However, it should be emphasized that stock markets may be impacted by a variety of unforeseeable circumstances, such as geopolitical events and economic crises, despite the fact that ANNs can offer valuable forecasts. As a result, it's crucial to see projections produced by AI as a helpful tool rather than as a perfect answer. For navigating the complicated and dynamic stock market, a strategy based on a combination of AI and human knowledge frequently produces the greatest results (Bahrammirzaee, 2010).

In this chapter, we developed an ANN model and Support Vectors Machines (SVM) to forecast the value of stock market index in a few African nations during a specified time period. Our approach involves emulating the steps taken by data mining techniques, beginning with pre-processing and exploratory analysis of the data, followed by the definition of the parameters of the approach used to build our model using the proportion of training data, and finally the evaluation of the model using the proportion of test data using the various metrics of validation derived from the confusion matrix (Safari and Ghavifekr, 2021).

Moreover, the main contribution of this chapter lies in the application of ANNs to predict the value of stock market index, specifically focusing on African countries. Also, by utilizing this advanced technique, the research provides valuable insights into the performance and dynamics of African stock markets, which have been relatively underexplored in previous studies.

DOI: 10.1201/9781032667478-4

The application of ANNs to the prediction of stock market index values is arousing keen interest among investors and researchers alike (Chen et al., 2018). The central problem lies in the ability of these methods to provide reliable and accurate forecasts, enabling market players to make informed decisions.

In order to answer the exposed question, we propose to explore for the first time the various theoretical debates realized. Then, we will approach the aspect of the stock exchange index as well as the methods of the ANN and SVM. Finally, we will explore and interpret the different results achieved.

3.2   RELATED WORKS

In order to provide a scientific value to our studies, it is necessary to explore the different results obtained in the different scientific papers published in relation to the different axes of our paper.

However, several studies have revealed the importance of exploiting new methods from AI to obtain accurate prediction results in an attempt to outperform conventional linear and non-linear approaches.

Due to the non-stationarity and volatility of stock market data, it is commonly thought that predicting the future movement of a stock price is a challenging undertaking (Esfahanipour and Aghamiri, 2010). The stock market's chaotic behavior and dynamic price variations have made price prediction a more challenging problem. Investors now face greater challenges in making quick investment decisions due to the stock market's inherent non-linear, dynamic, and complex domain. This necessitates the use of new AI techniques. Therefore, AI in business can be defined as the theory and advancement of computer systems capable of performing tasks that typically require human intelligence (Deloitte, 2017).

However, IBM (2020) defined AI as a field that leverages computing with robust data sets to improve problem solving and business decision-making.

According to Paquet (1997), there are two basic reasons why researchers have become interested in AI techniques and specifically in ANN:

- Since there are no presumptions regarding the functional structure of the link between the features and the likelihood of default, ANNs are more adaptive than certain conventional statistical approaches.
- Since the nature of the link between the variables under study cannot be predicted a priori, ANNs are an effective tool for handling complicated, unstructured issues.

In addition, this overview of the review has led to the conclusion that this area of research is receiving continued attention from researchers and that the literature is becoming increasingly specific and in-depth.

However, for the study conducted by Guresen et al. (2011) as well as the one conducted by Safari et al. (2021), the overall results show that the ANN model performs best in forecasting time series and obviously for stock market index, while the hybrid methods failed to improve the forecasting results.

However, SVM, which was developed by Cortes and Vapnik (1995) and defined by Noble (2006) as an algorithm that learns from historical data to assign labels to new anonymous data, has been the subject of several research studies, such as Pławiak et al. (2019), who showed that the SVM approach is a very powerful technique for default probability prediction.

The use of kernels for non-linearly separable data was made by Suykens and Vandewalle (1999) and Amzile and Amzile (2021), who concluded that an SVM with an RBF kernel has excellent performance and low computational cost.

The use of SVM to calculate customer scores is reported by Huang et al. (2007), who concluded that the score based on the SVM approach correctly classifies credit applications, while Guyon et al. (2002) concluded that the SVM method is very efficient for classification with an accuracy of 98%.

Kara et al. (2011) used two AI methods, ANN and SVM, to predict the annual trend of the Istanbul Stock Exchange index. They obtained an ANN model with an accuracy of 76.74% when tested with their test data set, while their SVM model was 71.52% accurate. Furthermore, according to Kara, ANN has certain limitations because stock market data contains a huge amount of noise, non-stationary characteristics, and complex dimensionality.

3.3 ILLUSTRATION OF ARTIFICIAL NEURAL NETWORKS (ANN)

A typical ANN is made up of a number of layers, each of which gets its inputs from the outputs of the layer before it (Atsalakis and Valavanis, 2009). The neurons that make up the layer below supply information to the N_i Neurons (nr) that comprise each layer I. The first layer is known as the input layer; the last layer, which is made up of a single neuron, is known as the output layer; and the middle levels are known as hidden layers, as shown in Figure 3.1.

3.3.1 AN ARTIFICIAL NEURAL NETWORK'S ARCHITECTURE AND FUNCTIONING

A device that accepts input from other neurons and gives it genuine weights, known as synaptic coefficients or synaptic weights, is referred to as an artificial neuron.

Consider the neuron j of a layer i. Let us note $x_1^i, x_2^i, \ldots, x_{N_{i-1}}^i$ the N_{i-1} inputs from the layer $i-1$ to the neuron j of the layer i. We also consider the N_{i-1} weights denoted $w_{1j}^i, w_{2j}^i, \ldots, w_{N_{i-1}j}^i$. The neuron j calculates the sum of its inputs weighted by the respective synaptic coefficients, to which it adds a constant term called the bias b_j^i. This gives the formula:

$$S_j^i = \sum_{k=1}^{N_{i-1}} w_{kj}^i x_{kj}^i + b_j^i$$

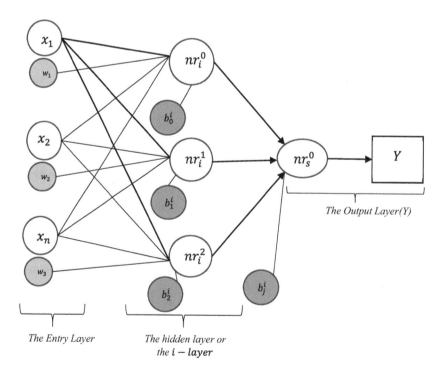

FIGURE 3.1 An artificial neural network's architecture.

The bias is an external parameter of the neuron j. It can be integrated into the weighted sum as the signal x_0^i that takes the value 1, weighted by the weight w_{0j}^i whose value is equal to the bias b_j^i:

$$\begin{cases} x_{0j}^i = 1 \\ b_j^i = w_{0j}^i \end{cases}$$

The sum S_j^i can thus be written as:

$$S_j^i = \sum_{k=0}^{N_{i-1}} w_{kj}^i x_k^i$$

To this sum S_j^i, the neuron applies activation or transfer function φ to obtain an output y_j^i (Figure 3.2).

$$y_j^i = \varphi\left(S_j^i\right) = \varphi\left(\sum_{k=0}^{N_{i-1}} w_{kj}^i x_k^i\right)$$

The output y_j^i (output) of the neuron j neuron in the i layer is sent to other neurons or to the outside.

3.3.2 MATRIX WRITING

We consider the layer i composed of M_i neurons.

For any neuron j with $1 \leq j \leq M_i$, we put:

$$X^i = \begin{pmatrix} x_0^i \\ x_1^i \\ \vdots \\ x_{N_{i-1}}^i \end{pmatrix} \quad W_j^i = \begin{pmatrix} w_{0j}^i \\ w_{1j}^i \\ \vdots \\ w_{N_{i-1}j}^i \end{pmatrix}$$

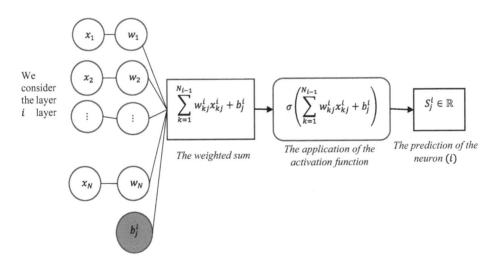

FIGURE 3.2 The mathematical formulation of an artificial neurons.

So:

$$S_j^i = \sum_{k=0}^{N_{i-1}} w_{kj}^i x_{kj}^i = \begin{pmatrix} w_{0j}^i & w_{1j}^i & \cdots & w_{N_{i-1}j}^i \end{pmatrix} \cdot \begin{pmatrix} x_0^i \\ x_1^i \\ \vdots \\ x_{N_{i-1}}^i \end{pmatrix} = {}^T W_j^i . X^i$$

We pose:

$$S^i = \begin{pmatrix} S_1^i \\ S_2^i \\ \vdots \\ S_{M_i}^i \end{pmatrix}$$

So:

$$S^i = \begin{pmatrix} S_1^i \\ S_2^i \\ \vdots \\ S_{M_i}^i \end{pmatrix} = \begin{pmatrix} w_{01}^i & w_{11}^i & \cdots & w_{N_{i-1}1}^i \\ w_{02}^i & w_{12}^i & \cdots & w_{N_{i-1}2}^i \\ & & \vdots & \\ w_{0M_i}^i & w_{1M_i}^i & \cdots & w_{N_{i-1}M_i}^i \end{pmatrix} \cdot \begin{pmatrix} x_0^i \\ x_1^i \\ \vdots \\ x_{N_{i-1}}^i \end{pmatrix} = \begin{pmatrix} {}^T W_1^i \\ {}^T W_2^i \\ \vdots \\ {}^T W_{M_i}^i \end{pmatrix} \cdot X^i$$

We put:

$$W^i = \begin{pmatrix} w_{01}^i & w_{02}^i & & w_{0M_i}^i \\ w_{11}^i & w_{12}^i & & w_{1M_i}^i \\ \vdots & \vdots & \cdots & \vdots \\ w_{N_{i-1}1}^i & w_{N_{i-1}2}^i & & w_{N_{i-1}M_i}^i \end{pmatrix} = \begin{pmatrix} w_{kj}^i \end{pmatrix}_{\substack{0 \le k \le N_{i-1} \\ 1 \le j \le M_i}}$$

So:

$$S^i = {}^T W^i . X^i$$

The outputs of the M_i neurons in the layer are then written (Figure 3.3):

$$Y^i = \begin{pmatrix} y_1^i \\ y_2^i \\ \vdots \\ y_{M_i}^i \end{pmatrix}$$

So:

$$Y^i = \begin{pmatrix} y_1^i \\ y_2^i \\ \vdots \\ y_{M_i}^i \end{pmatrix} = \begin{pmatrix} \varphi(S_1^i) \\ \varphi(S_2^i) \\ \vdots \\ \varphi(S_{M_i}^i) \end{pmatrix} = \varphi \begin{pmatrix} S_1^i \\ S_2^i \\ \vdots \\ S_{M_i}^i \end{pmatrix} = \varphi(S^i)$$

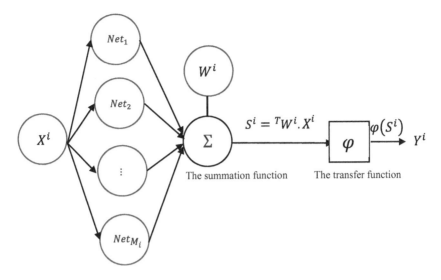

FIGURE 3.3 Architecture and functioning of ANN.

3.3.3 THE TRANSFER FUNCTIONS

Information is propagated from layer to layer via a function known as the transfer function, also known as the activation function or thresholding function. Table 3.1 contains a list of the most popular functions mentioned in the literature:

3.3.4 ERROR FUNCTIONS

The difference between the predicted output and the output generated by the network must be calculated in order to determine the proper weights (parameters). There are several ways to calculate the error:

- *R–square:*

$$R^2 = 1 - \frac{\sum_{i=1}^{n}\left(y_i - \widehat{y}_i\right)^2}{\sum_{i=1}^{n}\left(y_i - \overline{y}\right)^2}$$

With:

y_i : *the exact value*

$\overline{y}$: the average of the values of y_i

$\widehat{y}_i$: *the value we have predicted*

- *Mean Absolute Error «MAE»:*

$$\text{Error} = \frac{1}{m}\sum_{i=1}^{m}\left|y_i - \widehat{y}_i\right|$$

m The quantity of subjects, objects, or observations that must be predicted.

TABLE 3.1
The Activation or Propagation Functions

The Function Title	The Function	The Graphic Representation
Sigmoid	$\sigma(x)=\dfrac{1}{1+e^{-x}}$	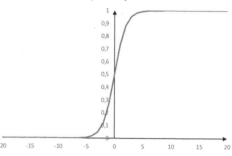
Hyperbolic tangent	$\text{Tanh}(x)=\dfrac{e^{x}-e^{-x}}{e^{x}+e^{-x}}$	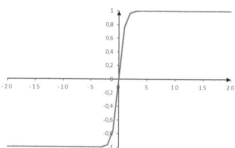
ReLu	$\text{ReLu}=\text{Max}(0,x)$	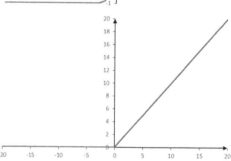

Source: Author.

- *Mean Squared Error «MAE»:*

$$\text{Error} = \frac{1}{2m}\sum_{i=1}^{m}\left(y_i - \widehat{y}_i\right)^2$$

3.3.5 LEARNING THE ARTIFICIAL NEURAL NETWORK

A "training" procedure is found in the great majority of neural networks, and it consists of changing the synaptic weights in response to a collection of input data (Louzada et al., 2016). The neural network is being trained with the intention of helping it learn from the instances (Knoll and Houts, 2012). The network can provide output answers that are remarkably similar to the original values of the training dataset if the training is done correctly (Figure 3.4). However, the capacity of neural networks to generalize from the test set is what makes them interesting.

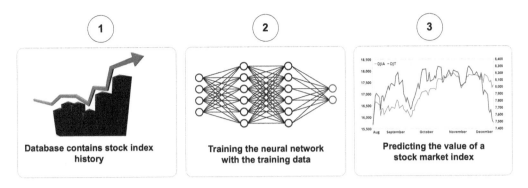

FIGURE 3.4 The modeling procedure using artificial neural networks.

So, it is feasible to build a memory using a neural network; this is called neural memory. When the network is compelled to converge to a certain end state when it is presented with a pattern, supervised learning takes place (Yao et al., 2019).

In contrast, when a pattern is provided in unsupervised learning, the network is free to converge to any final state.

ANN learning can be achieved, among other things, by:

i. Adjusting weights;
ii. Altering the network structure (adding or removing neurons, connections, or layers);
iii. Using suitable attractors or other suitable steady-state locations;
iv. The selection of transfer functions.

Since backpropagation training is a gradient descent process, it can get stuck in local minima in this weight space. It is because of this possibility that neural network models are characterized by high variance and instability.

- **Backpropagation:**

Backpropagation entails passing on a mistake made by a neuron to the neurons linked to its synapses. For neural networks, the backpropagation of the error gradient is typically used, which corrects errors based on the significance of the components that have actually contributed to these errors: the synaptic weights that contribute to generating a large error will be modified more significantly than the weights that have subsequently generated a marginal error.

How to decide how many layers and neurons to use:

The performance of an ANN in terms of prediction quality is directly influenced by the number of neurons and layers. In fact, we may use a method that involves beginning with a single hidden layer and modifying it until we achieve the perfect design to figure out how many hidden layers there are.

So, if one layer doesn't yield satisfying results, we must immediately consider adding another layer until we do. The number of neurons is also subject to modification in order to achieve the desired outcomes. Each layer's number of neurons cannot be greater than its number of input variables. To create an ANN that is relevant and effective in terms of accuracy in predicting the output variables, you must thus consider doing a number of experiments.

However, while the network's capacity increases with more layers, you run the danger of *overfitting* if you overestimate the number of layers or neurons. Conversely, if you underestimate the number of layers, you run the risk of *underfitting*.

We aim to partition the data into four parts and try to alternate the combinations between these parts in order to avoid the problems of *underfitting* and *overfitting*. By using this method, we will have a flawless test of the data because all components will be utilized.

3.4   SUPPORT VECTOR MACHINES (SVM)

The methodology presented in this section uses SVM modeling to build a model capable of predicting stock prices over the course of a trading session.

Let $x_i = (x_{1i}, x_{2i}, x_{3i}, \cdots x_{pi}) \in X \subset IR^p \in$ an observation of X concerning action (i), $x_i \epsilon IR^p$ The decision function (separator) is defined by:

$$F \; : \; IR^p \to \{-1, +1\}$$

With $\{-1\}$ means a decrease in the share price and $\{+1\}$ an increase in price.

In terms of probability, this amounts to minimizing the probability of error of the decision function to assign to the action x_i:

$$P\big(F(x) \neq y_i / x_i\big)$$

With P an unknown distribution defined on $\big(IR^p, \{-1, +1\}\big)$.

To produce the decision function, we need to use the training data $\{(x_i, y_i)\} i = 1 \ldots n$ with $x_i \in X$ and $y_i \in \{-1, +1\}$.

SVM is used to determine this decision function from the observations collected and to update this function according to the new data. This paragraph presents the theoretical aspect of this method, including the definition of the discrimination problem to be solved and the approach to be followed.

A discrimination problem is linearly separable if there exists a linear decision function F called a linear separator of the form

$$\begin{cases} F(X) = \text{sign}\big(h(X)\big) \\ h(x) = w^T X + w_0 \end{cases}$$

With $w \in IR^p$ and $w_0 \epsilon IR$.

Correct classification of the learning set $\{F(x_i) = y_i; i = 1, \ldots, n\}$.

The decision boundary associated with the decision function is defined as follows:

$$S(w, w_0) = \big\{x \epsilon IR^p / w^T x + w_0 = 0\big\}$$

S is a hyperplane $h(x)$ of equation $w^T x + w_0 = 0$:

$$F(x_i) = \begin{cases} \text{Augmentation du prix} & Si \; w^T x_i + w_0 \geq 0 \\ \text{Diminution du prix} & Si \; w^T x_i + w_0 < 0 \end{cases}$$

The point above or on the hyperplane will be classified in class +1, and the point below the hyperplane will be classified in class +1.

Either $\{(x_i, y_i)\}$, $i = 1 \ldots n$ with $x_i \in IR^p et \; y_i \in \{-1, +1\}$ An SVM separator is a linear discriminator of the form $F(x) = \text{signe}\big(w^T x_i + w_0\big)$ where $w \in IR^p$ and $w_0 \in IR$ are given by the solution of the following optimization system:

$$\begin{cases} \text{Min} \dfrac{\|w\|^2}{2} \\ \text{with} \quad y_i\big(w^T x_i + w_0\big) \geq 1, \; i = 1, \ldots, n \end{cases} \tag{3.1}$$

where $\|w\| = \sqrt{w_1^2 + w_2^2 + \ldots + w_p^2}$

Problem (3.1) is a convex quadratic problem under linear constraints whose objective function must be minimized. The Lagrangian of this problem is written:

$$\mathcal{L} = \frac{\|w\|^2}{2} - \sum_{i=1}^{n} \lambda_i \left(y_i \left(w^T x_i + w_0 \right) - 1 \right) \tag{3.2}$$

With:

$\lambda_i, i = 1, \ldots, n$ Lagrange multipliers.

To solve problem (3.1), the Lagrangian $\mathcal{L}$ must be minimized with respect to w and w_0 and maximized with respect to the variables λ_i. In this case, the saddle point (minimum with respect to variables and w_0 and maximum with respect to variables λ_i) must satisfy the "Karush-Kuhn-Tucker (KKT)" conditions.

Consequently, the separation hyperplane is defined by:

$$h(x) = \sum_{i=1}^{n} \lambda_i^* y_i (x_i, x) + w_0^* \tag{3.3}$$

Determining the hyperplane defined by formula (3.3) enables us to define the classification rule for a new observation (x), which is as follows:

$$F(x) = \text{sign} \left(\sum_{i=1}^{n} \lambda_i^* y_i (x_i, x) + w_0^* \right) \tag{3.4}$$

With (x_i, x) is the scalar product

When it is impossible to completely separate the data with a hyperplane, the data are non-linearly separable. In this case, the data must be processed to obtain a separable representation. Since SVMs are not capable of handling such a problem, the processing to be carried out consists in using techniques that transform the data, making them linearly separable after transformation.

In fact, the transformation is performed by the function ψ defined by:

$$\psi : \mathfrak{R}^m \rightarrow \mathfrak{R}^d$$

$$x \rightarrow \psi(x)$$

Therefore, to find the separation hyperplane, we use the same line of equation presented in the previous case, but replace the x_i by $(x_i), i = 1, \ldots, n$. This allows us to determine the classification function and the separation hyperplane from formulas (3.3) and (3.4):

$$h(x) = \sum_{i=1}^{n} \lambda_i^* y_i \left(\psi(x_i), \psi(x) \right) + w_0^* \tag{3.5}$$

$$F(x) = \text{sign} \left(\sum_{i=1}^{n} \lambda_i^* y_i \left(\psi(x_i), \psi(x) \right) + w_0^* \right) \tag{3.6}$$

Formulas (3.5) and (3.6) contain a scalar product that we need to define. $\left(\psi(x_i), \psi(x) \right)$ which we need to define, using the kernel method.

TABLE 3.2
Explanatory Variables Used

1	Date	Price	Open	High	Low	Change (%)
2	31-juil	11,947.29	11,818.46	11,953.54	11,796.27	1.09
3	24-juil	11,818.46	11,923.99	11,944.48	11,758.41	−0.89
4	17-juil	11,923.99	11,830.07	12,072.68	11,819.56	0.78
5	10-juil	11,831.42	11,701.65	11,834.60	11,701.65	1.03
6	03-juil	11,711.37	11,710.27	11,812.12	11,518.12	−0.28
7	26-juin	11,744.22	12,013.69	12,096.00	11,744.22	−2.25
8	19-juin	12,014.91	12,082.59	12,198.75	11,930.64	−0.57
9	12-juin	12,083.57	12,564.76	12,613.70	12,083.57	−3.83
10	05-juin	12,564.76	12,595.67	12,728.19	12,531.87	−0.36
11	29-mai	12,609.89	12,334.93	12,609.89	12,298.49	2.23
12	22-mai	12,334.93	12,626.59	12,660.52	12,306.00	−2.31
13	15-mai	12,626.59	12,682.21	12,749.42	12,625.57	−0.44
14	08-mai	12,682.21	13,051.01	13,060.85	12,568.50	−2.83
15	01-mai	13,051.01	13,136.65	13,173.92	13,046.44	−0.65
16	24-avr	13,136.65	13,169.67	13,183.37	13,065.96	−0.25
17	17-avr	13,169.67	12,999.31	13,169.67	12,990.96	1.31
18	10-avr	12,999.31	12,946.49	12,999.31	12,923.04	0.41
19	03-avr	12,946.49	12,843.52	12,953.94	12,803.95	0.80

The kernel K is a symmetrical, positive, two-variable function that defines a scalar product in the transformation space:

$$K(x_i, x_j) = \langle \psi(x_i), \psi(x_j) \rangle$$

The choice of kernel has an impact on the prediction performance of SVMs (Caner Savas et al., 2019). The literature suggests some kernels whose K function is defined as follows:

The linear kernel is defined by: $K(x_i, x_j) = (x_i^T x_j)$

The polynomial kernel: $K(x_i, x_j) = (x_i^T x_j + 1)^d$

RBF Gaussian kernel (radial basis function)[1]: $K(x_i, x_j) = e^{-\frac{\|x_i - x_j\|^2}{\gamma}}$

Inverse Multi-Quadratic: $K(x_i, x_j) = \dfrac{1}{\sqrt{(x_i - x_j)^T (x_i - x_j) + \beta}} = \dfrac{1}{\sqrt{\|x_i - x_j\|_2^2 + \beta}}$

3.5   DATA

In the data preparation stage, we used weekly historical stock market index data for the different selected African countries, which spans a time interval [January 2022–July 31, 2022], and we have downloaded the data from the website (ww.investing.com) (see Tables 3.2 and 3.3).

TABLE 3.3
The Stock Market Index Used

Country	Stock Market Index	Interval	Frequency
IvorCoast	BRVM 10	[January 2022–31 July 2022]	Weekly
Egypt	EGX 30	[January 2022–31 July 2022].	Weekly
Morocco	MASI	[January 2022–31 July 2022].	Weekly
Kenya	Kenya NSE 20	[January 2022–31 July 2022].	Weekly
Nigeria	NSE 30	[January 2022–31 July 2022].	Weekly

Source: Author.

3.6   RESULTS

a. ANN

We started by getting the data ready for modeling before using the ANN. For this, we established the parameters and layer count required to construct our network.

Our model is trained using the function *fit*(), which enables us to select the ideal values for the weight matrix W. Based on the error functions, the computations are carried out using the gradient descent method. On X_train (beginning values) and Y_train (anticipated arrival values), the training data are kept. Figures 3.5 and 3.6 depict the development of the Moroccan stock market index (MASI) model's accuracy and error (loss) during the training period.

As the learning algorithm iteratively changes the weights and biases in the neural network in accordance with the training data, we can observe in Figure 3.5 that the error lowers and the accuracy increases with iterations. We see in Figure 3.6 that the error (loss) curves for the test and train data both converge to 0, demonstrating the model's effectiveness and its potential for forecasting future MASI values (Nair and Mohandas, 2014).

As a result, we calculated the accuracy metric for the training and test data, and utilizing the test data, we were able to acquire a better accuracy of 97.48% for the MASI Stock Market Index.

b. SVM method

After preparing the data for modeling, we proceeded to apply the SVM method to the various historical stock market index data. However, we obtained the following results for the MASI index:

Linear kernel result:
Precision score:
0.6843768995757146

	Precision	recall	f1-score	support
0	0.00	.00	0.00	100
1	0.68	1. 00	0. 78	317

Accuracy 0.68

RBF kernel result:
Precision score:
0.785571442857143

	Precision	recall	f1-score	support
0	0.65	0.42	0. 51	100
1	0.790	.88	0.80	317

Accuracy 0.79

Poly Kernel result:
Precision score:
0.718693009118541

	Precision	recall	f1-score	support
0	0.75	0. 100	.18	100
1	0.72	0.98	0. 79	317

Accuracy 0.72

According to the results obtained when running the different kernels, we can say that the model obtained by the RBF kernel represents a fairly high level of accuracy. Table 3.4 shows the accuracy of the different indexes used.

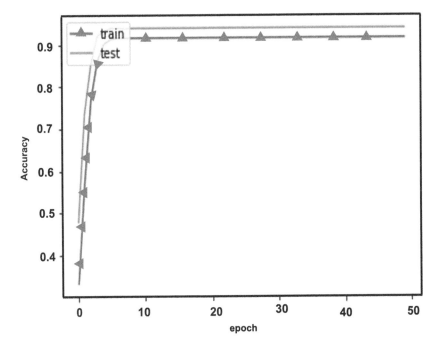

FIGURE 3.5   Accuracy curve of the resulting model.

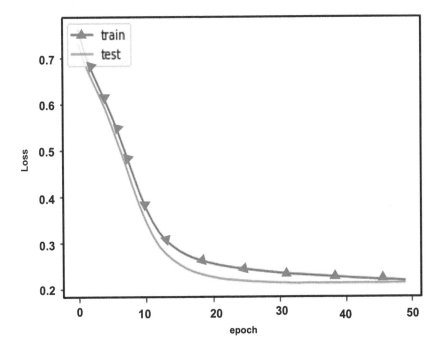

FIGURE 3.6   Error curve of the resulting model.

TABLE 3.4
Accuracy for the Different Indexes Used

Country	Stock Market Index	Accuracy ANN(%)	Accuracy SVM_RBF(%)
	BRVM 10	95,03	73.10
Ivory Coast			
	EGX 30	94,30	60.16
Egypt			
	MASI	97,48	79.12
Morocco			
	Kenya NSE 20	96,87	70.21
Kenya			
	NSE 30	95,41	63.59
Nigeria			

We can conclude that the models developed by ANN have a reasonably high level of predictability, which will assist us in making precise short- and long-term projections of the value of the stock market index based on the accuracy values obtained for the various African nations (Table 3.4).

3.7   DISCUSSION AND CONCLUSION

The rapid growth of information technology, including the Internet and other means of telecommunication, is contributing to the rapid development of computational methods. As a result, this paper is variable in content and objectives, but it remains true to one purpose, namely to study the performance of AI techniques in predicting the value of stock market index.

Data mining and AI allow us to make predictions with nearly perfect accuracy. In order to build a model that might anticipate the values of the stock market index, we thus sought to determine the significance of employing these strategies. We relied on the ANN and SVM methods to do this; thus, we took advantage of historical stock market index data for a few African nations (from January 2022 to July 31, 2022).

The accuracy attained at the end of the modeling process for the various models obtained by ANN, which exceeds 97% for the model obtained for the modeling of the MASI, which reveals the robustness of the model obtained, reflects the performance of the AI techniques as well as the process adopted by the data mining techniques. These results provide further evidence that deep learning and more specifically AI techniques are effective methods for predicting stock prices and indeed stock market indexes.

The results obtained in this chapter are remarkable, with an accuracy of 97% of the ANN model in predicting the value of the stock market index. This demonstrates the effectiveness of ANNs in analyzing African stock market data and their ability to identify patterns and trends specific to this region. This high level of accuracy can have significant implications for investors and traders, enabling them to make more informed decisions and adjust their strategies accordingly.

In considering the significance of this chapter in relation to previous research, it is important to emphasize that the inclusion of African stock market index enables a better understanding of these specific markets. African markets have unique characteristics in terms of economic, political, and investor behavioral factors, which can have a significant impact on stock index performance. By using African data in the analysis, this chapter contributes to bridging the existing gap in academic literature and provides valuable insights into trends and patterns specific to African markets.

This chapter varies in both content and objectives, but it focuses on two fundamental contributions, namely the evaluation of the effectiveness of AI systems in predicting the value of stock market indices in a number of African countries, while the second contribution is the comparison of the level of accuracy of a number of AI methods in predicting stock market index value.

However, this chapter had two significant flaws that might be fixed in follow-up investigations. Our chapter was based on one of numerous AI techniques, and it had two limitations: first, it was only able to analyze weekly stock market index data. However, it should be noted that the volatility of financial markets can have a significant impact on prediction results. Stock markets can be subject to rapid and unpredictable fluctuations, making it difficult to accurately forecast index prices.

In future studies, it is essential to further investigate the application of ANN in stock index value prediction. We also advocate a comparative study of the different methods emanating from intelligence using daily data of different stock index prices. Similarly, it would be interesting to explore more complex network architectures, integrate additional data such as macroeconomic data or financial news, and improve existing models for greater accuracy.

NOTE

1   With "γ" controls the bandwidth of the Gaussian: the narrower the Gaussian, i.e., the smaller the distance between x_i, x_j must be for the kernel to be different from 0.

REFERENCES

Alsheibani, S.A., Cheung, Y.P., & Messom, C.H. (2018). Artificial Intelligence Adoption: AI-readiness at Firm-Level. *PACIS*. https://aisel.aisnet.org/pacis2018/37.

Amzile K., et al. (2022). Towards a digital enterprise: The impact of Artificial Intelligence on the hiring process. *Journal of Intelligence Studies in Business* 12(3), https://doi.org/10.37380/jisib.v12i3.894.

Amzile, K., & Amzile, R. (2021). L'utilisation de la méthode KPV émanant de l'intelligence artificielle pour la prédiction de la solvabilité des clients bancaires. https://doi.org/10.5281/zenodo.5806789.

Anggoro, D.A. (2020). The implementation of subspace outlier detection in k-nearest neighbors to improve accuracy in bank marketing data. *International Journal of Emerging Trends in Engineering Research* 8(2), 545–550, févr. 2020. https://doi.org/10.30534/ijeter/2020/44822020.

Atsalakis, G.S., & Valavanis, K.P. (2009). Forecasting stock market short-term trends using a neuro-fuzzy based methodology. *Expert Systems with Applications* 36, 10696–707.

Bahrammirzaee, A. (2010). A comparative survey of artificial intelligence applications in finance: Artificial neural networks, expert system and hybrid intelligent systems. *Neural Computing and Applications* 19, 1165–95.

Cortes, C., & Vapnik, V. (1995). Support vector machine. *Machine learning* 20(3), 273--297.

Chen, L., Qiao, Z., Wang, M., Wang, C., Du, R., & Stanley, H. E. (2018). Which artificial intelligence algorithm better predicts the chinese stock market? *IEEE Access* 6, 48625–48633. https://doi.org/10.1109/ACCESS.2018.2859809.

Cybenko, G. (1989). Approximation by superpositions of a sigmoidal function. *Mathematics of Control, Signals and Systems* 2, 303–314.

Deloitte (2017). "Enquête sur les technologies cognitives 2017". https://www2.deloitte.com.

Dwivedi, Y.K., et al. (2019). Artificial Intelligence (AI): Multidisciplinary perspectives on emerging challenges, opportunities, and agenda for research, practice and policy. *International Journal of Information Management*. https://doi.org/10.1016/j.ijinfomgt.2019.08.002.

Esfahanipour, A., & Aghamiri, W. (2010). Adapted Neuro-Fuzzy Inference System on indirect approach TSK fuzzy rule base for stock market analysis. *Expert Systems with Applications* 37, 4742–4748. https://doi.org/10.1016/j.eswa.2009.11.020.

Gandhmal, D.P., & Kumar, K. (2019). Systematic analysis and review of stock market prediction techniques. *Computer Science Review* 34, 100190.

Guresen, E., Kayakutlu, G., & Daim, T.U. (2011). Using artificial neural network models in stock market index prediction. *Expert Systems with Applications* 38(8). https://doi.org/10.1016/j.eswa.2011.02.068.

Guyon, I., Weston, J., Barnhill, S., & Vapnik, V. (2002). Gene selection for cancer classification using support vector machines. *Machine Learning* 46, 389–422.

Huang, C.L., Chen, M.C., & Wang, C.J. (2007). Credit scoring with a data mining approach based on support vector machines. *Expert Systems with Applications* 33(4), 847–856.

IBM (2020) "Qu'est-ce que l'intelligence artificielle (IA)?". https://www.ibm.com/fr-fr/cloud/learn/what-is-artificial-intelligenceWeb. 3. June. 2020.

Inthachot, M., Boonjing, V., & Intakosum, S. (2015). Predicting SET50 index trend using artificial neural network and support vector machine. In M. Ali, Y. S. Kwon, C.- H. Lee, J. Kim, & Y. Kim (Éds.), *Current Approaches in Applied Artificial Intelligence* (vol. 9101, pp. 404–414). Springer International Publishing. https://doi.org/10.1007/978-3-319-19066-2_39.

Kara, Y., Acar Boyacioglu, M., & Baykan, Ö.K. (2011). Predicting direction of stock price index movement using artificial neural networks and support vector machines: The sample of the Istanbul Stock Exchange. *Expert Systems with Applications* 38, 5311–5319.

Knoll, M.A.Z., & Houts, C.R. (2012), The financial knowledge scale: An application of item response theory to the assessment of financial literacy. *Journal of Consumer Affairs* 46, 381–410. https://doi.org/10.1111/j.1745-6606.2012.01241.x.

Louzada, F., Ara, A., & Fernandes, G.B. (2016). Classification methods applied to credit scoring: Systematic review and overall comparison. *Surveys in Operations Research and Management Science* 21(2), 117–134, déc. 2016, https://doi.org/10.1016/j.sorms.2016.10.001.

Nair, B.B., & Mohandas, V.P. (2014). Artificial intelligence applications in financial forecasting - a survey and some empirical results. *Intelligent Decision Technologies* 9(2), 99–140. https://doi.org/10.3233/IDT-140211.

Noble, W.S. (2006). What is a support vector machine?. *Nature Biotechnology* 24(12), 1565–1567.

Obthong, M., Tantisantiwong, N., Jeamwatthanachai, W., & Wills, G. (2020). A survey on machine learning for stock price prediction: Algorithms and Techniques. Paper presented at the FEMIB 2020-2nd International Conference on Finance, Economics, Management and IT Business, Prague, Czech Republic, May 5-6; pp. 63–71.

Paquet, P. (1997). "L'utilisation des réseaux de neurones artificiels en finance", (Laboratoire Orléanais de Gestion - université d'Orléans). https://econpapers.repec.org/paper/logwpaper/1997-1.htm.

Pławiak, P., Abdar, M., & Acharya, U.R. (2019). Application of new deep genetic cascade ensemble of SVM classifiers to predict the Australian credit scoring. *Applied Soft Computing* 84, 105740.

Qiu, M., Song, Y., & Akagi, F. (2016). Application of artificial neural network for the prediction of stock market returns: The case of the Japanese stock market. *Chaos, Solitons & Fractals* 85, 1–7. https://doi.org/10.1016/j.chaos.2016.01.004.

Safari, A., & Ghavifekr, A.A. (2021). International stock index prediction using artificial neural network (ANN) and python programming. *2021 7th International Conference on Control, Instrumentation and Automation (ICCIA)*, 1–7. https://doi.org/10.1109/ICCIA52082.2021.9403580.

Safari, A., et al. (2021). International stock market index prediction using artificial neural network (ANN) and python programming. *7th International Conference on Control, Instrumentation and Automation (ICCIA)*.

Singh, S. & Kumar, V. (2013a). Analyse de la performance des étudiants en génie pour le recrutement à l'aide de techniques d'exploration de données de classification. *Revue internationale des sciences, de l'ingénierie et de la technologie informatique* 3(2), 31.

Singh, S., & Kumar, Dr. V. (2013b). Performance analysis of engineering students for recruitment using classification data mining techniques. *Samrat Singh et al | IJCSET |*, 3(2), 31–37.

Suykens, J. A., & Vandewalle, J. (1999). Least squares support vector machine classifiers. *Neural Processing Letters* 9, 293–300.

Yao, J., Pan, Y., Yang, S., Chen, Y., & Li, Y. (2019). Detecting fraudulent financial statements for the sustainable development of the socio-economy in china: A multi-analytic approach. *Sustainability* 11(6), 1579, mars 2019, https://doi.org/10.3390/su11061579.

4 Machine Learning Clustering
Application on Aggregated and Non-Aggregated Financial Data

El Aidouni Salma, Benhouad Mohamed,
Mestari Mohammed, and El Mansouri Adnane

4.1 INTRODUCTION

Financial systems are known for their extreme complexity. Fluctuations and changes in these systems are difficult to predict because of their complex interconnectedness. Researchers in the financial field try to understand these systems by fitting them into mathematical models where the effects of parameters can be separated and analyzed. 'Structural' approaches focus on theory and generate algebraic representations of phenomena, while 'non-structural' approaches refer to more data-driven studies where there is little or no financial/economic theory behind the selection of variables and the process of adjusting models. From another point of view, econometrics represents the empirical analysis and statistical modeling of economic phenomena. Theoretical (structural) approaches and data-based (non-structural) approaches are accepted and widely applied in the field [14].

The lack of data has been common in many econometric studies [2]. The quality and availability of data have drastically varied among countries and different economic phenomena. Macroeconomic studies illustrate this example; indeed, looking at regressing a variable against a chosen set of variables in an exploratory way will not be an easy task. Observations are limited by the number of years with concise data collection; they may have different limits, and data quality varies across countries and regions. Thus, building a robust statistical model and taking into account all pertinent information becomes continuously more difficult. Despite the rise of the literature that analyzes the effect of the financial system's development on several macroeconomic variables, financial indicators are often unsuited to international comparisons. Various specific measures of the maturity level of these systems are available, but they are generally specific to certain countries or groups of countries [1]. This peculiarity of data makes international comparisons complicated, and authors use a wide range of indicators, often incomplete or unsuitable for the intended theoretical purposes. This article bypasses this peculiarity of financial systems analysis and proposes a new index sensitive to several aspects of financial development. The main component analysis technique is used to group three indicators and build a composite index of financial development [9]. The proposed indicator provides a financial measure for 79 countries between 1985 and 2016. On the basis of this composite index, a comparison via clustering is possible among the panel of countries. To do this, we used the method of ascending hierarchical classification (AHC), k-means. A second approach was also adopted; it consists of grouping all non-aggregated observations in a global database and subsequently applying the two clustering methods mentioned above. The choice of indicators has been subjected to the identification of the efficiency with which financial systems manage to deal with private credit in the economy and monetary policy, considering financial sector characteristics, especially size and bank sector depth. Among the selected indicators, private domestic credit to GDP indicates the amount of private credit in the economy; broad money to DGP is closely related to monetary policy and the demand for money; and it fluctuates for reasons that are unrelated to

DOI: 10.1201/9781032667478-5

financial development. Domestic credit supplied by the financial sector as a part of GDP measures financial sector development and banking sector depth. The aim of this chapter is to grasp the difference between a parametric classification method and a hierarchical method, on the one hand. On the other hand, it is a question of concluding on the quality of the results obtained, taking into account the nature of the database (aggregated or non-aggregated).

4.2 STYLIZED FACTS ON FINANCIAL DEVELOPMENT

For a long time, the economic literature has been discussing the determinants of financial system development and its effects on economic activity. The history and evolution of economic systems over time are at the origin of the multitude of financial systems encountered in the world. This natural development, which goes in tandem with the evolution of societies, has promoted the growth of common practices and, subsequently, the development of institutions. In each country, this historical development has favored the development of distinct economic and institutional systems with their own characteristics. A multitude of legal, regulatory, and tax systems have thus developed. This set of factors explains the strong international variations observed in terms of financial development [4].

The ultimate objective of the financial system is the intermediation between agents who show a surplus of capital and those who show a deficit of capital. This intermediation, however, encounters a multitude of obstacles that hinder, on the one hand, the formation of savings and, on the other hand, the financing of investment projects. By improving friction in the market, financial systems act on the allocation of resources in time and space. Financial development is thus identified by the reduction of these frictions and the enhancement of intermediation [12].

Precisely, there are six important functions of systems that affect the fluidity of financial intermediation. Efficiency in the performance of these functions characterizes financial development, i.e., the improvement of the capital allocation process:

1. Mobilization of capital;
2. Production and dissemination of information on economic agents;
3. Control and sharing of risk;
4. Investment monitoring (so that capital is used optimally) and corporate governance;
5. Reduction of transaction costs;
6. The liquidity of financial investments.

While the financial sector in each country executes the allocation of resources between savers and capital seekers, there are huge gaps in the performance of each of these six functions. The more the friction, the less the effective financial intermediation. And the extent of the financial friction encountered by the agents defines the level of maturity of each system [15].

By reducing friction, financial systems act on the allocation of resources among agents. As an example, the development of a banking network facilitates the recognition of the owners of capital and reduces the costs of mobilizing savings. This improves the allocation of credit in the economy. In the same way, the appearance of institutions that produce information on firms and households minimizes financial friction. Strengthening contracts increases investor confidence and encourages increased savings. The supply of capital for corporate finance increases when the governance of firms favors the interests of the owners of capital at the expense of managers [5].

Financial development takes place when financial institutions minimize the existing friction and smooth the path of resource allocation. Specifically, financial development must reflect the efficiency of the system in the performance of each of the six major financial functions, thereby providing an optimal allocation of resources.

Today, there is considerable debate about financial development and its impact on economic growth. According to Mckinnon and Shaw (1973), proponents of financial liberalization, the stunting of developing countries is strongly linked to the low level of development of their financial system caused by the strong intervention of the state in the financial system, particularly through the setting of interest rates, the credit framework, or the taxation of financial intermediaries (by obliging banks to hold a certain percentage of their deposits in the form of minimum reserves).

These constraints imposed on the financial system have created several distortions, such as low interest rates, the discouragement of savings, the bad choice of investment projects, and the financing of unprofitable government projects. According to these two authors, the financial reforms, which are mainly based on the liberalization of interest rates, the suppression of credit control, the facilitation of access to loanable funds, and the establishment of a new banking regulation and privatization of the banking system, can stimulate the accumulation of savings and consequently the growth of productive investment, which can in turn contribute to economic growth.

4.3   METHODOLOGY

We note that our data base is provided by the World Bank Data precisely from the section on financial sector data.

4.3.1   Principal Component Analysis

In order to build the index, we performed a principal component analysis (PCA), which is a factorial method to represent our variables as accurately as possible in a small space. This amounts to determining the (factorial) axes that pass through the maximum number of points. In the sense of the OLS (Ordinary Least Squares), these axes are the eigenvectors of the correlation matrix associated with the largest eigenvalues.

In our case, we are interested in the information provided by the first principal component, i.e., the composite indicator; the greater the information or the variance, the greater the confidence in our index.

The PCA technique is used for the construction of the composite financial development index proposed in the paper. This is a widely used technique and one of the most used in multivariate data analysis. Moreover, this methodology is among the oldest of multivariate statistical analysis, being introduced initially by Pearson (1901) and Hotelling (1933).

Moreover, this methodology is among the oldest in multivariate statistical analysis, having been introduced initially by Everitt and Dunn (2001).

Thus, when the first components of the data matrix are responsible for a large part of the total variation of the sample, the set of observations can be represented by a reduced number of dimensions. In the article, this method is used to reduce the amount of information in a database composed of observations on the financial systems of 79 countries in the 31 years between 1985 and 2016. Each country is represented by three variables, which measure several aspects of financial development.

4.3.2   Hierarchical Classification

For a given level of precision, two individuals can be confused in the same group, while at a higher level of precision, they will be distinguished and belong to two different subgroups. The result of a hierarchical classification is not a partition of all individuals. It is a hierarchy of classes such that:

- Any class is not empty.
- Every individual belongs to one (and even several) classes.
- Two distinct classes are disjoint, or verify an inclusion relation.
- Any class is the meeting of classes that are included in it.

The advantage of this method is that it is not subject to any particular parameter initialization, which makes it deterministic, and furthermore, the class number does not have to be fixed a priori.

However, this type of method requires the calculation of the distance matrix of all the observation points with all the others, and this mass of calculations is much too important given the time we want to devote to this step. One of the most commonly used unsupervised methods is the AHC approach.

In fact, the AHC allows the building up of an entire hierarchy of objects following the form of a "tree" in the ascending order. We begin by treating each individual as a class and attempt to merge two or more appropriate classes (according to a similarity) to form a new class. The process is repeated until all individuals are in the same class. This classification generates a tree that can be cut at different levels to produce a larger or smaller number of classes.

Different interclass distance measurements can be used: the Euclidean distance, the lower distance (which favors the creation of low inertia classes), the higher distance (which favors the creation of larger classes of inertia), and so on [7].

4.3.3 K-MEANS

K-means is a method whose purpose is to divide observations into k partitions, in which each observation belongs to the score with the closest average. Using the dynamic cloud method, the problem is the search for a partition in k (fixed) classes of a set of n individuals. It's an iterative algorithm.

Let I be a population of individuals. This population is representable on R and forms a cloud of n points. We seek to constitute a partition in k classes on I. Each class is represented by its center, also called the kernel, consisting of a small subset of classes that minimize the dissimilarity criterion.

On the assumption that there are K distinct classes. In the beginning, we designate K centers of classes $\mu_1,\ldots,\mu_K$ among the individuals. These centers can be chosen by the user "Representatively", or randomly selected. The following two steps are then carried out iteratively:

- For each individual who is not a class center, it is assigned to the cluster with the nearest class center. Thus, we obtain K classes $C_1,\ldots,C_K$, where $C_i =$ {set of the points closest to the center μ_i.
- In each new class C_i, the new class center is considered to be the center of gravity of the points of C_i.

The algorithm stopped according to a stop criterion set by the user, which can be chosen from among the following: either the limit number of iterations is reached or the algorithm has converged, that is to say that between two iterations the classes formed remain the same; in other words, the intra-class inertia will no longer vary between two iterations [7].

4.4 FINANCIAL INDICATORS

This session presents the three indicators used as part of this chapter.

4.4.1 DOMESTIC CREDIT TO THE PRIVATE SECTOR OF GDP

The first financial indicator measures the quantity of credit engaged in the private sector, relative to the size of the economy. In particular, the domestic credit to private sector variable measures the total private resources mobilized to finance the private sector, divided by GDP. In this variable, private resources are made up of loans to individuals, the purchase of securities (without participation rights), trade credits, and other debits that establish a right of repayment.

This indicator is the main measure of financial development and the most commonly used in the literature. It is the most direct aggregated indicator of the amount of financial intermediation in the private sector. One of its advantages is its opposition to credits issued by public institutions and the financing of government expenditures. The main virtue of the variable is thus the isolation of the private sphere and its measure of the credit constraint experienced by private agents outside government [6].

In other words, domestic credit to private GDP indicates the amount of private credit in the economy. A significant proportion of private credit in GDP indicates the intense activity of intermediaries. In this case, companies experience low financial constraints. Increasing this variable reflects financial development [10].

Private sector development and investment are essential for alleviating poverty. In parallel with the public sector, private investment, especially in competitive markets, has important potential to enhance economic growth. Private markets are the engine of productivity growth and job creation. Furthermore, with the government playing a complementary role in regulation, funding, and service provision, private initiative and investment can help provide the basic services and conditions that empower poor people by improving health, education, and infrastructure [8].

The variable domestic credit to the private sector in our data varies greatly between two years, but the main variations occur in cross-country comparisons. The overall average of the sample is 45.78%, which indicates that the total credit offered by financial intermediaries to the private sector equates, on average, to just about half of GDP. Inequalities are, however, strong. Domestic credit to the private sector varies between 0.16%, recorded by Liberia in 1988, and 312.12%, the amount of credit compared to the GDP of Iceland in 2006. The overall standard deviation is 42.77%.

4.4.2 Domestic Credit Provided by the Financial Sector of GDP

Domestic credit mobilized by the financial sector includes all credit to different sectors on a gross basis, with the exception of credit to the government, which is net. The financial sector includes monetary authorities, deposit banks, and other financial intuitions where data is available (including financial institutions that don't accept transferable deposits and although incur liabilities such as time and savings deposits, and the size of the financial sector). Domestic credit mobilized by the financial sector as a share of GDP measures the size of financial sector development and banking sector depth.

A higher level of domestic credit provided by the financial sector implies a greater degree of dependence on the financial system. Also, this fact implies financial development since the financial system is more likely to ensure the main financial functions, as it can facilitate hedging and trading, mobilize savings, reduce asymmetric information, diversify and pool risk, monitor managers, exert corporate control, acquire information about investments, and allocate resources. This variable is essential to measuring the level of development, providing also information about the system's size and performance [13].

In fact, both financial systems and banking promote growth, which is the primary driver of poverty reduction, whereas domestic stock markets tend to become more active and efficient at higher levels. The mobility and size of international capital flows make it important to monitor the resilience of financial systems because a robust financial system can enhance economic activity, but financial instability can disrupt financial activity and lead to an economic downturn [3].

The variable domestic credit provided by the financial sector varies greatly between years, but the main variations occur in cross-country comparisons. The overall average of the sample is 59.65%, which indicates that the total credit provided by the financial sector equates, on average, to just over half of GDP. Inequalities are, however, strong. The domestic credit supplied by the financial sector exhibits a range from a low of −79.09%, reported by Botswana in 1998, to a high of 2066.18%, representing the credit amount relative to Liberia's GDP in 1996. The overall standard deviation is 68.85%.

4.4.3 BROAD MONEY OF GDP

Broad money is the total of currency held outside banks; demand deposits other than those of the central government; the time; foreign currency deposits of resident sectors other than those of the government; savings; bank and traveler's checks; and other securities such as certificates of deposit and commercial paper. It is the most inclusive method of calculating a given country's money supply. The ratio of money to GDP is associated with monetary policy and the demand for money, and it changes for numerous reasons that are unrelated to financial development. In other words, because cash can be exchanged for a variety of financial instruments and held in a variety of restricted accounts, economists face a difficult task in determining how much money is currently in a given economy. Therefore, the money supply is measured in many different ways. Economists use the capital letter "M" followed by a number to refer to the calculation that they are using in a given context. [16]. For the measurement of money supply, M0, M1, M2, and M3 are the most commonly used money supply measures. These measurements vary according to the liquidity of the accounts included. M0 includes only the most liquid instruments, such as cash or assets that can be converted into currency quickly, making it the most limited definition of money. M3 includes both liquid and less liquid instruments and is thus considered the most comprehensive measure of money. In general, M3 is referred to colloquially as broad money [11].

In simple terms, if there is more money to go around, the economy tends to accelerate, with businesses having easy access to financing. If there is less money in the system, the economy cools off and prices may drop or stop rising. In this context, broad money is one of the measures that central bankers use to determine what interventions, if any, they choose to make in the economy.

The variable broad money varies greatly between years, but the main variations occur in cross-country comparisons. The overall average of the sample is 52.78%, which indicates that total broad money equates, on average, to just over half of GDP. Inequalities are, however, strong. Domestic broad money varies between 0.43%, recorded by Liberia in 1985, and 242.87%, the amount of broad money compared to the GDP of Japan in 2016. The overall standard deviation is 37.62%.

4.5 RESULTS

For analysis, we proceeded with two approaches. The first one consists in analyzing the global non-aggregated data by applying ascendant hierarchical classification and k-means methods. The second approach applies the PCA to the non-aggregated data for each country to obtain coefficients for the composite financial index, and then it regroups the data constituted by these composite financial indices using ascendant hierarchical classification and k-means methods.

4.5.1 FIRST APPROACH

Our non-aggregated data is composed of 7872 observations; we dispose of three variables for 82 countries from 1985 to 2016.

The first step is centering and reducing the data to avoid having variables with strong variances weigh unduly on the results. Under R Studio, we adopt the criterion of Ward using the square of the Euclidean distance and Ward's minimum variance method, which minimizes the total within-cluster variance. At each step, the pair of clusters with the minimum between-cluster distance is merged. The dendrogram suggests a division into four groups, as presented in the following Figure 4.1.

Results show that 56 countries have their three variables in the same cluster, which represents 68.29% of total observations. The other countries exhibit heterogeneous behavior in terms of clustering; for example, some countries belong to one cluster considering one of the three variables and belong to another cluster by reasoning with the other two variables.

In this case, the clusters are in the decreasing order from left to right. On the left end, we find cluster 4 corresponding to highly developed countries, followed by developed countries in cluster 3,

Optimal number of clusters

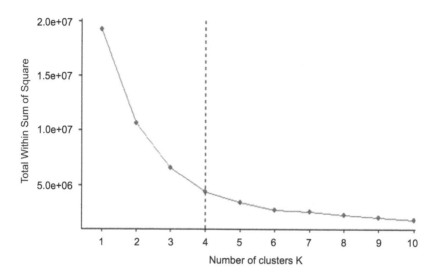

FIGURE 4.1   Elbow method.

then moderately developed countries in cluster 2, and finally low-developed countries in cluster 1 at the right end.

In short, for our case, 56 countries, namely 68.79%, have three weights in the same cluster (Morocco, Malaysia, Japan, etc.); 25 countries, i.e., 30.48%, have two weights in one cluster and one weight in another (India, Iceland, Liberia); and one country (the United States) illustrates the extreme case of having a weight in three different clusters.

K-means method: In this phase, we will apply k-means method to non-aggregated data. The k-means method requires the determination of the parameter k to launch the algorithm. In order to have correct and unbiased results, we will first determine the optimal number of clusters (K). To do so, we adopt:

- The elbow method, which is a direct method that consists of optimizing the within-cluster sums of squares criterion.
- The gap statistic, which is a statistical testing method that involves comparing evidence to the null hypothesis.

According to these methods, the optimal number of clusters in our case is $k=4$.

Figures 4.1–4.3 illustrate this choice. With four clusters, we obtain the following graphics.

Results show that 48 countries have their three variables at the same cluster which represents 58.54% of total observations. The other countries exhibit heterogeneous behavior in terms of clustering, as seen above with the AHC method.

According to the graphical representation above, there is a small difference in clusters' numbering between the AHC and k-means methods. In fact, there is an inversion between clusters 3 and 4. For k-means, cluster 1 corresponds to low-developed countries, cluster 2 coincides with moderately developed countries, cluster 3 conforms to highly developed countries, and cluster 4 agrees with developed countries.

To sum up, 48 countries, namely 58.54%, have three weights in the same cluster (Morocco, Malaysia, Japan, etc.), and 34 countries, i.e., 41.46% of observations, have two weights in one cluster and one weight in another (India, Iceland, the US).

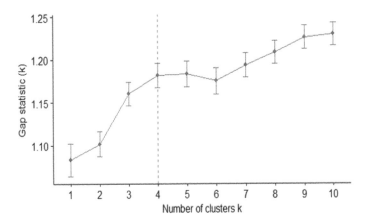

FIGURE 4.2   Gap statistic method.

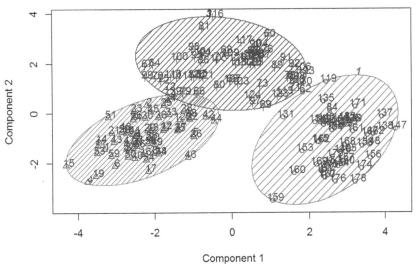

FIGURE 4.3   K-means clusters – non-aggregated data.

Also, the total variance explained by clustering is 78.3%. This means that by assigning the samples to 4 clusters rather than n ($n \geq 4$ is the number of samples) clusters, the achieved reduction in sums of squares is about 75.5%.

4.5.2 SECOND APPROACH

The second approach differs from the first in terms of the used data. In this second case, we proceed differently. Firstly, we applied a PCA across all three variables for each country separately to obtain the coefficients determining the composite index of financial development. Then we apply the AHC method and k-means method on the basis of data consisting of all the composite indices obtained in the first step of this approach for the 79 countries over the period from 1985 to 2016.

PCA method: As known, PCA is a tool for compressing and synthesizing information. It is a factorial analysis in that it produces factors (or principal axes), which are linear combinations of the initial variables that are hierarchical and independent of each other. We met two conditions in the application of the ACP (example of Morocco in the Appendix), namely:

- the correlation between variables greater than 0.5
- the proportion explained by the retained axis greater than 80%

AHC method: As preceded in the first approach, we started with the standardization of data, and then adopted the criterion of Ward using the square of the Euclidean distance and Ward's minimum variance method, which minimizes the total within-cluster variance. The dendrogram suggests a division into four groups. Results show that countries are divided into four clusters according to the level of financial development based on the composite indexes of financial development calculated in the first stage of this session. We distinguish between low-developed, moderately developed, developed, and highly developed countries. Table 4.1 gives the number of countries according to the level of development.

We notice that 44.3% of the chosen sample of countries are low developed, 31.65% are moderately developed, 22.78% are judged developed, and just 1.26% are considered highly developed in terms of financial development.

K-means method: Unlike the first approach, in this phase we will apply the k-means method to the aggregated data. To determine the optimal number of clusters, we used the elbow method and the NbClust package on R, which provides 30 indices for determining the number of clusters and recommends the best clustering scheme from the various results obtained by varying all combinations of cluster number, distance measures, and clustering methods. According to these methods, the optimal number of clusters is $k=4$. Figure 4.4 illustrate this choice.

The k-means method provides these results with four clusters (Table 4.2).

We notice that 35.44% of the chosen sample of countries are low developed, 31.65% are moderately developed, 25.32% are judged developed, and 7.59% are considered highly developed in terms of financial development.

Also, we dispose of a measure of total variance in our data set that is explained by the clustering. K-means minimize the within-group dispersion and maximize the between-group dispersion. By assigning the samples to four clusters rather than n ($n \geq 4$ is the number of samples), clusters achieved a reduction in sums of squares of 75.5%.

Within-cluster sum of squares by cluster: [1] 196.74983 171.01365 188.45978 54.54049
(between$_{SS}$/total$_{SS}$=75.5%)

With $k=4$, the graphical representation of clusters is provided in Figure 4.5.

TABLE 4.1

Countries by Development Cluster – Hierarchical Classification Method

Development Level	Countries
Low developed	35
Moderately developed	25
Developed	18
Highly developed	1

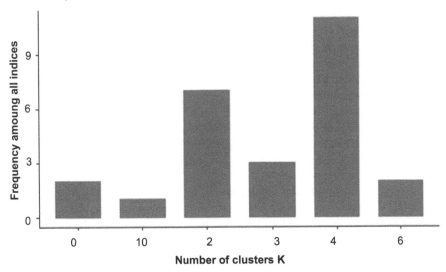

FIGURE 4.4   Indices method.

TABLE 4.2

Countries by Development Cluster – K-Means Method

Development Level	Countries
Low developed	28
Moderately developed	25
Developed	20
Highly developed	6

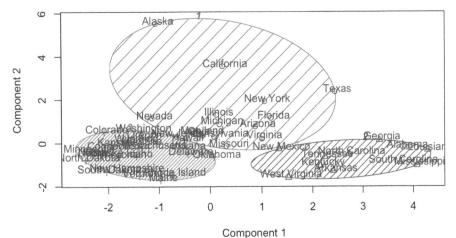

FIGURE 4.5   K-means clusters – aggregated data.

4.6 CONCLUSION

The differences between k-means and hierarchical clustering are significant, ranging from how the algorithms are implemented to how the results are interpreted. The k-means algorithm is parameterized by the value k, which is the desired number of clusters. The algorithm begins by creating k centroids. It then iterates between an assign step where each sample is assigned to its closest centroid and an update step where each centroid is updated to become the mean of all the samples that are assigned to it. This iteration continues until some stopping criteria are met.

Agglomerative hierarchical clustering, instead, builds clusters incrementally, producing a dendrogram. The algorithm begins by assigning each sample to its own cluster (top level). At each step, the two clusters that are the most similar are merged; the algorithm continues until all of the clusters have been merged. Unlike k-means, there is no need to specify a k parameter. Once the dendrogram has been produced, the user can navigate the layers of the tree to see which number of clusters makes the most sense for his particular application.

To summarize, the most important difference is the hierarchy. Actually, there are two different approaches that fall under this name: top-down and bottom-up. We divide the data into k clusters using top-down hierarchical clustering. Then, for each cluster, we can repeat this process until all of the clusters are either too small or too similar for further clustering to make sense, or we reach a predetermined number of clusters.

In bottom-up hierarchical clustering, each data item has its own cluster to begin with. Then we look for the two items that are the most similar and group them together in a larger cluster. We continue until all of the clusters we have left are too dissimilar to be assembled together, or until we reach a predetermined number of clusters (Table 4.3).

In terms of approaches, the first approach is beneficial for countries with heterogeneous clustering behavior because it allows a closer look at the variable that needs to be addressed in order to improve the positioning of a given country in terms of financial development, taking into account measures considered in this work. Meanwhile, the second approach gives just a classification on the scale of financial development; it does not make it possible to detect variables that need to be addressed so as to improve the classification of a given country. We can say that the k-means method applied to non-aggregated data provides relevant results that can be used as decision-support tools. In fact, the non-aggregation of data makes it possible to keep the entire information without losing any residuals, and the k-means method, via the procedure of determining the optimal K, gives an idea beforehand of the clusters to be considered, which facilitates the interpretation of results. These aspects are both potentially necessary and sufficient conditions for obtaining accurate results.

In brief, cluster analysis, financial system classification, and convergence analysis are different facets of the global financial system. Low convergence implies a lower probability of countries moving in one direction financially. In its turn, it means that we can observe many different types of financial systems, which may have little in common. It can obviously be an obstacle to forming adequate policies for international entities. One reflection of such heterogeneity is the debt crisis in Europe, where different countries coexist. From a more applied point of view, knowing which country refers to which type of financial system can facilitate the right decision-taking and prevent governments from conducting destructive reforms. If we find that there are several prevailing

TABLE 4.3
Mean Differences between K-Means and AHC Methods

	K-Means	AHC
Partitioning	Single partitioning	Different partitioning depending on the desired level of resolution
Number of clusters	Needs to be specified	Doesn't need to be specified
Run time	Efficient	Moderately slow

models of financial systems in the world, it may serve as a proof that successful policies conducted in one country can be adopted in another with the same structure of financial system. On the other hand, the major difference of one financial system from others may provide some caution in adopting other countries' practices.

REFERENCES

1. Cezar, R.: Newtoning financial development with heterogeneous firms. In: *61'eme Congr'es de l'AFSE.* p. 31 (2012).
2. Einav, L., Levin, J.: Economics in the age of big data. *Science* 346(6210), 1243089 (2014).
3. Emmanuel, E.: African business and development in a changing global political economy: Issues, challenges and opportunities. In: *International Academy of African Business and Development (IAABD). Peer-Reviewed Proceedings of the 13rd Annual International Conference*, vol. 13, pp. 173–183 (2012).
4. Estrada, G.B., Park, D., Ramayandi, A.: Financial development and economic growth in developing Asia. Asian Development Bank Economics Working Paper (233) (2010).
5. Fafchamps, M., Biggs, T., Conning, J., Srivastava, P.: Enterprise finance in Kenya (1994).
6. Fisman, R., Love, I.: Financial development and intersectoral allocation: A new approach. *The Journal of Finance* 59(6), 2785–2807 (2004).
7. James, G., Witten, D., Hastie, T., Tibshirani, R.: *An Introduction to Statistical Learning*, vol. 112. Springer, Cham (2013).
8. Krattiger, A.F.: Public-private partnerships for efficient proprietary biotech management and transfer, and increased private sector investments: A briefings paper with six proposals commissioned by UNIDO, vol. 4. Citeseer (2002).
9. Lenka, S.K.: Measuring financial development in India: A PCA approach. *Theoretical & Applied Economics* 22(1), 187–198 (2015).
10. Manova, K., Zhang, Z.: China's exporters and importers: Firms, products and trade partners. Tech. rep., National Bureau of Economic Research (2009).
11. McLeay, M., Radia, A., Thomas, R.: Money creation in the modern economy. *Bank of England Quarterly Bulletin* 54, 14–27 (2014).
12. Oyebola Fatima, E.M.: The effect of financial market development on capital and debt maturity structure of firms in selected African countries/Oyebola Fatima Etudaiye-Muhtar. Ph.D. thesis, University of Malaya (2016).
13. Reinhart, C.M., Tokatlidis, I.: Financial liberalisation: The African experience. *Journal of African Economies* 12(suppl 2), ii53–ii88 (2003).
14. Reiss, P.C., Wolak, F.A.: Structural econometric modeling: Rationales and examples from industrial organization. *Handbook of Econometrics* 6, 4277–4415 (2007).
15. Vaknin, S.: A critique of Piketty's capital in the twenty-first century (2014). *A Narcissus Publications Imprint*, Harvard University Press, ISBN: 978-0674430006.
16. Zingales, L., Rajan, R.: The great reversals: The politics of financial development in the 20th century. National Bureau of Economic Research (2001).

5 Candlestick Patterns Recognition in Bitcoin Price Action Graph Chart Using Deep Learning

*Abdellah El Zaar, Nabil Benaya,
Abderrahim El Allati, and Toufik Bakir*

5.1 INTRODUCTION

The stock market is an environment where regular activities like buying and selling are conducted. Though it is called a stock market, it is primarily known for trading stocks. Operations in stock markets are made in a secure and controlled environment. Nowadays, the stock markets operate electronically on well-known platforms designated for trading. These platforms bring together hundreds of thousands of market participants who wish to buy and sell shares. The trading platforms contain all the information and tools that the trader needs to analyze the different markets. Traders also refer to the trading chart, which displays information that can help them decide when to enter and exit a position in the live market. There are many kinds of trading charts, like bar charts, line charts, but the most commonly used is the candlesticks chart (see Figure 5.1). Each candlestick provides market information within a selected time interval; it also gives important visual indicators for market movements, the things that make traders who trade several markets at the same time unable to control and analyze the market movement, and therefore they risk losing market deals worth millions of

FIGURE 5.1 Candlestick chart.

DOI: 10.1201/9781032667478-6

dollars [1]. For this reason, we use deep learning to recognize the candlestick patterns in the trading chart graph. Computer vision and deep learning will complete the human vision, so our work aims to investigate the deep Convolutional Neural Network (CNN) to help traders take more accurate and effective action and analysis while trading in live markets and also when analyzing historical data. We exploit the power of the CNN algorithm to perform the training process and recognize multiple candlestick patterns in the price action chart of Bitcoin cryptocurrency. The graph chart contains several types of candlesticks such as Dojis, Big Candles, Gravestone, Shooting Star, and Hammer. In addition to that, these candlesticks appear in two principal colors: red and green, which represents an important visual indicator. The main goal of our work is to encode, recognize, and identify these powerful candlestick patterns. We used two different representations of the same data to train different architectures of CNN and to make a comparison between all the trained architectures. The first data form is the 1D dataset which contains (OCHL) Open, Close, High, Low of the Bitcoin price in a specific time interval (see Table 5.2), and the second is the 2D dataset, which contains images of Hammer and shooting-star candlesticks (see Figure 5.2). These two patterns have a great role in spotting the trend of Bitcoin cryptocurrency.

The collected datasets are fed to our CNN algorithm for training. Two different CNN strategies are used in order to make a comparison between performances. The first one consists of training CNN from scratch, and the second one is based on transfer learning and uses a combination of the CNN algorithm and the Support Vector Machine (SVM) algorithm. For the OCHL dataset, we implemented a one-dimensional CNN that contains 1D layers. On the other side, for image data, we implemented two-dimensional CNN with 2D layers.

The graph shows the price action in a one-hour time interval. The duration of each candlestick is one hour.

To test our proposed models and strategies, we split the collected datasets into the training set and the test set. The test set represents unseen data for the models, and it is used to test our model behavior after the training process. Figure 5.2 presents two different parts of the graph chart of Bitcoin cryptocurrency price action. The red candlesticks show that the stock closed lower and signals are selling signals, and the green ones indicate that the stock closed higher and signals are buying signals. We observe in Figure 5.2a that after the downtrend of the market, the Hammer candlestick gives a buy signal. On other side, in Figure 5.2b, we observe that after the uptrend of the market with a shooting-star candlestick at the end, a downtrend of the market is coming. The shooting-star candlestick pattern provides a good sell signal. Traders call these observations: Technical analysis [2]. The goal of our work is also to facilitate technical analysis for Bitcoin cryptocurrency traders by implementing deep learning and computer vision algorithms. This will allow them to predict Bitcoin cryptocurrency trends and signals. The rest of the chapter is organized as follows: Section 5.2 presents related work done recently in this field. Section 5.3 describes the proposed approach and experiments. The achieved results are discussed in Section 5.4. Conclusion and perspectives are given in Section 5.5.

5.2 RELATED WORK

In this section, we will explore some of the state-of-the-art work done in the field of stock market analysis using deep learning. We will also compare these works with our proposed research work. The majority of works done in this area of research do not provide effective and accurate results, but they give an important study about the technical analysis of the market using machine learning algorithms. For example, in [3], J. Hao Chen and Y. Cheng Tsai proposed a two-step approach based on the GAF-CNN algorithm to recognize candlestick patterns automatically. They were able to identify eight types of candlesticks patterns with 90.7%. Kusuma et al. [4] used the CNN to perform candlestick analysis. Their method provides a satisfactory result with a recognition score

(a)

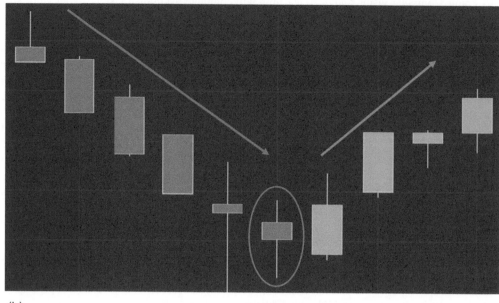

(b)

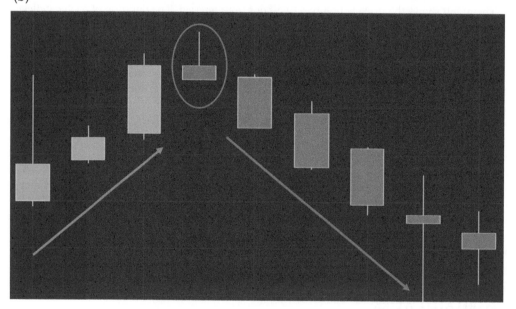

FIGURE 5.2 Trends of the Bitcoin cryptocurrency after Hammer and shooting-star candlesticks: (a) Hammer pattern and (b) shooting-star pattern.

of 92%. A. Andriyanto, A. Wibowo, and N. Z. Abidin [5] presented a CNN approach to identify the strength of a trend pattern in the movement of the stock market. The proposed approach produces an accuracy of 99% with remarkable noise during the training process. This problem is generally behind the noisy dataset and a non-suitable CNN strategy during the training process. In [6], J. H. Chen et al. provide an approach called adversarial attacks algorithm based on the local search to predict the patterns of candlesticks. The applied strategy gives good results with an attack ratio of 64.36%.

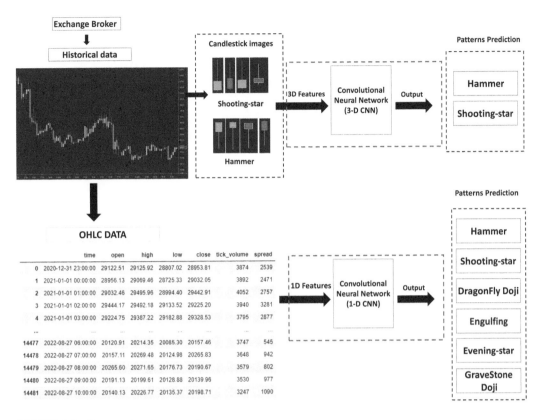

FIGURE 5.3   Proposed system overview.

5.3   PROPOSED APPROACH

In our case study, we used the power of a CNN algorithm to encode and recognize the most powerful candlestick patterns in the price action trading chart of Bitcoin cryptocurrency. The most powerful candlesticks in Bitcoin chart price movement are Evening Star, Engulfing, Dragonfly Doji, Gravestone Doji, Shooting Star, and Hammer (see Table 5.1). Our proposed system can also be used to perform technical analysis of the Bitcoin cryptocurrency market using historical data [7]. The technical analysis is an important study that gives the market context, an overview, and reports. It is the most important thing that traders refer to, to make a decision on the market. An overview of our system is illustrated in Figure 5.3.

5.3.1   DATA

One of the key characteristics of the CNN algorithm is that it can provide high-accuracy results when using a large amount of data. For that reason, we collected two years historical Bitcoin data (see Figure 5.4). The collected data contains 14,481 Candlesticks with a one-hour time frame (Figure 5.4). Each candle provides four price features: OHLC (Open, High, Low, Close) within a one-hour time interval (see Table 5.2). We focused on identifying six classes of the most powerful and influential candlesticks on the Bitcoin trend signal: Evening Star, Engulfing, Dragonfly Doji, Gravestone Doji, Shooting Star, and Hammer (see Table 5.1). The body of the candlestick is the difference between the opening and closing prices. If the closing price is higher than the opening price, the body is green, which signals rising prices. If the opening price is higher than the closing price, the body is red, which signals falling prices (see Figures 5.5 and 5.7). Our proposed approach is also used to recognize images of Hammer and shooting star. We used two different data representations in order to analyze and compare our model performance when using the same data with two different representations [8].

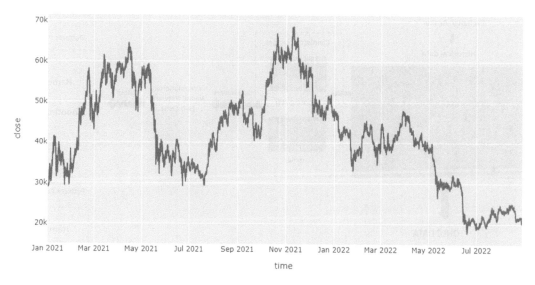

FIGURE 5.4 Bitcoin closed prices from 01-01-2021 to 27-08-2022.

5.3.2 PROPOSED CNN ARCHITECTURE

Our CNN model is built around the Visual Geometry Group Network (VGGNet) architecture with a little modification in the number of Conv-pool layers and the number and size of filters. There are many versions of VGGNet architectures, like VGG-11, VGG-16, and VGG-19. VGG-11 contained 8 convolution layers, VGG-16 had 13 convolution layers, and VGG-19 had 16 convolution layers. All of them ended the same with three fully connected layers [9]. In our case, we used a less complex architecture with four convolutional layers, two pooling layers, and three fully connected layers.

For the 1D collected dataset, we used a 1D CNN with 1D convolution layers, filters, and Maxpooling layers. For 2D image data, we used 2D convolutions, filters, and Maxpooling layers. In addition to that, we perform the recognition process using two learning strategies: CNN trained from scratch strategy and combined CNN-SVM strategy. When using the CNN from scratch learning strategy, the data is fed from the input layer to the output layer, traversing the convolutional layers and the Maxpooling layers. For the combined CNN-SVM strategy, CNN is used as a feature extractor, and the SVM algorithm is used as a multiclass classifier (see Figure 5.6).

Before feeding our 3D dataset images to the 2D CNN, we reshape and normalize them to the size of $32 \times 32 \times 1$. Each convolution layer is equipped with filters and a Rectified Linear Unit activation function. For the first two convolution layers, we used 16 filters of size 3×3. In the next two convolution layers, we doubled the number of filters and kept the same size. The convolution operation is a dot product between an input layer and the filter. The filters (Kernels) are three-dimensional parameters $(K_n \times K_n \times d_n)$, which represents the network parameters. The dimensions (length and width) of the output layer after performing the convolution operation are as follows:

$$L(n+1) = L_n - K_n + 1 \tag{5.1}$$

$$W(n+1) = W_n - K_n + 1 \tag{5.2}$$

To avoid the well-known overfitting problem and to maintain the effectiveness of our model, we used batch normalization and dropout optimization.

TABLE 5.1

Candlestick Patterns

Candlestick Patterns	Names	Candlestick Patterns	Names
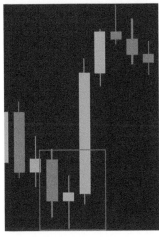	Hammer	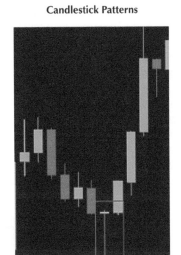	Dragonfly doji
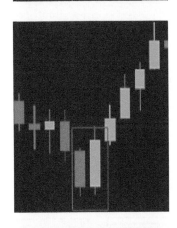	Engulfing	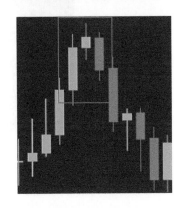	Evening star
	Shooting star	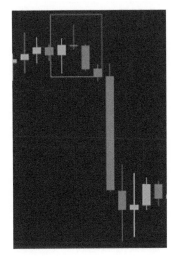	Gravestone doji

TABLE 5.2

Bitcoin Historical Data from 09-02-2018 to 31-08-2022 One-Hour Time Frame

	Time	Open	High	Low	Close	Tick Volume	Spread
0	2018-02-09 10:00:00	8251.64	8273.12	8234.30	8259.18	543	1494
1	2018-02-09 11:00:00	8254.44	8292.32	8249.11	8258.06	886	1529
2	2018-02-09 12:00:00	8258.23	8270.17	8193.92	8204.09	913	1519
3	2018-02-09 13:00:00	8207.18	8207.72	8201.21	8201.89	96	1546
4	2018-02-10 20:00:00	8199.83	8202.86	8187.04	8199.31	73	1725
5	2018-02-10 21:00:00	8194.33	8278.17	8187.15	8278.17	648	1344
6	2018-02-10 22:00:00	8282.15	8294.33	8281.21	8294.30	515	1143
7	2018-02-11 01:00:00	8294.21	8296.99	8266.99	8269.80	505	1410
8	2018-02-11 02:00:00	8269.99	8287.60	8261.33	8284.18	319	1344
9	2018-02-11 12:00:00	8279.78	8296.61	8238.42	8242.40	828	1671

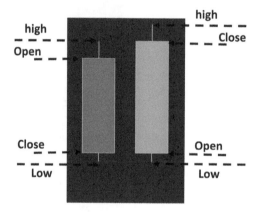

FIGURE 5.5 Candlestick composition.

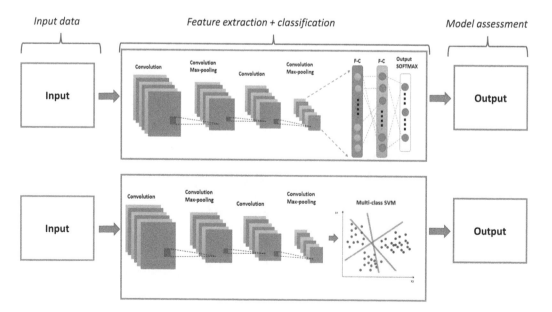

FIGURE 5.6 CNN from scratch and CNN-SVM strategies.

5.4 EXPERIMENTS, RESULTS, AND DISCUSSION

In this section, we will discuss the experiments and the results achieved using our system approach. We will also analyze the performances during the training process using our collected dataset.

5.4.1 EXPERIMENTS

The CNN algorithm was implemented using the Python language. We used several libraries that contain visualization and image processing functions, such as *Tensorflow*[TM], *Keras*[TM] Models, and *Sklearn*[TM]. We run our CNN model on a Microsoft Arure virtual machine with six core processors and 56 GB of RAM. Working with CNN with a high number of epochs and visualizing a large amount of data can be a time-consuming process. For this reason, we used a TESLA K80 NVIDIA GPU. GPU boost gives superior performance to our model and accelerates libraries and training processes.

5.4.2 RESULTS AND DISCUSSION

Figure 5.7 shows the performances of our 2D CNN model using an image dataset during the training process, while Figure 5.8 illustrates 1D CNN model training performance using an OHLC dataset.

It can be seen that our 2D CNN behavior during the training process is perfect. With 100 epochs, the system provides high-accuracy and effective results from our collected dataset. Our proposed 2D CNN model shows an error rate of 0.01%. For the 1D dataset, the 1D CNN proposed model also provides perfect results with the challenging OHLC collected dataset. We trained our collected datasets using the SVM algorithm combined with CNN to analyze and compare the performances of the collected candlestick datasets. We used CNN as a feature extractor and the SVM algorithm for classification and prediction. We provide also a comparative study of performances of the two used methods with preprocessing step and without preprocessing step.

In Table 5.3, we observe that the accuracy of both methods is higher when applying the preprocessing step compared to without preprocessing. Particularly in the second method based on CNN-SVM, the accuracy is significantly lower when the preprocessing step is not utilized. This problem is due to the inability of traditional machine learning algorithms like SVM to give high-accuracy results without the preprocessing and feature extraction steps. The major advantage of the CNN algorithm is that it can provide high-accuracy results and does not require any feature extraction or preprocessing steps. The first method, based on CNN and trained from scratch, offers a dual advantage: it presents a less complex architecture while delivering high and effective results. In the case of the 1D CNN and 1D OHLC datasets, we obtained satisfactory results, especially with such a challenging dataset, to the best of our knowledge. Our results are the best and most accurate compared with the state-of-the-art works in this field (Table 5.4).

5.5 CONCLUSION

In this chapter, we proposed a CNN architecture to encode and recognize candlestick patterns in the graph chart of Bitcoin cryptocurrency. We used two different representations of the same dataset in order to detect the candlestick patterns with high accuracy. We collected a large historical data dataset of 14,481 candlesticks that contains the Open, High, and close prices of each candle to train the 1D CNN. For the 2D CNN, we collected over 900 images of the two most powerful candlesticks that have had a great impact on the Bitcoin market trend. Our approach can give a decision-making signal to multi-market traders using computer vision. We conclude that developing deep learning and computer vision trading agents can also help stock market traders achieve accurate trades. These agents can also provide a real-time technical analysis of different markets.

(a)

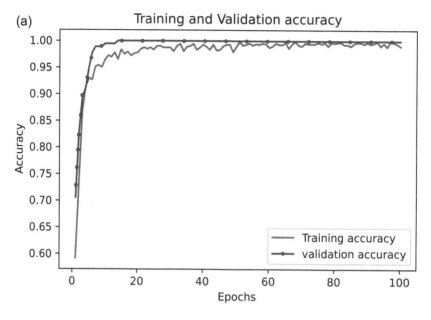

(b)

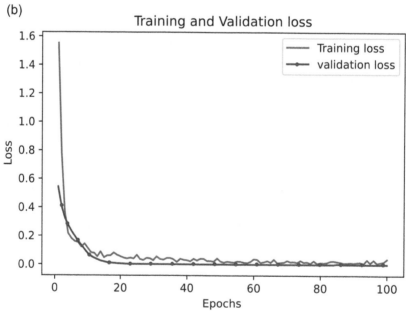

FIGURE 5.7 Proposed approach performances for 2D CNN: (a) accuracy and (b) loss.

(a)

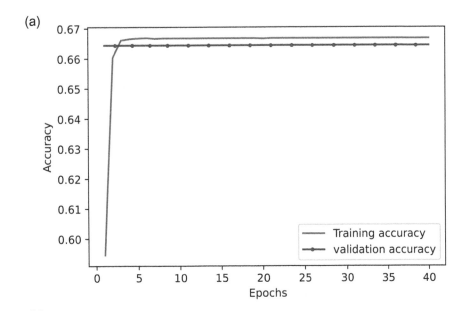

(b)

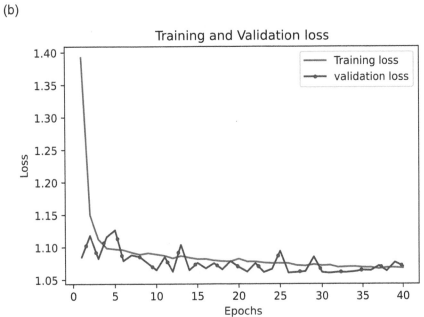

FIGURE 5.8   Proposed approach performances for 1D CNN: (a) accuracy and (b) loss.

TABLE 5.3

Comparison of Training Accuracy of CNN and CNN-SVM Methods

Method	With Preprocessing Accuracy (%)	Without Preprocessing Accuracy (%)
2D-CNN trained from scratch (Extracted images dataset)	99	98.46
Combined 2D-CNN-SVM (Extracted images dataset)	99	97
1D-CNN trained from scratch (Extracted OHLC dataset)	83	75
Combined 1D-CNN-SVM (Extracted OHLC dataset)	86	75

TABLE 5.4

Comparison of Test Accuracy of CNN and CNN-SVM Methods

Method	With Preprocessing Accuracy (%)	Without Preprocessing Accuracy (%)
2D-CNN trained from scratch (Extracted images dataset)	98	95
Combined 2D-CNN-SVM (Extracted images dataset)	96	92
1D-CNN trained from scratch (Extracted OHLC dataset)	82	73
Combined 1D-CNN-SVM (Extracted OHLC dataset)	83	72

REFERENCES

1. M. Wang and Y. Wang, "Evaluating the effectiveness of candlestick analysis in forecasting us stock market," in *Proceedings of the 2019 3rd International Conference on Compute and Data Analysis*, pp. 98–101, 2019.
2. J. Chen, *Essentials of Technical Analysis for Financial Markets*. John Wiley & Sons, New York, 2010.
3. J.-H. Chen and Y.-C. Tsai, "Encoding candlesticks as images for pattern classification using convolutional neural networks," *Financial Innovation*, vol. 6, pp. 1–19, 2020.
4. R. M. I. Kusuma, T.-T. Ho, W.-C. Kao, Y.-Y. Ou, and K.-L. Hua, "Using deep learning neural networks and candlestick chart representation to predict stock market," *arXiv preprint arXiv:1903.12258*, 2019.
5. A. Andriyanto, A. Wibowo, and N. Z. Abidin, "Sectoral stock prediction using convolutional neural networks with candlestick patterns as input images," *International Journal*, vol. 8, no. 6, pp. 2249–2252, 2020.
6. J.-H. Chen, S. Y.-C. Chen, Y.-C. Tsai, and C.-S. Shur, "Explainable deep convolutional candlestick learner," *arXiv preprint arXiv:2001.02767*, 2020.
7. Y.-J. Goo, D.-H. Chen, and Y.-W. Chang, "The application of japanese candlestick trading strategies in taiwan," *Investment Management and Financial Innovations*, vol. 4, no. 4, pp. 49–79, 2007.
8. F. Han and H. Liu, "Scale-invariant sparse pca on high-dimensional meta-elliptical data," *Journal of the American Statistical Association*, vol. 109, no. 505, pp. 275–287, 2014.
9. K. Simonyan and A. Zisserman, "Very deep convolutional networks for large-scale image recognition," *arXiv preprint arXiv:1409.1556*, 2014.

6 Prediction of Stock Markets Using Deep Learning Architectures

Khalid Bentaleb, Benhouad Mohamed,
and Mohammed Mestari

6.1 INTRODUCTION

A stock market is a place where buyers and sellers meet to trade shares of public companies. As discussed in [1], stock market forecasting is a challenging task because of the volatile, noisy, and nonlinear nature of the financial equity markets. However, the stock price is influenced by various economic factors such as political conditions, general economics, investors' expectations, commodity price index, and movements in other equity markets. Therefore, the prediction of stock market values and stock market movements is one of the most challenging problems for investors and researchers [2]. There are two types of stock price prediction. Technical analysis forecasts prices based on historical stock prices, using technical indicators like the exponential moving average, momentum, and moving average convergence divergence [3]. On the other hand, fundamental analysis looks at the deep-seated value of stocks, the economy, the performance of the industry, the political climate, and financial factors that influence a business [4]. The principal difference between these two methods of analyzing financial markets is the time period taken into account by the investment strategies. Technical analysis considers shorter time frames, while fundamental analysis focuses on the next long time frame. A number of traditional machine learning models, such as k-nearest neighbors, support vector machines, and random forests, have been used for stock market forecasting. However, it is difficult for these models to produce reliable stock predictions with high precision because they are powerful in many problems, but in such highly volatile and nonlinear problems, they suffer from stability issues. This is one of the most important reasons for the difficulty of stock market forecasting. The increasing eminence of deep learning in various industries has enlightened many researchers to apply deep learning algorithms to the financial sector, and some of them have yielded quite promising results and better performances [5]. In this study, we propose classification models of deep learning for forecasting the price movement on the Stock Exchange of Casablanca using long short-term memory (LSTM), bidirectional LSTM (BiLSTM), gated recurrent units (GRU), and convolutional neural networks (CNN). The models have used as input data a set of 83 technical indicators retrieved from the financial dataset: open, high, low, close, and volume. Furthermore, the results are based on four different assessment metrics, such as accuracy, precision, F1-score, and recall. Our objective is to compare the predicted movement results of stock prices obtained with different deep learning–based classification models. The main contribution of our work is to demonstrate and test the abilities of deep learning models to predict the stock market with a very high degree of accuracy.

The rest of this chapter is divided into five sections. A short overview of related work is presented in Section 6.2. Section 6.3 provides the methodology of techniques applied. Section 6.4 discusses the result. Finally, conclusions and future work are presented in Section 6.5.

DOI: 10.1201/9781032667478-7

6.2   LITERATURE SURVEY

In the past few decades, there has been a lot of research available in the areas of deep learning and stock price forecasting. Some research was implemented based on predicting financial data, while others was based on the use of technical indicators and financial news to improve predicting performance. Bhardwaj [6] constructed a CNN model and used 1D time-series data for the model. The purpose of this research was to categorically predict whether the price would go up or down. The accuracy of this pattern was 50.3%. Gavriel [7] created and deployed several LSTM model architectures, namely: stacked LSTM, shallow LSTM, four-layer LSTM model with dropouts, and BiLSTM. The results showed that the stacked LSTM and the shallow LSTM yielded almost the same results, with a normalized RMSE of 0.0247 and 0.0230, respectively. In 2021, Wu et al. [8] suggested an algorithm for forecasting the stock market that combines CNN and LSTM. The results indicate that CNN-LSTM achieved better performance than traditional CNN, LSTM, and statistical methods. Komori [9] built a CNN by applying image recognition to 2D candlestick stock charts. The results were expressed in terms of accuracy; the 3-day-ahead prediction gave the best accuracy with 50.0%. In 2019, Hoseinzade and Haratizadeh [10] proposed two CNN approaches to predict future trends in share prices on the basis of technical indicators and macroeconomic variables. Their results demonstrate that the CNN model can point out future trends as a classification model. Tsai et al. [11] suggested the use of CNN in combination with Gramian Angular Field, a time-series image representation technique, for the classification of the chart patterns. These patterns are commonly used to determine turning points in share prices and can indicate a buy/sell transaction. The results showed that the accuracy was around 80%. Fischer and Krauss [12] concentrated on the performance of the LSTM model. They found that LSTM performance exceeds traditional classifiers of machine learning like deep neural networks, logistic regression, and random forests. However, the RAF performed better in one case, during the global financial crisis. In 2017, Vargas et al. [13] used CNN and LSTM for a NLP approximation. The results showed that CNN may be better at capturing text semantics than LSTM, whereas LSTM is better at capturing contextual information.

6.3   METHODOLOGY

6.3.1   DATA PREPARATION AND PREPROCESSING

Historical data is collected from https://www.cdgcapitalbourse.ma. The dataset contains information on the stock price, such as the open, high, low, close, and volume of the trades for each trading day. We have considered banking sector stocks, namely Bank of Africa, Banque Populaire (BP), CIH Bank, Banque Marocaine pour le Commerce et Industrie (BMCI), and Attijari Wafa Bank (AWB). The collected dataset includes between 4000 and 6000 days. From these datasets, we determine a target that we are trying to predict. Our focus will be on whether the price will go up or down the next day. If the price has gone down, the target will be 0, and if it has gone up, the target will be 1. Also, we need to extract technical indicators, as these indicators hold valuable hidden information about prices. There are 83 different indicators that have been created. These indicators have been used as features of the models. 80% of the dataset is used for training and 20% for testing. Before training, we are cleaning and processing data. Missing and redundant data are removed from the dataset, and input values are normalized at the [0, 1] range. The expected outputs are adjusted to 0 or 1.

6.3.2   CLASSIFICATION METHODS

Deep learning [14] is an advanced type of machine learning that is basically made up of multiple layers of artificial neural networks. Deep learning has made its way into financial institutions for its power in forecasting time-series data with a high level of accuracy, and it is used to determine if the stock price will go up or down the next day based on all historical data.

6.3.2.1 Long Short-Term Memory (LSTM)

LSTM is a variant of the RNN. It was introduced by Hochreiter and Schmidhuber [15] in 1997. These networks can learn and memorize long-term dependencies in sequence prediction problems. It's useful in time-series prediction because it can store past information that is important and forget the information that is not. Figure 6.1 shows a representation of an LSTM cell.

The structure of an LSTM model consists of the following cells:

- **The input gate:** The input gate adds information to the cell status.
- **The forget gate:** It decides what information should be discarded or kept.
- **The output gate:** This gate controls the value of the next hidden state.

6.3.2.2 Bidirectional LSTM (BiLSTM)

A BiLSTM is the process that makes any neural network have sequential information in both forward (past to future) and backward (future to past). BiLSTMs effectively improve the context available to the algorithm and increase the amount of information available to the network.

6.3.2.3 Gated Recurrent Unit (GRU)

GRU [16] is a gating mechanism in RNNs. It's similar to LSTM but without an output gate. GRU can be considered a variation of the LSTM because both have a similar design and give equal results in some cases. Fewer parameters mean that GRUs are generally faster and easier to train than their LSTM counterparts. Figure 6.2 shows a representation of a GRU cell.

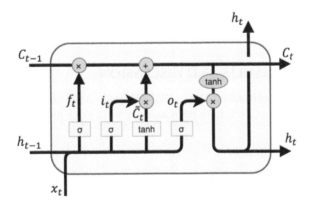

FIGURE 6.1   The architecture of the LSTM cell.

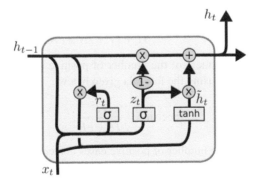

FIGURE 6.2   The architecture of the GRU cell.

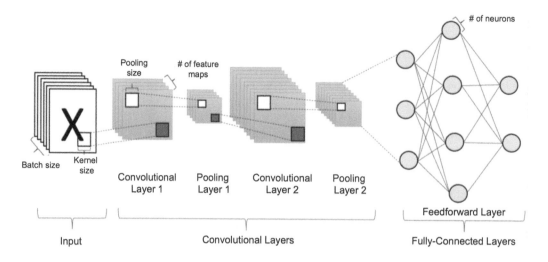

FIGURE 6.3   Basic convolutional neural network architecture.

6.3.2.4   Convolutional Neural Network (CNN)

A CNN is a network architecture for deep learning proposed by Lecun et al. [17] in 1998. CNN precisely learns from data, eliminating the need for extracting features manually. It consists of a collection of neurons that are organized in layers, each with their own learnable weights and biases. Figure 6.3 shows a basic CNN architecture.

6.4   EXPERIMENTAL RESULTS AND DISCUSSION

In this research, five datasets (Bank of Africa, BP, CIH Bank, AWB, and BMCI) and four classifiers (LSTM, BiLSTM, GRU, and CNN) have been used for stock price prediction.

6.4.1   Performance Evaluation

As our problem is a classification problem, the models will provide the result of whether the equity market will go up or down the next day based on all the past data. We have used precision, accuracy, recall, and F1-score to evaluate the performance of deep learning models.

Accuracy: Accuracy is the base metric used for classification model evaluation. It indicates how close the measurements are to a specific value. The formula used for accuracy in this case is:

$$\text{Accuracy} = \frac{TP + TN}{TP + FP + TN + FN} \tag{6.1}$$

Precision: Precision is a measurement of the number of positive predictions that are correct (true positives), and it is calculated by using the formula given below:

$$\text{Precision} = \frac{TP}{TP + FP} \tag{6.2}$$

Recall: Recall is a measure of the number of positive cases correctly predicted by the classifier for all positive cases in the data. It is calculated as per the formula given below:

$$\text{Recall} = \frac{TP}{TP + FN} \tag{6.3}$$

F1-score: The F1-score is the harmonic mean of precision and recall scores. It may be explained by weighted average precision and recall. The formula for calculating the F1-score is given below:

$$F1\text{ -score} = \frac{2 * \text{Precision} * \text{Recall}}{\text{Precision} + \text{Recall}}$$

Where TP is the number of true positives, FP is the number of false positives, TN is the number of true negatives, and FN is the number of false negatives.

6.4.2 Experimental Results

The experiments are conducted on five stocks. For this, we used four types of deep learning models named LSTM, BiLSTM, GRU, and CNN. In this chapter, we have considered between 4000 and 6000 prediction days. Each model is assessed separately; we have used accuracy, precision, recall, and F1-score. The motive behind this was to compare the performance of LSTM, BiLSTM, GRU, and CNN. The results obtained are presented in Tables 6.1–6.5.

The models classified as having the highest to lowest accuracy are GRU (92%), LSTM (91%), BiLSTM (81%), and CNN (88%). These results clearly indicate the superiority of RNN variants over CNN. For the BP dataset, the highest accuracy of 92% belongs to GRU. The second highest belongs to BiLSTM (91%) and the lowest belongs to CNN (87%). For the Bank of Africa dataset, among all the classifiers, BiLSTM has the highest accuracy of 90%. The second highest comes from LSTM at 89%. The lowest accuracy of 88% belongs to GRU and CNN. For the AWB dataset, the highest accuracy of 88% belongs to LSTM and GRU. The second highest (83%) belongs to BiLSTM, and the lowest belongs to CNN (81%). For the CIH Bank dataset, the highest accuracy of 86% belongs to LSTM, BiLSTM, and CNN. The lowest belongs to GRU (83%). Also, for the BMCI dataset, LSTM has the highest accuracy of 91%. BiLSTM and GRU have the second highest of 90%, and CNN has the least accuracy CNN (87%).

These results indicate that our models perform better than the proposals submitted by Naik and Mohan [18], by Patel et al. [19], and by Wang [20] and provide satisfactory performance. In choosing the right algorithm, performance is the most important consideration, along with ease of use and interpretation. Due to the variety of related work in the literature, providing a completely fair and objective comparison is very challenging.

TABLE 6.1
Metrics Comparison for Banque Populaire

Model	Precision	Recall	F1-Score	Accuracy
LSTM	0.91	0.90	0.90	0.90
BiLSTM	0.91	0.91	0.91	0.91
GRU	0.92	0.92	0.92	0.92
CNN	0.87	0.87	0.87	0.87

TABLE 6.2
Metrics Comparison for Bank of Africa

Model	Precision	Recall	F1-Score	Accuracy
LSTM	0.90	0.89	0.89	0.89
BiLSTM	0.91	0.91	0.91	0.90
GRU	0.89	0.89	0.89	0.88
CNN	0.89	0.88	0.88	0.88

TABLE 6.3
Metrics Comparison for Attijari Wafa Bank

Model	Precision	Recall	F1-Score	Accuracy
LSTM	0.90	0.88	0.88	0.88
BiLSTM	0.87	0.83	0.82	0.83
GRU	0.89	0.88	0.88	0.88
CNN	0.86	0.81	0.80	0.81

TABLE 6.4
Metrics Comparison for CIH Bank

Model	Precision	Recall	F1-Score	Accuracy
LSTM	0.89	0.86	0.86	0.86
BiLSTM	0.88	0.86	0.86	0.86
GRU	0.86	0.83	0.83	0.83
CNN	0.87	0.86	0.86	0.86

TABLE 6.5
Metrics Comparison for BMCI

Model	Precision	Recall	F1-Score	Accuracy
LSTM	0.91	0.91	0.91	0.91
BiLSTM	0.90	0.90	0.90	0.90
GRU	0.90	0.90	0.90	0.90
CNN	0.88	0.88	0.87	0.87

6.5 CONCLUSION

Predicting the stock market is critically important to financial analysts and investors. As a financial series, the equity market is nonlinear, noisy, and chaotic. So, it's very hard to predict exactly the stock market. In this chapter, four deep learning architectures were adopted for efficient forecasting of the stock market. Using five banking stocks in Stock Exchange of Casablanca, we trained and tested our models to show that LSTM can achieve excellent forecasting performance for almost all stocks as well as comparable proposals in the literature. This study contributes to enriching the predictive area of the financial series. It may be a useful tool to help financial analysts and investors forecast the equity market. In addition to historical prices, there is more information that could impact stocks, such as politics, social media, economic growth, and financial news. Research has shown that sentiment analysis has a significant impact on future prices. In future work, we are going to use deep learning models for NLP to predict the stock price based on fundamental analyses.

REFERENCES

1. Chandar, K., Hybrid models for intraday stock price forecasting based on artificial neural networks and metaheuristic algorithms. *Pattern Recognition Letters*, 2021, vol. 147, pp. 124–133.
2. Chen, W., Jiang, M., Wei-Guo, Z., et al., A novel graph convolutional feature based convolutional neural network for stock trend prediction. *Information Sciences*, 2021, vol. 556, pp. 67–94.
3. Chen, Y. and Hao, Y., A feature weighted support vector machine and K- nearest neighbor algorithm for stock market indices prediction. *Expert Systems with Applications*, 2017, vol. 80, pp. 340–355.
4. Khan, W., Malik, U., Ghazanfar, M. A., Azam, M. A., Alyoubi, K. H., and Alfakeeh, A. S., Predicting stock market trends using machine learning algorithms via public sentiment and political situation analysis. *Soft Computing*, 2019, vol. 24, pp. 11019–11043.
5. Hiransha, M., Gopalakrishnan, E. A., Menon, V. K., and Soman, K. P., NSE stock market prediction using deep-learning models. *Procedia Computer Science*, 2018, vol. 132, pp. 1351–1362.
6. Bhardwaj, K., Convolutional neural network in stock price movement prediction. *Neural and Evolutionary Computing*, 3 Jun 2021. doi: 10.48550/arXiv.2106.01920.
7. Gavriel, S., Stock market prediction using long short-term memory (B.S. thesis). University of Twente.
8. Wu, J.M.T., Li, Z., Herencsar, N. et al. A graph-based CNN-LSTM stock price prediction algorithm with leading indicators. *Multimedia Systems* 2023, vol. 29, pp. 1751–1770. https://doi.org/10.1007/s00530-021-00758-w
9. Komori, Y., Convolutional neural network for stock price prediction using transfer learning. *SSRN Electronic Journal*, 2020. doi: 10.2139/ssrn.3756702.
10. Hoseinzade, E. and Haratizadeh, S., CNNpred: CNN-based stock market prediction using a diverse set of variables. *Expert Systems with Applications*, 2019, vol. 129, pp. 273–285.
11. Tsai, Y.-C., Chen, J.-H., and Wang, C.-C., Encoding candlesticks as images for patterns classification using convolutional neural networks. *Financial Innovation*, 2020, vol. 6.1, pp. 1–19.
12. Fischer, T. and Krauss, C., Deep learning with long short-term memory networks for financial market predictions. *European Journal of Operational Research*, 2018, 270(2), 654–669.
13. Vargas, M. R., De Lima, B. S., and Evsukoff, A. G., Deep learning for stock market prediction from financial news articles. In *2017 IEEE International Conference on Computational Intelligence and Virtual Environments for Measurement Systems and Applications (CIVEMSA)*, pp. 60–65. IEEE, 2017.
14. Sarker, I. H. Deep learning: A comprehensive overview on tech- niques, taxonomy, applications and research directions. *SN Computer Science*, 2021, vol. 2, no. 6, pp. 420.
15. Hochreiter, S. and Schmidhuber, J., Long short-term memory, *Neural Computation*, 1997, vol. 9, no. 8, pp. 1735–1780.
16. Chung, J., Gulcehre, C., Cho, K., and Bengio, Y., Empirical evaluation of gated recurrent neural networks on sequence modeling, ArXiv Prepr. ArXiv14123555.
17. Lecun, Y., Bottou, L., Bengio, Y., and Haffner, P., Gradient-based learning applied to document recognition, *Proceedings of the IEEE*, 1998, vol. 86, no. 11, pp. 2278–2324.
18. Naik, N., and Mohan, B. R., "Stock price movements classification using machine and deep learning techniques-the case study of indian stock market." *Engineering Applications of Neural Networks*, edited by John Macintyre et al., Springer International Publishing, Cham, 2019, pp. 445–52.
19. Patel, J., et al., Predicting stock and stock price index movement using trend deterministic data preparation and machine learning techniques. *Expert Systems with Applications*, Jan. 2015, vol. 42, no. 1, pp. 259–68.
20. Wang, Y., Stock price direction prediction by directly using prices data: An empirical study on the KOSPI and HSI. *International Journal of Business Intelligence and Data Mining*, 2014, vol. 9, no. 2, p. 145.

7 Improving Credit Card Fraud Detection with Distributed Machine Learning and Bfloat-16

Bushra Yousuf, Rejwan Bin Sulaiman, and Musarrat Saberin Nipun

7.1 INTRODUCTION

The banking industry is significant to everyone's daily lives because of the services it provides. We utilize banking services and products, including internet banking and physical credit and debit cards. When someone takes your actual credit card or gets the data of your credit card or bank account from you with dishonest intentions over the phone or online, this is considered fraud.

When you reveal your credit card information to a dishonest individual, you risk being a victim of online fraud, phishing, and spam. Because of the massive number of transactions that are taking place, both the client and the bank have a very high risk of falling victim to fraud since criminals are more likely to succeed in their schemes [1].

Fraud may come in many forms, such as insurance fraud, credit card fraud, and accounting fraud, all of which lead to monetary loss for the victims, who might be customers or financial institutions. Consequently, it is of the utmost importance to uncover schemes of this nature. Examining signs that are both evident can help in the investigation and identification of accounting crimes. Transactions of high value in odd locations or with uncommon merchants require further verification. Traditional approaches involve the use of rule-based procedures to identify fraudulent operations. These methods do not concentrate on catastrophic circumstances or an excessive imbalance of positive and negative examples. Rule-based fraud systems use methods that run various checks to identify distinct fraud frameworks. These frameworks are manually constructed by fraud analysts and then used by rule-based fraud systems.

About 400 unique approaches are used by the existing conventional systems whenever a transaction has to be validated [2]. The algorithms must have additional cases physically added and can hardly distinguish between correlated and uncorrelated interactions. In addition, rule-based software typically uses antiquated software that is barely capable of processing real-time data, which is essential for the present market. Detecting fraudulent activity on credit cards is a complicated process due to the following two reasons: (i) the behavior of fraudsters often varies from one instance to the following (ii) the data are not balanced, which means that the actual set of data samples outnumbers the limited dataset samples, which means that the genuine fraudulent instances are underrepresented in the data. When machine learning ML is fed severely imbalanced data as an input, the resulting model is biased in favor of the original dataset. As a direct result, it has a greater propensity to display a fake record while giving the impression that it is an actual record [3].

DOI: 10.1201/9781032667478-8

7.1.1 ML Algorithms

There are some hidden and subtle occurrences in a user's behavior that might nonetheless suggest probable fraud, although these events are not immediately visible. We can design algorithms that can handle large datasets that contain a variety of factors, and these algorithms can assist us in determining the hidden connections that exist between the activities of the user and fraudulent behavior by utilizing machine learning. Compared to traditional rule-based systems, one of the primary benefits of machine learning is that it is far faster at processing data and requires significantly less labor on the user's part.

For example, machine learning algorithms integrate effectively with behavior analytics; as a result, the number of verification stages may be reduced. Large financial institutions have already begun to employ the technology of machine learning to combat fraudsters. For instance, MasterCard uses a combination of artificial intelligence and machine learning to analyze and keep tabs on various factors, including time, transaction size, location, purchase data, and device. The technology above then analyzes the performance of the bank account during each set of operations. It provides real-time reasoning, which determines whether the transaction that took place was fraudulent or legitimate. This initiative hopes to accomplish its goal of lowering the percentage of erroneous declines that occur during the processing of merchant payments [4].

According to recent studies, the cost incurred by retailers due to false decreases is around $118 billion annually, whereas the loss incurred by customers is approximately $9 billion annually [5]. This is a significant sector where fraudulent activity occurs in the financial services industry. As a result, combating fraudulent activity is one of the essential things for payment companies and banks to consider. This demonstrates that there is a significant requirement for developing an effective fraud detection system for credit cards based on machine learning.

Machine learning requires enormous datasets to train a model, and the training process might take a significant amount of time. A real-time fraud detection system at a bank calls for a more effective system because of the constant stream of fraudulent activity it uncovers. Additionally, a model can be too large to fit within the working memory of the training device. This would be the case if the model were too detailed.

Even if we purchase a large system with significant memory and processing power, utilizing several smaller machines will be less expensive than purchasing one large machine. To put it another way, vertical scaling is a costly endeavor. Because of this, there is a demand for a scalable machine learning technique that is both effective and affordable.

The main topic of discussion in this chapter is how to effectively achieve scalability in machine learning for fraud detection in credit card transactions by financial institutions.

7.1.2 Problem Statement

The machine learning technique that is now accessible primarily emphasizes distributed computing, such as the utilization of multicore processors, GPUs, or HPCC, all of which are quite pricey options. In addition, the detection system for credit card fraud used by banks must be able to do real-time data processing for anomaly instances on big datasets. This need makes machine learning systems necessary.

Because a fraudster will not adhere to the same pattern more than once, we must train our system with a fresh set of examples relatively quickly. The currently available methods for machine learning call for a significant amount of time to be spent "training" a dataset, followed by "testing" the findings and "implementing" the technique. In addition, the focus on developing machine learning algorithms to detect bank fraud on credit cards is primarily on raising the accuracy of the algorithms, not on increasing their scalability.

The accuracy will indeed be impacted if the processing time is shortened; this may be accomplished by scaling up the system. Nevertheless, the primary objective is to train the model in a shorter time to increase the likelihood that it will be able to recognize a proportion of frauds in the interim that would have been used up in training the model. According to the findings of recent studies, fraudulent activities that go undetected can quickly cost financial institutions millions or even billions of dollars [5].

This suggests that the time spent training the models on big datasets for the machine learning system for fraud detection might cause a significant loss to the bank for the amount of time allowed to go undiscovered.

7.1.3 RESEARCH OBJECTIVE

In recent years, the identification of fraud in credit card transactions has significantly grown, attracting the attention of most academics and researchers. The implementation of machine learning in real time is hampered by a number of methodological hurdles, which researchers are working to overcome. Numerous studies have been conducted in a variety of fields, including the following: the detection of abnormal patterns [6], biometric identification [7] the prediction of diabetes [8], the detection of anomalies [9], the detection of pneumonia [10], and the prediction of informational efficiency using deep neural networks [11]. Despite these limitations, academics are working hard to develop machine learning systems to identify fraudulent activity.

This chapter aims to offer a solution for machine learning scalability in banks' fraud detection systems for credit cards. Specifically, the solution should be able to process the data quickly and effectively while taking up as little time as possible.

Currently, the only comparison of available machine learning algorithms focuses solely on the accuracy of the various methods:

* to research the many strategies for scalable machine learning that are currently accessible.
* to put out an innovative solution for the scalability of machine learning for banks to use in the detection of fraud in credit cards, one that requires less time to analyze the data and helps support the efficient detection of fraud in real time.
* to put the method into practice and present the test results alongside a discussion on the scalability of distributed computing.

7.2 LITERATURE REVIEW

7.2.1 INTRODUCTION

The machine learning algorithms, which are computer programs, are constantly modifying their internal workings to provide ever-improved outcomes whenever new data is introduced into the mix. In machine learning, the term "learning" refers to how computer programs gradually transform how they process data over time. This mirrors how people acquire new knowledge through the process of learning.

Scalability in machine learning refers to the capacity to manage large amounts of data and carry out a variety of calculations in a manner that is both extremely cost-effective and very time-saving. The following is a list of the intrinsic benefits of devoting priority to scalability:

* Productivity: Currently, machine learning is carried out through experiments, such as locating a one-of-a-kind problem with a one-of-a-kind design (algorithm). We can attempt more things and be more creative if we have a pipeline that facilitates speedy executions of each stage of the process (training, assessment, and deployments, respectively).
* Portability: It will be more advantageous if other teams can employ the training outcomes and the trained model for improved results. This will make it possible for more people to benefit from the training.

- Cost Reduction: Optimizing for expenses is always an excellent choice to consider because it may reduce costs. If we scale, it will help us make the most efficient use of the available resources. In addition to this, it will result in a compromise between the level of precision and the marginal cost.
- Automation: The amount of human involvement is reduced by automating the algorithms as much as is practically possible. This will allow humans to take it easy for a while and concentrate on other matters.

For instance, 25 % of engineers at Facebook are responsible for training models, accomplishing this task at a rate of 550 thousand models per month. Their online prediction service is capable of producing 5,000,000 forecasts every second. The training of the Baidu Deep Search model leverages the processing capacity of 200 TFLOP/s distributed over a cluster of 128 GPUs. Because of this, it is easy to comprehend why it is critical for businesses of this nature to grow their operations effectively and why scalability in machine learning is important at the moment [12].

The primary purpose of this literature review is to investigate the present algorithms used for machine learning in the credit card fraud detection system used by banks and to investigate the current possibilities available for machine learning scalability.

The following is an overview of some of the machine learning algorithms most commonly used for fraud detection in banks.

7.2.2 Logistic Regression

It is a supervised classification method that yields the possibility of a dependent binary variable that is guessed based on the quantity of an independent variable in a dataset. The logistic regression predicts the likelihood of an associated result that can take on two distinct values, such as 0 or 1, affirmative or no, and true or false.

Comparable to statistical regression, logistic regression generates a curve rather than a line as the output, in contrast to the line provided by statistical regression. Using one or more predictors or independent variables is contingent upon the prediction type; logistic regression generates curves that plot the values between 0 and 1, depending on the range of those values. When the value of the dependent variable is known in advance, a regression model may be used to investigate the relationship between several independent variables [13] (Figure 7.1).

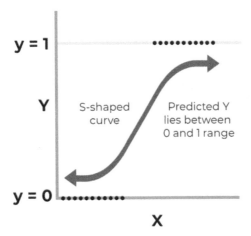

FIGURE 7.1 Logistic curve.

The distinction between logistic regression and linear regression is illustrated in the preceding graph, which demonstrates that logistic regression displays a curve whereas linear regression illustrates a straight line.

7.2.3 DECISION TREE

It is an algorithm that gets closer and closer to goal functions with discontinuous values. It is shown by a function that is acquired through learning. Such categories of algorithms are commonplace when inductive learning is being carried out. They have proven to be useful in various applications, which is rather impressive.

We apply a name label to a recent transaction and state whether or not it is genuine, while indicating that the class label for this transaction is unknown. After that, the value of the transaction is validated using the decision tree. Then a path, beginning at the root node and leading to the output for that particular transaction, is identified. The decision rules determine the outcome of the information included in the leaf node (Figure 7.2).

The decision tree helps determine the best, anticipated, and worst values for several scenarios, simplifies the information to know and comprehend, and enables the addition of additional potentially occurring scenarios. The first step in developing a decision tree is to determine the entropy of each characteristic by using the dataset associated with the issue at hand. After that, the dataset is subdivided into many sets by employing an attribute's greatest gain or minimal entropy to construct a decision tree node comprising that dataset's attributes. Those nodes are then used to split the dataset. In the conclusion, and as a last step, recursion is carried out on subsets by using lingering attributes to construct a decision tree [14].

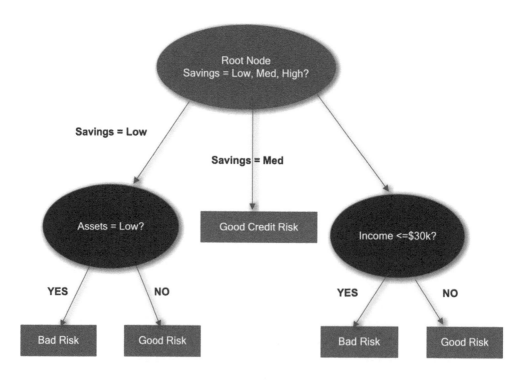

FIGURE 7.2 Decision tree.

7.2.4 SUPPORT VECTOR MACHINES

In addition, it is a supervised machine learning approach that may be used for regression difficulties or classification issues. It employs the "kernel trick," which modifies your data. Subsequently, based on these modifications, it finds an optimal boundary among the feasible outputs.

Put another way, it executes a highly complex data transformation and then determines how to divide the data depending on the labels or outputs you supply.

The model represents a large number of classes in the space of a hyperplane with many dimensions. After that, the hyperplane is constructed in a repeating manner with the use of SVM to bring the error rate down. This SVM approach aims to divide the datasets into sets of classes so that a maximum marginal hyperplane may be discovered [15].

7.2.5 RANDOM FOREST

A regression and classification technique known as random forest is described here. In a sense, it is a collection of different classifiers based on decision trees. Because it corrects the pattern of overfitting in their training set, it has a significant benefit over DT, which stands for decision trees. To train every single tree and produce a decision tree, a random subset of the training set is sampled and then sampled again to sample a random subset of the complete set. Each node in the tree then separates a feature selected from the random subset of the full set.

Because each tree is trained independently from the others throughout the training process, the process is exceptionally effective even when used with much larger datasets containing various attributes and data (Figure 7.3).

In addition, the random forest method provides an accurate assessment of both the generalization problem and overfitting problems. The random forest provides a more organic method for ranking the variables' relevance in a classification or regression task [15].

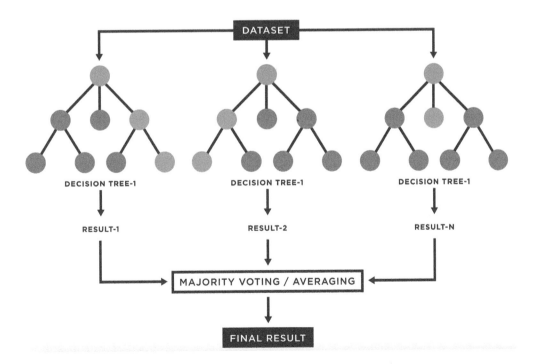

FIGURE 7.3 Random forest classifier.

We observed that the overall accuracy of a random forest is 76%, which is higher than the accuracy of a decision tree. Both the chances of a false positive and a false negative are lower than those associated with decision trees [14]. The likelihood of a false positive is 2.51%, while the likelihood of a false negative is 0.33%.

7.3 RESEARCH METHODOLOGY

The purpose of this part is to talk about the many stages that are included in the research process. It will supply facts that will help select this chapter's study strategy and research technique. In addition to that, it will include an in-depth discussion of the analysis approach and tool that were utilized. The primary objective here is to offer a strategy for scaling up machine learning in financial institutions' credit card fraud detection systems [16].

Since the outcomes of the study will be presented in the form of numerical data, the method of research that has been applied in this work is known as quantitative empirical research. The quantitative empirical research approach that was applied here will assist us in creating a statistical analysis that will compare the results of previously developed algorithms currently available for machine learning scalability in banks' credit card fraud detection systems with the solution that we are presently proposing [17].

The fact that quantitative empirical research can be turned into actionable data is the primary justification for its utilization. It takes a more organized approach, and the statistics produced from this technique will help us calculate the chance our proposed solution will succeed. This strategy also has greater structure. It will also assist us in tracking any changes that may have occurred during the project as a result of statistical data. It will support testing the hypothesis, where we believe that the suggested approach would boost the scalability of machine learning in credit card fraud detection systems at the price of losing some accuracy. This will support the testing of this hypothesis [18].

7.3.1 RESEARCH METHOD

Experiments are carried out on the machine learning algorithms, and the results of those experiments are compared with the answers presented in this work. This is the research approach that is employed. In the process of comparing, statistical inference will be utilized to establish how one approach is superior to the other. The purpose of the experiments is to determine the link that exists between the many variables. Because we are employing the quasi-experimental methodology, we will not be using the random assignment of variables for the system; in other words, we will be keeping it the same.

7.3.2 DATA COLLECTION

The information needed for this system is obtained from a pre-existing database of credit cards that is controlled by the financial institutions. This database will include "true positive" data, in which fraudulent activity was discovered, and "false negative" data, in which no fraudulent activity was discovered. A machine learning model can also be trained to recognize false positives. The information is kept in a file format known as CSV.

7.3.3 DATA ANALYSIS

In the program called TensorFlow, schema-based validation is used to do the analysis of the data. After that, the findings are displayed in the form of graphs and charts to compare the two approaches utilized for machine learning scalability in the credit card fraud detection systems of banks.

7.3.4 ASSUMPTIONS

- Supervised data is utilized as the input during the machine learning model's training process.
- We are operating under the assumption that the data utilized for this system is massive and that it has sufficient information to successfully train the machine learning model used for credit card fraud detection.
- The dataset utilized for machine learning does not represent the whole.
- The data now accessible will be compatible with all the tools that will eventually be utilized for the system.

7.3.5 CONSTRAINTS AND LIMITATIONS

- Training on big datasets requires a lot of processing power, such as GPUs, supercomputers, or high-performance computing centers (HPCC). Because of financial and temporal restraints, employing such a high level of computer capacity is impossible.
- It will not be possible to build up storage capable of retaining such a vast dataset since the dataset necessary for training such a model is so large.
- In order to correctly train the model, the data supplied for machine learning ought to contain examples of real fraud taking place and situations in which fraud has not taken place.
- Collecting millions of data records to correctly train the machine learning model is impossible. This is because it would take too long.
- Because they are only expressed in numerical terms, the delivered results have some restrictions.
- Because we only have a limited amount of computing power available, training the model might sometimes take longer.
- Because the findings depend on the data used to train the model, there is a possibility that the findings will be skewed.

7.3.6 PROPOSED METHOD

The plan is to utilize a training approach with a low degree of accuracy. In most cases, the default settings of machine learning frameworks call for using 32-bit floating-point precision when it comes to inference and training the models. Some data may be used to justify reducing the numerical precision (for example, using 16 bits for training and 8 bits for inference), although at the expense of some accuracy.

We are using the float16-bit data type, representing the floating-point value using 16 bits. In certain circles, it is also referred to as a half-precision float. The first bit is known as the sign bit; the following five bits are used to represent an exponent; and the next ten bits are used to represent a fraction, or mantissa. Python's numpy package provides ready access to it, so you can use it immediately.

When the accuracy is decreased, the immediate result will be a smaller memory footprint, improved bandwidth consumption, enhanced caching, increased throughput by a factor of two, and accelerated model performance.

7.4 SYSTEM DESIGN

In the next part, we will concentrate on the primary design and implementation of the machine learning system to detect credit card fraud. In addition, we will demonstrate the concept by constructing a prototype of the system to explain how it will work.

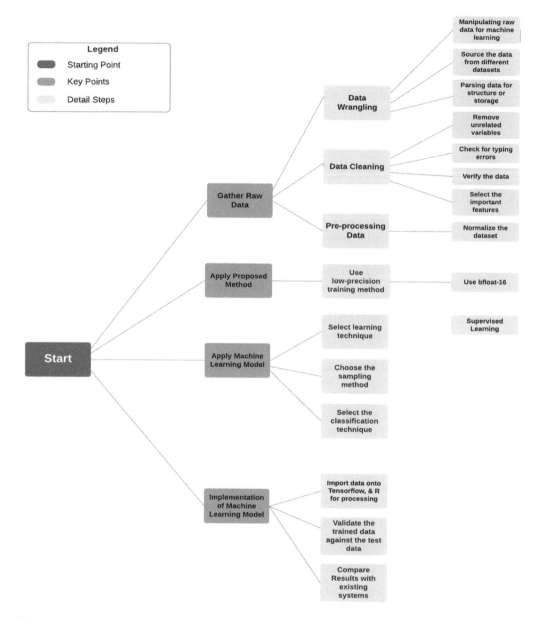

FIGURE 7.4   ML scalability process for credit card fraud detection.

7.4.1 PROTOTYPE

The first thing that must be done is to collect raw data so that the machine learning model may be trained on it later, as illustrated in Figure 7.4. The flowchart of the procedure is broken down into the following sections:

- The first stage is called "wrangling the data." It consists of locating and sourcing the data from the databases and reducing the set to just include the essential data.
- In addition, parsing must be done so that the data may be organized into a structure and saved for later use.
- We need to do some sort of transformation on the raw data to prepare it for consumption by the machine learning model.

Stage 1: Data Cleaning
- We must verify that the various datasets gathered are all in the same format. This means that the data obtained from the photographs, audios, and texts must all be converted into the same format.
- At this point in the process, the data must be double-checked for any typographical errors, variables that aren't needed, and rows or columns with a significant number of missing values or none.
- The data have to be checked and double-checked to ensure that they are accurate.
- We must pick the most significant features from the various aspects we need for prediction. Doing so will help us do computations more effectively and reduce the amount of memory we need to use.

Stage 2: Pre-processing:
- At this phase, we have to normalize the data.
- It is cleaned up to prepare the data for inclusion in a machine learning model and then uniformly dispersed throughout all fields.
- Method of Low Precision, which is the second stage:
- We are going to use the approach suggested on this dataset by making use of float16, and we will implement it on this dataset in TensorFlow with the assistance of libraries.
- Following this step, the machine learning model will be trained using data converted from 32-bit or 62-bit formats into 16-bit formats.

Stage Three:

Select Learning Technique:
- We will go with the supervised learning strategy since it is simple to put into action. This is since the data may be labeled and separated into several categories. Only courses 0 and 1 will be available to students. We will train our dataset to differentiate between fraud cases and cases that do not include fraud.
- Because the data that was supplied is not balanced, the number of fraudulent instances would not be the same as the number of cases that were not fraudulent, so we will need to balance the data before we can train our model.

Select Sampling Method:
- There are various options for sampling techniques, but we will use one called SMOTE (synthetic minority oversampling technique). With this method, we will artificially make the number of fraudulent and non-fraudulent transactions in our dataset equivalent.

Select Classification Technique:
- The random forest classifier, an advanced version of decision trees, will be the one we go with. The rationale behind this decision was established in the literature review because it is the most accurate algorithm out of all the others used for machine learning.

Stage Four:

Process the Data:
- TensorFlow will then be used to prepare our data for processing when loaded.
- In addition, statistical analysis will be performed in R.
- Execute Model and Conduct Validation Tests: After executing our model, we will conduct validation tests of the trained model compared to the test model to locate the confusion matrix.

Compare Results:
- We will demonstrate the results of the random forest technique after it has been implemented on the same data. Additionally, we will compare the amount of time needed to train the same model using 32 bits, as well as the resources and confusion matrix that were used.
- Because of this, we will better understand how scalability was attained in machine learning.

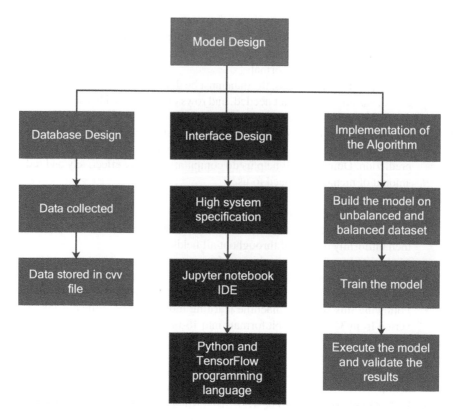

FIGURE 7.5   Model design flow process for machine learning scalability for credit card fraud detection.

7.4.2   MODEL DESIGN

This section will provide an in-depth discussion of the many aspects of model design. The database that the training model will be applied to is the most significant consideration since this work is centered on the applicability of machine learning to the detection of credit card fraud on a large scale. This unit will emphasize database design and interface design, i.e., determining the environment in which the method will be implemented using various programming languages.

Figure 7.5 illustrates the first stage, which is creating the database design, which has to be saved in a format that the programming can handle. This completes the first step. In addition to this, we will go through each component in further depth.

7.4.2.1   Database Design

The dataset includes transactions that credit card holders across Europe made during September 2013. This dataset contains transactions that took place over two days, and out of a total of 284,807 transactions, there were 492 instances of fraudulent behavior.

Data Analysis
   To investigate the data, we will first demonstrate how the data are represented and then list the computations for the various components. The dataset is skewed, with just 0.172% of all transactions belonging to the positive class (frauds).
   As a result of applying a PCA (principal component analysis) transformation, it comprises solely numerical input variables as its only constituents.
   Because of concerns about privacy and the need to protect confidentiality, the original characteristics of the data and other background information are not being disclosed.

TABLE 7.1

Statistical Insight into the Dataset

	Time	Amount	Class
Count	284807.000000	284807.000000	284807.000000
Mean	94813.859575	88.349619	0.001727
Std	47488.145955	250.120109	0.041527
Min	0.000000	0.000000	0.000000
25%	54201.500000	5.600000	0.000000
50%	84692.000000	22.000000	0.000000
75%	139320.500000	77.165000	0.000000
Max	172792.000000	25691.160000	1.000000

The major components derived with PCA are the features V1, V2,..., V28. The only features not converted with PCA are 'Time' and 'Amount,' the only features. The 'Time' feature stores the number of seconds that have passed since the beginning of the dataset for each subsequent transaction in the dataset. This feature, which may be utilized for example-dependent cost-sensitive learning, is referred to as the transaction amount and is referred to as "Amount." The answer variable is called Feature 'Class,' and it has a value of 1 if there was fraud and a value of 0 otherwise (Table 7.1).

As was just demonstrated, only three features of the dataset have not been subjected to the PCA transformation. This is because these features include crucial information used to construct and train the machine learning model. The most that may have been fraudulently obtained through fraudulent behavior is 25691 US dollars. If there was no fraud, the value is 0; if there was fraud, the class value is 1. If there was fraud, the fraud class is determined by the numbers 0 and 1.

Synopsis:

- The sum involved in the transaction is not very large. Approximately 88 dollars is the mean price of all of the mounts that have been produced.
- As a result of our investigation, we discovered that there are no "Null" values; hence, we do not need to change or add any values.
- In the data period, fraud occurred just 0.17% of the time out of 280,000 transactions. In contrast, most transactions were legitimate (99.83% of the time), making up 99.83% of the total. This demonstrates a significant gap between the ratio of non-fraud to fraud in the data.

Data Coherency

We have to check to make sure that the dataset does not contain any null values, since they might mess with computations and cause inaccurate results to be shown. The dataset is saved in a comma-separated values (csv) file format. Table 7.2 demonstrates that the dataset does not include any null values, which assures that the data is coherent.

7.4.2.2 Interface Design

In this section, we will discuss the system specifications required for the program to work properly. The platform or platforms on which the programming language is executed.

The graphic and specification below show that the interface is being run on a 64-bit Windows 11 with an Intel Core i7 processor. Because of limitations in both resources and costs, the system is presently being operated on the CPU rather than the GPU.

TABLE 7.2

Non-Null Values of Dataset

6	V6	284807	non-null	float64
7	V7	284807	non-null	float64
8	V8	284807	non-null	float64
9	V9	284807	non-null	float64
10	v10	284807	non-null	float64
11	v11	284807	non-null	float64
12	v12	284807	non-null	float64
13	v13	284807	non-null	float64
14	V14	284807	non-null	float64
15	v15	284807	non-null	float64
16	v16	284807	non-null	float64
17	V17	284807	non-null	float64
18	V18	284807	non-null	float64
19	v19	284807	non-null	float64
20	v20	284807	non-null	float64
21	v21	284807	non-null	float64
22	v22	284807	non-null	float64
23	v23	284807	non-null	float64
24	V24	284807	non-null	float64
25	v25	284807	non-null	float64
26	v26	284807	non-null	float64
27	v27	284807	non-null	float64
28	v28	284807	non-null	float64
29	Amount	284807	non-null	float64
30	Class	284807	non-null	int64

dtypes: float64 (30), int64 (1).
memory usage: 67.4   MB.

7.4.2.3   Model Implementation

This section will focus mostly on implementing the bfloat-16 methodologies for the scalability of credit card fraud detection machine learning. Before executing the solution given, we will also talk about the pre-processing processes that need to be carried out.

Data Visualization

To examine the dataset's additional characteristics, we are using the data visualization tools that Python offers. The distribution chart for each color and its associated amount and time are displayed below in graph form (Figure 7.6).

Because the information in the other columns has been concealed as a result of the PCA transformation, we can concentrate on more explicitly displaying the Time and Amount columns (Figure 7.7).

We can draw the conclusion that fraudulent transactions have a more equal distribution than transactions that do not include fraud. There is a balance of fairness in the timing of fraudulent transactions (Figure 7.8).

Fraudulent transactions (amount) against time. The time that is presented is in seconds that are calculated by subtracting the current time from the total number of seconds (48 hours, over two days). This illustrates that the majority of fraudulent operations take place during the night rather than during the daytime.

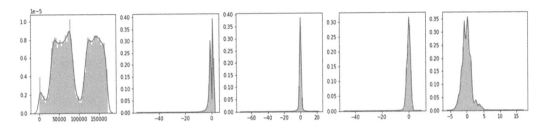

FIGURE 7.6 Graphs depicting the columns of dataset.

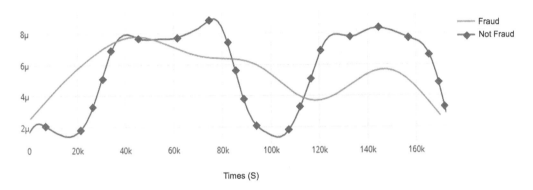

FIGURE 7.7 Credit card transactions of fraud and non-fraudulent transactions.

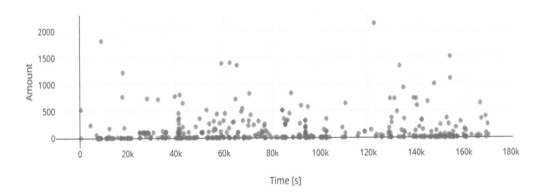

FIGURE 7.8 Fraudulent transactions.

Supervised ML – Unbalanced Data

Because our dataset is already segmented into multiple groups and the values are already known, we will deploy a supervised machine learning approach. A model is trained on a collection of transactions that have been "labeled" correctly. Every transaction is categorized as either fraudulent or legitimate activity. The quality and quantity of the training data used in supervised machine learning directly affect the accuracy of the resulting models.

The following distribution graph shows that the data is severely lopsided, as we can see. We have roughly 99% of instances that are not fraudulent, represented by class 0 and colored blue. We have 1% of fraudulent cases, represented by class 1 and colored red (Figure 7.9).

To build the confusion matrix for imbalanced data, we will utilize the classification technique known as random forest (Tables 7.3 and 7.4).

The genuine negative value is rather large, as we can deduce from the confusion matrix, which provides us with this information. This is because the dataset we are working with is not balanced.

Supervised ML – Balanced Data

When dealing with unbalanced data, there are two different sorts of resampling procedures available, the first of which is sampling and the second of which is oversampling.

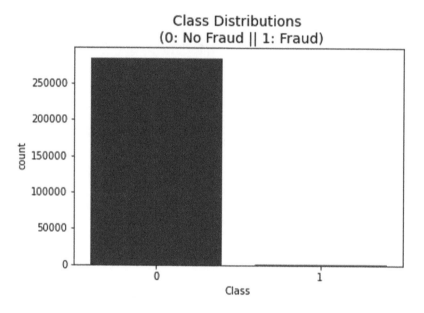

FIGURE 7.9   Distribution of fraud and non-fraud classes.

TABLE 7.3
Confusion Matrix for Unbalanced Data

Confusion Matrix – Unbalanced Data

Actual		
	85286	9
	32	116
	Not fraud	Fraud
	Predicted	

TABLE 7.4
Random Forest Classifier for Unbalanced Dataset

Random Forest Classifier

Accuracy	0.999520148
Precision	0.99962493
Recall	0.99989448

TABLE 7.5

Confusion Matrix for Unbalanced Data

		Confusion Matrix	
Actual	Not fraud	85293	2
	Fraud	31	117
		Not Fraud	Fraud
		Predicted	

TABLE 7.6

Calculation of Accuracy and Other Parameters

Random Forest Classifier (SMOTE)

Accuracy	0.9996137776
Precision	0.99963668
Recall	0.99997655

- Undersampling occurs when we take random values from observations that do not include fraud in order to match the number of observations that involve fraud. We are putting in a lot of data and information in a random order right now. In certain circles, it is also known as Random Undersampling.
- The term "oversampling" refers to the process by which we enhance the number of fraud samples contained in our data by taking random values from incidents of fraud and copying these observations.
- SMOTE: This technique allows us to adjust the data imbalance by oversampling the minority observations (fraud cases). Rather than simply copying the minority samples, we create new synthetic fraud cases by using the nearest neighbors of the existing fraud cases to create new synthetic fraud cases.

In order to balance our data, we are employing the SMOTE approach, and after that, we will determine the confusion matrix and the accuracy. The SMOTE methodology, which generates new synthetic values as an alternative to the standard practice of duplicating them, is the method that offers the greatest choice (Tables 7.5 and 7.6).

We have higher accuracy and precision as compared to the unbalanced dataset.

7.5 RESULTS AND COMPARISON

Now, we will discuss the outcomes of using the algorithm and compare those outcomes with those from before. It will comprehensively explain how adding the float16 data type may make a system more scalable.

Confusion Matrix for Unbalanced Dataset

As can be seen in the following graph, the confusion matrix's value shifted when we tested the model on data that was not evenly distributed after we had used the float16 data type. Because the dataset was imbalanced, the confusion matrix has been affected by the minor rise in the negative value. Additionally, converting the bits to float16 has affected the matrix (Table 7.7).

Confusion Matrix for Balanced Dataset – SMOTE

We have utilized the oversampling strategy known as SMOTE to verify that the dataset is representative of its entirety. The proportion of unfavorable aspects is significantly lower than average (Table 7.8).

Accuracy, Precision, and Recall

We will calculate the accuracy, precision, and recall for both unbalanced and balanced data, demonstrating the difference between using float16 bits and not using them (Table 7.9).

TABLE 7.7

Confusion Matrix with Float 16 on Unbalanced Data

	Confusion Matrix	
Not fraud	85287	8
Fraud	34	114
	Not Fraud	Fraud
	Predicted	

(Actual on vertical axis)

TABLE 7.8

Confusion Matrix with Float16 on Balanced Data

	Confusion Matrix	
Not fraud	85275	20
Fraud	27	121
	Not fraud	Fraud
	Predicted	

(Actual on vertical axis)

TABLE 7.9

Different Models Results

Model	Accuracy	Precision	Recall
Random forest classifier (Float16)	0.999508444225975	0.9996015	0.99990621
Random forest classifier (Without Float16)	0.999520148	0.99962493	0.99989448
Random forest classifier (SMOTE – Float16)	0.99944992568144	0.99968348	0.99976552
Random forest classifier (SMOTE)	0.9996137776	0.99963668	0.99997655

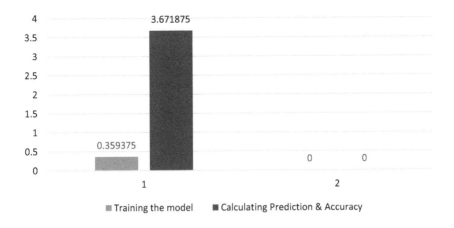

FIGURE 7.10   CPU Execution Time on graph for training and calculation on unbalanced data.

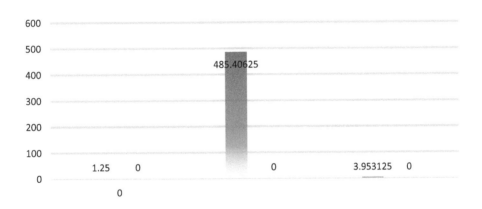

FIGURE 7.11   CPU Execution Time on graph for training and calculation on balanced data.

CPU Execution Time – Before Float16 Implementation
     We will display graphs demonstrating the difference between the amount of time it takes for the CPU to execute on imbalanced data and balanced data (Figures 7.10 and 7.11).
CPU Execution Time – After Float16 Implementation
     We will provide graphs demonstrating the difference in the amount of time it takes for the CPU to execute imbalanced data and balanced data while using float16 bits (Figure 7.12).
     As seen above, using float16 reduces the time needed for the CPU to complete an operation by approximately 55% compared to not using float16 bits (Figure 7.13).
     The above graph demonstrates that we may conclude that utilizing float16 bits can help achieve scalability in machine learning by minimizing the amount of time that the CPU spends executing code. This, in turn, leads to quicker processing and the use of fewer resources.

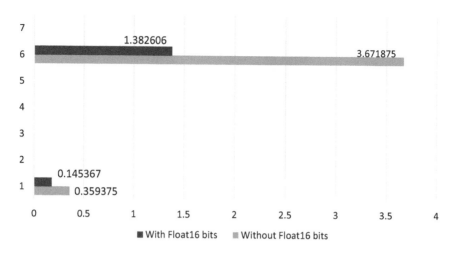

FIGURE 7.12 Depiction of CPU execution time with float16 (unbalanced data).

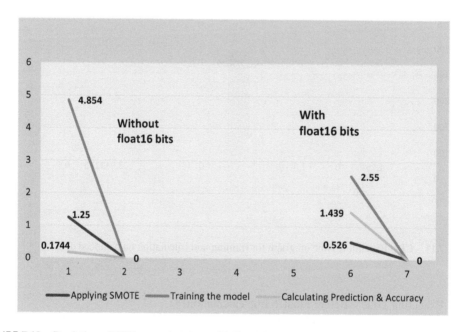

FIGURE 7.13 Depiction of CPU execution time with float16 (balanced data).

7.6 CONCLUSION AND FUTURE WORKS

This chapter aims to provide a scalable system for detecting credit card fraud in financial institutions based on machine learning. Including float16 bits in the data led to a reduction of between 50% and 60% in the amount of time it required to carry out the operation. After introducing float16 bits, the findings reveal that the accuracy was significantly decreased by 0.0002% for both balanced and unbalanced data. This was the case regardless of whether the data was balanced or not.

It is realistic to expect that accuracy can be sacrificed in the process of detecting credit card theft, given the urgent requirement for immediate action to be taken to detect credit card fraud. It is more vital to swiftly identify the fraudster before they steal the money. A fraudster can completely deplete a credit card in seconds before the crime is discovered. This method can help save millions of dollars by quickly identifying fraudulent credit card activity.

We recommend using two extra methods to increase the identification of fraudulent activity using credit cards. To begin, we can use the brain floating-point format, also known as bfloat16 or BF16. This format uses 16 bits to represent floating-point numbers, with 1 bit denoting the sign, 8 bits used for the exponent, and 7 bits used for the fraction, or mantissa. Bfloat16 is superior to the conventional fp16 format since it has eight exponent bits instead of just five, appropriately represents all integers, and is simple to convert from the IEEE-754 32-bit format. Second, because we were short on time and resources for this study, the dataset we used for implementation was quite tiny. On the other hand, if processing billions of transactions in real time is essential, we advise utilizing the TPUs (Tensor Processing Units) made available by GCP (Google Cloud Platform). TPUs are not available to all users and require money to use, but major businesses and banks can profit from their use since it speeds up data processing.

REFERENCES

1. Rambola, R., Varshney, P. and Vishwakarma, P., 2020. Data Mining Techniques for Fraud Detection in Banking Sector – IEEE Conference Publication. [online] Ieeexplore.ieee.org. Available at: https://ieeexplore.ieee.org/document/8777535/.
2. Oualid, A., Maleh, Y., and Moumoun, L. "Federated learning techniques applied to credit risk management: A systematic literature review." *EDPACS* (2023): 68, 1–15.
3. Kansal, S., 2020. Machine learning: Why scaling matters. [online] Codementor.io. Available at: https://www.codementor.io/blog/scaling-ml-6ruo1wykxf [Accessed 19 March 2020].
4. Tietoevry.com. 2020. Robust and Scalable Machine Learning Lifecycle I Tietoevry. [online] Available at: https://www.tietoevry.com/en/blog/2019/12/robust-and-scalable-ml-lifecycle-for-a-high-performing-ai-team/ [Accessed 19 March 2020].
5. Kansal, S., 2020. Machine Learning: How To Build Scalable Machine Learning Models. [online] Codementor.io. Available at: https://www.codementor.io/blog/scalable-ml-models-6rvtbf8dsd [Accessed 19 March 2020].
6. Jakaite, L., Schetinin, V., Hladůvka, J., et al. Deep learning for early detection of pathological changes in X-ray bone microstructures: Case of osteoarthritis. *Sci Rep* 11, 2294 (2021). https://doi.org/10.1038/s41598-021-81786-4.
7. Selitskaya, N., et al. (2020). Deep learning for biometric face recognition: Experimental study on benchmark data sets. In: Jiang, R., Li, CT., Crookes, D., Meng, W., Rosenberger, C. (eds.) *Deep Biometrics*. Unsupervised and Semi-Supervised Learning. Springer, Cham. https://doi.org/10.1007/978-3-030-32583-1_5.
8. Hassan, M. M., Peya, Z. J., Mollick, S., Billah, M. A., Shakil, M. M. H., and Dulla, A. U., Diabetes prediction in healthcare at early stage using machine learning approach. In: *12th International Conference on Computing Communication and Networking Technologies (ICCCNT)*, 2021, pp. 01–05. https://doi.org/10.1109/ICCCNT51525.2021.9579869.
9. Bin Sulaiman, R., Schetinin, V., and Sant, P. Review of machine learning approach on credit card fraud detection. *Hum-Cent Intell Syst* (2022). https://doi.org/10.1007/s44230-022-00004-0
10. Kareem, A., Liu, H., and Sant, P. Review on pneumonia image detection: A machine learning approach. *Hum-Cent Intell Syst* (2022). https://doi.org/10.1007/s44230-022-00002-2
11. Sulaiman, R.B. and Schetinin, V. (2022). Deep neural-network prediction for study of informational efficiency. In: Arai, K. (eds.) *Intelligent Systems and Applications. IntelliSys 2021. Lecture Notes in Networks and Systems*, vol. 295. Springer, Cham. https://doi.org/10.1007/978-3-030-82196-8_34.
12. Csail.mit.edu. 2020. Re-Thinking Machine Learning Algorithms for Scalability I MIT CSAIL. [online] Available at: https://www.csail.mit.edu/event/re-thinking-machine-learning-algorithms-scalability [Accessed 19 March 2020].
13. Pathmind. 2020. Machine Learning Algorithms. [online] Available at: https://pathmind.com/wiki/machine-learning-algorithms [Accessed 19 March 2020].
14. Codes, C., 2020. Essentials of Machine Learning Algorithms (With Python And R Codes). [online] Analytics Vidhya. Available at: https://www.analyticsvidhya.com/blog/2017/09/common-machine-learning-algorithms/ [Accessed 19 March 2020].

15. Javatpoint.com. 2020. Support Vector Machine (SVM) Algorithm – Javatpoint. [online] Available at: https://www.javatpoint.com/machine-learning-support-vector-machine-algorithm [Accessed 19 March 2020].

16. Medium. 2020. Random Forest Simple Explanation. [online] Available at: https://medium.com/@williamkoehrsen/random-forest-simple-explanation-377895a60d2d [Accessed 19 March 2020].

17. Analytics India Magazine. 2020. Distributed Machine Learning Is the Answer to Scalability and Computation Requirements. [online] Available at: https://analyticsindiamag.com/distributed-machine-learning-is-the-answer-to-scalability-and-computation-requirements/. [Accessed 19 March 2020].

18. TensorFlow. 2020. TensorFlow. [online] Available at: https://www.tensorflow.org/ [Accessed 19 March 2020].

8 A Proposed Semi-decentralized Approach of Federated Learning for Enhancing CCFD Using CS-SVM and K-Means Algorithms

Rejwan Bin Sulaiman

8.1 INTRODUCTION

Banks and financial firms have continuously researched and devised solutions to detect and prevent credit card fraud. It is one of the primary focuses for researchers and other research institutions to find possible ways to tackle credit card fraud anomalies. Each year, the world faces the loss of billions of pounds caused by fraudulent transactions, with an estimated 31.67 billion dollars of loss in 2020 [1]. The risk of online credit card fraud has been widely increased due to the advancement of e-commerce. If handed over to a third party, consumers' credit card details are more vulnerable to exploitation. To avoid this, two main mechanisms are in use: fraud detection and fraud prevention. The machine learning (ML) approach is used to detect fraud, as it is an effective way to train the system to perform predictions based on the data. However, the problem exists as the dataset for credit card transactions is too sensitive to be shared. Secondly, the credit card dataset is skewed, which means the fraudulent transaction dataset is highly unbalanced compared to non-fraudulent transactions [2,3]. While considering these datasets' concerns, our research is based on devising a model that could potentially get the trained model from the real-time datasets while ensuring the privacy of the customers' data and enhancing the fraud detection accuracy by having the trained model from a variety of datasets [4,5]. Many methodological limitations are still expected to be overcome in ML applications. Some attempts have been undertaken to find efficient solutions in different domains, such as the detection of abnormal patterns [6–12], biometric identification [13,14], trauma severity evaluation [15–18], collision avoidance at Heathrow [19], and early detection of bone pathologies [20].

8.1.1 MOTIVATION

Several factors have motivated us to conduct research on credit card fraud detection. Every day, millions of people experience credit card fraud, which is increasing day by day. The firms' and individuals' financial consequences are one of the driving forces to work in this research area. There is currently a less efficient system for detecting credit card fraud, which gives hackers a favorable environment to abuse the system. It has led to our motivation to introduce a model that can potentially fulfill the limitations of the current credit card fraud detection system (FDS) and bring up an efficient model that can never be breached.

DOI: 10.1201/9781032667478-9

8.1.2 AIMS AND OBJECTIVES

The following are some of the aims and objectives of our research work:

- Our research will focus on the predictive analysis of credit card transactions, whether fraudulent or non-fraudulent.
- The credit card transactions' prediction will be based on ML techniques, both supervised and unsupervised.
- A prediction model will be used to classify standard and fraudulent transactions.
- One of the research's primary objectives is to devise a model capable of ensuring the privacy of the customer's data when using it as a dataset for research purposes.
- We aim to use the federated learning model to ensure the data's privacy.
- We use CS-SVM as a supervised learning model to detect fraudulent credit card transactions.
- We use K-means as an unsupervised learning model to detect credit card anomaly–based fraudulent transactions.
- The primary goal is to build up a hybrid approach of FL, supervised (CS-SVM), and unsupervised (K-means) to improve credit card fraud detection accuracy while ensuring the privacy of customers' data according to GDPR.

8.1.3 RESEARCH HYPOTHESIS

The hypothesis of this research work is based on combining FL with CS-SVM and K-means techniques to formulate a model effective for detecting fraudulent transactions. As mentioned in the literature review (next chapter), we have seen that SVM performs better in credit card fraud detection accuracy [21]. Similarly, K-means performs better in dataset clustering. We have chosen these two algorithms based on our observations from the previous literature and concluded that combining these algorithms with FL can enhance fraud detection accuracy, which has never been researched before. This hybrid approach will increase accuracy more than traditional credit card fraud detection models [22].

8.1.4 PROBLEM STATEMENT

The following are some of the problems and challenges we have discovered in the literature review for credit card fraud detection:

- Credit card transactions are considerable in quantity and are diversified in kind because users of credit cards utilize credit cards for various reasons based on geographical regions and currencies [23]. Fraudulent credit card transactions are pretty varied.
- Detecting fraudulent activity is another example of a multi-objective job. Customers of banks and other financial institutions must always receive quality service and an enjoyable experience. Therefore, using the consumer datasets for experimental purposes while protecting their privacy and maintaining service availability is hard.
- A significant obstacle in CCFD is represented by the wide variety of fraudulent transactions and unbalanced datasets [24].
- It may be pretty challenging to obtain datasets of real-time credit card transactions [25]. Because of GDPR, banks and other financial institutions do not make their customers' personal information publicly available. As a result, it presents a problem for the researchers to compile the information necessary for identifying instances of credit card fraud.
- Customer behavior diversity and the imbalance between fraudulent and genuine transactions are among the hurdles in CCFD.
- Various financial firms and banks lack labeled data, which creates a challenge for the supervised learning model.

8.1.5 Research Questions

The following are the research questions for our research work:

- RQ1

  Can various banks and financial institutions use real-time datasets mutually to detect fraudulent transactions while ensuring customers' privacy?
- RQ2

  Is it possible to enhance the detection of fraudulent transactions with a hybrid ML approach in real-time datasets?

8.1.6 Potential Contribution

Our research's significant contribution will be focused on understanding the implications of the federated learning model in CCFD for the privacy of data and keeping the accuracy of fraud detection high. Another significant contribution of our research is focused on using a supervised and unsupervised model of SVM and K-means, respectively, for training the datasets in collaboration with the FL model. Per our hypothesis, it could give a productive output regarding credit card fraud detection accuracy and the customers' data security.

8.2 LITERATURE REVIEW

This section of the report constitutes related work done in the past to enhance credit card fraud detection accuracy. We have analyzed comparative technologies and techniques recently adapted and accepted by financial firms. After doing the comparative analysis, we have identified the gap concerning CCFD and selected techniques to adopt for our proposed work.

8.2.1 Hybrid Models Used in CCFD

Researchers have been trying to identify credit card fraud using the hybrid approach of ML. We have analyzed various research done in CCFD by using the hybrid model as follows.

The traditional way of detecting transactions has been replaced by ML methods, which have improved CCFD performance to a greater extent. Wen et al. have proposed a credit card loan fraud detection method using ML techniques [26]. In the experiment, the authors have used the extreme gradient boosting (XGBoost) model and data mining techniques motivated by effective credit card fraud detection performance. The research work involves the preservation of meaningful information while keeping that information without knowing about it.

To make it possible, the authors have used a hybrid approach by combining supervised and unsupervised learning (PK-XGBoost) concerning the XGBoost algorithm and kernel principal component analysis (kernel PCA) [27]. By performing the tests, it has been proven that PK-XGBoost performs better than using just XGBoost [28]. It is more effective for detecting credit card fraud while maintaining the users' privacy. Although it is a great leap toward fraud detection while ensuring the clients' privacy details, this model has limitations in providing client data privacy. However, this model has limitations regarding the enormous number and variety of transactions.

Furthermore, XGBoost overfits the datasets while processing, requiring a combination of the parameters to achieve higher accuracy. While considering this, it is necessary to have a common and effective solution that can promise better fraud detection while ensuring privacy and confidentiality. To solve the issue, our research work will be based on a federated learning mechanism to fix the privacy concern for training the model, which is incorporated with CS-SVM and K-means for producing effective credit card fraud detection.

8.2.2 RANDOM FOREST AND ISOLATION FOREST

A hybrid strategy for detecting credit card fraud using random forest and isolated forest has been presented by Vynokurova et al. This approach is utilized to identify anomaly-based transactions [29].

The model that has been suggested is composed of two basic sub-systems. One of the things that one is concerned about is anomaly detection software that uses unsupervised learning. The second possibility involves an interpretation that takes into account the anomalous kind. It makes use of supervised learning as its foundation. The speed of the data, which functions efficiently when examined with the hybrid model and applied to real-time data [29,30], is the work's key focus. The system was tested to determine its capability of determining users' geolocation while they carried out transactions for detection. This hybrid model does not consider the anomaly level in its calculations.

On the other hand, it is determined by the type of abnormality. An FDS based on anomalous transactions can use geolocation to identify fraudulent activity. Despite this, the findings of this study effort do not adequately protect users' privacy and confidentiality, which is problematic given that identifying fraudulent transactions requires access to real-time data. The researcher did not specify any hashing or encryption methods to protect the user's data from being made public. Therefore, to meet the requirements of this obstacle, there is a requirement to guarantee the privacy of the data provided by users of credit cards for research. The study did not explain how geolocation spoofing techniques might be combated to stop fraud. Our contribution will center on considering the geolocation and temporal characteristics to identify fraudulent activity. Specifically, we will combine the SVM and K-means clustering algorithms with a federated learning technique to protect the confidentiality of the data.

8.2.3 CCFD FOCUSING ANN

Dubey et al. have suggested a model for the identification of credit card fraud that is based on an artificial neural network (ANN) [31] and backpropagation [21,31]. The process requires the client's information, including the customer's name and the transaction I, D, and time. The author performed experiments with 80% of the data for training, 10% of the data for testing, and 10% for validation. A significant result has been achieved using the model presented for detecting fraudulent transactions in real-time data. For the purpose of evaluation, the authors have employed the use of a confusion matrix as well as recall, accuracy, and precision. The ANN and backpropagation model that have been proposed to be used may be shown below [31], together with the user's personal information that was used to train the model.

By carrying out this experiment, the level of accuracy that can be attained has increased to 99.96%, which is better than the prior model when considering the real-time data. Even though it has generated positive outcomes, this study activity does not include a viable solution to the problem of data threats posed by the researcher or even by an individual bank worker. This is a limitation that affects both training and research. As a result, it is necessary to have a solution capable of meeting all of the criteria for maintaining the confidentiality of customer information and the integrity of bank credit card transactions. The authors have not mentioned data confidentiality, even though it is being used for training. This includes the participant's name, age, and gender. In light of this, the study that we have suggested would use a federated learning model to protect the confidentiality of the data while it is being trained to detect credit card fraud.

Li et al. have proposed an innovative technique that has enhanced the capability of SVM [23,32]. It was based on using cuckoo search (CS) algorithms that cause an enhancement in the kernel function of SVM [32]. The combination of CS-SVM has produced higher accuracy, precision, and other parameters than SVM alone. It has revolutionized CCFD, where the data is complex and noisy, whereas simple SVM possesses limitations. However, the proposed method of combining the CS with SVM has achieved an accuracy of 98%, which is considered one of the most efficient ways to use a supervised learning approach for CCFD. Our research is motivated by this approach's future direction. We have considered using CS-SVM as a supervised learning method for research based on monitoring the capability of time and location parameters of credit card transactions.

8.2.4   CCFD Focusing SVDD

Khedmati et al. have proposed a method to combat the challenge of getting samples of the fraud detection dataset. The method uses only one-class classifier, support vector data description (SVDD), based on SVM and the REDBSCAN algorithm to reduce the need for training samples [33,34]. This method has helped to achieve better performance and speed in analyzing fraud. The advantage of using a one-class classifier is that it requires only one data class for training. It does not need to take fraudulent data, which is challenging.

The experiment results were compared with the SVM, and various components like precision, recall, and f-measure were obtained separately. It has been shown that SVDD performs well even in terms of speed at which the REDBSCAN algorithm is used [33]. Although the result obtained by SVDD compared with SVM is better in speed and performance, it requires a large volume of samples to be solved by our novel method proposed using the federated learning method on end devices in our research work. We consider federated learning with SVM and CS, as combining the two will produce higher accuracy than any existing state-of-the-art techniques.

8.2.5   CCFD Focusing SVM

Naufal Rtayli and colleagues have developed a method for calculating credit card fraud risk for higher-dimensional data. This approach makes use of the classification techniques of random forest classifiers (RFC) [35] and support vector machines (SVM) [35,36]. The concept emerged due to the feature selection of fraudulent transactions in the vastly skewed dataset. Because of their low volume, fraudulent transactions are becoming increasingly difficult to identify. The model's creator has employed assessment measures like accuracy and recall, as well as the area under the curve, to evaluate the model. A dataset containing 284,807 transactions was utilized for this investigation, 492 of which were fraudulent transactions.

Based on SVM, while utilizing RFC stated that it had created an accuracy of 95%, false-positive transactions are minimized by raising the sensitivity to 87%, resulting in improved fraud detection in the enormous dataset and unbalanced data [35,36]. Additionally, this model has increased the performance of categorization. This model limits the transaction's privacy when completing the accuracy and recall assessment metrics, although the approach delivers efficient matching output for fraud detection when employing classification characteristics. As a result, to address privacy concerns, we are using a federated learning model that instructs data on a local level. We are also combining it with CS and SVM. RFC performs slowly when dealing with large datasets. It requires high processing power, whereas CS is less intensive and uses global search parameters to produce effective results.

8.2.6   CCFD Focusing Deep Ensemble Algorithm

CCFD has been challenging because the conventional model does not involve the identification of hidden patterns. According to Vitaly, Selitskaya, et al. [37], the regularization of the deep learning neural network can be attained by reducing the parameters that are followed in other multidisciplinary fields. While considering this challenge, Arya et al. have proposed a method based on categorical accuracy, which comprises the overall accuracy for the imbalanced datasets [38]. While considering this, the authors have introduced a predictive framework based on deep ensemble algorithm learning [38,39] to detect fraudulent transactions that comprise imbalanced data. It also explores hidden patterns by observing the user's spending behavior. The experiment was performed in Python with Sci-Kit Learn, Google TensorFlow, and the Keras deep learning library.

For the purpose of the experiment, a dataset of 284,807 transactions was used, of which 492 were fraudulent and 284,315 were standard transactions. The results have been compared with multi-layer perception, CNN, and two other MK models. It has produced a significant impact when checked with a confusion matrix. Although this framework has performed well in terms of precision and recall for the CCFD in the real data stream, it still needs improvement for next-generation computing (serverless). The data collection methodology needs data confidentiality assurance for the research to proceed. One of the techniques to improve the current system is to use a decentralized serverless method like blockchain. However, it possesses limitations in terms of the anonymity of the fraudulent data. Therefore, we propose the semi-decentralized model of a federated learning approach to solve this problem, potentially ensuring data authenticity and efficiency in detecting fraudulent transactions.

8.2.7 CCFD Focusing Weighted Extreme Learning

The classification of the imbalanced dataset is one of the significant challenges in credit card fraud detection. Zhu et al. have proposed a method based on the weighted extreme learning machine (WELM) to cope with this challenge [37]. The work focused on using smart methods to optimize WELM on imbalanced datasets. The authors have used the WELM algorithm due to its fast performance. For improvement in the algorithm, three dandelion algorithms were proposed with the probability-based mutation [37,40–42]. They applied it to detect credit card fraud, which has resulted in better classification. The experiment was done on the three imbalanced datasets of credit cards. It has produced improved detection while comparing precision and recall. Although the results achieved by this method are effectively improved, the classification performance can be further enhanced by considering more parameters in WELM. Using the distributive computing platform and data confidentiality are also the primary considerations not covered in the proposed method. Our proposed model will fix this gap by using the federated learning privacy approach and combining it with existing supervised and unsupervised ML, i.e., CS-SVM and K-means.

8.2.8 CCFD Focusing HMV Technique

Lucas et al. proposed a framework while considering the sequence of credit card transactions by using the Hidden-Markov model (HMV) [41,43,44].

Lucas et al. experimented based on three transaction parameters: the genuineness of the transaction (fraudulent or authentic), the amount spent based on timings concerning the last transaction, and either the transaction from fixed cardholders or the fixed terminal. With the combination of these parameters in a binary sequence, it gives eight sets of sequences. To experiment with this set of sequences, the cardholder transaction history is observed using the time and amount of credit card transactions via the HMV model. It proved efficient for anomaly detection [45]. This model can associate the likeliness of a transaction with the previous transaction. The authors used the random forest to detect and classify fraudulent transactions in this experiment. The method proposed by the authors is only concerned with the features of time and transactional behavior. However, the authors cannot consider the transactions' geographical location, which is essential for identifying fraudulent transactions. We believe using features like time and location combined with a supervised and unsupervised ML approach will ensure higher detection and accuracy.

8.2.9 CCFD Focusing K-Means

According to Carcillo et al., credit card fraud is complicated because customer behavior frequently alters [46]. It changes the fraudsters' capability to use novel techniques to commit fraud. Therefore, the authors have seriously considered finding a technique for detecting anomaly-based attacks. So, in this context, the supervised learning approach possesses limitations. It is required to consider unsupervised learning methods in CCFD. To make the CCFD more accurate based on the anomaly, the authors have used a hybrid approach combining the supervised and unsupervised learning approaches. The experiment was performed on the dataset based on customer history transactions. The clustering approach was adapted as an unsupervised (K-means) learning method [47]. The outcomes have shown that combining unsupervised ML with a supervised ML approach has greatly enhanced the accuracy of CCFD based on the classification anomaly. We have used this approach in our research work while combining the unsupervised method with supervised learning in a federated learning model. We chose K-means for our hybrid model because it can be executed on a large dataset and easily control the aggregation.

8.2.10   CCFD Focusing Meta-Learning Ensemble

Olowookere et al. have proposed a framework for effective credit card fraud detection based on the cost-sensitive meta-learning ensemble technique [48,49]. The framework follows the base classifier's adjustment with the ease of cost-sensitive learning combined with ensemble learning to regulate the meta-classifier while not allowing cost-sensitive learning on any base classifier. The authors have evaluated the framework by the area under the receiver operating characteristics curve [49]. The experiment was followed by the use of three datasets comprising various proportions of fraudulent transactions. During the experiment phase, the cost-sensitive ensemble framework was observed to keep a sufficient area under the receiver operating characteristics curve record while having an efficient result across the datasets. It has been shown that using a cost-sensitive ensemble technique has produced good results for detecting fraudulent transactions in the datasets. The framework's cost-sensitive classifiers are useful in detecting fraudulent transactions equally for all available datasets, irrespective of their numbers. It can be seen that the use of ensemble learning patterns with meta-learning patterns produces a significant result for imbalanced datasets; however, when it comes to the training, there is a lack of research to ensure the confidentiality of real-time data for analysis. This research also emphasizes the strength of cost-sensitive classifiers in processing highly skewed data sets [48]. Our proposed work promises federated learning technology for the privacy of real-time datasets while considering time and location elements.

8.2.11   Detecting Fraudulent Transactions by Blockchain Technology

Several applications built on blockchain technology have garnered significant interest from the general public. It is predicated on the idea that it exceeds the capabilities of centralized servers found in establishments such as banks and other organizations. In its place, it offers a decentralized method in which the behavior of individual users is contingent on the characteristics of the blockchain technology. Imagine a piece of malicious software capable of causing fraud in blockchain wallets being installed on an edge device. In blockchain technology, Ostapowicz et al. have presented a solution that utilizes supervised ML [50]. The authors have used this method on the blockchain of Ethereum. The experiment was carried out on a sample size of three hundred thousand accounts, and the findings were analyzed using random forest, SVM, and XGBoost [51]. The experiment's results led them to conclude that different transaction settings can affect the value of accuracy and recall. They have also argued that blockchain technology is capable of performing its maintenance. Particularly in finance, this over-reliance on it poses the possibility of danger. As a result, the foundation of our study is a more practical technique known as federated learning. This method is semi-decentralized, and it guarantees both effectiveness and confidentiality simultaneously.

8.2.12   Why Not Blockchain?

ML applications are revolutionizing people's lives and developing rapidly every day to make our surroundings more pleasant. The varied and complicated training data represent the most significant obstacle in ML. Crowdsourcing is one method that may be utilized to collect data for the central server; nevertheless, it has certain restrictions regarding the confidentiality of the data [52]. One of the new technologies that provides a decentralized platform and potentially increased data security is blockchain [53]. As a result, it could be possible to see it as the medium via which data is collected for CCFD in terms of how data is safely transmitted across banks and other financial organizations.

However, the technology suffers from several shortcomings and limits that make it less effective for data transfer. In addition, because of GDPR, there are privacy issues regarding data exchange. When considering CCFD, the following are some of the drawbacks of using blockchain technology:

- When there are an excessive number of users on a network, the procedure becomes more laborious.
- It is more difficult to scale up the data on a blockchain since it uses a consensus approach.
- It needs a larger use of energy.
- When it comes to its operations, blockchain may occasionally be inefficient.
- Users are required to save their information in wallets.
- The technology comes at a high price.
- It is not standardized in every region of the world.

Because of the abovementioned problems with blockchain technology, academic institutions and researchers are hesitant to implement CCFD. The federated learning methodology is semi-decentralized and will be used by our proposed study to resolve this issue. This would result in quicker processing capabilities and more data scalability than blockchain technology. It would also enable higher levels of efficiency since the participants would train their models locally, which would retain security.

8.2.13   USE CASES OF FL TECHNOLOGY

In 2017, Google introduced the world to the technology of federated learning for the first time. They have implemented it in their Gboard, which has improved the user experience while interacting with the Gboard [54]. The technology works by learning the typed words or phrases from the users, combining NLP and federated learning. The idea was to keep the user data local to their devices; however, with federated learning technology, the learned model was sent across to the Google platform, where it was shared with other users. This way, the data remains on the individual user devices; only the trained model is shared with Google. It has resulted in learning new patterns of words and phrases; in this way, the next word or phrase is predicted, enhancing the user experience. The proposed work uses federated learning, where data security is promised.

Google Home has also adopted federated learning technology to improve the user experience in terms of interaction. It follows the training procedure for the user's voice commands (using voice recognition) on the devices without sharing data with the Google central server [55]. The model gets trained locally on users' own devices, and afterward, the trained model is sent to Google, where it collaborates with other trained models. By this way, it is shared across different devices, resulting in a better user experience. We have adopted this method in our proposed hybrid framework to enable the data security feature in CCFD.

The technology of federated learning has also been adopted in Gboard by Google, where the emoji was predicted based on the text by users. Once the user types the "word" or "sentence", the emoji is predicted, and the model is trained locally on user devices. The trained model is then centrally shared to enhance the prediction of emoji based on user text [56].

Recently, Facebook tried to use federated learning technology in its application. The idea is to observe the user behavior while interacting with the Facebook application, which includes the navigation to the advertisement bar, clicking behavior, and frequency of visiting certain pages or profiles. This information is trained on the users' devices, and the model is sent to Facebook to enable them to understand the overall user interaction with their system and improve the user experience simultaneously [57]. We combine our approach of supervised and unsupervised ML algorithms with a federated learning model to attain a higher accuracy of credit card fraud in a privacy-preserving manner.

8.3 METHODOLOGY

The methodology is one of the principal components of a research project that helps understand the theoretical and systematic concepts. The methodology includes building a model, collecting data, analyzing it, representing it, and assessing it. For this research, it is important to explore defined tools and methods to solve the identified problem. Therefore, it is required to have methods that can promise efficient accuracy while keeping data privacy assured.

8.3.1 PROBLEMS IN THE CONVENTIONAL APPROACH OF CCFD

The primary concern is the customer's data which must be regulated under the GDPR 2018 Act. According to the GDPR, "Personal Data must be available to the only authorized body and must be kept confidential" [58].

In compliance with the GDPR, no financial institution can share the data with other institutions or researchers for research purposes. Therefore, as researchers, we do not have access to data to devise a model that could effectively optimize the detection of fraudulent transactions. Therefore, to solve this issue, we have proposed a research work that could ensure the model's training for banks without accessing each other's data. The proposed work constitutes the following research question in contrast to the conventional model:

8.3.2 JUSTIFICATIONS FOR THE RESEARCH QUESTION

This section focuses on the justification of our proposed research questions based on the analysis from the literature review as follows:

RQ1: Can real-time datasets be used by various banks and financial institutions mutually to detect fraudulent transactions while ensuring the customers' privacy?

The data involved in banks and other financial firms are susceptible. It is stored and processed locally on their servers. So, banks usually train the ML model locally in their database to detect fraudulent activities. It is known as a centralized approach that involves ML techniques. It keeps data on local servers along with the model.

In the literature review, we observed the work identifying fraudulent and non-fraudulent transactions. Multi-layer perception is a practical approach to identifying the nature of transactions. The major problem with the centralized technique is that various banks have seen various fraudulent transactions that differ from each other. It prohibits the effectiveness of fraud detection by all banks. So, the solution to this problem involves all banks' mutual corporations participating and bringing up a shared model that incorporates all banks' various transactions. It is indeed a sensitive concern for banks and financial firms. They do not want to expose themselves to the fraudulent transactions they encounter.

The above research question is based on federated learning. It is responsible for achieving security, data privacy, and a fixed and transparent system that ensures data integrity compared to conventional centralized techniques and methods. FL involves a semi-decentralized structure far more enhanced for data privacy than traditional centralized servers. The FL model can help banks and other financial firms adopt a model that provides privacy protection to their customers.

The FL model will incorporate a central platform for all banks to share their models but keep the data unexposed. Individual banks will be involved in bringing up their trained model from their datasets, and that model without data is shared centrally. Every bank will add its model to the central platform. That model will be trained by the mutual collaboration of all banks while taking the average of the trained models. Ultimately, this main model is shared among all banks, which would cause a practical and broader approach to fraudulent transactions. The whole process of FL does not involve data sharing at any stage. No sensitive information is shared, like in the centralized technique. So, the ultimate model will be sufficient to detect a wider range of fraudulent transactions than any individual bank.

We will use K-means, the SVM CS algorithm [23], and the federated learning algorithm to make a hybrid approach. Based on the tests and results, this research aims to determine whether the K-means, CS-SVM, and FL hybrid approaches are practical for the CCFD. By understanding the limitations, scope, and privacy and security issues, we will compare and contrast the results with the other models to evaluate our model.

RQ2: Is it possible to enhance the detection of fraudulent transactions with a hybrid ML approach in real-time datasets?

The scope of our research is based on detecting fraudulent transactions, which will effectively categorize the transaction type. Based on the analysis of the various state-of-the-art in CCFD, our proposed solution would potentially solve the customers' data privacy and the accuracy of detecting fraudulent transactions while combing the supervised and unsupervised models. The primary outcome of this research is based on the fact that it will be able to attain high-speed data processing in real time. The FL approach will achieve the model from real-time data, improving accuracy and data security.

The abovementioned research involves applying a hybrid approach of the FL algorithm to the clustering location of transactions and transaction time based on customers' previous history. This clustering technique uses features of time and geolocation of a historical transaction to provide real-time data on the customers' behavior in the banking system and train the model. Designing a model that keeps the customers' data private while using historical transactions' geolocation is essential.

Our research will be focused on customer transaction timing, behavior analysis, and geolocation of the historical transaction. We will use a supervised (CS-SVM) algorithm for accuracy alongside the unsupervised learning algorithm (K-means) that will enable us to classify the diverse transactions of various customers based on classification and clustering. A federated learning technique will make the users' data decentralized and keep the data secure.

As mentioned in the research paper titled "Application of Credit Card Fraud Detection Based on CS-SVM", the paper discusses the performance of Support Vector Machines (SVM) in Credit Card Fraud Detection (CCFD). Comparing the SVM with the other classifier techniques can potentially solve the problem of linear and nonlinear binary classification issues that could differentiate the input data. The kernel function is one of the mediums that incorporates the classification of SVM. The SVM method is based on the grid search method. However, it is not the best solution in the category of an optimal local solution. We will use the CS algorithm to solve this problem, incorporating the more considerable step into smaller steps that cause SVM to prohibit local-optimum. It can significantly increase the performance of SVM in terms of its classification by utilizing the optimizing components of SVM. We have decided to use CS-SVM in our research because researchers achieve higher accuracy, and the maximum achieved accuracy is 98% [23].

In the research paper "Combining unsupervised and supervised learning in credit card fraud detection", the authors have used the K-means algorithm in CCFD [47]. The authors mentioned that fraudulent transactions could be identified using the number of transactions and values in the customers' transaction history. The paper has experimented with that clustering technique on the geolocation of historical transactions, and customer transaction behavior (time) has produced good results. We have decided to use the K-means clustering algorithm for our research since it is simple to interpret. It performs quickly on more extensive databases and helps control data aggregation rapidly. It is an important consideration to choose the correct feature for clustering. We have chosen geolocation of the historical transaction and customer payment behavior (time) for our research. After analyzing the literature review, we have proposed a hybrid model that incorporates the features of time and location while using the K-means algorithm and SVM for classification to achieve effective results for CCFD while combining with FL.

8.3.3 THE DATASET USED IN CCFD

We will be using the dataset for our research from the ULB ML group, which can be found at www.kaggle.com/mlg-ulb/creditcardfraud. This dataset constituted credit card transactions by European cardholders back in 2013. It has 284,807 transactions that happened over the two days. The fraudulent transaction was 0.172% of the total transaction. This data is quite imbalanced concerning a fraudulent transaction. It constitutes numerical inputs that have produced 28 principal components based on PCA. In this work, 30 various features have been analyzed. Due to privacy reasons, formal details cannot be exposed. The time feature shows the elapsed seconds between transactions, and the amount feature shows the total transaction amount. Finally, the feature class shows binary classifications. It takes "1" for the fraudulent transaction and "0" for the non-fraudulent transaction.

8.3.4 PROPOSED MODEL DESIGN

The centralized approach is a strategy that is often used to identify fraudulent activity on credit card accounts. An FDS will perform less effectively when there are just a few datasets. Because of GDPR, banks and other financial centers are prohibited from sharing their data on a centralized server. Even if the "anonymized" dataset is stored locally on servers, the user's privacy may still be at risk because it is possible to do reverse engineering. As a result, to meet this obstacle's demands, we have incorporated FL into our research model. This provides the capability to train the real-time data locally on edge devices. The trained model is shared with other financial institutions and research centers. This helps to improve the accuracy with which fraudulent transactions are identified.

Second, in our research model, we will utilize the K-means method to cluster the client data depending on the client's location and the transaction time. This will be done in conjunction with the support vector machine to achieve greater accuracy. Furthermore, after implementing this hybrid strategy, we will employ FL, which plays a crucial role in achieving user data privacy in the provided hybrid model approach. This is because FL is one of the hybrid models that play a role in the given hybrid model approach.

8.3.5 PROPOSED MODEL WITH FEDERATED LEARNING, SVM, AND K-MEANS

In our FL model, the following steps are involved in training the model until all participants achieve the complete transition:

- **Clients' selection**
  The server chooses which clients will participate in the activity based on the criteria used to determine eligibility.
- **Broadcasting**
  At this point, the selected client connects to the server to retrieve our model, a combination of the SVM and the K-means clustering algorithm, as well as the training criteria.
- **Computation phase**
  At this point, all participating devices will execute the application supplied by the server to compute the model update.
- **Aggregation**
  At this point, the server is responsible for doing the aggregate of the updates that were received from the device.
- **Model update**
  As part of this process, the shared server will aggregate the updates made by the clients locally and will also update the shared model.

8.3.5.1 Model Outline

The federated learning model with CS-SVM and K-means that has been presented may be broken down into three parts, each of which is carried out one after the other until the last phase is finished, at which point the cycle begins again.

Step 1

In this stage, our model, consisting of supervised and unsupervised components (CS-SVM and K-means, respectively), is sent from the centralized server to the various correspondent banks and financial institutions. The figure that follows shows it as the "Black Brain" symbol. As soon as the multiple banks acquire the model, they begin training it with the datasets readily available in their respective locations. The training process is depicted in the following image, in which the trained model is shown to be separated from the bank using different colors (A-purple, B-blue, C-green, and D-red). The number "1" represents the beginning of the first part of delivering our model to the banks (Figure 8.1).

Step 2

After the first phase has been completed, step two will immediately begin, which will involve the transmission of a trained model from banks to the central server of the federated learning model. On the server, all of the models from the various banks are merged to create an "upgraded model," as seen in Figure 8.2.

Step 3

Step 3 is the final step in our proposed model, and it reflects the transmission of the "upgraded model" to the individual bank independently denoted by the digit "3." This "upgraded model" is generated by the mean average of all related trained models from various banks. In addition, as soon as the model is received from the banks, it immediately undergoes step 1 of the local training. After the training has been finished, the data is then uploaded to the server. The procedure will be carried out cyclically until the desired result is achieved (Figure 8.3).

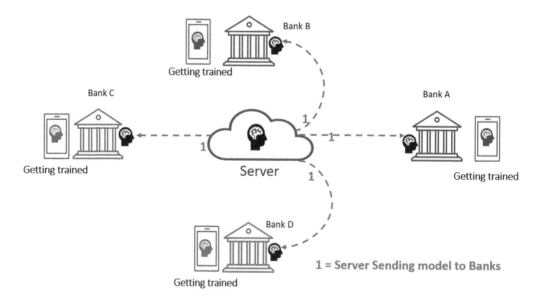

FIGURE 8.1   Step 1 of the proposed model.

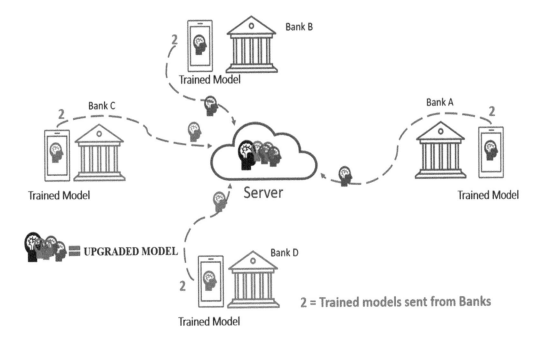

FIGURE 8.2   Step 2 of the proposed model.

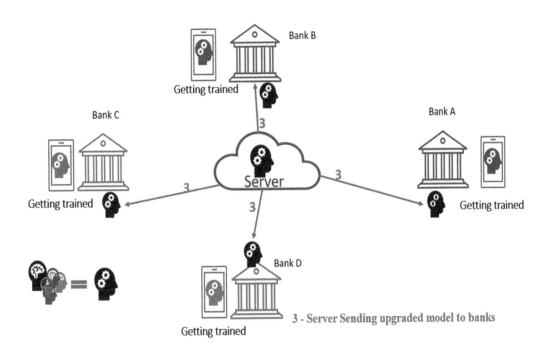

FIGURE 8.3   Step 3 of the proposed model.

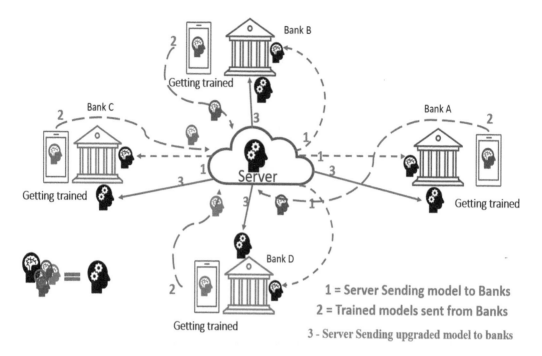

FIGURE 8.4   Full model architecture.

8.3.5.2   Cycle Repetition

The procedure is continued after step 3, and the trained model is uploaded to the server in the same manner that step 1 was described above. Again, the server calculates the mean and sends it along to each bank in the same form. Our hypothesis suggests that a better level of accuracy in CCFD may be achieved through a process of recurrent training. The steps involved in the complete procedure are shown in Figure 8.4.

By carrying out the measures outlined above, FL is able to improve the accuracy of the fraud detection process considerably. The paradigm in which the data is stored locally in each institution's database is one that financial institutions and other research organizations routinely and cooperatively share. On the other hand, just the trained model and not the actual data are made available to all participants. Each participant will train the central server in turn, leading to improved classification compared to the results obtained by the individual models trained locally. To put it more succinctly, the pattern is learned on each client side, and the central server then aggregates all of the patterns that have been learned. It is trained based on the collaborative efforts of all of the participants. All participants have access to this central model, and fraud detection is carried out in accordance with it.

8.3.6   Limitation

Although it is an excellent solution for the banks to have a shared model, the limitation is based on the fact that the various datasets used by the banks could be complicated in a way that they use various features and other tags that can make the training model complex. To solve this issue, we would be considering only two features in our research work: time and location. We would use these two features based on supervised and unsupervised learning mechanisms. For classification and clustering, CS-SVM and K-means will be used, respectively. Using the FL, which will include the data following GDPR, simultaneously protects the confidentiality of the customer's personal information.

8.4 CONCLUSION

The proposed research reflects introducing a new model based on federated learning. The previous study included the limitations in data confidentiality and data perseverance. Some researchers devised the idea of using blockchain to tackle this limitation. Blockchain technology resolves data security and data alteration. Still, other limitations (discussed in the literature review) make it inefficient to use in the FDS. After considerable research, our proposed model combines federated learning with supervised (SVM) and unsupervised (K-means) [47] ML algorithms to fill the bank's data privacy gap and keep the detection of fraudulent transactions high. We have concluded that using FL with the CS-SVM [23] and the K-means model can effectively increase the capability to detect fraudulent transactions while considering the elements of time and location. This model can produce a shared approach for all banks and financial institutions to adopt. It can significantly increase the capability of detecting fraudulent transactions while giving the opportunity to train a model from real-time data where privacy is ensured simultaneously. It also helps banks and financial institutions develop a more complicated but accurate model with massive data collection from end-users.

Our research model can benefit banks and financial corporations by encouraging them to collaborate for a mutual connection. It helps analyze fraud detection individually and produce the corresponding solution while training the model from various real-time datasets and following the GDPR guidelines. The proposed technique can alter the way financial banking is done in terms of fraud detection. It can bring collective benefits, which include data privacy and higher accuracy in fraud detection. The research can open doors for the finance industry to a new dimension of data security and fraud detection using the FL approach.

REFERENCES

1. Li, Z., Liu, G., and Jiang, C. (2020). Deep representation learning with full center loss for credit card fraud detection. *IEEE Transactions on Computational Social Systems*, 1–11. https://doi.org/10.1109/tcss.2020.2970805.
2. Hegedűs, I., Danner, G., and Jelasity, M. (2021). Decentralized learning works: An empirical comparison of gossip learning and federated learning. *Journal of Parallel and Distributed Computing*, 148, 109–124.
3. Danner, G., Berta, Á., Hegedűs, I., and Jelasity, M. (2018). Robust fully distributed minibatch gradient descent with privacy preservation. *Security and Communication Networks*, 2018, 6728020. https://doi.org/10.1155/2018/6728020.
4. Oualid, A., Maleh, Y., and Moumoun, L. (2023). Federated learning techniques applied to credit risk management: A systematic literature review. *EDPACS*, 68, 1–15.
5. Bahnsen, A.C., Stojanovic, A., Aouada, D., and Ottersten, B. (2014). Improving credit card fraud detection with calibrated probabilities. In: *Proceedings of the 2014 SIAM International Conference on Data Mining* (pp. 677–685). SIAM.
6. Jakaite, L., Schetinin, V., and Maple, C. (2012) Bayesian assessment of newborn brain maturity from two-channel sleep electroencephalograms. *Computational and Mathematical Methods in Medicine*, 1–7. https://doi.org/10.1155/2012/629654.
7. Jakaite, L., Schetinin, V., Maple, C., and Schult, J. (2010). Bayesian decision trees for EEG assessment of newborn brain maturity. In: *The 10th Annual Workshop on Computational Intelligence UKCI 2010.* https://doi.org/10.1109/ UKCI.2010.5625584.
8. Jakaite, L., Schetinin, V., and Schult, J. (2011). Feature extraction from electroencephalograms for Bayesian assessment of newborn brain maturity. In: *Proceedings of the 24th IEEE International Symposium on Computer-Based Medical Systems* (pp. 1–6). https://doi.org/10.1109/ CBMS.2011.5999109.
9. Nyah, N., Jakaite, L., Schetinin, V., Sant, P., and Aggoun, A. (2016). Evolving polynomial neural networks for detecting abnormal patterns. In: *2016 IEEE 8th International Conference on Intelligent Systems (IS)* (pp. 74–80). https://doi.org/10.1109/IS.2016.7737403.

10. Nyah, N., Jakaite, L., Schetinin, V., Sant, P., and Aggoun, A. (2016). Learning polynomial neural networks of a near-optimal connectivity for detecting abnormal patterns in biometric data. In: *2016 SAI Computing Conference (SAI)* (pp. 409–413). https://doi.org/10.1109/SAI.2016.7556014.

11. Schetinin, V., and Jakaite, L. (2012). Classification of newborn EEG maturity with Bayesian averaging over decision trees. *Expert Systems with Applications*, 39(10), 9340–9347. https://doi.org/10.1016/j.eswa.2012.02.184.

12. Schetinin, V., and Jakaite, L. (2017). Extraction of features from sleep EEG for Bayesian assessment of brain development. *PLOS ONE*, 12(3), 1–13. https://doi.org/10.1371/journal.pone.0174027.

13. Schetinin, V., Jakaite, L., Nyah, N., Novakovic, D., and Krzanowski, W. (2018). Feature extraction with GMDH-type neural networks for EEG-based person identification. *International Journal of Neural Systems*. https://doi.org/10.1142/S0129065717500642.

14. Selitskaya, N., Seliski, S., Jakaite, L., Schetinin, V., Evance, F., Conrad, M., and Sant, P. (2020). Deep learning for biometric face recognition: Experimental study on benchmark data sets. In: R. Jiang, C. Li, D. Crookes, W. Meng, C. Rosenberger (eds.) *Deep Biometrics* (pp. 71–970). Springer. https://doi.org/10.1007/978-3-030-32583-15.

15. Schetinin, V., Jakaite, L., Jakaitis, J., and Krzanowski, W. (2013). Bayesian decision trees for predicting survival of patients: A study on the US national trauma data bank. *Computer Methods and Programs in Biomedicine*, 111(3), 602–612. https://doi.org/10.1016/j.cmpb.2013.05.015.

16. Schetinin, V., Jakaite, L., and Krzanowski, W. (2018). Bayesian averaging over decision tree models: An application for estimating uncertainty in trauma severity scoring. *International Journal of Medical Informatics*, 112, 6–14. https://doi.org/10.1016/j.ijmedinf.2018.01.009.

17. Schetinin, V., Jakaite, L., and Krzanowski, W. (2018). Bayesian averaging over decision tree models for trauma severity scoring. *Artificial Intelligence in Medicine*, 84, 139–145. https://doi.org/10.1016/j.artmed.2017.12.003.

18. Schetinin, V., Jakaite, L., and Krzanowski, W.J. (2013). Prediction of survival probabilities with Bayesian decision trees. *Expert Systems with Applications*, 40(14), 5466–5476. https://doi.org/10.1016/j.eswa.2013.04.009.

19. Schetinin, V., Jakaite, L., and Krzanowski, W. (2018). Bayesian learning of models for estimating uncertainty in alert systems: Application to air traffic conflict avoidance. *Integrated Computer-Aided Engineering*, 26, 1–17. https://doi.org/10.3233/ICA-180567.

20. Jakaite, L., Schetinin, V., Hladuvka, J., Minaev, S., Ambia, A., and Krzanowski, W. (2021). Deep learning for early detection of pathological changes in x-ray bone microstructures: Case of osteoarthritis. *Scientific Reports*, 11. https://doi.org/10.1038/s41598-021-81786-4.

21. Patidar, R. and Sharma, L. (2011). Credit card fraud detection using neural network. *International Journal of Soft Computing and Engineering (IJSCE)*, 1, 32–38.

22. Irny, S.I. and Rose, A.A. (2005) Designing a strategic information systems planning methodology for Malaysian institutes of higher learning (isp- ipta). *Issues in Information System*, VI(1), 325–331.

23. Li, C., Ding, N., Dong, H., and Zhai, Y. (2021, January). Application of credit card fraud detection based on CS-SVM. *International Journal of Machine Learning and Computing*, 11(1), 34–39.

24. Zhu, H., Liu, G., Zhou, M., Xie, Y., Abusorrah, A., and Kang, Q. (2020). Optimizing weighted extreme learning machines for imbalanced classification and application to credit card fraud detection. *Neurocomputing*, 407, 50–62.

25. Modi, K. and Dayma, R. (2017). Review on fraud detection methods in credit card transactions. In: *2017 International Conference on Intelligent Computing and Control (I2C2)* (pp. 103–107). IEEE, Coimbatore, India.

26. Wen, H. and Huang, F. (2020, May). Personal loan fraud detection based on hybrid supervised and unsupervised learning. In: *2020 5th IEEE International Conference on Big Data Analytics (ICBDA)* (pp. 339–343). IEEE.

27. Hu, J. and Lyu Z. (2017). Model validation method with multivariate output based on kernel principal component analysis. *Journal of Beijing University of Aeronautics and Astronautics*, 43(7), 1470–1480.

28. Zieba, M., Tomczak, S. K., and Tomczak, J. M. (2016). Ensemble boosted trees with synthetic features generation in application to bankruptcy prediction. *Expert Systems with Applications*, 58(C), 93–101.

29. Vynokurova, O., Peleshko, D., Bondarenko, O., Ilyasov, V., Serzhantov, V., and Peleshko, M. (2020, August). Hybrid machine learning system for solving fraud detection tasks. In: *2020 IEEE Third International Conference on Data Stream Mining & Processing (DSMP)* (pp. 1–5). IEEE.

30. Rai, A.K. and Dwivedi, R.K. (2020, July). Fraud detection in credit card data using machine learning techniques. In: *International Conference on Machine Learning, Image Processing, Network Security and Data Sciences* (pp. 369–382). Springer, Singapore. https://doi.org/10.1007/978-981-15-6318-8_31.

31. Dubey, S.C., Mundhe, K.S., and Kadam, A.A. (2020, May). Credit card fraud detection using artificial neural network and backpropagation. In: *2020 4th International Conference on Intelligent Computing and Control Systems (ICICCS)* (pp. 268–273). IEEE.

32. Devi, K. N., Bhaskaran, V. M., and Kumar, G. P. (2015). Cuckoo optimized SVM for stock market prediction. In: *Proceedings of 2015 International Conference on Innovations in Information, Embedded and Communication Systems* (pp. 1–5).

33. Khedmati, M., Erfani, M., and GhasemiGol, M. (2020). Applying support vector data description for fraud detection. arXiv preprint arXiv:2006.00618.

34. Hines, C. and Youssef, A, (2018). Machine learning applied to rotating check fraud detection. In *2018 1st International Conference on Data Intelligence and Security (ICDIS)* (pp. 32–35). IEEE, 2018.

35. Xuan, S., Liu, G., Li, Z., Zheng, L., Wang, S., and Jiang, C. (2018, March). Random forest for credit card fraud detection. In: *2018 IEEE 15th International Conference on Networking, Sensing and Control (ICNSC)* (pp. 1–6). IEEE.

36. Rtayli, N. and Enneya, N. (2020). Selection features and support vector machine for credit card risk identification. *Procedia Manufacturing*, 46, 941–948.

37. Selitskaya, N., Vitaly, S., et al. (2020). Deep learning for biometric face recognition: Experimental study on benchmark data sets. In: Jiang R., Li CT., Crookes D., Meng W., Rosenberger C. (eds.) *Deep Biometrics.* Unsupervised and Semi-Supervised Learning. Springer, Cham. https://doi.org/10.1007/978-3-030-32583-1_5.

38. Arya, M. and Sastry, H.G. (2020). DEAL-'Deep Ensemble Algorithm' framework for credit card fraud detection in real-time data stream with google tensorflow. *Smart Science*, 8(2), 71–83.

39. Roy, A., Sun, J., Mahoney, R., et al. (2018). Deep learning detecting fraud in credit card transactions. In: *2018 Systems and Information Engineering Design Symposium (SIEDS)* (pp. 129–134). IEEE, Charlottesville, VA; 27 April 2018.

40. Huang, G.B., Zhou, H., Ding, X., and Zhang, R. (2011). Extreme learning machine for regression and multiclass classification. *IEEE Transactions on Systems, Man, and Cybernetics*, 42(2) 513–529.

41. Schetinin, V., Jakaite, L., and Krzanowski, W. (2018). Bayesian averaging over decision tree models: An application for estimating uncertainty in trauma severity scoring. *International Journal of Medical Informatics*, 112, 6–14.

42. Husejinovic, A., (2020). Credit card fraud detection using naive Bayesian and C4. 5 decision tree classifiers. *Periodicals of Engineering and Natural Sciences*, 8, 1–5.

43. Lucas, Y., Portier, P.E., Laporte, L., He-Guelton, L., Caelen, O., Granitzer, M., and Calabretto, S. (2020). Towards automated feature engineering for credit card fraud detection using multi-perspective HMMs. *Future Generation Computer Systems*, 102, 393–402.

44. Yadav, S. and Siddartha, S. (2018). Fraud detection of credit card by using HMM model, *International Journal of Engineering Research & Technology*, 6(1), 41–46.

45. Agbakwuru, A.O. and Elei, F.O. (2020), Hidden Markov Model Application for Credit Card Fraud Detection Systems. *International Journal of Innovative Science and Research*, 5(1).

46. Carcillo, F., Le Borgne, Y.A., Caelen, O., Kessaci, Y., Oblé, F., and Bontempi, G. (2019). Combining unsupervised and supervised learning in credit card fraud detection. *Information Sciences*, 557, 317–331.

47. Rai, A.K. and Dwivedi, R.K. (2020, July). Fraud detection in credit card data using unsupervised machine learning based scheme. In: *2020 International Conference on Electronics and Sustainable Communication Systems (ICESC)* (pp. 421–426). IEEE.

48. Olowookere, T.A. and Adewale, O.S. (2020). A framework for detecting credit card fraud with cost-sensitive meta-learning ensemble approach. *Scientific African*, 8, e00464.

49. Bahnsen, A.C., Villegas, S., Aouaday, D., and Ottersten, B. (2017). Fraud detection by stacking cost-sensitive decision trees. *Data Science for Cyber-Security*, 1–15. https://doi.org/10.1142/9781786345646_012.

50. Ostapowicz, M. and Żbikowski, K. (2020, January). Detecting fraudulent accounts on Blockchain: A supervised approach. In: *International Conference on Web Information Systems Engineering* (pp. 18–31). Springer, Cham. https://doi.org/10.1007/978-3-030-34223-4_2.

51. Carneiro, N., Figueira, G., and Costa, M. (2017). A data mining based system for credit-card fraud detection in e-tail. *Decision Support Systems*, 95, 91–101. https://doi.org/10.1016/j.dss.2017.01.002.

52. Xu, J.J. (2016). Are blockchains immune to all malicious attacks? *Financial Innovation*, 2(1), 1–9.

53. Ostapowicz, M. and Żbikowski, K. (2020, January). Detecting fraudulent accounts on Blockchain: A supervised approach. In: *International Conference on Web Information Systems Engineering* (pp. 18–31). Springer, Cham. https://doi.org/10.1007/978-3-030-34223-4_2.

54. Federated Learning: Collaborative Machine Learning without Centralized Training Data. https://ai.googleblog.com/2017/04/federated-learning-collaborative.html [accessed on: 07/6/2020].

55. Nishio, T. and Yonetani, R. (2019, May). Client selection for federated learning with heterogeneous resources in mobile edge. In: *ICC 2019-2019 IEEE International Conference on Communications (ICC)* (pp. 1–7). IEEE.

56. Ramaswamy, S., Mathews, R., Rao, K., and Beaufays, F. (2019). Federated learning for emoji prediction in a mobile keyboard. arXiv preprint arXiv:1906.04329.

57. How federated learning could shape the future of AI in a privacy-obsessed world https://venturebeat.com/2019/06/03/how-federated-learning-could-shape-the-future-of-ai-in-a-privacy-obsessed-world/ [accessed on 20/11/2020].

58. Data Protection Act 2018. https://www.legislation.gov.uk/ukpga/2018/12/contents/enacted [accessed on 11/12/2020].

9 Medical-DCGAN
Deep Convolutional GAN for Medical Imaging

Rguibi Zakaria, Hajami abedlamjid,
Dya Zitouni, and Amine ElQaraoui

9.1 INTRODUCTION

Machine learning has seen widespread adoption in the last decade, due to the advent of deep neural networks and their state-of-the-art results on a variety of medical imaging tasks. Deep neural networks are composed of many layers, each of which can be trained to recognize certain features in an image. This allows for more accurate recognition than traditional machine learning algorithms [1].

Medical imaging is one area where deep neural networks have seen significant success. They have been used to improve the accuracy of tumor detection and classification, as well as the detection of other abnormalities such as blood clots. In addition, they have been shown to be effective at predicting patient outcomes after surgery or radiation therapy [2,3].

The use of deep neural networks for medical imaging is still relatively new, and there is still much research that needs to be done in this area. However, the results that have been achieved so far are very promising and suggest that these techniques will play an important role in future medical care [2–4].

DCGAN, or Deep Convolutional Generative Adversarial Networks, one of the most significant early advancements in Generative Adversarial Networks (GANs) since the invention of the technology two years earlier was GAN. While a team of academics had previously attempted to employ ConvNets in GANs before Alec Radford, Luke Metz, and Soumith Chintala's introduction of the DCGAN in 2016, it was the first time they had really been successful in doing so [5].

DCGAN networks are composed of two subnets: a generator network (G), which produces synthetic data samples; and a discriminator network (D), which evaluates how real these data samples are compared to actual data. The key innovation that allowed DCGAN networks to outperform other methods was their use of convolutional layers within both the generator and discriminator nets. This gave these networks an advantage in terms of speed and accuracy over traditional GAN models [5].

Since its beginning, DCGAN has become one of the most widely used techniques for generating artificial images and videos. In addition, because DCGAN networks are relatively easy to train compared to other deep learning models, they have become popular tools for teaching deep learning concepts.

An artificial intelligence system known as a GAN puts two neural networks against one another in a "zero-sum game" to boost the performance of both. In the context of medical imaging, GANs can be used to generate realistic images from training data, which can then be used for diagnostic purposes or as part of machine learning models [6].

Another advantage of using GANs for medical imaging is that they can help reduce the amount of noise present in images. This is because GANs are able to learn how to suppress noise while still preserving important image features. As a result, this could lead to improved diagnosis accuracy and reduced scan times for patients.

DOI: 10.1201/9781032667478-10

The main goal of our work is to generate high-quality images as close as possible to the original images based on the DCGAN model. We have been successful in generating some very photorealistic images and are continuing to work toward making even more improvements.

The rest of this chapter is organized as follows: the background and understanding of generative and discriminative models is discussed in Section 9.2. Then, Section 9.3 covers the GAN for medical imaging. The conclusion of this work is presented in Section 9.4.

9.2 UNDERSTANDING GENERATIVE AND DISCRIMINATIVE MODELS

GANs are a special kind of deep neural network that can produce new data that resembles the data it was trained on. Two primary blocks of GANs compete with one another to make creative works of art. The first block, called the generator, takes a set of training data and produces new data samples. The second block, called the discriminator, evaluates these new samples to see how similar they are to the original training set. If the discriminator can't tell the difference between generated and real data samples, then the generator is considered successful.

GANs were developed in 2014 by Ian Goodfellow [7] and his teammates as a way of improving on traditional deep learning methods. Since their inception, GANs have been used for tasks such as creating realistic images and generating text descriptions of objects [8]. Some recent successes include generating photorealistic faces and synthesizing speech sounds that sound like natural human speech (Figure 9.1).

GAN may be thought of as the interaction of two separate models: the generator and the discriminator [9,10]. As a result, each model will have its own loss function. Let's try to motivate an intuitive comprehension of the loss function for each in this section. The generator attempts to produce images that are as realistic as possible, while the discriminator attempts to discern between genuine and generated images [11]. The purpose of the training procedure is to find a Nash equilibrium between these two models.

In their most basic formulation, GANs are trained to optimize the following value function [7]:

$$\min_G \max_D E_{x \sim p}\text{data}(x)\big[\log D(x)\big] + E_{z \sim p}z\ (z)\big[1\ \log D\big(G(z)\big)\big] \tag{9.1}$$

Let's create some notation that we'll utilize throughout this essay to avoid misunderstanding.

- x: Real data
- z: Latent vector
- $G(z)$: Fake data
- $D(x)$: Discriminator's evaluation of real data
- $D(G(z))$: Discriminator's evaluation of fake data

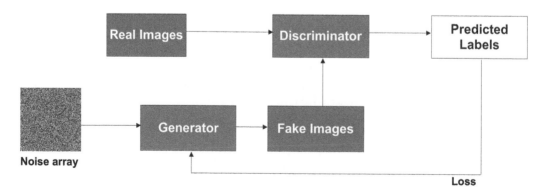

FIGURE 9.1 An illustration of the structure of a Generative Adversarial Network.

For the generator, reducing the loss is identical to minimizing $\log(1 - D(G(z)))$ since the generator cannot directly influence the $\log(D(x))$ term in the function.

9.3 GENERATIVE ADVERSARIAL NETWORK FOR MEDICAL IMAGING

9.3.1 CONSTRUCTING A DATASET AND DATA PROCESSING

Medical imaging is one of the most important tools used in the diagnosis of diseases because of its speed in terms of time compared to other tools such as medical analysis. In this sense, our work is to develop a useful tool that allows us to make a classification of medical images to detect the presence or absence of a disease in order to help doctors make the best decision and thus meet the needs of patients. Furthermore, due to the use of CNN and resent blocks, the proposed model is a good choice for this kind of task [12].

As we know, one of the most challenging aspects of working with medical imaging data is the lack of datasets available. Developing an algorithm that can give us a better patient outcome is essential to improving patient care. However, this process is not easy; it requires significant time and effort. In our work, we will use chest X-ray datasets (Figure 9.2).

Whichever technique you choose, it's important to make sure your dataset is representative of real-world patients if you want your results to be meaningful. The first data is about 5960 images from patients without and with pneumonia [13], where 1690 are normal patients and 4270 are pneumonia, and all images have different 224 pixels, and all images are resized to a resolution of 224×during the algorithm evaluation (see Figure 9.2). No additional image from other sources has been used in resolution, and no data augmentation has been used during the training time.

9.3.2 ARCHITECTURE AND CONFIGURATIONS

DCGANs are a recent development in the field of deep learning that have successfully combined CNNs with GANs, resulting in a more suitable solution for image generation. DCGANs are composed of two main parts: the generator and the discriminator. The generator is responsible for creating synthetic images, while the discriminator is used to distinguish between real and generated images. The advantage of using a DCGAN over other methods, such as standard GANs or CNNs alone, is that it can learn to generate high-quality images without any prior information about what those images should look like. This makes it an attractive option for medical imaging applications [9,10].

Grid of X-Ray NORMAL images

Grid of X-Ray PNEUMONIA images

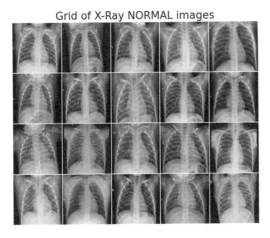
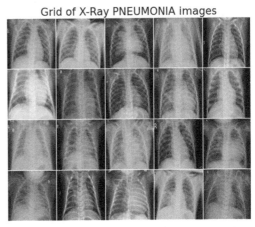

FIGURE 9.2   X-ray normal and Pneumonia images.

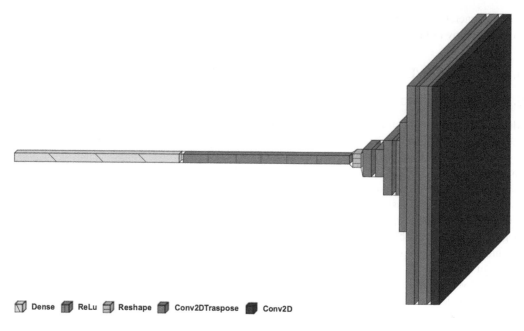

Dense **ReLu** **Reshape** **Conv2DTraspose** **Conv2D**

FIGURE 9.3 Discriminator architecture.

In the discriminator, we receive an image as an input and apply to the input 5 convolutional lay-ers with 128 filters with a kernel size of 5×5, a 2×2 stride, and the same padding, each followed by a leaky relu activation function. After that comes a dropout layer with a rate of 0.5. Finally, the activation passes through a sigmoid activation function to produce the output. We fixed the number of filters to 128 instead of increasing/decreasing them layer by layer since we found it gave us the best results (Figure 9.3).

In the generator, we receive a noise vector and apply a relu activation function to it right away and then we reshape it into an image. We then apply four transformed convolution layers with 128 filters with a kernel size of 4×4, a 2×2 stride, and the same padding, each followed by a relu activation function. And finally, a convolutional layer with three filters with a kernel size of 5×5 and the same padding, followed by a tanh activation function, produces our intended output.

9.3.3 RESULTS AND EVALUATION

DCGANs are a powerful tool for generating high-quality images, but the quality of the generated images is often difficult to evaluate [14]. This is because different people may have different opin-ions on what constitutes a "good" image. In addition, the quality of a DCGAN generator model can vary depending on the parameters used to train it (Figure 9.4).

One way to evaluate a DCGAN generator model is to compare its output with that of other mod-els. This can be done by using an evaluation metric such as mean squared error. Another approach is to ask people to rate the quality of the generated images from best to worst. However, this approach can be subjective and unreliable.

A better way to evaluate DCGANs is by measuring how well they reproduce certain aspects of natural images. For example, one could measure how well a generator model reproduces edges or textures in an image. By doing this, one can get an idea of how good the generator model is at pro-ducing realistic-looking images.

In the appendix section of our chapter and Figures 9.5 and 9.6, we present our findings from the experiment. There is now no way to objectively assess the effectiveness of training and the relative or absolute quality of the model using loss alone since there is yet no objective loss function used to train

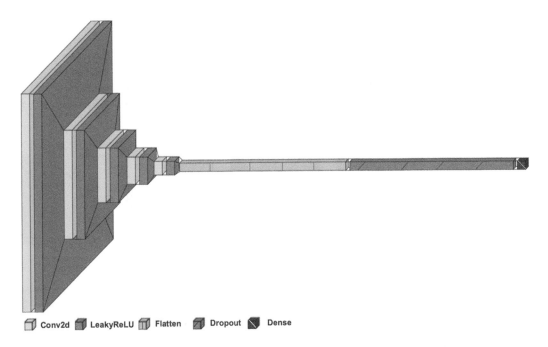

🔲 Conv2d 🔲 LeakyReLU 🔲 Flatten 🔲 Dropout ◨ Dense

FIGURE 9.4 Generator architecture.

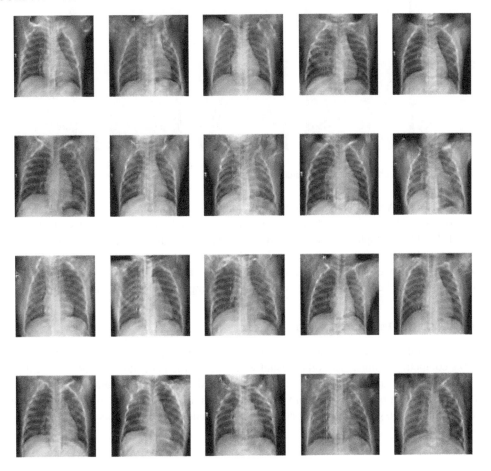

FIGURE 9.5 Normal chest images generated.

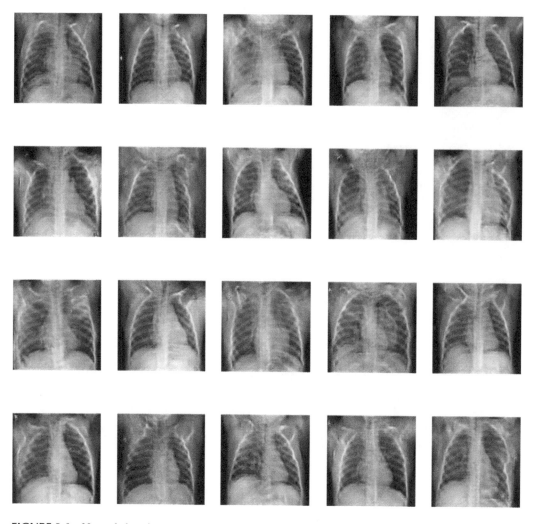

FIGURE 9.6   Normal chest images generated.

the generator of a GAN. Instead, a variety of qualitative and quantitative methods have been developed to assess a GAN model's performance based on the standard and variety of synthetic images generated [14]. In our chapter, we evaluate our model based on its fidelity; in other words, the fidelity of the produced samples is a measure of their realism. The better the quality, the more lifelike the visuals. Consider how dissimilar each false sample is from its nearest genuine sample. We hope that this information will be helpful to those who are considering similar projects in the future.

9.3.4   Research Opportunities and Challenges

In the field of machine learning, a GAN is a type of artificial neural network composed of two networks competing against each other, usually one called the generator and the other called the discriminator. The goal of GANs is to generate realistic data samples. Recently, GANs have been used to generate high-quality images and videos. However, it is still difficult to evaluate GAN models because there is no gold standard for comparison. In some cases, evaluators may be fooled by the generated data and give inaccurate results. Additionally, different evaluators may produce different results for the same GAN model due to their own biases. Therefore, it is important that we develop better ways to evaluate GAN models so that we can trust their outputs more accurately.

Transformers inside generator models are the future of our work. Transformers today have better results in the field of imaging, which is why we should focus on this technology for our future work.

The use of transformers inside generator models has already been proven to be more effective than traditional transformers. For example, a study by researchers at MIT showed that using transformer-based generators produced higher-quality images with less noise and distortion than traditional generator designs. This is due to the fact that transformer-based generators are able to produce more consistent current flow, which leads to clearer images.

Not only do transformer-based generators produce better-quality images, but they are also more efficient than traditional designs. A study by Sandia National Laboratories found that transformer-based generator systems were up to 30% more efficient than traditional systems! This means that we could save a significant amount of energy by using transformer-based generators in our equipment.

Based on these findings, it is clear that using transformers inside generator models is the future of our work. Transformers today have better results in the field of imaging, and they are also more efficient than traditional designs. We should focus on this technology for our future work in order to achieve superior results for our customers [11,12].

9.4 CONCLUSIONS

Medical imaging is an important field of study that helps medical professionals diagnose and treat patients. However, one challenge in medical imaging is the lack of data. This can be due to a number of factors, such as the need for specific equipment or expertise to generate images. To address this issue, we are developing an algorithm based on DCGANs that can generate high-quality images from limited data sets.

DCGANs have been found to be efficient at producing high-quality pictures from small input sets. They function by pitting a generator network and a discriminator network against one another in a neural network competition. As the discriminator network tries to tell the difference between produced and actual pictures, the generator network makes new images. Even with very little training data, DCGANs can learn how to produce realistic pictures by continuously modifying their parameters. We believe that our approach (with high accuracy and high image quality) could help improve medical imaging in fields where data is scarce. We are currently developing an algorithm to automate this process, depending on the input data set.

ACKNOWLEDGMENTS

I would like to thank my company, Procheck, and its executive director, Zineb Oudghiri, for supporting my research. The resources and the time they gave me allowed me to grow as a researcher. I am grateful for the opportunities they have provided me and look forward to continuing our partnership in the future.

APPENDIX

This section demonstrates the evolution of image quality in our algorithm training. As you can see, the image quality has improved significantly over time. We believe that this progress will continue as we continue to refine our algorithms (Figures 9.7–9.10).

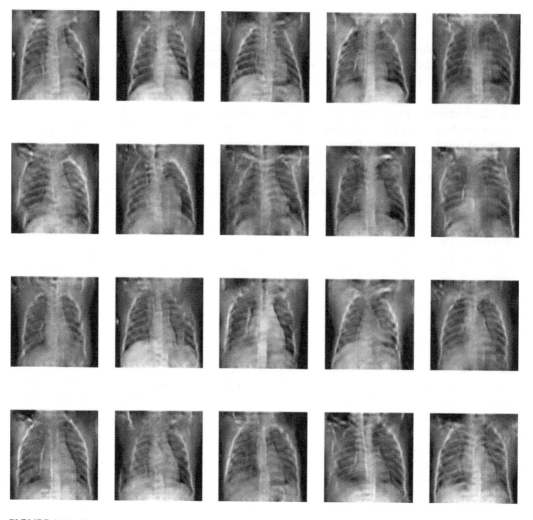

FIGURE 9.7   Normal chest images generated the first 100 epochs.

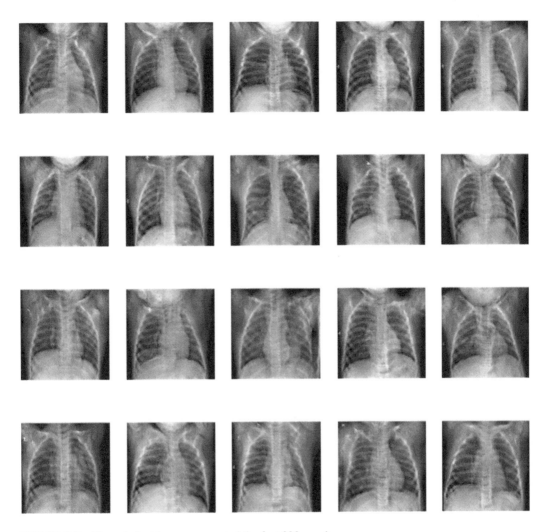

FIGURE 9.8 Normal chest images generated the first 200 epochs.

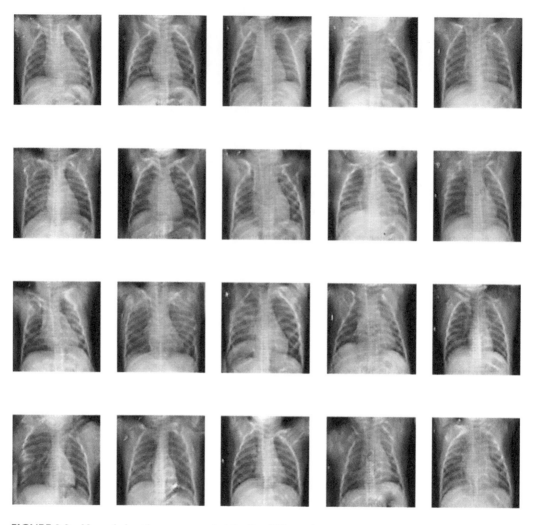

FIGURE 9.9   Normal chest images generated the first 300 epochs.

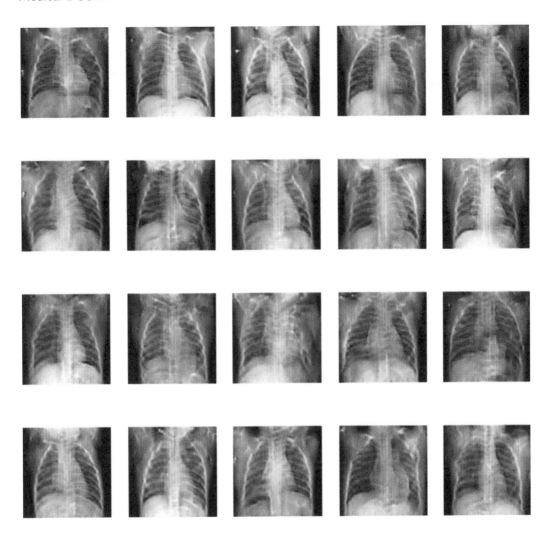

FIGURE 9.10 Normal chest images generated: final results.

REFERENCES

1. Sarker, I. H. (2021). Machine learning: Algorithms, real-world applications and research directions. *SN Computer Science*, 2(3), 1–21.
2. Zakaria, R., Abdelmajid, H., Zitouni, D. (2022). Deep learning in medical imaging: A review. *Applications of Machine Intelligence in Engineering*, 131–144. doi: 10.1201/9781003269793-15.
3. Aggarwal, R., Sounderajah, V., Martin, G., Ting, D. S., Karthikesalingam, A., King, D.,... Darzi, A. (2021). Diagnostic accuracy of deep learning in medical imaging: A systematic review and meta-analysis. *NPJ Digital Medicine*, 4(1), 1–23.
4. Zhou, S. K., Greenspan, H., Davatzikos, C., Duncan, J. S., Van Ginneken, B., Madabhushi, A., ..., Summers, R. M. (2021). A review of deep learning in medical imaging: Imaging traits, technology trends, case studies with progress highlights, and future promises. *Proceedings of the IEEE*, 109(5), 820–838.
5. Radford, A., Metz, L., Chintala, S. (2015). Unsupervised representation learning with deep convolutional generative Adversarial Networks.
6. Creswell, A., White, T., Dumoulin, V., Arulkumaran, K., Sengupta, B., Bharath, A. A. (2018). Generative adversarial networks: An overview. *IEEE Signal Processing Magazine*, 35(1), 53–65.
7. Goodfellow, I., Pouget-Abadie, J., Mirza, M., Xu, B., Warde-Farley, D., Ozair, S., ... Bengio, Y. (2014). Generative adversarial nets. *Advances in Neural Information Processing Systems*, 27.

8. Saxena, D., Cao, J. (2021). Generative adversarial networks (GANs) challenges, solutions, and future directions. *ACM Computing Surveys (CSUR)*, 54(3), 1–42.

9. Karras, T., Laine, S., Aila, T. (2019). A style-based generator architecture for generative adversarial networks. In *Proceedings of the IEEE/CVF Conference on Computer Vision and Pattern Recognition* (pp. 4401–4410).

10. Wang, K., Gou, C., Duan, Y., Lin, Y., Zheng, X., Wang, F. Y. (2017). Generative adversarial networks: Introduction and outlook. *IEEE/CAA Journal of Automatica Sinica*, 4(4), 588–598.

11. Shamshad, F., Khan, S., Zamir, S. W., Khan, M. H., Hayat, M., Khan, F. S., Fu, H. (2022). Transformers in medical imaging: A survey. arXiv preprint arXiv:2201.09873.

12. Parvaiz, A., Khalid, M. A., Zafar, R., Ameer, H., Ali, M., Fraz, M. M. (2022). Vision Transformers in Medical Computer Vision-A Contemplative Retrospection. arXiv preprint- arXiv:2203.15269.

13. Wang, X., Peng, Y., Lu, L., Lu, Z., Bagheri, M., Summers, R.M. ChestX-ray8: Hospital- scale Chest X-ray Database and Benchmarks on Weakly-Supervised Classifica- tion and Localization of Common Thorax Diseases. IEEE CVPR 2017, ChestX- ray8Hospital-ScaleChestCVPR2017 paper.pdf.

14. Borji, A. (2019). Pros and cons of gan evaluation measures. *Computer Vision and Image Understanding*, 179, 41–65.

Section 2

Financial Inclusion, Literacy, and Behavior

10 Financial Inclusion and Economic Growth in the Presence of a Cross-Sectional Dependence

Huynh Thi Thanh Truc

10.1 INTRODUCTION

In the socio-economic development of countries, it is indispensable to contribute to the financial market. In the traditional sense, financial markets play an important role in the allocation of capital in the economy in terms of the connection between saving and investment. Basically, the financial markets significantly develop, transaction costs in the economy can be reduced, and businesses can access low-cost loans, which therefore increases corporate financial efficiency and economic performance in the future.

The technological revolution has taken place rapidly and has made leaps and bounds since 2014 until now. The achievements of this revolution have brought about many changes in the banking system. Banks in countries can apply technology such as digital banking and e-banking, promotion of online payments, and ATMs. Moreover, thanks to the application of technology, banking products and services can serve the majority of people and businesses, even when they are located far from the city center. It is difficult for traditional banks to supply financial products and services in remote areas due to high transaction costs and low-population density. Traditionally, it is difficult to ensure the break-even point when setting up a branch there. However, developing digital and electronic banking can help banks provide products and services to remote areas without setting up branches, or it can be said that the financial system has been providing financial inclusion in the economy.

Research on the impact of financial inclusion on economic growth has been carried out through a number of recent studies, and all show that financial inclusion has a positive impact on economic growth, typically the study of Ong et al. (2023), Daud and Ahmad (2023), or Ozili et al. (2022). However, Dabla-Norris et al. (2021) argued that there is a trade-off between financial inclusion, economic growth, and income distribution within the country, which shows that financial inclusion does not always have a positive impact on growth, and this is also confirmed by the study of Ozili et al. (2022). That is the main reason for our study. Moreover, countries in the region are often interdependent, so a change in financial inclusion policy or growth policy in this country can have an impact on the others, which makes this study unique.

The countries of Southeast Asia have a long tradition of cooperation, complementarity, and interdependence in economic, cultural, political, social, and investment fields. Studying the influence of financial inclusion on economic growth in the context of cross-sectional dependence, especially in the case of Southeast Asia, lacks previous evidence. That is why we selected this topic for this analysis.

DOI: 10.1201/9781032667478-12

In addition to the introduction, the rest of the study is presented as follows: Part 2 is a system of previous studies. Section 10.3 collects data and research methods. Sections 10.4 and 10.5 discuss the results of the study and the general conclusions of the study.

10.2   LITERATURE REVIEW

Financial inclusion is the ability to provide people and businesses with affordable access to financial products and services. As the banking system develops, the application of technology in banks can bring more experiences to users at low costs and thus increase the opportunities for consumers to access financial products and services. Due to economic needs, people and businesses always need the banking system to be able to conduct transactions, make payments, borrow capital, or save money. With the development of comprehensive finance, it is possible to gain access to more financial products and services.

Research by Dabla-Norris et al. (2021) argued that there is a trade-off between financial inclusion, economic growth, and income distribution within the country. This means that financial inclusion has not really brought about immediate economic benefits, but in the long run, it may have a positive impact on growth. Therefore, Dabla-Norris et al. (2021) also emphasized the comprehensive impact on the economy influenced by other factors and interwoven relationships that play a major role in promoting economic growth. Complementing this statement, Ong et al. (2023) argued that financial inclusion can lead to higher economic growth. In particular, the digitization of business processes can promote financial inclusion in ASEAN countries. Digitize business processes modeled with fixed high-speed mobile, mobile, and broadband subscriptions. The study also suggested that digitization has a significant effect on private enterprises with domestic credit provided by commercial banks.

Extending the study to a large sample, Daud and Ahmad (2023) studied panel data across 84 countries and confirmed that financial inclusion has a positive impact on economic growth, especially digital technology, which has the potential to increase this effect. It shows that financial inclusion is supported by digital technology and promotes economic growth, thus improving the financial ecosystem only through investment in digital technology infrastructure. Moreover, it can be said that the role of information and communication technology (ICT) has also become more important in the current technological revolution. Wang et al. (2023) argued that ICT plays an important role in promoting financial inclusion to improve inclusive growth in the current period. Research conducted in Africa and the GMM regression method for the period 2000–2020 suggest that ICT and financial inclusion come together to improve inclusive growth in each country. Therefore, governments should set up measures to develop ICT, open up trade, and attract international capital flows to improve financial inclusion, thereby creating momentum for inclusive growth in the host economy.

The study by Ozili et al. (2022) stated that most studies show a positive effect of financial inclusion, and very few studies show a negative effect of this relationship. In the most common sense, financial inclusion positively impacts growth due to greater access to financial products and services offered by financial intermediaries, which translates into economic growth. However, if financial inclusion develops only in quantity and there is little improvement in quality, it is unlikely to benefit the economy in the long term and may have negative effects. Furthermore, Ifediora et al. (2022) studied 22 sub-Saharan African countries between 2012 and 2018 and used the GMM method, suggesting that the availability aspect of financial inclusion, the penetration dimension, and the aggregate financial inclusion have a significant and positive impact on economic growth. In addition, the number of bank branches and ATMs has a positive effect on economic growth, reflecting increased coverage of financial products and services that can drive economic growth. In addition, the study also found that the mobile money index weakens economic growth while mobile money accounts and mobile money transactions promote economic growth, but this effect is weak. Thereby, it is suggested that African countries should pursue policies to improve financial literacy education so that people better understand the benefits of using banking services and develop financial inclusion.

10.3 DATA COLLECTION AND METHODOLOGY

10.3.1 DATA COLLECTION

In this study, we collected data from each country's statistical bureau. The data were collected between 2002 and 2019 in the ASEAN6 countries in Southeast Asia, including Indonesia, Vietnam, the Philippines, Thailand, Malaysia, and Singapore.

10.3.2 METHODOLOGY

We use the original model from the study of Ifediora et al. (2022) and make adjustments to suit the conditions of this study. The regression equation is as follows:

$$\text{GROWTH}_{it} = \beta_0 + \beta_1 \text{INCLUSION}_{it} + \beta_2 FD_{it} + \beta_3 \text{SPENDING}_{it} + \beta_4 \text{STOCK}_{it} + u_{it}$$

where GROWTH_{it} is the economic growth rate in country i, time t; INCLUSION_{it} is the natural logarithm of the number of ATMs per 100,000 inhabitants in country i, time t, which is a representative parameter for the development of financial inclusion; FD_{it} is the financial development of country i, time t, as measured by the broad money relative to GDP; SPENDING_{it} is government spending in country i, time t; STOCK_{it} is the stock index showing the stock market of each country, measured through the log of the stock market index in country i, time t; β_0 is the intercept; β_i is the estimated coefficients; and u_{it} is the white noise.

10.4 RESULTS AND DISCUSSIONS

10.4.1 DESCRIPTIVE STATISTICS

Table 10.1 presents the results of the descriptive statistics for the variables in the model. For economic growth, this index reaches an average value of 5.30%, which is a relatively good growth rate. Economic growth in some countries even reaches 14.52%, which is very high. In terms of access to financial inclusion, the average number of ATMs reached 42.4 machines per 100,000 people and improved continuously between 2002 and 2019, reflecting people's ability to access financial products and services when their population is increasing.

TABLE 10.1
Descriptive Statistics

Variable	Mean	Std. Dev	Min	Max
GROWTH	5.306823	2.131159	−1.513529	14.52564
INCLUSION	42.47762	29.03265	1.409589	117.7916
FD	96.39653	36.41437	36.00172	164.8682
SPENDING	10.60218	2.90556	5.465202	17.12117
STOCK	2297.078	1960.273	153.1892	8558.42

Source: Author's analysis.

TABLE 10.2
Correlation Analysis

Variable	GROWTH	INCLUSION	FD	GOV	STOCK
GROWTH	1.0000				
INCLUSION	−0.3384	1.0000			
FD	−0.1439	0.4901	1.0000		
SPENDING	−0.3800	0.6329	0.2460	1.0000	
STOCK	0.0511	0.2648	−0.4163	0.0764	1.0000

Source: Author's analysis.

TABLE 10.3
VIF Analysis

Variable	VIF	1/VIF
INCLUSION	3.26	0.306495
FD	2.41	0.415689
STOCK	1.98	0.505720
SPENDING	1.79	0.558931
Mean VIF	2.36	

Source: Author's analysis.

Table 10.2 analyzes the correlation between the variables in the estimated model. Research results show that all pairs of variables have low correlation and are all less than 0.8, which reflects the unlikely possibility of multicollinearity. Table 10.3 also confirms again that there is no possibility of multicollinearity when the VIF coefficient is small.

10.4.2 Estimated Results

In this study, we used panel data estimation, including pooled OLS, FEM, and REM, and evaluated the dependence relationship between countries through panel-corrected standard errors regression. Based on the F-test, it shows that OLS pooled regression has better results than REM. The Hausman test shows that FEM regression is better than REM, so it can be confirmed that pooled OLS regression can show good results. Through testing of autocorrelation and heteroskedasticity, it shows that the pooled OLS model does not have a defect problem. The results are shown in Table 10.4.

Research results show that:

Financial inclusion has not had a positive impact on economic growth. It means that countries in the region that have developed financial inclusion through the expansion of ATM locations have not yet brought about economic growth. This study is set out in the context of interdependence between countries, but the research results still show that there is no positive impact between financial inclusion and economic growth. Ozili et al. (2022) argued that most studies show a positive effect of financial inclusion on growth, but there are also some studies showing a negative effect of this relationship, because financial inclusion that only develops in terms of increasing access to financial products and services without their quality may only have a positive impact in the short term, but in the long run, the positive impact is reduced and may have an adverse effect on the economy.

TABLE 10.4

Estimated Results–Dependent Variable: GDP

Variable	Pooled OLS	FEM	REM	PCSE
INCLUSION	-2.1226^b	-3.1246^b	-2.1226^b	-2.1226^c
	(0.046)	(0.036)	(0.043)	(0.000)
FD	0.0109	-0.0277	0.0109	0.0109
	(0.228)	(0.166)	(0.225)	(0.112)
SPENDING	-0.1676^a	-0.5913^b	-0.1676^a	-0.1676^c
	(0.084)	(0.028)	(0.081)	(0.006)
STOCK	1.4242^a	6.2694^c	1.4242^a	1.4242^b
	(0.083)	(0.001)	(0.079)	(0.029)
_cons	4.5622^a	-1.4540	4.5622^a	4.5622^a
	(0.095)	(0.739)	(0.091)	(0.065)
F-test	$F(5, 88)=1.90$			
	Prob>F=0.1022			
Hausman test		chi2(4)=9.54		
		Prob>chi2=0.0490		
Autocorrelation (tự tương quan)	H0: no first-order autocorrelation $F(1, 5)=1.086$ Prob>F=0.3451			
Breusch-Pagan / Cook-Weisberg test for heteroskedasticity	Ho: Constant variance Variables: fitted values of GROWTH chi2(1)=2.45 Prob>chi2=0.1175			

Note: [a, b, c] represent the significance level of 10%, 5% and 1%, respectively.
Source: Author's analysis.

The research results also show that government spending has a negative impact on economic growth, thereby reflecting that countries in the region increase government spending that may not bring benefits to economic development. It can be explained that an increase in government spending may lead to a decrease in the private sector's access to capital as interest rates rise and the private sector's inability to access capital increases, thus not increasing economic growth. This requires countries to make appropriate their government spending to ensure economic growth and private economic development. However, the study also confirmed the positive impact of the stock market on economic growth. The expansion of the stock market makes it possible for businesses to raise long-term capital to meet their investment and development needs and ultimately improve economic growth.

10.5 CONCLUSIONS

The development of the fourth industrial revolution has brought many great benefits to economies. The countries of Southeast Asia have a long tradition of cooperation and interdependence in economic, cultural, political, social, and investment fields. Studying the effect of financial inclusion on economic growth in the context of the interdependence effect among countries in the same region, the research results show that financial inclusion has not had a positive impact on economic growth in Southeast Asia. The study also confirms that stock market development has a positive effect on growth, but an increase in government spending has a negative impact on growth.

REFERENCES

Dabla-Norris, E., Ji, Y., Townsend, R. M., & Filiz Unsal, D. (2021). Distinguishing constraints on financial inclusion and their impact on GDP, TFP, and the distribution of income. *Journal of Monetary Economics*, *117*, 1–18. https://doi.org/10.1016/j.jmoneco.2020.01.003.

Daud, S. N. M., & Ahmad, A. H. (2023). Financial inclusion, economic growth and the role of digital technology. *Finance Research Letters*, *53*, 103602. https://doi.org/10.1016/j.frl.2022.103602.

Ifediora, C., Offor, K. O., Eze, E. F., Takon, S. M., Ageme, A. E., Ibe, G. I., & Onwumere, J. U. J. (2022). Financial inclusion and its impact on economic growth: Empirical evidence from sub-Saharan Africa. *Cogent Economics & Finance*, *10*(1), 2060551. https://doi.org/10.1080/23322039.2022.2060551.

Ong, H.-B., Wasiuzzaman, S., Chong, L.-L., & Choon, S.-W. (2023). Digitalisation and financial inclusion of lower middle-income ASEAN. *Heliyon*, *9*(2), e13347. https://doi.org/10.1016/j.heliyon.2023.e13347.

Ozili, P. K., Ademiju, A., & Rachid, S. (2022). Impact of financial inclusion on economic growth: review of existing literature and directions for future research. *International Journal of Social Economics*, *ahead-of-print*(ahead-of-print). https://doi.org/10.1108/IJSE-05-2022-0339.

Wang, W., Ning, Z., Shu, Y., Riti, M.-K. J., & Riti, J. S. (2023). ICT interaction with trade, FDI and financial inclusion on inclusive growth in top African nations ranked by ICT development. *Telecommunications Policy*, *47*(4), 102490. https://doi.org/10.1016/j.telpol.2023.102490.

11 The Role of Tax Policy in Stimulating and Encouraging Investment in Jordan

Yaser Arabyat and Hussam-eldin Daoud

11.1 INTRODUCTION

The study, which depends on a descriptive research approach and an analytical fieldwork, aims to learn about the role of tax policy and its effects on investment in Jordan. A comprehensive questionnaire survey was carried out for all members of the sample community, which consists of foreign and domestic investors who benefit from the investment promotion law of the Jordan Investment Commission.

A random sample was chosen from among (109) participants in the sample community. The results of the study showed that despite the positive impacts on the following sectors (tourism, transport, and communication), it was found that the tax incentives for these three sectors are not efficient. This is because of the relative importance of incentives that are given to these sectors, whereas they are lacking in other sectors, including the industrial sector. The study also indicated that the investor is satisfied with tax incentives, tax exemptions, and their decreasing rate, but he/she is not satisfied with the ease of practical enforcement or the continuous modification of tax laws.

The study reached many recommendations, such as regularly reviewing of the regulations and legislation that organize the investment process in Jordan, working to simplify and standardize the investment procedures, and doing similar studies to include other environments and societies. In addition, it recommends searching for and studying the reasons that affected investment in some sectors, especially the industrial sector. Accordingly, the researcher recommends creating a new comprehensive tax law for this sector.

11.1.1 STUDY PROBLEMS

With the declining role of grants and foreign assistance and the accompanying economic, social, and political consequences, it has affected the Jordanian economy, creating the challenge of adopting the necessary policies to create an appropriate environment for investment by activating the role of tax policy as one of the most important instruments of the state's fiscal policy under the policy of incentives, tax exemptions, and their rates of decline, which have become factors influencing the attraction and stimulation of national and foreign investments in order to stimulate economic growth and development. It was therefore necessary to demonstrate the role that tax policy had played with its benefits, incentives, and tax exemptions in attracting and encouraging national and foreign investment through Jordan's investment promotion laws as well as amendments to those laws. The problem of the study can therefore be summed up by the following question:

DOI: 10.1201/9781032667478-13

Has Jordan's tax policy played its role in attracting and encouraging national and foreign investment?

11.1.2   THE IMPORTANCE OF STUDY

Taxes affect the ability and readiness of different individuals and entities to form savings, thus investing in various and different areas. This capacity depends on the tax policy, types and rates of taxes, the nature of the sectors on which different taxes are levied alone, as well as the nature of the state economy, the types of investment activities in circulation, the progress or backwardness of this economy, and many of the basic determinants with which tax policy deals, which impose themselves on the relationship between this system and domestic investment.

The Jordanian economy has witnessed a series of different economic phenomena accumulated from economic stagnation and decline in investment. Jordan, like other developing countries, is working in a race to stimulate national investment and attract foreign investors by providing many advantages and providing the investment climate with its legal, political, social, and economic elements, so this study seeks to highlight the role of tax policy and its impact on domestic and foreign investment and the effectiveness of tax reforms in Jordan in stimulating and localizing national and foreign companies. In addition, this topic was chosen because of the importance of the role that tax policy can play as an encouraging tool for national and foreign investments in line with Jordan's economic development plans, which have become a tool for providing incentives and benefits that stimulate investment.

11.1.3   STUDY OBJECTIVES

The study aims to identify the tax policy adopted by Jordanian governments in investment laws to further increase economic growth and attract investment in various economic fields and activities. The objectives of the study can therefore be summarized by the following points:

1. To see how tax policy variables affect investment performance in Jordan at the macro and sectoral level from the investor point of view.
2. To clarify the variables resulting from the change in tax policy laws used to stimulate investment.
3. To identify the appropriateness of the tax policies applied to economic conditions and identify the problems they face in terms of planning and implementation.

11.2   TAX POLICY AND ITS IMPACT ON INVESTMENT DECISIONS

The concept of tax policy has been associated with the development of societies and the complexity of their economic, social and political construction, so taxation has become one of the most important instruments of fiscal policy adopted by states to fill the gap of local resources and the needs of citizens, and an indicator in the expression of economic transformations and their political and social effects on the state and society, so taxes occupied a distinct position in financial studies, as the tax and Investment and its decisions have become important topics in the priorities of economic, financial, banking and administrative studies that are concerned with structural developments in developed societies, accompanied by a similar development in the field of investment study beside it in kind and financial, and after the sources of investment financing became not limited to local, global and traditional sources of investment finance. It has evolved and developed tools, mechanisms, and an information and communication network in global markets that have provided investment opportunities and benefits that were not previously available (Momani, 2002).

First: The impact of tax policy on investment decisions (Al-Hanawi, 2003):

The foundation's choice to invest and select among investment projects relies on various criteria, with the most frequently utilized being the present net value standard, the payback period for invested capital, and the internal rate of return criterion. Tax policy directly influences these standards in the following ways:

defined as the period of time required to recover the initial cost of investment from the net cash flow generated by it, and the investment project acquires greater importance and lower risk whenever this period is short, which is what the tax policy seeks to reach by reducing tax rates, granting exemptions that would raise the value of cash flows for investment, thereby recovering the initial investment cost in As little as possible, this encourages investment.

1. The impact of tax policy on the internal rate of profits: The internal rate of profit criterion is one of the most important criteria used in investment decision-making, defined as "the rate at which total cash flows enter equal to total cash flows out of domestic values".

   Corporate profits taxes reduce the value of current cash flows, which makes the internal rate of profits decline, thereby negatively affecting investment. In order to raise the value of the internal rate of investment, reductions and tax exemptions are granted in relation to corporate profits taxes within the state's expectations to invest in an activity, thereby increasing current cash flows and encouraging desirable investment.

2. The role of financial incentives in capital-exporting countries to invest in developing countries: Capital-exporting countries may contribute to encouraging their citizens to invest their money in developing countries for what they see as their interest, so capital-exporting countries may give their citizens who invest their money in developing countries tax incentives and benefits, such as being completely or partially tax-free from profits to their country, or taking into account in one way or another somehow. They paid him taxes abroad in order to avoid double taxation, and there is no doubt that when deciding to invest outside the borders of his country, the investor takes into account the tax burden that he will bear in his country compared to the tax benefits he can enjoy in the host country for his investment, he may prefer not to enjoy these tax benefits if the cost of investing without them is lower than his cost if he is exempted from the taxes of the host state, and the investor even if he sees that his profits abroad will be taxed in his country, which may lead him to refrain from investing abroad, so the financial incentives established in capital-exporting countries to invest in developing countries are factors influencing the investment decision (Mabrouk, 2007).

Second: The impact of financial incentives on encouraging investment decisions

The polarized state of investment in general aims to provide tax incentives and benefits to achieve goals in multiple dimensions, perhaps the most important of which is (Mabrouk, 2007):

Encouraging investors to set up investment projects of economic feasibility in a country or to continue and expand the list.

Urge foreign investors to reinvest returns (profits) in the State rather than offshore them.

3. Encouraging foreign businesses to operate in accordance with the requirements of the National Economic Development Plan, whether by directing them to certain sectors of the national economy or to specific areas of the State territory as required by the Economic Development Plan.

   Financial incentives may be aimed at encouraging investment projects capable of absorbing a large amount of national employment, resulting in increased technical skills for such employment, the revitalization of domestic markets, and the reduction of unemployment.

In order for tax incentives and benefits to have implications for encouraging investment, the legal regulation that determines such incentives and benefits should take into
account the following:

1. Prior to determining tax incentives and benefits, the polarized State must conduct
   multiple studies to demonstrate the relationship between the cost and benefit of such
   incentives on the one hand and the stages and requirements of economic development
   on the other. The introduction of random tax credits without prior consideration has
   a bad impact on the State in its quest to attract foreign capital, as it creates a sense of
   un-seriousness in the investor.
2. The State can use financial incentives as a tool to channel foreign investment to important priority economic sectors in the national development plan, generally by granting
   greater financial exemptions and incentives to investment projects that go to those economic sectors, and preferential tax treatment can play an influential role in promoting
   export industries by providing distinctive facilities for their inputs and outputs.

In light of the above, it is clear that tax policy is very important in the investment process, as investment is the vital and effective element of achieving the process of economic and social development,
and if we take into account that any initial increase in investment will lead to increases in income
through the investment multiplier, any increase in income must go partly to increase investment
through the so-called accelerator.

It should be noted that tax is not only one of the main sources of funding but also an effective
strategic means that enables the State to intervene positively in economic, social, and political life,
where tax-derived funds contribute to the financing of state expenditures for all sectors such as
health, education, and other services, upgrading and strengthening infrastructure, stimulating economic policies such as support for certain economic goods, sectors, or activities, or improving individual income, social welfare programs, social security, and insurance, and providing employment
opportunities, which will lead to increased investments in the country (Obayami, 2001).

To achieve these data and reach the strategic objectives desired in stimulating investment,
the state needs an efficient and professional tax system with sound and transparent systems and
procedures, accuracy in the data adopted during tax collection to see the results and assess the
success or failure of tax policy, or its compatibility with the needs of development plans, and
the ability of taxpayers to deal with the vocabulary of tax laws and adapt their institutions and
projects to the benefit of society, hence the direct relationship between tax policy and investment
(Hohots, 2000).

11.2.1 State Economic Objectives in Light of Tax Exemptions

1. Achieving economic growth: The rate of economic growth depends on the amount of
   investment and its distribution among the different branches of economic activity. The
   tax affects the tendency to invest; when a low rate is taxed on an economic sector or the
   exemption is wide, financiers direct their investments toward the sector because it is more
   profitable (Ezzedine, 2010).
2. Using tax policy to encourage certain productive activities: Each country in the world
   seeks to achieve different economic objectives using taxes, depending on its geographical
   location. There are countries specific to a particular sector, such as industrial, agricultural,
   and tourism countries, where they exempt the privileged sector from any tax, either in full
   or in part, and the State, through tax policy, encourages investment in these productive
   activities, and the State may seek to attract foreign capital for investment through incentives and exemptions. Tax that encourages investment (Shamia, 1997).
3. The use of tax policy to prevent the concentration of economic projects: the phenomenon
   of concentration is a feature of the current era, and this phenomenon is addressed using

tax policy by increasing the tax rate in the stages of production of the commodity for the company heading toward concentration, which increases its expenditures and therefore the lack of direction toward concentration (Khatib, 1997).

4. Incentives and their impact on savings and investment: this is done through investment-encouraging tax exemptions and no tax on development bonds issued by the State or joint stock companies (Al Mawla).

5. Using tax policy to finance development processes: Tax policy is an important tool to serve development purposes, where tax is an important item of public revenue in the state budget, which allocates its proceeds to spending on development purposes, and tax policy is an important way to help stimulate the company's ability to increase output and investment and lead to increased employment opportunities (Ionel Report, 2013).

6. Use of tax policy as a tool to address economic underdevelopment: this is done through the development of a comprehensive economic plan using economic policy instruments to address this phenomenon, and one of the tools used by the State is tax policy, which is an important part of fiscal policy since it is known that the tax structure is influenced by the economic structure of the State, especially in the manner of development, but the tax can also affect the economic structure by using a successful tax policy and cooperating with all tools for development and push forward economic growth (Hamdallah, 2005).

Encouraging investment is a major goal and one of the priorities of the economic work of all successive Jordanian governments, without exception. In order to increase the pace of economic growth, many legislation and laws regulating the FDI process have been enacted in proportion to their objectives to support the national economy, provide it with foreign currency, create jobs, and achieve comprehensive economic development. Jordan has carried out numerous economic, administrative, and legislative reforms to remove restrictions on the movement of its foreign trade, the movement of capital and investment, and to enter global markets. It has succeeded in achieving good growth rates as a result of these reforms, recognizing the importance of investments as a key tool for correction and economic openness. In order to achieve the desired objectives of these investments, the Jordanian government has taken many corrective measures by enacting laws and legislation. This has been done with the aim of providing the right investment climate to attract investments, especially in the context of competition posed by neighboring countries (Al-Maliki, 2001).

There are many previous studies that have talked about the subject of tax policy and its impact on investment, including the Hanish study (1992). The authors studied and analyzed the role of tax in promoting private national sector investments in Algeria. The study found that social development is linked to economic development, which is linked to investments and the development and tax it achieves for the State, and considers that the tax has two effects: a positive impact for the institution and a negative impact on states, the institution can benefit from these funds payable, on the other hand he believes that the tax has two effects: a positive impact on the institution and a negative impact on states. The State has lost amounts of money that should have been seized as a result of tax evasion. The State's tax intervention takes many forms, including reduction, increase, or exemption; flexibility in treatment is intended to encourage or limit certain economic activities; and the tax-investment relationship is an old complementary one.

The Hijazin study (1996) aimed to identify Jordanian IPAs and their role in attracting foreign direct investment. The study emphasized the existence of security and political stability in Jordan, that Jordanian society welcomes direct investments and stressed that the Jordanian economic environment needs great efforts to raise its level in accordance with the needs of the investor. This may have prompted Momani (2002) to study the role of incentives and tax measures in encouraging investment, where the researcher presented taxes and investment incentives addressing investment promotion institutions and administrative procedures in investment, and concluded that the Jordanian government is improving the working environment in order to encourage investment at the local and security levels. Addressing the government's great efforts to create a suitable and competitive investment climate at the regional level by introducing new investment policies with many advantages and competing with international standards satisfied some of the hypotheses, which serves his study explaining the impact of income tax on increasing private investment and GDP, which has not been fully demonstrated.

The Msimi study (2006) aimed to identify the tax policy in Palestine and its role in the development of the Palestinian economy, finding that tax exemptions play an important role in achieving economic objectives, that they increase per capita income and thus increase the ability to save and increase what is allocated to consumption, and that the privileges granted in the Investment Promotion Act encouraged enterprises coming from abroad while internal projects did not benefit from these privileges, which in turn will affect the Palestinian economy. In the event of the exit of projects coming from abroad and their withdrawal from the markets, especially under the current economic conditions, the study also concluded that investment incentives, if exploited, would support the national economy and that indirect taxes in the financing of public revenues contribute to GDP and national income by a greater proportion of direct taxes.

The Voget study (2010) also examined the impact of taxes on FDI, finding that tax cuts are an option to attract more investment by multinational companies, so in order to increase foreign investment, taxes, particularly corporate taxes, must be reduced.

This may be why it is believed that a country's economic development and tax level affect the level of investments, particularly foreign direct investment, and that there is a strong link between corporate income tax revenues and FDI, as well as other factors such as the market, production prices, labor prices, and transport prices affecting FDI (Nistor and Pzun, 2013).

In the same vein, Rahal (2011) analyzed the impact of the Temporary Income Tax Act (28) of 2009 in encouraging investment in Jordan, finding that there was no effect of the Temporary Income Tax Act on increased investment in Jordan, found that there was no effect of the Temporary Income Tax Act in terms of incentives in the Investment Promotion Act in Jordan, and recommended that tax procedures be simplified and as far from complications as possible to contribute to encouraging investment in Jordan.

The Habib study (2012) also analyzed tax policy and its impact on Algeria's economy and found that taxes in Algeria were an essential means for the government to encourage and sustain investment, especially in recessions and economic crises, as there could be no economic reform and investment without a modern tax system and in keeping with global economic development. The study also emphasized that the tax is a stable and effective source of government revenue to reduce reliance on oil resources and fluctuations in the oil market, and that the tax system is an effective means for the State.

One study that emphasized the impact of taxes on investment was a study (Jacob et al., 2013), which analyzed the effects of consumer taxes on corporate investment, finding that investment companies respond to changes in consumption taxes, which is economically important and stronger for companies with less flexible supply, more flexible demand, and less uncertainty, as increased consumption taxes lead to lower investments by companies with less flexible supply and more flexible demand.

The Mahlawi Study (2015) aimed to demonstrate the role of tax policy in stimulating investment, and it became clear that there was a positive impact of the Tax Act (8) of 1997 on the total investments made on GDP for all sectors. It was found in the multiple linear equation that there was a positive impact on tax incentives on both the industrial and tourism sectors, while there was a negative impact on the agriculture, construction, and communications sectors.

Suleiman's study (2016) analyzed the impact of indirect tax exemptions on investment in Sudan, where the study found a correlation between the size of tax exemptions and the size of investment and that tax exemptions are used as a tool to direct investments toward a particular sector, projects, or geographical area, and the study emphasized that double taxation leads to the flight of domestic and foreign capital.

The tax system and its impact on encouraging investment in Jordan Jordan's tax policy is a plan to achieve objectives that represent the state's policy and therefore the conduct of the State in accordance with a plan it sets out to pursue in its tax affairs in order to achieve its objectives. The most important investment incentives, including tax exemptions, are derived from Jordan's investment promotion laws:

> Jordan's main tax law is income tax law No. 57 of 1985, and many amendments have been made to the law since it came into force. According to the law, income generated or considered to be originating in Jordan (Jordan's source of income) is taxable, and in order to determine the taxable income of the taxpayer, all expenses are deducted entirely and exclusively or incurred in income production during the year. The company's expenses for training, marketing, research, and development are tax-free.
>
> (Momani, 2002)

Jordan has recognized the need for encouraging business frameworks with strong incentives for entrepreneurial initiatives, and His Majesty King Abdullah II has identified increased investment as a national goal as FDI, together with domestic investment, is the surest engine for sustainable development.

Through the establishment of the Investment Promotion Corporation in 1995 and other practical steps aimed at strengthening the investment environment and opening up Jordan's economy, it liberalized trade and foreign exchange and implemented ambitious privatization programs as part of its quest to integrate into the global economy.

The IPA offers attractive and generous incentives to Jordanian and non-Jordanian investors alike, namely duty-free exemptions, income and social services exemptions, and unrestricted transfers of capital and profits.

Accordingly, a series of laws and legislation have been passed in Jordan to encourage investment since the early fifties, which have been keeping pace with the economic and social development of the country, including these laws and legislation.

1. Investment Promotion and Industry Guidance Act No. 27 of 1955.
2. Foreign Capital Promotion and Employment Act No. 28 of 1958.
3. Investment Promotion Act No. 1 of 1967.
4. Temporary Investment Promotion Act No. 53 of 1973.
5. Laws amended by Law No. 53 of 1995.
6. Investment Promotion Act No. 11 of 1987.
7. Investment Promotion Act No. 16 of 1995.
8. Arab and Foreign Investment Promotion Act No. 27 of 1995.
9. Investment Promotion Act No. 68 of 2003.
10. Investment Promotion Act No. 30 of 2014.

The above-mentioned laws included a large number of facilities for investors to work to provide the right investment climate, unhindered. These include tax and customs exemptions, tax exemptions that also concern income, profits, social service taxes, buildings and land, incentives for the transfer of capital, salaries and wages for foreign workers, and capital returns. However, these laws met the needs of investors only limitedly, and foreign investment was not at a safe level. A new law had to meet the needs of investors more and have a significant impact on encouraging foreign investment, and accordingly, the Investment Promotion Act No. 16 (1995) and the Arab and Foreign Investment Promotion Act No. 27 (1995) were enacted as an attempt to fill and cover the weaknesses of previous laws, as the new law granted larger customs and tax exemptions for longer periods of time and gave the investor the right to own the project and manage it as it wanted without restriction or a condition.

11.2.2 BENEFITS AND INCENTIVES PROVIDED BY JORDANIAN INVESTMENT LAWS: 1995, 2003, AND 2014

Interest in taxes and their role in economic, social, and political life began to increase their impact on the process of economic growth and the importance of taxation as an encouraging tool for the growth of the private industrial sector and the revitalization of the national economy as a whole by rationalizing tax exemptions and encouraging investment through its financial incentives (Qasim, 2003). These financial incentives are exemptions and facilities granted in accordance with the provisions of the law in order to stimulate, encourage investments, attract, employ, and direct capital in various sectors of the national economy in order to meet the requirements of the construction plans in those countries to reach their economic objectives, so that tax policy becomes a financial economic means of implementing a country's policy aimed at encouraging productive activity (Yadk, 2006).

Jordan Investment Promotion Act No. 16 and its amendments for the year 2003 are an appropriate legislative framework for attracting foreign investment and stimulating domestic investment, as they are competitive with the advantages, incentives, and guarantees of investment laws at the regional level, including those benefits and incentives related to:

1. Customs exemptions

   Fixed assets (machinery, equipment, machinery, and the number allocated for use exclusively in the project, furniture, furniture, and supplies for hotels and hospitals) are exempt from customs duties and taxes.

   Spare parts are exempt from customs duties but do not exceed 15% of the value of the fixed assets required by these parts.

   The project's fixed assets are exempt from customs duties and taxes for the purposes of expansion, modernization, or renewal if this leads to an increase of at least 25% of the project's production capacity.

   Hotel and hospital projects grant additional fee and tax exemptions for purchases of furniture, furnishings, and supplies for modernization and renovation purposes at least every seven years.

   The increase in the value of fixed assets imported for the project account is exempt from duties and taxes if the increase results from the high prices of those assets in the country of origin, the high freight rate, or a change in the transfer price.

2. Tax facilities

   Income tax is met from taxable income for companies operating within the mining industry, hotels, hospitals, transportation, and construction contracting at 15%, 35% for companies within the banking and financial sectors, and 25% for other companies, and income tax for agricultural projects is zero.

   The Kingdom has been divided into three development zones (A, B, and C) according to the degree of economic development so that investment projects within the sectors stipulated in the Investment Promotion Act in any of these areas enjoy exemptions from income and social services taxes in their ratios described below for ten years from the start-up date of service projects or the actual production of industrial projects as described in Table 11.1.

The period of exemptions shall be based on the committee's decision (ten years), starting from the date of commencement of service projects or from the date of commencement of production for industrial projects. The Commission grants additional exemption if the project has been expanded, developed, or modernized and has increased its production capacity. The additional

TABLE 11.1

Development Zones and Exemptions from Income and Social Services Taxes

Category	Area	Reduction Time (years)	Reduction Rate (%)
A	The following brigades (Northern Valleys, Deir Alla, Southern Shona, Southern Valleys, Ruweished, Northern Badia, North West Badia, and Giza, with the exception of the borders of the Municipality of Giza Al-Jadida, which is revered except for the borders of the Municipality of Al-Muqar), include the Aqaba governorate, except for the Aqaba special economic zone, and the District of Azraq.	20	100
B	Includes provinces (Maan, Tafila, Karak, and Ajloun).	20	80
C	The provinces include Jerash, Mafraq, and Irbid, except for the borders of the greater Irbid municipality.	20	60
D	The provinces (the capital except the municipality of Amman, Zarqa, with the exception of the borders of Zarqa municipality, Madaba, and Balqa) include the borders of the municipality of Al-Rsaifeh.	20	40

exemption period for one year for each increase in production capacity is at least 25% and for a maximum of four years.

1. Investment guarantees

   The Non-Jordanian Investment Promotion Act treats a Jordanian investor as granting him the right to invest in the Kingdom by owning, participating in, or contributing to any economic project in accordance with the ratios specified in the Non-Jordanian Investment Regulation System No. (54) of 2000 and to have the freedom to fully own projects within the industry, information technology, agriculture, hotels, hospitals, shipping and railways, amusement and tourist cities, conference and exhibition centers, and services for the extraction, transportation, and distribution of water, gas, and derivatives. Oil pipelines, as well as many other economic sectors.

   The minimum non-Jordanian investment in any project (50,000) is 50,000 dinars, excluding investment in public joint stock companies.

   The investor has the right to take out the foreign capital he has entered into the Kingdom to invest in it, the proceeds or profits he has earned from his investment, and the proceeds of liquidating his investment or selling his project, stake, or shares.

   For technical and administrative workers in any project, transfer their salaries and compensation out of the Kingdom.

   No project may be expropriated or subject to any procedures leading to this except by acquiring the requirements of the public interest, provided that the investor is paid fair compensation so that compensation is paid in convertible currency.

   Investment disputes between investors and Jordanian government institutions are settled amicably, and if the dispute is not resolved within a period of not more than six months, either party may resort to justice or refer the dispute to the International Center for Settlement of Investment Disputes, although Jordan ratified the accession agreement in 1972.

   The investor will re-export, sell, or waive the exempted fixed assets to another investor who is a beneficiary or is not benefiting from exemptions in accordance with special provisions (Investment Promotion Act No. 16, 2003).

In 2003, the Investment Promotion Act No. 68 was passed, which was one of the most important features of the Act (Al Momani, 2002):

A. Any non-Jordanian person has the right to invest in the Kingdom by participating or contributing, provided that the sectors and percentage of the non-Jordanian investor are entitled to participate or contribute to their borders.
B. Non-Jordanian investors are treated as Jordanian investors.
C. The investor has the right to manage the project in the way he sees it and by the persons of his choice.
D. The non-Jordanian investor has the right to take out his capital, which he has entered into the Kingdom for investment purposes, transfer the returns and profits of his investment out of Jordan, liquidate his investment, or sell his project, stake, or shares without delay, provided that he has fulfilled all his obligations to third parties, and non-Jordanian technical and administrative workers in any project are entitled to transfer their salaries outside the Kingdom and in accordance with the legislation in force.

Concrete efforts have been made by the Jordanian authorities to attract foreign investment by modernizing foreign investment laws and establishing the Investment Promotion Corporation. In 1995, the Investment Promotion Corporation was established by the Jordanian government, which provides attractive and encouraging incentives to Jordanian and non-Jordanian investors alike in terms of duty exemptions, exemptions from income taxes and social services, and unrestricted capital and profit transfers. On May 16, 2004, the investment window was introduced to simplify the procedures for obtaining approval, with the volume of foreign investment in projects amounting to 136.9 million dinars.

Jordan is also linked to several agreements, such as the Jordan-U.S. Partnership Agreement and the Jordan-Europe Partnership Agreement, to stimulate Jordan's economy and investment and to increase the volume of foreign direct investment through the preferential facilities granted by these agreements to Jordanian goods (Abed Rabbo, 2008).

In addition, since 2003, the Jordanian government has introduced a number of economic reforms in many areas with the aim of attracting investment, creating jobs, and integrating into regional and global markets, including the establishment of specialized government institutions to encourage investment and support export growth, as well as the use of economic zones. To achieve these objectives, the Jordan Investment Authority was formed in May 2014 as part of the Jordanian government's efforts to simplify and standardize key government investment institutions. The new committee is therefore the successor to the three previous institutions: the Export Promotion Directorate of the Jordan Foundation for Economic Project Development, the Free Development Zones Authority, and the Investment Promotion Corporation, so that the Jordan Investment Authority has everything necessary to make enormous gains for this country as well as the flexibility needed to respond to changes and patterns in the international and local economic investment environment.

Also, under the income tax reduction system in the less developed regions, the tax due on industrial and craft economic activities and economic activities related to schedule (3) above is reduced, with income tax due to any activity after the reduction not less than 5% of taxable income (Income Tax Act No. 28, 2009).

The new Investment Act (2014) integrated development zones, free zones, and industrial cities, the Investment Promotion Corporation, and the part on the establishment of commercial centers, representative offices, and exhibitions into a single joint body called the Investment Authority, which aims to reduce bureaucracy toward the investor and provide all solutions and facilities supporting his investment.

In addition, the Powers of the Jordan Investment Authority were strengthened by the adoption of Investment Law No. 30 of 2014, which enabled the Authority to become the only government entity responsible for attracting investments, and supporting exports, and providing a safe and stable investment environment.

The law gives the Jordan Investment Authority the authority and powers to find centralization in all investment-related procedures and accelerate them, including the establishment and regulation of special economic development zones, and the Investment Act provides for a framework for incentives and benefits available to investors and investment projects both inside and outside the current development zones or free zones, with the exception of the Aqaba Special Economic Zone.

Perhaps the most prominent feature of the new law (Investment Act 2014) is the creation of a unified investment window, which aims to provide a single place of service to license economic activities in the Kingdom, simplify its procedures, and answer investors' inquiries by various means and methods, including electronic methods.

Despite these continued efforts by the Jordanian government, there are some obstacles that are considered to be obstacles to Jordan's access to its ambitions, including: the existence of administrative bureaucracy, the multiplicity of entities that the investor must review, the monopoly of meaningful investment opportunities by a specific group of investors, the lack of a marketing program compatible with the Jordanian market, the small size of the market, the rise in income tax, sales tax, and energy prices with no stability and stability of investment policies, and the political situation. Unstable in the region, particularly in Jordan's Arab neighbors, lack of investment opportunities and weak competitiveness in many economic sectors, high cost of production and financing, control of enterprises by small and medium-sized industries, low income and wage levels, and the emergence of unemployment (Al Jamil, 2002).

11.2.2.1 Summary of Incentives and Benefits Provided by the Jordan Investment Promotion Act

Investment promotion laws reflect a strong awareness of the benefits of attracting FDI and include provisions that encourage local entrepreneurs, and Jordan targets most economic sectors to give them favorable tax treatment and customs duties, which benefit from:

1. The fixed assets of the project are exempt from fees and taxes and will be entered into the Kingdom within three years of the date of the committee's decision to approve the fixed asset lists of the project, and the committee may extend this period if it finds that the nature of the project and the volume of work in it require it.
2. Imported spare parts for the project are exempt from fees and taxes, provided that the value of these parts does not exceed 15% of the value of the fixed assets required by these parts to be entered into the Kingdom or used in the project within ten years of the date of start of production or work by a decision of the Committee to approve the lists and quantities of spare parts.
3. The Committee exempts the fixed assets needed to expand, develop, or update the project from fees and taxes if this increases at least 25% of the project's production capacity.
4. The Commission exempts from fees and taxes the increase in the value of fixed assets imported for the project account if the increase is the result of higher prices for such assets in the country of origin, higher freight rates, or a change in the transfer price (IPA 1995).
5. The project is exempt from any of the sectors or branches depending on the development area in which it is located, as follows:
   1. 25% if the project is in the development zone of category A
   2. 50% if the project is in the development zone of category B
   3. 75% if the project is in the development zone category C
6. The Committee grants additional exemption if the project is expanded, developed, or modernized and increases its production capacity by one year for each production increase of at least 25%, provided that the additional exemption period under this paragraph does not exceed four years.

7. Hotel and hospital projects grant additional exemptions from fees and taxes for purchases of furniture, furnishings, and supplies for modernization and renovation purposes at least once every seven years, to be introduced into the Kingdom and used in the project within four years of the committee's decision to approve the lists and quantities of purchases.

8. If the project is transferred during the period of exemption granted from a development area to another development area, the project will be treated for the purposes of exemption during the rest of the period, with the development area projects transferred to it being treated with the institution's information (Investment Promotion Act No. 16 of 1995).

11.2.2.2   Recommendations

Based on previous findings, the current study makes the following recommendations:

1. Work on an ongoing review of the regulations and legislation governing the investment process in Jordan to address the imbalance in the texts and articles of tax legislation.

2. Examining what shortcomings have affected the attraction of investments in some sectors, particularly the industrial sector, and what incentives are required for these sectors and what are absent from them, the study also recommends the development of a comprehensive new tax law for this sector with the inclusion of financial incentives as a tax exemption rate for different types of industries and their importance in the Jordanian economy.

3. Work to unify and simplify investment procedures in Jordan, especially since there are many parties (ministries, institutions, and government agencies) related to the investment process in terms of licenses and the application of regulations and laws other than those related to tax.

4. Further scientific studies similar to the subject of the current study should include other environments and communities with the aim of taking advantage of their findings and circulating their recommendations.

5. In order to achieve Arab economic integration and in line with Arab national objectives, Arab legislation should concern Arab investment with preferential treatment than those it decides for foreign investments other than Arab, and on the other hand, Arab joint projects, in particular productive projects that satisfy basic needs on other projects, should be distinguished by special treatment in terms of expanding benefits and tax facilitation in support of the Arab national trend aimed at expanding the productive base on an integrative basis.

6. The application of the indirect tax system in the event of difficulty in collecting direct tax from within, direct taxes on the part of governments are seen as optional while they are compulsory, but on the other hand, this is contrary to the principle of income distribution, which assumes that income tax is a compulsory tax collected from high-income people to act on projects and services that serve the lower income class and contribute to support for the poor.

REFERENCES

Abed Rabbo, Reda (2008). "Determinants of Foreign direct investment in the age of globalization", a comparative study of experiences for east and Southeast Asia and Latin America with application to Egypt, University House, Alexandria.

Al Habib, Mashri (2012). Tax Policy and Its Impact on Investment – The Case of Algeria – Ll.M., Specialty business law, Faculty of Law and Political Science, Mohamed Kheder University, Bskra.

Al-Hanawi, Said (2003). Tax systems are an analytical introduction to the knowledge, Cairo: University House.

Al-Maliki, Hessen (2001). "The importance of foreign investment in Jordan and its basic determinants", *Ennahda Magazine, Faculty of Economics and Political Science*, 2(4), 21–40.

Ezzedine, Walid (2010). *Investment Incentives in Accordance with the Latest Economic Legislation*, Cairo: Arab Renaissance House.

Hamdallah, Muhammad (2005). "Foreign direct investment in Islamic countries in light of the Islamic economy", Al Nafis Publishing and Distribution House, Amman.

Hanish (1992). Tax and its role in promoting investments in the private national sector, University of Algeria-Institute of Economic Sciences.

Hijazin, Jihad (1999). *Public Finance and Tax Legislation*, Oman: Wael Publishing and Distribution House.

Hohots, Viktor (2000). The Role of the tax System of Ukraine, Rusia. and Kazakhstan in attracting foreign Investments, Canada: Queens University at Kingston.

Investment Promotion Act No. 30 of 2014.

Investment Promotion Act No. 15 of 2003.

Income Tax Act No. 28 of 2009.

Ionel Report (2013). Corporate Income Taxation Effects on Investment Decisions in The European Union.

Jacob, Martin; Michaely, Roni, Muller, Maximilian A. (2016). The Effects of Consumption Taxes on Corporate Investments.

Khatib, Mohamed (1997). *Photo Tax Evasion*, Cairo: Arab Renaissance House.

Mabrouk, Samer (2007). The impact of income tax on private sector investment decision in the West Bank for the period 1994–2005, unpublished master's letter, National University of Success, Nablus.

Mahlawi, Ali (2015). The role of tax policy in stimulating investment, unpublished master's thesis, University of Menoufia.

Momani, Jibril (2002). The role of incentives and tax procedures in encouraging investment, unpublished master's thesis, University of Jordan, Amman.

Msimi Yahya (2006). The Role of Tax Policy in Encouraging Investment in Developing Countries – The Case of Algeria – A Supplementary Memorandum for a Master's Degree in Accounting and Finance, Faculty of Economics, Commercial Sciences and Management Sciences, Arab University Ben Mehidi.

Nistor, Loan-Alin, Paun, Dragos (2013). Taxation and Its Effect on Foreign Direct Investement – The Case of Romania, Universitatea Babes-Bolyi.

Obayami, Oluwole (2001). Nigeria's new investment laws revisited, *The Journal of Business Law* 2001, pp. 209–216.

Rahal, Mustafa (2011). Impact of the Temporary Income Tax Act (28) of 2009 in encouraging investment in Jordan, Business School, Accounting Department, Middle East University.

Shamia, Atef (1997). *Encouraging Investment and Security Requirements*, Palestine: Al Haq Publishing and Distribution Foundation

Suleiman, Hyam (2016). Indirect tax exemptions on investment in Sudan 2001–2013, unpublished master's letter, Sudan.

Voget, Johanees (2010). The Effect of Taxes on Foreign Direct Investment, University of Mannheim.

12 Financial Development and Economic Growth
An Experimental Evidence Based on Quantile Regression

Van Chien Nguyen

12.1 INTRODUCTION

The financial market is the place that provides capital for individuals, organizations, and businesses in the economy. The financial market has an intermediary role to connect a party with excess financial resources and a party that needs financial resources to carry out consumption and investment. Thus, developed financial markets are likely to be of great help to economic growth in most countries. Indeed, developed financial markets have the ability to reduce transaction costs, thereby lowering interest rates for investors when borrowing and thereby making investment projects more efficient.

Vietnam, a country that has grown strongly thanks to the effects of trade, has received many international investment flows, including FDI inflows. There are some sources of FDI capital invested in the financial market through buying shares from banks and credit institutions, or foreign investors become strategic shareholders of domestic banks. Through these investment sources, domestic banks have the ability to update technology, develop digital banking and e-banking, and thus increase financial products and services in the market. This change helps to reduce transaction costs and increase the business efficiency of banks and the economy. Usually, developed economies always require a developed financial market and, therefore, a well-developed, networked banking system.

The goal of economic development is one of the top goals on the agenda of countries in general and also in Vietnam. In recent years, Vietnam has always maintained a high level of economic growth, and therefore, the per capita income has increased rapidly and will reach 4,100 USD/person/year by 2022. It is predicted that Vietnam can achieve a high average income of about 7500 USD/person/year in 2035 and become a developed country in the following years. To achieve that milestone, it is indispensable for the financial market's contribution to the provision of financial resources and the allocation of financial resources to investment projects with high economic efficiency. However, in order to better assess the relationship between financial development and growth, in this study, we estimate this relationship, thereby assessing the role of finance in economic development.

12.2 LITERATURE REVIEW

Financial markets play an important role in connecting savings and investments. This connection process is associated with the allocation of financial resources from the surplus to the deficient side, and there is a need to use these financial resources for investment and consumption. In the usual sense, financial flows are always circulating between countries or within countries to meet

DOI: 10.1201/9781032667478-14

consumption and investment needs, and this capital flow often flows to places with higher real interest rates, better profitability, and therefore higher socioeconomic efficiency.

Research by Singh et al. (2023) argues that countries benefit from advanced financial systems that enable their economies to generate high growth and reduce poverty. Furthermore, a well-developed financial infrastructure will promote economic progress by reducing transaction costs in the connection between saving and investment and improving labor productivity. In addition, financial development has the potential to encourage economic progress by increasing competition in business and promoting innovation and economic efficiency. Therefore, the authors emphasized that financial development can promote growth through more efficient use of resources, but too much finance can be detrimental to the economy. In the case of India, it is found that there is a cointegration and asymmetric relationship between the variables, which suggests that there is a negative impact of financial development and FDI inflows on the country's economic growth in the long and short term, while trade liberalization and technology development have a positive effect. Therefore, the study suggests that it is necessary to be cautious in the process of trade liberalization in India in order to improve the quality of this country's economic growth.

Golder et al. (2023) argued that economic growth is an important economic indicator for policymakers. Considering that Bangladesh has recorded rapid economic growth over the past 20 years and is supported by the benefits of demographics, commodity exports, and inflows, the macroeconomy is stable. Maintaining a stable economic growth rate, the country has been achieving the goal of sustainable development by 2030 and middle-income and emerging countries by 2041. Golder et al. (2023) argued that financial progress through increased financial activity makes the financial sector more efficient, thereby eliminating financial recession and improving financial structure through innovation and diversification. Economic growth can be further boosted through financial development and investment by the private and public sectors. Given that Bangladesh's financial system is largely bank-dependent while the stock market is still in its infancy, financial sector credit access has the potential to boost private investment and national economic growth. The study covers the period from 1988 to 2020 and uses a nonlinear autoregressive distributed lag model. The study suggests that there is an asymmetric effect of financial progress and remittances on economic growth and the relationship in the long run. The study also confirms the positive and negative effects of financial progress and remittances on growth in Bangladesh. Therefore, the study argues that it is necessary to enact policies to support financial progress by ensuring the sustainable development of financial institutions and financial markets.

Jerónimo et al. (2023) argued that in the first decade of the 21st century, the world realized that the effects in the financial economy do not stay in the financial economy but also have a specific impact on the real economy. Financial markets have a very important role in most developed countries, connecting all countries and thus increasing liquidity and currency movement between countries. On the other hand, global financial linkages develop and increase systemic risk and financial oversensitivity, or may cause misallocation of resources in financial markets. Jerónimo et al. (2023) also argued that the financial economy and the real economy are connected through different transmission mechanisms, which are complex and developed, have a degree of coverage, and are comprehensive.

The financial sector plays an increasing role in global economic growth and development. Financial activities often rely on financial institutions and often require capital from banking and finance, while developed countries often rely on financing in financial markets. Research by Nguyen et al. (2022) argues that although there is an increasing amount of research on the relationship between finance and growth, there are not many similar studies in emerging countries. After researching 22 emerging markets between 1980 and 2020, the study finds that financial development has a positive effect on growth and that their relationship is linear. The study also finds a bidirectional Granger causal relationship between financial development and economic growth in all representations of financial development. Another study, Ze et al. (2023), also argues that financial globalization can stimulate economic growth and sustainable growth. Or Liu et al. (2023) affirm that the trend of developing green finance is gradually gaining popularity around the world.

12.3 DATA AND REGRESSION EQUATION

The data sources used in this study were collected in Vietnam for the period from 1991 to 2022. The data collected includes financial development, per capita income, and other indicators such as foreign direct investment, carbon emissions, and trade openness. The data are processed for errors, then used for regression analysis.

The regression is shown below:

$$GDP_t = \beta_0 + \beta_1 FD_t + \beta_2 CARBON_t + \beta_3 FDI_t + \beta_4 TRA_t + \varepsilon_t$$

where GDP_t is the per capita income (USD/person/year); FD_t is the country's level of financial development, as measured by the expansion of the money supply relative to GDP; $CARBON_t$ is carbon emissions per capita; FDI_t is net FDI attraction, calculated as % GDP; TRA_t is the trade openness, and is calculated according to % GDP.

12.4 RESULTS AND ANALYSIS

12.4.1 THE ANALYSIS OF DESCRIPTIVE STATISTICS

Table 12.1 shows that Vietnam's per capita GDP has improved rapidly between 1991 and 2022, from a very low level of 5.96 USD/person/year in 1991 to 4163.51 USD/person/year. This result reflects the great achievements of Vietnam's economy, from a very low-income country to a middle-income country, and is likely to achieve a higher-income country in the near future. This is also confirmed by Figure 12.1, GDP per capita has increased strongly from 2008 to now and the increase is higher than the period before 2008.

TABLE 12.1
Descriptive Statistics

	GDP	FD	CARBON	FDI	TRA
Mean	1458.257	75.69751	1.498968	5.536982	124.6814
Median	851.9524	78.79791	1.185215	4.812211	127.9877
Maximum	4163.514	146.2925	3.676440	11.93948	186.4682
Minimum	5.960844	19.56649	0.289598	3.390404	66.21227
Std. Dev.	1301.925	41.90126	1.080836	2.173468	32.77186
Skewness	0.644451	0.027764	0.807845	1.372480	−0.170335
Kurtosis	1.952706	1.702264	2.517657	3.923419	2.181626
Jarque-Bera	3.677457	2.249604	3.790812	11.18334	1.047722
Probability	0.159020	0.324717	0.150257	0.003729	0.592230
Sum	46664.24	2422.320	47.96696	177.1834	3989.804
Sum Sq. Dev.	52545230	54427.19	36.21440	146.4428	33293.84
Observations	32	32	32	32	32

Source: Author's analysis.

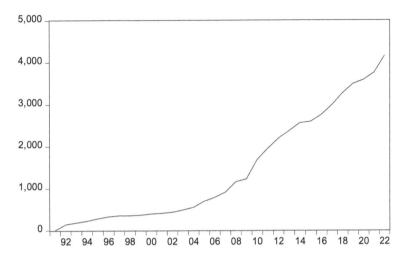

FIGURE 12.1   Trends of GDP.

Financial development, as measured by broad money as of % GDP, shows an expansion of financial development, from 19.56% of GDP in the early period to 146.29% of GDP. This reflects the development level of the financial market in Vietnam in recent years. Indeed, Vietnam's financial system has improved dramatically since the Ordinance on Banking was issued in 1990, and there is a clear division between the central bank and commercial banks. The process of trade liberalization has led to reforms in the financial market, and with the introduction of the stock market in 2000, it has created more long-term capital mobilization channels for the economy. It can be said that Vietnam's financial market is relatively modern and has many products and services supplied to the economy. The development of the financial market is associated with the development of digital banking, e-banking, and financial inclusion, thereby increasing the opportunities for beneficiaries to access financial services and products. Figure 12.2 also confirms the trend of financial development in Vietnam in recent times.

The regression results in Tables 12.2 and 12.3 show the following.

According to Table 12.2, financial development has a positive but not statistically significant impact on economic growth. The same thing happens for FDI when the regression coefficient of FDI is positive but not statistically significant. However, the study suggests that carbon emissions have a positive effect on economic growth. International economic integration through commercialization has a negative impact on growth.

Table 12.3 shows that financial development has a positive but not statistically significant effect at all percentiles. This result is similar to Table 12.2, which shows that there is no clear evidence of the impact of financial development on growth. This could explain that the financial development in Vietnam can have an impact on many other factors in the economy, such as stimulating investment, consumption, increasing access to financial products and markets, but these effects are not really large enough to have a change in economic growth.

The results of Table 12.3 also show that carbon emissions have a positive effect on growth in all quartiles, and this result is similar to the results of Table 12.2. It can be explained that increasing carbon emissions has a positive effect on the economy. It can be seen that Vietnam's level of economic development is low, so production activities often use outdated technology and have high carbon emissions. In addition, the economic structure is not modern and depends largely on energy-consuming production activities and high carbon emissions. In recent times, Vietnam's

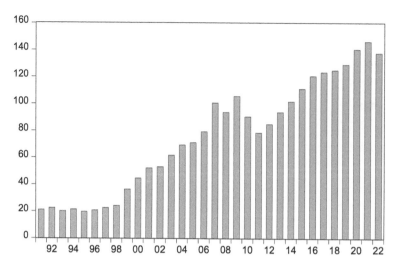

FIGURE 12.2 Trends of FD.

TABLE 12.2
OLS Regression

Dependent Variable: GDP

Method: Least Squares

Date: 07/23/23 Time: 22:15

Sample: 1991 2022

Included observations: 32

Variable	Coefficient	Std. Error	t-Statistic	Prob.
FD	7.363946	4.965374	1.483060	0.1496
CARBON	1173.898	125.6331	9.343862	0.0000
FDI	2.290011	22.89644	0.100016	0.9211
TRA	−9.880854	4.484906	−2.203135	0.0363
C	360.4679	354.5695	1.016636	0.3184
R-squared	0.964975	Mean dependent var		1458.257
Adjusted R-squared	0.959786	S.D. dependent var		1301.925
S.E. of regression	261.0818	Akaike info criterion		14.11015
Sum squared resid	1840420.	Schwarz criterion		14.33917
Log likelihood	−220.7623	Hannan-Quinn criter.		14.18606
F-statistic	185.9670	Durbin-Watson stat		0.714830
Prob(F-statistic)	0.000000			

Source: Author's analysis.

economic development has been accompanied by the expansion of energy production and consumption, and this process has improved labor productivity and efficiency in the economy. However, increasing carbon emissions increases the cost to the economy; when the cost of the economy is too high, it hinders economic growth. Therefore, the policy of increasing carbon emissions to promote growth is difficult to sustain in the long term, and then the Vietnamese economy needs to change its production activities and energy consumption habits to achieve higher levels of economic development in the long run.

TABLE 12.3
Quantile Regression

Quantile Process Estimates

Equation: UNTITLED

Specification: GDP FD CARBON FDI TRA C

Estimated equation quantile tau = 0.5

Number of process quantiles: 4

Display all coefficients

	Quantile	Coefficient	Std. Error	t-Statistic	Prob.
FD	0.250	−1.989672	8.036411	−0.247582	0.8063
	0.500	−2.265735	6.154949	−0.368116	0.7157
	0.750	15.34577	10.77128	1.424693	0.1657
CARBON	0.250	1184.406	183.4544	6.456130	0.0000
	0.500	1402.711	168.8760	8.306160	0.0000
	0.750	1263.318	201.5758	6.267209	0.0000
FDI	0.250	26.26305	27.69827	0.948184	0.3514
	0.500	10.32310	21.12505	0.488666	0.6290
	0.750	15.14041	33.26614	0.455130	0.6527
TRA	0.250	−2.115477	6.647349	−0.318244	0.7527
	0.500	−4.167247	5.261198	−0.792072	0.4352
	0.750	−22.75693	12.62472	−1.802570	0.0826
C	0.250	−255.9146	509.8369	−0.501954	0.6198
	0.500	−43.76012	388.6965	−0.112582	0.9112
	0.750	1350.929	963.1677	1.402589	0.1721

Source: Author's analysis.

Table 12.3 also shows that FDI has no clear impact on economic growth at all quintiles, and this result is similar to Table 12.2, so the impact of FDI on growth cannot be fully confirmed. However, Table 12.3 confirms that there is an impact of international trade on growth at the high percentile, and this result is similar to Table 12.2. The results suggest that a larger extent of international trade expansion is likely to have a negative effect on growth, and conversely, a lower degree of economic integration has a more positive effect on economic growth. This result supports Vietnam's trade liberalization policy, indeed, Vietnam's gradual loosening of policies on international trade has created many opportunities for businesses and investors to invest in the country and expand international trade, these policies have brought benefits to the economy such as promoting economic growth and development. However, excessive trade expansion can create risks for the economy.

12.5 CONCLUSION

Studying the impact of financial development on economic growth in Vietnam is set in the context that this country has had a deep integration into the global economy. The starting point was a closed economy with a very low level of development. The economy was heavily affected by the war and needed a long time to recover from the war's consequences and integrate into the international economy. The financial market reform process has promoted the development of the banking system and capital market, the stock market, and thereby brought opportunities to access financial services and products for subjects, businesses, and investors. Research results show that financial development has a positive impact on economic growth, but this effect is not statistically significant. Therefore, more time is needed to re-evaluate this relationship in order to better clarify the impact of financial development on the economy. The study also suggested that excessive trade liberalization can have a negative impact on the economy. Therefore, it is necessary to carry out appropriate trade liberalization in order to take advantage of the benefits of international trade, minimize risks, and strengthen the economy's ability to withstand external risks.

The study has several policy implications for Vietnam. Firstly, Vietnam continues to improve financial market performance in the direction of efficiency, thereby reducing transaction costs in the economy and increasing the efficiency of the economy. Second, Vietnam needs to have a direction to control carbon emissions, although in the short term, carbon emissions still create benefits for the economy. However, increasing emissions is unlikely to generate benefits or the overall prosperity of the economy in the long run as the economy incurs increased social costs. Third, in parallel with the process of opening the economy, Vietnam needs to have appropriate policies to minimize risks due to impacts from external factors that may affect the economy. Therefore, measures to improve risk management and improve internal resources in the economy need to be improved.

REFERENCES

Jerónimo, J., Azevedo, A., Neves, P. C., & Thompson, M. (2023). Interactions between financial constraints and economic growth. *The North American Journal of Economics and Finance*, *67*, 101943. https://doi.org/10.1016/j.najef.2023.101943.

Golder, U., Rumaly, N., Hossain, M. K., & Nigar, M. (2023). Financial progress, inward remittances, and economic growth in Bangladesh: Is the nexus asymmetric? *Heliyon*, *9*(3), e14454. https://doi.org/10.1016/j.heliyon.2023.e14454.

Liu, X., Zhao, T., & Li, R. (2023). Studying the green economic growth with clean energy and green finance: The role of financial policy. *Renewable Energy*, *215*, 118971. https://doi.org/10.1016/j.renene.2023.118971.

Nguyen, H. M., Thai-Thuong Le, Q., Ho, C. M., Nguyen, T. C., & Vo, D. H. (2022). Does financial development matter for economic growth in the emerging markets? *Borsa Istanbul Review*, *22*(4), 688–698. https://doi.org/10.1016/j.bir.2021.10.004.

Singh, S., Arya, V., Yadav, M. P., & Power, G. J. (2023). Does financial development improve economic growth? The role of asymmetrical relationships. *Global Finance Journal*, *56*, 100831. https://doi.org/10.1016/j.gfj.2023.100831.

Ze, F., Yu, W., Ali, A., Hishan, S. S., Muda, I., & Khudoykulov, K. (2023). Influence of natural resources, ICT, and financial globalization on economic growth: Evidence from G10 countries. *Resources Policy*, *81*, 103254. https://doi.org/10.1016/j.resourpol.2022.103254.

13 Importance of Financial Literacy Education and Financial Behavior for Developing Nations in the World

Venkata Vara Prasad Janjanam
and SubbaLakshmi A.V.V.S.

13.1 INTRODUCTION

Finance Literacy is a significant problem that every country is seeing. According to the S&P Global Financial Literacy survey in 2014 [1], only 33% of adults worldwide were financially literate (3.5 billion people). According to the map of the global variations of financial literacy, the highest financial literacy scores were recorded in Australia, Canada, Denmark, Finland, Germany, Israel, the Netherlands, Norway, Sweden, and the United Kingdom (65% were financial-literate). In major emerging economies like China, India, Brazil, the Russian Federation, and South Africa, less than 40% of adults are financially literate. Among the major emerging economies, India scored significantly lower. It reached below 24% for financially literate adults. Neighboring countries Pakistan, China, Thailand, Sri Lanka, and Bhutan scored higher than India. This is an awful result for India, the second most populous country in the world, with a younger people population.

These results were released in 2014, and after that, lot of economic changes happened in India: demonetization from November 8 to December 30, 2016, banning the currency notes 500rs and 1000rs in India due to lack of digitalization, more dependence on liquid cash payments, less usage of online payments and wallets, a smaller number of smartphone users, a low number of people using the internet, high internet service costs, lack of planning to face the emergency, etc. made the people tackle the currency ban very hard, destroying the businesses of small vendors and small and medium enterprises. Due to a low awareness of cashless payments, people forcefully make regular payments in small denomination notes like 10 rupees, 20 rupees, 50 rupees, and 100 rupees. This caused a lack of availability of liquid cash at banks and ATMs. During this period, the regular lives of common people were utterly disturbed, creating a complex environment for businesses to buy and sell goods and services. From just-born infants to old age, adults were affected by consumer needs and wants [2–4].

Now the number of smartphone users and cashless payments has increased at the highest rate. Hence, it shows that people lacking financial literacy awareness faced severe obstacles during the economic crisis.

Along with the rise of technology, the internet's availability exposed consumers to more financial products. Due to a lack of proper skills and awareness, people cannot use fine-quality products with more benefits. The market now has a pool of financial products and Fintech applications, and it is straightforward for the customer to get the product or service. But what is lacking is the proper skills to filter the exemplary product/service to get better benefits, make proper judgments and decisions, and focusing on extended stand usage and profits [5–7].

DOI: 10.1201/9781032667478-15

The organization of this chapter is as follows: Section 13.1 provides an introduction to the topic. Section 13.2 discusses the motivation for this research. Section 13.3 describes the research method used to conduct this study. Section 13.4 presents the literature review, which includes the key findings from various studies on financial literacy, financial behavior, and financial education. Finally, Section 13.5 provides a conclusion summarizing the chapter's main points and highlighting the research's implications.

13.2 MOTIVATION FOR THIS RESEARCH

As India is the second most populous country in the world [8], the number of people investing in the stock exchange, mutual funds, bonds, and commodity markets is shallow [9,10], and the number of farmers doing farm insurance is deficient [11]. Agriculture and farming were among the leading contributors to the nation's growth. Especially in Southern India, there are more natural disasters. Farmers who take loans from banks or informal lenders cannot pay or clear them within time due to a lack of awareness of various farm insurance policies [12]. People were attracted to easy money-making and fell into the scams of different fraudsters and institutes [13] (trade frauds, chit fund cheatings, famous financial frauds in India). Payday loan lenders charge high interest rates. Some people didn't know its compound interest rate nature, and with a lack of awareness, they lost their hard-earned money, assets, and even physical, sexual, and mental harassment and blackmailing by the lenders [14]. The number of digital lending app frauds was raised to a very high number. Out of 1100 Indian loan apps, nearly 600 apps were illegal loan apps (www.rbi.org.in, 2021) [15].

Hence, a high financial literacy rate and good financial behavior play a vital role in minimizing and protecting against all these problems. It helps to learn and acquire proper skills, behavior, and attitudes to make solid decisions and judgments for the growth and well-being of the consumer and the nation's economy. It raises awareness about various financial products, the compounding nature of money, financial security during emergencies, and risk management through diversification of portfolios; helps to build wealth; fulfills needs and wants; gives consumers the right to complain to regulating bodies for mishandling of user data, money, etc.; and allows consumers to buy authorized legal products. It also helps the individual achieve financial wellness. Finally, it makes the individual's life happier and more prosperous.

The financial literacy gap is more prevalent in the nation, so taking it as a severe research problem and knowing the main barriers to the low financial literacy rate, weird financial behavior to market inconstancy, and low attendance in different financial products. This research article features the financial literacy levels of India and acknowledges the results of splendid financial education and financial behavior papers.

13.3 RESEARCH METHODS

A systematic literature review methodology was adopted to enhance the analysis further and ensure that all relevant studies were included. This involved a comprehensive search of electronic databases, including Google Scholar, Scopus, and Web of Science, using relevant keywords and combinations of financial literacy, financial behavior, financial education, and developing countries.

The studies included in the analysis were predominantly quantitative, with a few qualitative studies that provided a deeper understanding of the experiences and perspectives of individuals regarding financial literacy and behavior. The data collected from these studies were analyzed using various statistical methods, including regression analysis, correlation analysis, and structural equation modeling, to identify the relationship between financial literacy, financial behavior, and other socioeconomic factors.

Furthermore, the limitations of the studies were also critically assessed and presented. These included self-reported measures, small sample sizes, and a lack of longitudinal data. Despite these limitations, the studies provided valuable insights into the factors that influence financial literacy and behavior in developing countries and the effectiveness of financial education programs in improving financial literacy and behavior. Overall, the research method adopted in this study was robust and comprehensive, ensuring that the analysis presented was based on the best available evidence.

13.4   LITERATURE REVIEW

13.4.1   FINANCIAL LITERACY (GLOBAL LEVEL)

The S&P Global Fin Lit survey provides insights into financial literacy around the world [16]. The map shows that most nations have below-average adult financial literacy rates. In this state, we must take financial literacy as a global problem and work together to get good results across nations. According to the results of the survey, India has a low adult financial literacy rate of 24%, as do all other emerging economies.

BRICS (Brazil, Russia, India, China, and South Africa) have an average adult financial literacy rate of 28%. A study on financial literacy was conducted in the most literate state of India, Kerala. The study consists of questions on essential knowledge, money management, savings and investments, risk management, perception and opinion, and the authors own questions. The results were not good. The mean of the sample on adult financial literacy was 43.3% for the most literate state. This shows a severe gap in the education system in India. Hence, the author stated there is a need for financial education awareness programs, especially for females and youth [17]. These results were obtained in another study on financial literacy attitudes, behaviors, and cognitive biases [18]. The author conducted the study in four waves in the urban slums of Ahmedabad in Gujarat state, India, with a sample size of 1200. The study aimed to improve financial literacy and financial behavior through video-based financial education, pay for performance in tests, counseling, and goal setting. The mean financial literacy rate was 1.59 out of 3, which shows half of the respondents have low financial literacy levels. The results showed that the treatment and pay-for-performance financial literacy treatment groups performed better than the control groups.

Another study on financial literacy among high school students in grades 10–12 in Tamilnadu, India, got identical results to the above studies. The survey results for basic financial literacy have a mean percentage of 44.5%, whereas the mean rate for sophisticated financial literacy is 44%. This shows that more than half of the students answered incorrectly. The students performed well in numeracy questions, but this skill was insufficient to achieve good financial literacy scores [19]. Another major national survey on financial literacy also showed approximate results, which was conducted by the National Center for Financial Education (NCFE), India. The financial attitude toward spending money was good; 47% of people showed a good attitude. And toward saving money, 40% of people showed a savings attitude for the long term. Finally, toward planning money, 40% of people showed a strong attitude toward planning money.

When it comes to financial behavior, 65% of the respondents prepare a budget. 40% of the people's income did not cover their living costs in the last 12 months, and 31% used the available credit to meet their living costs. Half of the respondents don't have active savings accounts that generate returns. 55% agreed to setting long-term financial goals and striving to achieve them. 70% of the respondents agreed to monitor their financial affairs. 80% of the respondents agreed on paying bills on time, and 84% stated that they buy only what is affordable.

Regarding financial knowledge, respondents showed strong results on the ability to divide, with 90% answering correctly. In contrast, 70% of them were unable to calculate compound interest. The respondents' mean value for basic knowledge was 58.5% [20].

Financial Literacy among Young Professionals and Retired Persons. Agarwalla et al. [21] surveyed financial literacy among students, employees, and retired people in India. Most of the respondents have average financial knowledge. For the correct answers to questions, more than half answered rightly to division, interest, risk and return, and inflation and prices. This shows participants were strong in mathematics, and this skill didn't help them to score highly in financial knowledge.

For gender, males were stronger than females among employees and retired people in terms of financial knowledge. Participants from rural areas beat the town and city participants in terms of high levels of financial knowledge. Income-wise, income greater than 30,000 rupees dominated the high-level financial knowledge category. The second part of the study is financial behavior. About 68% of employed and 75% of retired respondents have shown positive financial behavior. A total of eight dimensions were used to measure financial behavior. This respondent achieved a mean of 88% on self-assessment of affordability, 76.5% on bill payment behavior, 66.5% on monitoring personal finances, 58% on long-term planning, 75.5% on household budget planning, 100% on active saving behavior, 84% on evaluation of financial products, and 69.5% on reliance on borrowing.

Gender-wise, among employees, males showed more positive financial behavior than females, whereas, in the retired category, both males and females showed good positive financial behavior with a mean of 75%. Education-wise, employees who studied higher than higher/secondary education positively influenced financial behavior. For the retired category, there is little difference between highly and lowly educated people. Income-wise, low-earning employees up to 5000rs showed the highest positive financial behavior. This shows the low-income group is more careful about expenses, consumption, borrowing, savings, and financial planning. Three items were used to measure the financial attitude, i.e., the extent of belief in planning, the propensity to save, and the propensity to consume. 48% agreed with the extent of belief in planning, 51.5% had the propensity to save, and 54% had the propensity to consume. For the overall financial literacy score, an average mean of 13.7 has been observed in Indians of both groups, equal to the OECD average rate of 13 countries. This result shows that Indians are not very poor in financial literacy [21].

But contradicting results appeared just a few years later. Agarwalla et al. [22] conducted a study on working young in urban India in six metropolises. The results show that the young Indians achieved the highest mean value of 6 out of 8 on financial behavior, an average mean value of 4.2 on financial knowledge, and an excellent positive financial attitude. Agarwalla et al. [22] compared India's and other OECD 14 countries' average financial literacy rates. Out of the 19 questions on financial literacy, only five questions crossed the OECD average. For the remaining 14 questions, they were unable to beat the OECD average.

The results of the influence of socio-demographic variables on financial literacy on overall family income were higher than Rs. 50,000.00 monthly, which has a solid, significant positive impact on financial literacy. Excluding income and marriage, none of the variables significantly positively impact financial literacy.

Similarly, for financial behavior between good, average, and indifferent groups, family income, education, lack of financial planning, and gender show a statistically significant influence. Finally, among positive, average, and indifferent financial attitudes, family income, marriage, and gender have shown a significant positive impact. Females showed more positive financial attitudes than males, and persons living in joint families showed more negative financial attitudes [22].

13.4.2 FINANCIAL LITERACY AND FINANCIAL EDUCATION

Teaching finance education to people so they can make sound financial decisions and adequately use financial products is very significant. The main concern is how we educate people, which should help them remember it and practice it as much as possible for the long term. Financial education helps to learn basic and advanced financial literacy topics, and it can be taught in regular classes, seminars, just-in-time, workshops, etc.

And some authors used visual tools and narratives to teach financial education. Lusardi [23] chose videos with narratives explaining financial literacy topics in this paper. She explained clearly the importance of narratives in behavior changes and their studies in psychology, cognitive thinking, and information processing.

The results showed an increase in financial literacy, confidence, and knowledge and a reduction in the don't know answers of the participants exposed to the videos. She concluded that the videos with interactive tools were better than the written scripts [23]. Like these results found by Ameer [24], Education in Finance & Economics has a significant association with a high level of financial literacy and appropriate confidence levels.

In another paper, Laura [25] used the service learning teaching method to deliver personal finance education to college students through skits. She used the brochures and YouTube videos to play the skits on personal finance topics. The paper's results showed the audience's overall improvement in knowledge, positive attitude, behavioral changes, willingness to participate in various financial credit rating services, motivation to maintain budgets, and preparation for financial planning.

Rashmi [26] experimented at a university in Singapore by providing students with the freedom to select the financial education course in the curriculum through a bidding process. The students who had successful bids scored higher in financial knowledge and financial planning than those whose bids were unsuccessful.

From the data in the paper, we can say that the students interested in finance education have performed better than those with less interest. This financial education course helped them to improve their financial literacy levels [26].

Financial Education Impact. Lewis Mandell and Linda Schmid Klein [27] studied the impact of financial management courses on high school students. Lewis et al. used Jump $tart survey questions to calculate the financial literacy score. Surprisingly, those who have not taken the course have an average rate of 69.9%, whereas the students who have taken the course have an average rate of 68.7%. Data didn't support the conclusion that the students who took the course were more saving-oriented than those who had not. But marvelous results for financial education affect financial behavior for high school students in the United States. Carla et al. [28] study found that subjects with financial education have fewer defaults and higher credit scores.

In some countries, financial education is mandatory for high school students. Douglas [29] showed that the adults exposed to the mandates had increased their savings and wealth along with their age, and in the states that imposed orders, the adults had more protection and wealth than in the states that didn't set the mandates. And coming to student and parent relations, he categorized frugal and non-frugal parents. From the data, for the students whose parents were non-frugal, the mandates helped them to improve their savings and wealth, whereas for the adults whose parents were frugal, mandates delivered neutral effects [29].

Not only the current research data, but some authors also used National Government Financial surveys to know how education affected people's lives. Shawn Cole [30] used the three surveys: the US Census, the income and program participation survey, and the Federal Reserve Bank of New York consumer credit panel/Equifax data. Here Cole considered the US census data from the 1980, 1990, and 2000 censuses for the age group of 18–75 years. He analyzed people's demographic data, education, and income. Does education affect financial behavior or not? According to his results, the additional year of schooling has impacted credit outcomes like an increase in credit score, a rise in account balance, and a reduction in the number of quarters delinquent. And finally, the additional year of schooling helped with higher labor earnings, high asset accumulation, an increase in credit score and balance in the account, a reduction of chances of bankruptcy and foreclosure, the accumulation of wealth and earnings through financial investments and financial products, improved cognitive ability and financial behavior, better financial decisions, fewer negative returns, and less delinquency [30].

Is financial education making people wealthier? For this, Cole [31] published a paper. The results of the paper show that the percentage of people who had an education has increased significantly with an increase in age from 35 to 55. The rate of investment income positively correlates with education, and investment income is higher for people with the highest education qualifications.

The years of schooling have impacted the rise of income from investments and retirement savings by a significant rate. The compulsory schooling laws positively affected students' attendance year by year, with good growth in attendance. Cole [31] presented the tabular data of the states with no mandates and mandates implemented conditions. The states with no mandates have given negative investment returns from their income through investments and retirement savings.

The economic position of the states in the form of GDP growth rate was studied five years prior, and post-commerce financial education in high schools was analyzed. The results showed an abnormal GDP growth rate before and after mandates were implemented. Overall, the GDP growth rate after the mandates was higher than before the mandate period. The results show that ability and knowledge related to education have a positive relationship. With increased knowledge, the amount of income through investments improved, and the dependence on mortgages and loans for the first time was positive and the second negative. The final verdict the author has given is that the additional year of schooling benefits more [31].

Financial Education, Debt Literacy, and Debt Behavior. Regarding spending behavior, young people were spending more on unnecessary things by using credit cards and falling into debt traps. To evaluate the debt literacy of the students' various education programs were implemented, like financial education, economic education, financial literacy, consumer education, and math education, as analyzed by Brown [32]. Brown considered it four years before the reforms and four years after the reforms. For economic reforms, the following results were observed: the risk score started decreasing after the implementation of economic reforms, which means students were taking huge risks. And the worst part is that the delinquent balance, non-home debt, and home debt chances of bankruptcy in the next few months increased drastically after the implementation of economic reforms. Still, student economic reforms helped reduce student debt after implementing the reforms.

The results of the financial literacy reforms were the complete opposite of the economic reforms. The delinquency rate, chances of bankruptcy in the following months, and home and non-home debt decreased significantly after implementing the financial literacy reforms [32].

Lusardi [33] developed a simple question to evaluate the debt literacy of young people, and results showed females answering correct answers lower than men. And many people didn't know the time value of the money; the highest income group has a higher literacy rate than the middle and lower income groups. In the self-assessment of financial literacy, many people rated it higher than average [4.8/7]. This shows people think they have better financial literacy, but the real fact is they have low literacy rates.

To evaluate the financial experience, the author divided people into four categories based on their type of transactions: traditional borrowers, alternate financial services borrowing, saving and investing, and credit card users. Here, the author used cluster analysis and categorized the consumers into four clusters: alternate financial users [AFS users], pay-full, pay fees, and borrowers/savers. Debt literacy is very low for AFS users. The AFS users have low participation rates in investment products and high usage of payday loans, pawn loans, and tax-incurred loans. In contrast, the savers/borrowers group has the highest usage of financial investment products and more dependence on debt.

The results clearly show a positive relationship between financial literacy and debt behavior. Those with good literacy rates were very good users of financial products and had low debts and fines. Those with low literacy rates are not good at using financial products and use high-debt interest products from a traditional market like payday loans, pawn shops, etc. and incur more losses by paying the highest interest rates [33].

Evaluation and Reliability of Financial Literacy Education. Teaching financial literacy education by any means and maintaining proper evaluation metrics to know its outputs and increase the quality of program delivery is crucial to Lyons et al. [34]. In this paper, the author states the importance of evaluation and its necessity from its basic roots and also states the importance of adopting a standard methodology to measure the educational impact on the behavior, attitude, skills, stratification, and confidence levels of the subjects. And the major barrier to evaluating the education programs was getting the right atmosphere and attention of the people and getting the data, which has high quality in terms of reality and applicability [34]. Some authors want to evaluate the financial literacy training. Bruce [35] provided an 18-hour financial training to students and allowed access to a junior achievement financial park, where students were given fictitious identities and made financial decisions on 17 kiosks arranged at the park. The results showed that the students who went for the treatment were deciding to reduce their debts, spending very little on clothing, groceries, home improvement, sports activities, charity, low phone bills, etc. This shows that financial training has impacted student behavior toward unnecessary spending by paying less and more for the required ones, like taking high-cost health premiums, spending more on education, and dining out [35].

Not only the evaluation, but data reliability is critical. Lauren [36] mentions the barriers to financial literacy education misrepresentation and lack of evidence, especially in the data collection, and notes that education affects behavior in credit scores, bankruptcy, and mortgage payments due to a lack of proof. Lauren clearly said that the non-experiment studies on financial literacy education didn't have the proper evidence to justify the effects of financial literacy education. When compared with experiment papers, there were a lot of differences in the data. He cited various problems in the financial education papers but could not justify them with the proper method and data [36].

Against Financial Literacy Education. Not only the advantages of financial education, but some disadvantages are also listed by Lauren [36] in the paper. He mentioned the financial education drawbacks, difficulties consumers face while making a crucial decision, overloaded information, and inactive cognitive thinking and knowledge when making significant financial decisions. He talked about the various biases and illusions leading to bad financial behavior. He briefly identified multiple drawbacks in the financial education classes offered by National Government Institutes, Private financial firms like credit card companies. The ineffectiveness financial education and its causes and reasons were highlighted by Lauren [36].

Lauren [36] discussed the costs of effective financial education and the reasons why it is highly costly. Factor one is the extremely bad state of most consumers' financial literacy. The second factor is complexity in financial decision-making and the heterogeneity of consumer financial circumstances and values. The third factor is the high speed of product offerings and industry practice changes. The fourth factor is a lack of interest in or resistance to a participant. The fifth factor is that the industry has substantial resources to outsource education. The last factor is that even those who are financially knowledgeable and skilled make poor financial decisions. Watanapongvanich's [37] survey results clearly showed that elementary school financial education doesn't significantly reduce gambling behavior. But high financial literacy and financial knowledge usage in decision-making significantly impact avoiding gambling behavior.

Workplace Financial Education. Bernheim [38] et al. studies the impact of financial education on participants' savings, retirement wealth, and total wealth. From the sample, the highly educated were earning a high income, and those who made a high income had a higher 401k and thrift balance, or net wealth. Anshika et al. [39] revealed that small and medium enterprise entrepreneurs have higher financial literacy than their counterparts. The higher the gross profit ratio of the firm, the more likely they are financially literate. Javed et al. [40] conducted a study on the financial literacy and access to external finance of small and medium-sized enterprises in London. The results revealed that financial literacy positively impacts the Firm's Growth, access to external finance, and maintaining balance sheets perfectly within time. Jing [41] conducted the effects of financial education on the financial capability of American subjects in 2012 by using data from the National Financial Capability Study of approximately 25,509 American adults. From the results, a 3.06 mean out of 5 is observed for objective financial literacy, 5.18 out of 7 for subjective financial literacy, 5.74 out of 7 for perceived financial capability, and 7.82 out of 20 for desired financial behavior. Those who received financial education have higher mean values for financial literacy variables than those who didn't receive any at all. The overall Financial Capability Index is slightly higher for those who received workplace financial education than for those receiving financial education at high school or college [41].

Importance of Financial Literacy and Its Tool. Lewis [42] stated the financial literacy problem. It impacts the economy, people's wealth, savings, development of the nation, and welfare. Its consequences are foreign investor dominance, bankruptcy, misery, bad decisions, and a market bubble. He discussed the assessment of financial literacy, the drop of financially literate students from 57% in 1997 to 50% in 2002 of the Jump $tart survey of high school students, and concern about the lack of connection between education and financial literacy.

For improving financial literacy, Lewis [42] taught that playing stock market games makes students score high on financial literacy tests, and no penalties in games make students less risk-averse. The low-risk-averse skill is needed to save the economy.

Lusardi [43] raised the question of whether individuals are ready to make a financial decision for complex financial products. Due to the rise in technology, accessing credit is easier than before. Lusardi [43] suggested that every person needs a financial license to invest in sophisticated products. Programs should not only encourage savings and investment behavior but also borrowing behavior. The results show that half of the respondents are weak in numeracy, and more than 4/5 cannot answer compound interest. Along with this, Lusardi [43] used the advanced financial literacy questionnaire with eight questions for the sample of highly educated groups [44]. Results showed that even education didn't help the respondents calculate the relationship between the interest rate and bond prices. Finally, Lusardi [43] studied the relationship between financial literacy and retirement planning. Those who were strong in compound interest calculation mostly planned retirement. Hence, she suggested that financial literacy programs should be interactive, and one size fits all didn't work for everyone. Add technology to improve the participants' financial literacy. The saving and investing pyramid is needed for general guidelines on saving and investing for people.

Measured Financial Literacy and Manipulated Financial Literacy. Daniel [45] conducted a meta-analysis of financial literacy and its relationship with financial education and financial behavior in 168 papers, covering 201 prior studies. His study was about the effect size of r on measured financial literacy and manipulated financial literacy. The measure of financial literacy has a larger effect size than the manipulated financial literacy on financial behavior. The effect size for financial interventions has a minimum effect, $r^2 = 0.0011$. For measured financial literacy of 111 effect size, $r^2 = 0.0179$ is better than manipulated financial literacy. The author stated that the following were the main reasons for manipulated financial literacy having weaker effects than measured financial literacy:

Reason 1: The intervention decays over time. Daniel clearly shows a graph between the months since the intervention to effect the partial size r between financial literacy and financial behavior. The results were presented for the interventions as a function of the number of hours, from 1 to 24 hours of intervention, with a time interval of 6 hours. The decay rate is very strong for long hours of intervention.

Reason 2: Financial education interventions have weak effects on financial knowledge that intend to cause financial behavior.

Reason 3: Due to omitted variable bias in the effects of measured financial literacy studies.

In his study, Xiao Li [46] found that measuring financial literacy is challenging due to its qualitative nature and tough calculation due to time series change. Hence, his study proposed to add textual analysis to survey-based methods to measure financial literacy precisely.

13.4.3 FINANCIAL LITERACY AMONG VARIOUS DEMOGRAPHICS

Financial literacy among various demographics was discussed in this session. Lusardi's [47] results from the paper show that youngsters have standard knowledge of inflation and risk diversification topics. And students with high cognitive ability have high financial literacy levels. Regarding family characteristics, those students of the parents who are highly educated, actively participate in financial products, and have adequate wealth show higher financial literacy levels than others. One-size-fits-all education programs are ineffective for young groups. Parents' education and cognitive ability play a crucial role in the financial knowledge of young people. In another paper, Chen [48,49] found that the youngest people have a low level of knowledge in investments, insurance, savings, and borrowings. The students of business majors performed slightly better than the non-business majors, whereas per educational level, highly studied students performed at a reasonable rate than the lower-class rank students. Regarding gender, females performed marginally better or less than males, showing less interest in financial literacy subjects. And the significant evidence shows that those who have more experience, aged >40, and have high income levels of $50,000 or more performed well in the test.

Baker [50], in their survey paper on various models of financial literacy and demographic variables related to behavioral biases results, revealed that financial literacy has negative relation with disposition bias and herding bias, and positive and significant relation with mental accounting bias. Age, occupation, and investment experience were the crucial demographic variables related to behavioral biases. Lusardi [51] used data from a particular module devised from the 2004 Health & Retirement Study. From the data on financial literacy questions, only 61.9% of women correctly answered interest rate. For the inflation question, 70.6% answered correctly, and more than half of the respondents could not answer correctly to risk diversification. For retirement planning, the author categorized the respondents into three categories, simple planners, serious planners & committed planners. Lusardi [51] concludes that financial literacy has a significant impact on retirement planning in households. In investments and insurance products, the students with high personal finance knowledge outperformed those with low personal financial knowledge.

Agarwalla [52] found that young people in urban India have low financial knowledge, good financial behavior and a bad financial attitude. The results clearly show that young people with higher education, a higher family income >50,000, and who are married were financially literate, whereas females, and people living in joint families have low financial literacy levels. Teaching finance education along with activities and games of investment helps the students acquire more knowledge instead of traditional courses like personal finance, economics, etc. [53]. The results of the Jump $tart surveys over the years 1997–2006 show that the students whose parental income and education have a positive relationship with financial literacy. In the paper, Chen [48] says that most of the high schools running in his study were putting less effort into financial education. The findings of his study suggest that college students' knowledge of personal finance is inadequate, and their weakest area is investments. The reasons for this are a lack of sound personal finance education in college curricula and less exposure to financial issues due to the early financial life cycle.

And these young people were more focused on spending or consumption than investing. Coming to the results of his paper, he says that the students were more familiar with the auto insurance products among all financial products, and the participants with higher personal income, business majors, more experience, and higher-class ranks were more financially knowledgeable. Regarding gender and age, men were more financially knowledgeable than women, older participants had more financial knowledge than the younger ones, and these knowledgeable participants were making correct choices rather than the less knowledgeable ones. Maintaining personal financial records and spending less than their incomes were helpful for the students [46,53].

13.4.4 FINANCIAL LITERACY AND FAMILY COMMUNICATION

Family and parents are crucial in the child's financial literacy. It is essential to help the children learn financial literacy topics along with child development. More focus should be placed on the financial behavior and attitudes of the students, and it is crucial to develop programs for parent-teacher education. The more the parent learns and provides the best social environment and tools for his children, the easier it is for the children to be proficient and have good financial behavior [54]. The higher the mother's education level, the more financially literate the children will be [55].

Financial Socialization. Shim [56] conducted a cross-sectional study to test a conceptional financial socialization process model. The author used two theories to build the model name: the Theory of Planned Behavior [57] and the Theory of Consumer Socialization [58]. From the results, we can see that parental financial behavior and direct parental teaching have a strong impact on the adoption of parental financial role modeling and finance knowledge. At the same time, high school financial education and work experience significantly impact financial knowledge. Financial knowledge has a strong impact on healthy financial behavior and a moderate impact on perceived behavioral control and financial attitude, but it doesn't influence parents' subjective norms. Overall, for healthy financial behavior, financial attitude is very important. It has a stronger impact on it than other variables. Perceived behavior control is the only variable affecting financial satisfaction [57]. Deenanath's [59] model results showed that confidence in making money decisions and students' earned income were two factors that have a strong direct effect on financial knowledge and a huge indirect impact on financial behavior. Aisa's [60] paper on the financial literacy of high school students in the Netherlands revealed that students with low socioeconomic status, low mathematical ability, student's mothers without a college degree, and no family or peer communication about financial matters were showing lower levels of financial literacy. But contradicting the results found in the Ameer [22] study, financial socialization had no significant association with high financial literacy and confidence levels. The Federal Reserve Board's triennial Survey of Consumer Finances [61] data for the years 2016–2019 conducted in the United States of America clearly showed family income, wealth, credit use, and various financial outcomes. The mean income is highest for the consumers whose usual income was 90–100, highly educated (college degree or more), whites, who have their own house, live in a Metropolitan statistical area, and have the highest percentage of net worth among the sample. Regarding saving behavior, families have steadily given good results with the rise in the rate of savings from 2016 to 2019. Families with the highest usual income group (90%–100%) have an above-80% savings rate, whereas those below 50% of the usual income group have the lowest savings rate compared to the total mean value. Median wealth by parental education and own education shows that the median wealth for parents having a college degree in three times that of their own education parents. Families with direct and indirect holdings of stock in the 2013–2019 survey chart showed that consumers with a 90–100 percentile of income were holding four times more than the 0–49.9 percentile of income. Retirement plan participation is good for all years 2013–2019, highest for the usual income group 90–100 with a mean percentage of approximately 95%, and lowest at 45% for the usual income group of 0%–49.9% [61].

13.4.5 FINANCIAL LITERACY AND FINANCIAL INSTRUMENTS

Acquiring and improving financial literacy helps a lot in wealth generation, getting more profits, making better decisions, making the right decisions according to the various information shocks, etc. Those with high financial literacy performed well with different financial instruments and products.

Maarteen Van Rooij [44] showed that education level, income level, and wealth were directly proportional to financial literacy, and those subjects with high financial literacy levels had a higher participation rate in stocks. But financial literacy is hump-like for the age of the samples, and economic education has a low impact on stock market participation. Retirement wealth build-up and planning were significant for the smooth running of life since financial literacy teaches the right investment and doing the perfect planning and investing and helps to overcome risk-aversion behavior in crucial periods. Lusardi [62] compared two cohorts of 1992 and 2004 (baby boomers) in the USA on housing, non-housing, and total wealth among all races, from low to highly educated. The results have shown that financial and political literacy are strongly linked with planning and wealth building. And suggest that the one-size-fits-all program cannot help build wealth for an entire population. Lusardi [62] used the HRS (Health and Retirement Data of the United States). The data showed that people with more financial literacy could accumulate wealth than financial illiterates. Again, the results showed that the higher the level of education (university or more), the higher the financial literacy level. Similar results were found in Singapore's older people [25]. Results showed that financial knowledge helped with good retirement plans, savings, wealth accumulation, and better portfolio diversification and investments in complex non-pension funds.

The need for emergency funds in this modern world is quite important. To make people aware of the need for emergency funds, financial literacy helps a lot. The results of this paper show financial knowledge, higher education, vital income sources, and employment were strongly correlated with the maintenance of emergency funds [63]. Not all financial products are expensive, and not all expensive or premium products give tremendous performance and returns. In terms of expenses on the products, financial literacy helps a lot of consumers understand that fund fees and discounts are not the main deal if the product has all the features, and character to deliver the best performance. Results show that sophisticated investors choose funds based on precise fees and are not going to buy cheap funds. Highly knowledgeable customers choose funds based on data from various sources, whereas less knowledgeable customers directly pick the traditional routine method for the selection of the funds [64]. Trading is a fine art that requires hard work, study, patience, cognitive thinking, and self-discipline. It involves a lot of practice and groundwork before investing. This is an investing platform, not a gambling platform. According to the paper by Barber et al. [65], individual investors who engage in short-term, aggressive trading behaviors such as day trading often experience reduced returns, with an average reduction of nearly 3.8% annually. This is often due to a combination of factors such as an undiversified portfolio, a lack of proper financial knowledge, and inadequate training. It's important for individual investors to understand the risks involved in these types of investment strategies and to approach investing with caution, seeking out proper education and guidance.

From the data, institutions recorded the highest profits for passive and aggressive trade from one-day to long-term trading. Later, AMCs and foreigners recorded significant growth rates. Due to low financial literacy levels and knowledge, individual traders only lose so much money in trading for both aggressive and passive trading for long-term investment.

Financial literacy helps reduce the entry barriers to the participation of complex financial products. Hsiao [66] clearly showed that individuals with low-level financial literacy have low-level involvement in the derivative markets compared with those with high financial literacy. Hence, we can say that individuals with high-level financial literacy and low-level risk-averse attitudes have a maximum participation rate in complex needs.

13.4.6 FINANCIAL BEHAVIOR

Financial behavior is crucial for making good financial decisions in a lifetime and finically achieving financial wellness, a good financial behavior helps to achieve a great experience and success in financial product usage. Hence for good financial behavior, is financial literacy helps? Yes, financial literacy helps a lot for good financial behavior. Somtip's [36] survey in Japan revealed that those who use financial knowledge and have high financial literacy levels would make excellent financial decisions and were very unlikely to do the gambling.

Sam [67] investigated the effect of financial literacy on financial behavior for both actual and perceived financial literacy. For subjective and objective financial knowledge, the respondents were assigned to perceived-high/actual-high, perceived-high/actual-low, and perceived-low/actual-high based on the scores they achieved. From the tabular data, credit card behavior results, perceived-high/actual-high, were significant and better than others in good credit card behavior. Similarly, for investment behavior, perceived-high/actual-high has a significant and high percentage of holding the products in stocks, IRAs, and rebalancing their products. For loan behavior, the perceived-high/actual-high group has better and more significant results than other groups by holding most of their own houses, paying mortgage payments on time, avoiding late payments, choosing mortgage products, comparing the different lenders, and considering the most recent auto loan. For financial advice behavior, the perceived-high/actual-high group dominated the results. Maintaining good behavior and taking advice from a financial professional before using various financial products were significant. Julia [68] studied how financial game applications improve financial behavior. Findings revealed that those who use the financial app tend to exhibit higher subjective knowledge, common objective financial knowledge, and want social and economic features of the application.

Trading Behavior. Based upon the market trends and volumes in the share market, Tobias [69] designed a strategy for trading decisions based on the Google query volume search on Google Trends for the years 2004–2011. He analyzed a graph between the change in the index value of Dow Jones and the relative volume search. During the crisis in the market, consumers searched more on the internet about the terms. This generates the query volume, which is useful for future market falls and developing a strategy to gain profits in these conditions. To investigate changes in the information gathering behavior as captured by Google trends, later changes in the index were implemented by an investment strategy for a portfolio using search volume data called 'Google Trends Strategy'. The performance of the Google Trends strategy based on search term debt was analyzed using a purely random investment strategy. The portfolio yielded a return of 326% by using the Google Trends strategy using the search volume of the term debt. In contrast, using a purely random investment strategy for 10,000 simulations, the yield rate was only 16%. The investment strategy successfully worked for US investors in the USA but may not work for global investors [69].

Financial Development and Investment Behavior. Similarly, investments were required to create new research and development for the nation's development, and a lot of capital and investments were needed in the sector's growth to reach the global market's pinnacle. Inessa [70] conducted vector autoregression on firms from 36 countries to analyze the relationship between the firm's investment decision and the respected countries' financial development. The sample was divided into two groups: low and high financial development countries based on financial product measuring five indicators from Demirguc-Kunt [71]. A total of 8000 firms' data for the years 1988–1999 were used. A total of four variables were used, namely investment to capital ratio, sales to capital ratio (SKB), cash flow scaled by capital (CFKB), investment to capital ratio (IKB), and Tobin's q, which measures the market value of assets to the book value of assets (TOBINQ). The impact of the lagged cash flow on the level of investment is greater in countries with a low financial development index than in countries with the higher financial development index.

From the graphs of impulse response for 1 lag VAR of SKB, CFKB, IKB, and TOBINQ between the high financially developed countries and low financially developed countries, the lag is more for the low financially developed countries in two samples. Inefficient funds for allocating capital were one of the significant causes of slower growth rates in low financially developed nations. For achieving good financial behavior, not only is objective financial knowledge required, but subjective financial knowledge also plays a key role [70,71].

In addition to financial knowledge, financial attitude and various psychological traits significantly impact financial behavior. For this, Tang [72] studied the relationship.

From the total direct and indirect effects of self-esteem on financial behavior, self-esteem immediately affects positive and significant impacts on increasing savings and risky investments and a negative effect on credit card debt on financial assets. In contrast, the indirect impact of self-esteem on financial behavior also has a solid significance with savings and investments and a negative effectiveness with credit card debt/ financial assets [72].

Financial Literacy, Financial Behavior, and Psychological Factors. Sharon [73] investigated a study with a sample rate of 5,329 high school students in financial knowledge, self-efficacy, and behavior before and after the High School Financial Planning Program. The results of financial knowledge almost increased by 54.56% after the course. Self-efficacy increased by 42.2% and financial behavior increased by 35.13% after the program's implementation [73]. Yoshihiko's [74] study on determinants of financial literacy in Japan found that financial literacy was unrelated to financial satisfaction and didn't directly affect financial anxiety in the future. Melisa's [55] study on the role of cognitive ability on financial literacy proved experimentally that there is a strong relationship between them and individuals with higher cognitive ability who are highly financially literate.

Best Practice Behavior. Everything in this world can be achieved with a good amount of practice, and not giving up the failure and gaining money wasn't the only goal enough. Along with this financial independence and stratification, many other psychological traits should be fulfilled. Cliff [75] researched to examine the relationship between personal financial knowledge, financial satisfaction and best practice financial behavior. Results showed that financial knowledge has a weaker correlation with financial satisfaction, whereas best practices are strongly correlated with financial satisfaction. From the Multiple regression results, financial knowledge has a significant impact on financial behavior, income has the highest impact on financial behavior, and both financial satisfaction and financial confidence have a more significant effect on financial behavior than financial knowledge. From this, we can conclude that, for appropriate financial behavior, educating for only financial knowledge will not get the expected results [75].

Herd Behavior. The market never stays in a single trend, and the same strategy doesn't work for all conditions. Similarly, every individual should think and do proper analysis before investing. Sometimes uncertainties occur in the market, leading to recession or bearish trends, especially in this volatile situation. Following herd behavior is not the right thing to do. Hence, blindly following others will not always be beneficial to individuals. Sushil [76] clearly stated what herding means, who benefits from herding, why they choose to herd, and the steps that lead to wrong investment decisions. Hence, herds start and lead to bad decisions for all of them, which in turn causes a snowball effect [76].

Applying Behavioral Theories to Financial Behavior. Xiao [77] clearly stated the importance of financial education for improvement in positive financial behavior, and Shim [78] conducted SEM using TPB (Theory of Planned Behavior). The results showed that financial attitude, directly and indirectly, impacts financial behavior and intention by mediating variable financial knowledge. And few others studied the relationship between financial literacy and financial behavior with evidence [57,77].

Financial Literacy and Financial Risk Behavior. Antonia's [79] study showed financial literacy and numeracy significantly positively correlated, and both negatively and significantly correlated with risk aversion, whereas in the Mudzingiri [80] study, high-level financial literacy strongly and significantly correlated with high risk-taking behavior and the right amount of confidence. Saurabh [81] used financial risk attitude and financial behavior as mediating variables and found that both have a solid mediating relationship between financial knowledge and financial satisfaction and financial socialization and financial satisfaction. Similarly, the study by Aboagye [82] found that risk tolerance, appropriate spending behavior, and having an emergency fund and retirement plan significantly influence financial satisfaction. Financial literacy is negatively related to risk-aversion behavior [56,82]. The Washington State Department conducted a sponsored survey of financial knowledge, financial behavior, financial attitude, and experiences in 2003. The survey was divided into two groups: one victim group and a second general group. From the tabular results on the financial knowledge test, for the overall twelve questions, the mean knowledge score was 8.04, for the victim group 7.95, and for the general population 8.21. For the financial experiences of the total of 14 questions, 64.90% had correct answers, the victim pool answered 66.07% correctly answers, and the general population answered 61.95% correctly. For positive financial behavior, out of nine questions, total respondents answered 48.42% correctly, with 45.43% mean correct answers for the victim pool 52.94% mean correct answers for the general population. Hence, general group exhibits better positive financial behavior than others [83].

Consumer Financial Behavior. Vanessa [84] researched to examine the relationship between consumer financial knowledge, income, and locus of control on financial behavior. From the results, the external LOC was negatively correlated with financial behavior, whereas financial knowledge and income were positively correlated with financial behavior. From the regression results, financial knowledge and income over $35,000 significantly affect financial behavior. In contrast, external LOC has a negative effect on financial behavior, hence from the results we can say that LOC is successfully mediating the relationship between financial knowledge and financial behavior, between income and financial behavior.

13.5 CONCLUSION

This chapter clearly shows the current levels of financial literacy rate in India and presented the literature review of famous financial literacy and financial behavior papers. This chapter analyzed the results of various methods used, scientific results, measurement, interventions used, determinants of financial literacy, the relationship of financial literacy, financial knowledge, and financial behavior.

REFERENCES

1. Lauren. E. Willis, "Against Financial Literacy Education," *Penn Law: Legal Scholarship Repository*, pp. 1–56, Mar. 2008.
2. A. E. Faulkner, "Financial literacy education in the United States: Exploring popular personal finance literature," *Journal of Librarianship and Information Science*, Jan. 2016, https://doi.org/10.1177/0961000615616106.
3. L. Klapper, A. Lusardi, and P. Van Oudheusden, "Financial Literacy Around the World: Insights from the Standard & Poor's Ratings Services Global Financial Literacy Survey," S&P. World Bank Development Research Group, 2016.
4. IANS, "People faced serious problems due to note ban: Survey," *Business Standard India*, Nov. 07, 2017. Available: https://www.business-standard.com/article/news-ians/people-faced-serious-problems-due-to-note-ban-survey-117110701393_1.html.
5. S. Staff, "Reading list: How the common man had to cope with demonetisation," *Scroll.in*, Aug. 31, 2017. https://scroll.in/latest/849126/reading-list-how-the-common-man-had-to-cope-with-demonetisation.
6. V. Kaul, "Viewpoint: Why Modi's currency gamble was 'epic failure,'" *BBC News*, Aug. 30, 2017. Available: https://www.bbc.com/news/world-asia-india-41100610.
7. "DIGITAL GOVERNANCE IN INDIA Contribution of Smartphones to," India Cellular & Electronics Association, 2020.
8. Population, total Data, The World Development Report (WDR), 2019. https://data.worldbank.org/indicator/sp.pop.totl?most_recent_value_desc=true.
9. "Mutual funds add more than 81 lakh investor accounts in 2020-21," *mint*, Apr. 25, 2021. https://www.livemint.com/mutual-fund/mf-news/mutual-funds-add-more-than-81-lakh-investor-accounts-in-202021-11619329011346.html.
10. "Indian Mutual Fund Industry's Average Assets Under Management (AAUM) stood at ₹ 38.89 Lakh Crore (INR 38.89Trillion)," *Amfiindia.com*, 2020. https://www.amfiindia.com/indian-mutual.
11. V. Mahambare and S. Dhanaraj, "Has crop insurance helped Indian farmers? Many don't get payments on time," *ThePrint*, Oct. 28, 2021. https://theprint.in/opinion/has-crop-insurance-helped-indian-farmers-many-dont-get-payments-on-time/757815/.
12. A. Mani, S. Mullainathan, E. Shafir, and J. Zhao, "Poverty impedes cognitive function," *Science*, vol. 341, no. 6149, pp. 976–980, Aug. 2013. https://doi.org/10.1126/science.1238041.
13. S. Poonam, "Chinese instant-loan apps have exploded in India. As lockdown hurts salaries, borrowers say they are threatened with violence and fake police reports over repayment.," *Business Insider*, May 07, 2020. https://www.businessinsider.com/india-lending-apps-threats-google-play-store-2020-5?IR=T.
14. P. Mallikarjunan, "How app-based lenders are harassing, sucking borrowers dry," *Moneylife NEWS & VIEWS*, Jun. 12, 2020. https://www.moneylife.in/article/how-app-based-lenders-are-harassing-sucking-borrowers-dry/60621.html?__cf_chl_captcha_tk__=VYzKpV8.rfeawMZUhION7_Lt6a7NR5Mg85la1mFEMm8-1638094872-0-gaNycGzNBqU.
15. "ReserveBankofIndia-Reports," *www.rbi.org.in*, Nov.18, 2021. https://www.rbi.org.in/Scripts/PublicationReportDetails.aspx?UrlPage=&ID=1189.
16. OECD (2018), OECD/INFE Toolkit for Measuring Financial Literacy and Financial Inclusion.

17. A. L. Kiliyanni and S. Sivaraman, "The perception-reality gap in financial literacy: Evidence from the most literate state in India," *International Review of Economics Education*, vol. 23, pp. 47–64, Sep. 2016, https://doi.org/10.1016/j.iree.2016.07.001.

18. F. Carpena, S. Cole, J. Shapiro, and B. Zia, "The ABCs of financial education: Experimental Evidence on attitudes, behavior, and cognitive biases," *Management Science*, vol. 65, no. 1, pp. 346–369, Jan. 2019, https://doi.org/10.1287/mnsc.2017.2819.

19. J. D. Jayaraman and S. Jambunathan, "Financial literacy among high school students: Evidence from India," *Citizenship, Social and Economics Education*, vol. 17, no. 3, pp. 168–187, Oct. 2018, https://doi.org/10.1177/2047173418809712.

20. "Financial Literacy and Inclusion in India," National Insititute of Securities Market, Final Report on the Survey Results, Mumbai, India. https://www.ncfe.org.in/images/pdfs/reports/NFLIS_2019.pdf, 2019.

21. S. Agarwalla, S. Barua, J. Jacob, and J. R. Varma, "A Survey of Financial Literacy among Students, Young Employees and the Retired in India", https://faculty.iima.ac.in/iffm/literacy/youngemployeessandretired2012.pdf, Indian Institute of Management Ahmedabad, 2012.

22. S. K. Agarwalla, S. K. Barua, J. Jacob, and J. R. Varma, "Financial literacy among working young in Urban India," *World Development*, vol. 67, pp. 101–109, Mar. 2015, https://doi.org/10.1016/j.worlddev.2014.10.004.

23. A. Lusardi, A. S. Samek, A. Kapteyn, L. Glinert, A. Hung, and A. Heinberg, "Visual tools and narratives: New ways to improve financial literacy," *SSRN Electronic Journal*, no. 20229, 2014, https://doi.org/10.2139/ssrn.2585231.

24. R. Ameer and R. Khan, "Financial socialization, financial literacy, and financial behavior of adults in New Zealand," *Journal of Financial Counseling and Planning*, vol. 31, no. 2, p. JFCP-18-00042, Mar. 2020, https://doi.org/10.1891/jfcp-18-00042.

25. L. D. DeLaune, J. S. Rakow, and K. C. Rakow, "Teaching financial literacy in a co-curricular service-learning model," *Journal of Accounting Education*, vol. 28, no. 2, pp. 103–113, Jun. 2010, https://doi.org/10.1016/j.jaccedu.2011.03.002.

26. R. Barua, B. Koh, and O. S. Mitchell, "Does financial education enhance financial preparedness? Evidence from a natural experiment in Singapore," *Journal of Pension Economics and Finance*, vol. 17, no. 3, pp. 254–277, Aug. 2017, https://doi.org/10.1017/s1474747217000312.

27. L. Mandell and L. S. Klein, "The impact of financial literacy education on subsequent financial behavior," *Journal of Financial Counceling and Planning*, vol. 20, no. 1, pp. 15–24, 2009.

28. C. Urban, M. Schmeiser, J. M. Collins, and A. Brown, "The effects of high school personal financial education policies on financial behavior," *Economics of Education Review*, p. 101786, Mar. 2018, https://doi.org/10.1016/j.econedurev.2018.03.006.

29. B. Bernheim, D. Garrett, and D. Maki, "Education and saving: The long-term effects of high school financial curriculum mandates," *Journal of Public Economics*, vol. 80, pp. 435–465, 2001.

30. S. Cole, A. Paulson, and G. K. Shastry, "Smart money? The effect of education on financial outcomes," *Review of Financial Studies*, vol. 27, no. 7, pp. 2022–2051, Feb. 2014, https://doi.org/10.1093/rfs/hhu012.

31. S. Cole et al., "If you are so smart, Why aren't you rich? The Effects of Education, Financial Literacy and Cognitive Ability on Financial Market Participation", working paper, https://afiweb.afi.es/eo/FinancialLiteracy.pdf, 2008.

32. J. Grigsby, W. Van Der Klaauw, J. Wen, and B. Zafar, *Financial Education and the Debt Behavior of the Young*, Oxford University Press, Oxford, 2016.

33. A. Lusardi and P. Tufano, "Debt Literacy, Financial Experiences, And OverIndebtedness," *NBER Working Paper*, no. 14808, Mar. 2009.

34. A. C. Lyons, L. Palmer, K. S. U. Jayaratne, and E. Scherpf, "Are we making the grade? A national overview of financial education and program evaluation," *The Journal of Consumer Affairs*, vol. 40 no 2, pp. 208–235, 2006.

35. B. I. Carlin and D. T. Robinson, "What does financial literacy training teach us?," *The Journal of Economic Education*, vol. 43, no. 3, pp. 235–247, Jul. 2012, https://doi.org/10.1080/00220485.2012.686385.

36. L. Willis, "Evidence and Ideology in Assessing the Effectiveness of Financial Literacy Education," Penn Law: Legal Scholarship Repository, Pennsylvania, Apr. 2008.

37. S. Watanapongvanich, P. Binnagan, P. Putthinun, M. S. R. Khan, and Y. Kadoya, "Financial literacy and gambling behavior: Evidence from Japan," *Journal of Gambling Studies*, no. 37, pp. 445–465, Mar. 2020, https://doi.org/10.1007/s10899-020-09936-3.

38. B. D. Bernheim and D. M. Garrett, "The determinants and consequences of financial education in the workplace: Evidence from a survey of households," *Journal of Public Economics*, vol. 87, pp. 1487–1519, Sep. 2001, https://doi.org/10.2139/ssrn.451.

39. Anshika, A. Singla, and G. Mallik, "Determinants of financial literacy: Empirical evidence from micro and small enterprises in India," *Asia Pacific Management Review*, vol. 26, pp. 248–255, Apr. 2021, https://doi.org/10.1016/j.apmrv.2021.03.001.

40. J. Hussain, S. Salia, and A. Karim, "Is knowledge that powerful? Financial literacy and access to finance," *Journal of Small Business and Enterprise Development*, vol. 25, no. 6, pp. 985–1003, Nov. 2018, https://doi.org/10.1108/jsbed-01-2018-0021.

41. J. J. Xiao and B. O'Neill, "Consumer financial education and financial capability," *International Journal of Consumer Studies*, vol. 40, no. 6, pp. 712–721, May 2016, https://doi.org/10.1111/ijcs.12285.

42. L. Mandell, "Financial literacy: If It's so important, why isn't it improving?," *SSRN Electronic Journal*, pp. 1–11, 2006, https://doi.org/10.2139/ssrn.923557.

43. A. Lusardi, "Financial literacy: An essential tool for informed consumer choice?," *National Bureau of Economic Research Working paper*, no. 14084, pp. 01–30, Jun. 2008, https://doi.org/10.2139/ssrn.1336389.

44. M. van Rooij, A. Lusardi, and R. Alessie, "Financial literacy and stock market participation," *Journal of Financial Economics*, vol. 101, no. 2, pp. 449–472, Aug. 2011, https://doi.org/10.1016/j.jfineco.2011.03.006.

45. D. Fernandes, J. Lynch, and R. Netemeyer, "Financial literacy, financial education, and downstream financial behaviors," *Management Science*, vol. 60, no. 8, pp. 1861–1883, 2014, https://doi.org/10.1287/mnsc.2013.1849. (full terms and conditions of use: https://pubsonline.informs.org/page/terms-and-conditions Financial Literacy, Financial Education, and Downstream Financial Behaviors).

46. X. Li, "When financial literacy meets textual analysis: A conceptual review," *Journal of Behavioral and Experimental Finance*, vol. 28, p. 100402, Dec. 2020, https://doi.org/10.1016/j.jbef.2020.100402.

47. A. Lusardi, O. S. Mitchell, and V. Curto, "Financial literacy among the young," *Journal of Consumer Affairs*, vol. 44, no. 2, pp. 358–380, Jun. 2010, https://doi.org/10.1111/j.1745-6606.2010.01173.x.

48. H. Chen and R. P. Volpe, "An analysis of personal financial literacy among college students," *Financial Services Review*, vol. 7, no. 2, pp. 107–128, 1998, https://doi.org/10.1016/s1057-0810(99)80006-7.

49. H. Chen and R. P. Volpe, "Gender differences in personal financial literacy among college students," *Financial Services Review*, pp. 289–307, 2002.

50. H. K. Baker, S. Kumar, N. Goyal, and V. Gaur, "How financial literacy and demographic variables relate to behavioral biases," *Managerial Finance*, vol. 45, no. 1, pp. 124–146, Jan. 2019, https://doi.org/10.1108/mf-01-2018-0003.

51. A. Lusardi and O. S. Mitchell, "Planning and financial literacy: How do women fare?," *American Economic Review*, vol. 98, no. 2, pp. 413–417, Apr. 2008, https://doi.org/10.1257/aer.98.2.413.

52. S. K. Agarwalla, S. K. Barua, J. Jacob, and J. R. Varma, "Financial literacy among working young in Urban India," *World Development*, vol. 67, pp. 101–109, Mar. 2015, https://doi.org/10.1016/j.worlddev.2014.10.004.

53. L. Mandell, "Financial literacy of high school students," In Xiao, J.J. (ed.) *Handbook of Consumer Finance Research*, vol. 163, Springer, New York, 2008.

54. G. Van Campenhout, "Revaluing the role of parents as financial socialization agents in youth financial literacy programs," *Journal of Consumer Affairs*, vol. 49, no. 1, pp. 186–222, Feb. 2015, https://doi.org/10.1111/joca.12064.

55. M. Muñoz-Murillo, P. B. Álvarez-Franco, and D. A. Restrepo-Tobón, "The role of cognitive abilities on financial literacy: New experimental evidence," *Journal of Behavioral and Experimental Economics*, vol. 84, p. 101482, Feb. 2020, https://doi.org/10.1016/j.socec.2019.101482.

56. S. Shim, B. L. Barber, N. A. Card, J. J. Xiao, and J. Serido, "Financial socialization of first-year college students: The roles of parents, work, and education," *Journal of Youth and Adolescence*, vol. 39, no. 12, pp. 1457–1470, Jul. 2009, https://doi.org/10.1007/s10964-009-9432-x.

57. I. Ajzen, "The theory of planned behavior," *Organizational Behavior and Human Decision Processes*, vol. 50, no. 2, pp. 179–211, Dec. 1991, https://doi.org/10.1016/0749-5978(91)90020-T.

58. G. P. Moschis and G. A. Churchill, "Consumer socialization: A theoretical and empirical analysis," *Journal of Marketing Research*, vol. 15, no. 4, pp. 599–609, Nov. 1978, https://doi.org/10.2307/3150629.

59. V. Deenanath, S. M. Danes, and J. Jang, "Purposive and unintentional family financial socialization, subjective financial knowledge, and financial behavior of high school students," *Journal of Financial Counseling and Planning*, vol. 30, no. 1, pp. 83–96, Jun. 2019, https://doi.org/10.1891/1052-3073.30.1.83.

60. A. Amagir, W. Groot, H. M. van den Brink, and A. Wilschut, "Financial literacy of high school students in the Netherlands: Knowledge, attitudes, self-efficacy, and behavior," *International Review of Economics Education*, vol. 34, no. 100185, p. 100185, Jun. 2020, https://doi.org/10.1016/j. iree.2020.100185.

61. N. Bhutta, et al., "Changes in U.S. Family Finances from 2016 to 2019: Evidence from the Survey of Consumer FinancesChanges," Federal Reserve Bulletin, United States of America, Sep. 2020. Accessed: May 23, 2022. [Online]. Available: https://www.federalreserve.gov/publications/files.

62. A. Lusardi and O. S. Mitchell, "Financial literacy and planning: Implications for retirement wellbeing," *SSRN Electronic Journal*, May 2011, https://doi.org/10.2139/ssrn.881847.

63. P. Babiarz and C. A. Robb, "Financial literacy and emergency saving," *Journal of Family and Economic Issues*, vol. 35, no. 1, pp. 40–50, Aug. 2013, https://doi.org/10.1007/s10834-013-9369-9.

64. S. Müller and M. Weber, "Financial literacy and mutual fund investments: Who buys actively managed funds?," *Schmalenbach Business Review*, vol. 62, no. 2, pp. 126–153, Apr. 2010, https://doi.org/10.1007/ bf03396802.

65. B. M. Barber, Y.-T. Lee, Y.-J. Liu, and T. Odean, "Just how much do individual investors lose by trading?," *Review of Financial Studies*, vol. 22, no. 2, pp. 609–632, Apr. 2008, https://doi.org/10.1093/rfs/ hhn046.

66. Y.-J. Hsiao and W.-C. Tsai, "Financial literacy and participation in the derivatives markets," *Journal of Banking & Finance*, vol. 88, pp. 15–29, Mar. 2018, https://doi.org/10.1016/j.jbankfin.2017.11.006.

67. S. Allgood and W. B. Walstad, "The effects of perceived and actual financial literacy on financial behaviors," *Economic Inquiry*, vol. 54, no. 1, pp. 675–697, Sep. 2015, https://doi.org/10.1111/ ecin.12255.

68. J. Bayuk and S. A. Altobello, "Can gamification improve financial Behavior? The moderating role of app expertise," *International Journal of Bank Marketing*, Feb. 2019, https://doi.org/10.1108/ ijbm-04-2018-0086.

69. T. Preis, H. S. Moat, and H. E. Stanley, "Quantifying trading behavior in financial markets using google trends," *Scientific Reports*, vol. 3, no. 1, Apr. 2013, https://doi.org/10.1038/srep01684.

70. I. Love and L. Zicchino, "Financial development and dynamic investment behavior: Evidence from panel VAR," *The Quarterly Review of Economics and Finance*, vol. 46, no. 2, pp. 190–210, May 2006, https://doi.org/10.1016/j.qref.2005.11.007.

71. A. Demirguc-Kunt and R. Levine, "Stock market development and financial intermediaries: Stylized facts," *The World Bank Economic Review*, vol. 10, no. 2, pp. 291–321, May 1996, https://doi.org/10.1093/ wber/10.2.291.

72. N. Tang and A. Baker, "Self-esteem, financial knowledge and financial behavior," *Journal of Economic Psychology*, vol. 54, pp. 164–176, Jun. 2016, https://doi.org/10.1016/j.joep.2016.04.005.

73. S. Danes, H. Haberman, and M. Candidate, "Teen financial knowledge, self-efficacy, and behavior: A gendered view," *Journal ofFinancial Counseling and Planning*, vol. 18, no. 2, pp. 48–60, 2007.

74. Y. Kadoya and M. S. R. Khan, "What determines financial literacy in Japan?," *Journal of Pension Economics and Finance*, vol. 19, no. 3, pp. 353–371, Jan. 2019, https://doi.org/10.1017/s1474747218000379.

75. C. A. Robb and A. S. Woodyard, "Financial knowledge and best practice behavior," *Journal of Financial Counseling and Planning*, vol. 22, no. 1, 2011.

76. S. Bikhchandani and S. Sharma, "Herd behavior in financial markets," *Palgrave Macmillan Journals*, vol. 47, no. 3, pp. 279–310, 2000.

77. J. Xiao and J. Xiao, "Applying behavior theories to financial behavior," In Xiao, J.J. (ed.) *Handbook of Consumer Finance Research*, Springer, New York, 2008, pp. 69–81.

78. H.-M. Shih, B. H. Chen, M.-H. Chen, C.-H. Wang, and L.-F. Wang, "A study of the financial behavior based on the theory of planned behavior," *International Journal of Marketing Studies*, vol. 14, no. 2, p. 1, Jun. 2022, https://doi.org/10.5539/ijms.v14n2p1.

79. A. Grohmann, "Financial literacy and financial behavior: Evidence from the emerging Asian middle class," *Pacific-Basin Finance Journal*, vol. 48, pp. 129–143, Apr. 2018, https://doi.org/10.1016/j. pacfin.2018.01.007.

80. C. Mudzingiri, J. W. Muteba Mwamba, J. N. Keyser, and W. C. Poon, "Financial behavior, confidence, risk preferences and financial literacy of university students," *Cogent Economics & Finance*, vol. 6, no. 1, p. 1512366, Jan. 2018, https://doi.org/10.1080/23322039.2018.1512366.

81. K. Saurabh and T. Nandan, "Role of financial risk attitude and financial behavior as mediators in financial satisfaction," *South Asian Journal of Business Studies*, vol. 7, no. 2, pp. 207–224, Jun. 2018, https:// doi.org/10.1108/sajbs-07-2017-0088.

82. J. Aboagye and J. Y. Jung, "Debt holding, financial behavior, and financial satisfaction," *Journal of Financial Counseling and Planning*, vol. 29, no. 2, pp. 208–218, Nov. 2018, https://doi.org/10.1891/1052-3073.29.2.208.

83. D. Moore, *Survey of Financial Literacy in Washington State: Knowledge, behavior, Attitudes, and Experiences*, Social & Economic Sciences Research Center, Pullman, Washington, 2003. https://doi.org/10.13140/2.1.4729.4722.

84. V. G. Perry and M. D. Morris, "Who is in control? The role of self-perception, knowledge, and income in explaining consumer financial behavior," *Journal of Consumer Affairs*, vol. 39, no. 2, pp. 299–313, Sep. 2005, https://doi.org/10.1111/j.1745-6606.2005.00016.x.

14 A Literature Review on Financial Stability

El Mansouri Adnane, Benhouad Mohamed,
Mestari Mohammed, and El Aidouni Salma

14.1 INTRODUCTION

Since the recent financial crisis of 2007–2008, financial stability has become an ultimate objective of central banks, besides price stability. Today, after years of the recent financial crisis, there is no widely accepted definition of financial stability and, therefore, no accepted framework or clear policies to achieve this objective. In fact, contrary to a monetary policy framework that has a clear objective of price stability by targeting the inflation rate, financial stability is a difficult financial concept to properly achieve with no clear bottom line to target.

Moreover, monetary and fiscal policies and microprudential regulations alone did not prevent the buildup of systemic risk in the recent financial crisis. To overcome this issue, financial regulators have made important technical advances in setting up macroprudential toolkits that can complement traditional policies to enhance the resilience of the financial system. Widening the vision of financial regulation to capture interconnectedness between financial institutions and maintaining spillover effects have been the main concerns of regulation since the recent financial crisis.

Below, we present the (Section 14.2) methodology used in this chapter: the multiple definitions of financial stability (Section 14.3), measurement of financial instability (Section 14.4), framework for assessing financial stability (Section 14.5), monetary policies and financial stability (Section 14.6), and financial stability and macroprudential policy tools (Section 14.7).

14.2 METHODOLOGY OF THE LITERATURE REVIEW

In identifying sources for this literature review, multiple article databases were used. Initially, Web of Science, Science Direct, and Google Scholar were utilized to take an initial sample of what types of articles were available. Regarding Web of Science, Science Direct, and Google Scholar, broad search terms were initially used to establish a list of research articles that were primary sources and peer-reviewed. In the beginning, we used a basic search for financial stability. From the article titles and research data derived from Web of Science, Science Direct, and Google Scholar, we were able to use a better list of more refined terms when utilizing other databases.

The search terms selected for this literature review consisted of: financial stability, framework of financial stability, monetary policy and financial stability, measurement of financial stability, financial stability index, and macroprudential policies. These terms were combined in various ways with "AND" commands to obtain the most relevant and appropriate articles. In addition to the database searching, a number of articles were located using the Snowball method. Each of the search terms used was selected due to their appropriateness and relevance in consideration of the purpose of this literature review. The majority of cited references were within recent years. A few researchers were included from previous decades to establish foundational concepts that continue to this day.

DOI: 10.1201/9781032667478-16

14.3 DEFINITION OF FINANCIAL STABILITY

At first, there is no widely accepted detention of financial stability in the academic world or among central banks (Issing [1], Oosterloo and De Haan [2]), although most of the definitions share two major characteristics: (i) the financial system is examined with an overall view rather than looking at each financial institution alone and (ii) they make implicit reference to financial distress or crisis because it is more observable and useful in implementing a framework to monitor financial stability.

Mishkin [3] defines financial stability as "the prevalence of a financial system, which is able to ensure in a lasting way, and without major disruptions, an efficient allocation of savings to investment opportunities." This definition is quoted from Issing [1]. This definition of financial stability is very hard to observe in the real world and provides limited guidance to the institutions in charge of maintaining financial stability, although the definition is still accepted intellectually for the reason that it looks at the overall resilience of the financial system.

Mishkin [4] has also elucidated financial stability as a more practical and observable one. In fact, he defined financial stability by its opposite, financial instability, in which the instability of the financial system materializes when shocks lead to restricting the key function of the financial system in channeling funds to the real economy.

Schinasi [5] defined financial stability as the ability of the financial system to perform three functions: (i) to facilitate the allocation of economic resources and the effectiveness of other economic processes; (ii) to price and manage financial risks; and (iii) to maintain the ability to perform these key functions in the presence of external shocks or by a buildup of imbalances.

Allen and Wood [6] define financial stability as the situation in which the occurrence of an episode of financial instability is a tail event (i.e., unlikely to occur), so that the decisions of economic agents are not affected by the fear of materializing such an episode of instability. In the same context, Gadanecz and Jayaram [7] define financial stability as the absence of excessive volatility and stress.

Tommaso and Padoa-Schioppa [8] define financial stability as a condition where the financial system is able to bear shocks without impairing the process of channeling savings to investment and the processing of payments.

Haldane et al. [9] defined financial stability as (i) enabling individuals to smooth consumption across time (for example, by saving or borrowing) or across states of nature (for example, through insurance contracts) and (ii) the efficient financing of investment projects with saved resources.

Borio and Derhman [10] defined financial instability as a set of conditions that can lead to financial distress/financial crisis in response to a normal shock to the financial system. The shocks can originate from the financial system or outside of the financial system.

Also, the definition of systematic risk can be seen as a proxy definition of financial instability. Summer [11] and Bandt and Hartmann [12] defined a systemic risk as an event where the release of bad news about a financial institution, or even its failure, or the crash of a financial market leads to considerable adverse effects on one or several other financial institutions or markets.

Chant [13] defines financial stability as a controverse financial instability, which refers to a set of conditions in financial markets that can impair the performance of the economy through their impact on the working of the financial system. The imbalances can emerge from shocks that originate within the financial system or from the transmission of shocks that originate elsewhere by way of the financial system.

Houben [14] defines financial stability as the situation in which the financial system performs the three keys' elements: (i) allocating resources efficiently between activities and across time; (ii) assessing and managing financial risks; and (iii) absorbing shocks.

From the point of view of institutions, the European Central Bank (2007) defines financial stability as a condition in which the financial system – comprising financial intermediaries, markets, and market infrastructure – is capable of withstanding shocks and the unraveling of financial imbalances, thereby mitigating the likelihood of disruptions in the financial intermediation process that are severe enough to significantly impair the allocation of savings to profitable investment opportunities. This definition seems incomplete as a result of the unknown origin of the shock. The Financial Stability Board defines financial stability in the Financial Stability Surveillance Framework (2021) as the capacity of the global financial system to withstand shocks, containing the risk of disruptions in the financial intermediation process and other financial system functions that are severe enough to adversely impact the real economy. In our point of view, the most appropriate definition of financial stability is the absence of an episode of financial distress or crisis because of the difficulties associated with observing the characteristics of other definitions.

14.4   MEASUREMENT OF FINANCIAL INSTABILITY

Measuring financial stability is a challenging task because financial stability itself lacks observable characteristics. Thus, the most appropriate definition useful in the measurement of financial stability is financial instability. In this context, the ideal measurement of financial instability is the output of structured model mapping instrument to the objective assigned (Borio, Drehmann [10]). As a consequence, policymakers use different quantitative indicators that try to capture financial instability signals. Those tools can be divided into three types of indicators:

- **Composite financial indicators:** These indicators are statistics that compress a wide range of variables (capital adequacy, asset quality, profitability, liquidity, market risk sensitivity, financial intermediaries, money market funds, insurance corporations, pension funds, nonfinancial corporations, and households) that aim to monitor the resilience of the financial system. The IMF published a guide for policymakers in order to calculate those indicators, called Financial Soundness Indicators (FSIs, 2019). The FSIs methodology was reviewed to capture concentration and distribution measures that provide information on tail events and variations in distributions (IMF, 2022). Borio and Drehmann [10] suggested that those indicators can be used in a more detailed analysis of financial vulnerabilities.
- **Early Warning Indicators:** The main objective of Early Warning Indicators (EWI) is to capture stress signals in the financial system that can lead to a situation of financial distress/crisis. Borio and Drehmann [15] found that household and international indicators, in particular cross-border debt, contain useful information about the next banking crisis. The BIS (2018) identified seven variables as EWI. The selected variables are credit-to-GDP gap, total debt to service ratio (DSR), property price, household DSR, household credit-to-GDP gap, foreign currency debt to GDP, and cross-border claims to GDP.
- **Macro-stress testing:** Macro-stress testing is a toolbox of instruments that aims to stress the overall financial system rather than micro-stress testing relative to individual institutions (banks, insurance companies). The objective of the macro-stress test is to assist public authorities in the process of identifying vulnerabilities and overall risk exposures in a financial system that could lead to financial instability. Bunn et al. [16] describe some findings of the macro-stress test carried out as part of the Financial Stability Assessment Program for the United Kingdom.

14.5   FRAMEWORK FOR ASSESSING FINANCIAL STABILITY

The recent financial crisis of 2007–2008 has shown that prudential tools alone can't achieve the stability of the financial system. Also, the monetary policy that targets price stability failed to ensure financial stability. Furthermore, fiscal policies should be integrated into the policies that aim for financial stability (Hannoun [17]). Thus, there is a need for a monitoring framework for financial stability.

The elements discussed above suggest that there is no precise operational framework for measuring financial stability. In fact, the definition of financial stability has some elements that are very hard to measure. So, any measurement is, by nature, incomplete or fuzzy. Still, knowing the limitations of any framework, regulators and researchers have developed frameworks in order to adjust them to be helpful in policy guidance.

The fundamentals of the frameworks suggested by Hannoun [17] and Borio and Drehmann [10] have identified six features of any operational framework:

- The framework should be system-wide, and the orientation of the framework should be macroprudential rather than microprudential. Besides, it should take into account the inter-linkages and feedback between the financial system and the real economy.
- Some institutions have to be in the scoop of the regulators more than others, in particular high-leveraged institutions or institutions with large liabilities outstanding (Arnold, Borio, Ellis and Moshirian, 2013).
- The significance of financial institutions' exposure to systemic risk is notable. As exposure to systemic risks increases, so does the likelihood of posing a threat to the financial system.
- The procyclicality of the financial system should be dampened using macroprudential tools, i.e., the building of buffers during the expansion phase of the economy and activating the buffers in the downturns of economic activity. Introduction of automatic stabilizers in the framework. This feature of the framework avoids the fear of "going against the wind" in the case of building up systemic risk.
- Institutions in charge of implementing this framework should have control over the instruments of financial stability.

Tobias et al. [18] have identified a set of financial vulnerabilities in the con- text of global monitoring framework. Financial vulnerabilities identified are the following (Tables 14.1 and 14.2).

TABLE 14.1

Financial Sector Leverage and Maturity Assessment

	Leverage	Maturity
Banking Sector	Regulatory capital	Short-term funds ratio
	Stress test capital	Liquid asset ratios
	Market-based measures	Regulatory liquidity
	Off-balance-sheet	
Non-bank financial sector	Regulatory capital	Short-term wholesale funds ratio
	Cap on loan-to-value ratio	Open-end funds and
	Securitizations	ex-change-traded funds (with
	Off-balance-sheet assets and derivatives	less liquid assets)
	Lending standards	Debt with adjustable rates
Households sector	Credit to GDP	
	Debt service	
	Credit growth	
Corporate sector	Government debt to GDP	Short-term debt
	Debt growth	Adjustable-rate debt
	Off-balance-sheet liabilities	Liquid assets
Government Sector	Government debt to GDP	Short-term debt
	Debt growth	Short-term debt
	Off-balance-sheet liabilities	Liquidity and depth of market

TABLE 14.2
External Debt Claims and Interconnections in Different Sectors

	External Debt Claims	Interconnections
Banking Sector	• External funding needs • Cross-border funding	• Interbank claims • Non-bank financial claims • Cross-border activities • Price-based systemic risk measures
Non-bank financial sector	• Open-end and other funds • invested in foreign debt	• Claims on banks • Claims on other non-bank institutions
Households sector		• Debt overhang
Corporate sector	• Debt issued in foreign currencies	
Government Sector	• External debt • US dollar versus local currency debt • Short-term debt to foreign exchange re-serves • Capital flows	

14.6 MONETARY POLICIES AND FINANCIAL STABILITY

The recent financial crisis of 2008 led to a rethinking of central bank principles of price stability using monetary policies. In fact, a monetary policy framework with a low and stable inflation rate didn't prevent the buildup of systemic risk. Therefore, price stability isn't a sufficient condition for financial stability. This evidence accelerated the emergence of the macroprudential framework that aims to strengthen the resilience of the financial system (the macroprudential tools will be discussed in the next chapter). The assignment of monetary policy to price stability and macroprudential policy to the soundness of the financial system is in line with Tinbergen's (1952) principal, which states that we should have as many instruments as many objectives.

The relationship between monetary policies and financial stability objectives is still an ongoing debate. The monetary policy framework should focus only on price stability, or financial stability objectives should be integrated into this framework. There are two major views on this issue:

- The first views suggest that the monetary policy framework should have the exclusive objective of maintaining price stability, and macroprudential authorities should focus on the resilience of the financial system to shocks (Greenspan [19]). Therefore, the objective of price stability is separated from the objective of financial stability.
- The second view argues that the financial stability objective should be part of the secondary objectives of monetary policy (Borio and White [20], Roubini [21], and Woodford [22]).

Macroprudential policy and monetary policy are hardly to be seen as two separate policies. In fact, macroeconomic instruments can affect macroeconomic activity (price stability). Also, it can have an impact on the transmission of monetary policy. For instance, if new regulations are issued to constrain borrowers' access to funds, the credit channel of monetary policy is probably to become weaker. For its part, monetary policy can contribute to the buildup of systemic risk in the context of low interest rates for a long period, and that prompts economic agents to take excessive risks [22].

We argue that monetary policy and macroprudential policy can work together to achieve price stability and financial stability. In fact, during an economic boom that is characterized by high inflation and the risk of building up financial imbalances, simultaneous tightening of monetary policy and the activation of the contracyclical capital buffer can underpin the objectives of the two policies. On the other hand, in a downturn of economic activity, banking losses can be addressed by lowering the interest rate and releasing the contracyclical capital buffer.

14.7   FINANCIAL STABILITY AND MACROPRUDENTIAL POLICY TOOLS

The macroprudential framework has become a key tool introduced in the reforms of the global financial system after the recent financial crisis of 2008, designed to ensure financial stability. The macroprudential framework is defined as the set of prudential tools targeting systemic risk that mitigate its economic costs (Borio [15]). Caruana [23] defined the macroprudential framework as the use of prudential tools with the explicit objective of promoting the stability of the financial system as a whole, not necessarily of the individual institutions within it.

Macroprudential tools can be divided into two broad types of instruments: those that can be used to increase the resilience of banks (e.g., capital, liquidity-based tools, large exposure limits) and those that affect the credit terms offered to borrowers. The different instruments can be structured as follows (Table 14.3).

The different macroprudential tools aim to limit credit growth (smoothing the credit cycle) and enhance the resilience of the financial system. In fact, 22 of the macroprudential tools had the objective of increasing the resilience of the financial sector, using capital liquidity or provisioning requirements, while 78 had the purpose of dampening the credit cycle (Gambacorta et al. [24]).

Evaluating the impact of macroprudential policies on the stability of the financial system is an early stage, and more development needs to be done [25]. For instance, empirical evidence suggests that caps on DSTI and caps on loan-to-value ratio are more effective than capital requirements in containing asset growth, while buffer-based policies seem to have little impact on credit growth. Overall, there is little evidence that the effectiveness of these tools varies by the intensity of the cycle (Claessens et al. [26]). Cerutti et al. [27] found that macroprudential tools are frequently

TABLE 14.3
2021 IMF Macroprudential Policy Survey Compilation Guide

Sector	Macroprudential Instrument
Banking Sector	Countercyclical capital buffer
	Capital conservation buffer
	Limit on leverage ratio
	Forward-looking loan loss provisioning requirement
Household Sector	Limit on amortization periods
	Cap on loan-to-value ratio
	Cap on loan-to-income ratio
	Cap on debt-service-to-income ratio
	Restrictions on unsecured loans
Corporate Sector	Cap on loan-to-value ratio for commercial real estate
	Cap on debt-service coverage ratio for real estate
	Cap based on borrower leverage
Liquidity instruments	Liquidity coverage ratio
	Liquidity coverage ratio differentiated by currency
	Liquid asset ratio
	Liquid asset ratio differentiated by currency
	Net stable funding ratio differentiated by currency.
	Limits on maturity mismatches
	Loan-to-deposit ratio
	Net stable funding ratio

used in emerging economies, and in advanced economies, borrower-based policies (such as caps on loan-to-value and debt-service-to-income ratios) are commonly used. Moreover, borrower-based policies proved to be effective in limiting credit growth and house prices. However, macroprudential tools are known to be less efficient in economic downturns.

14.8   CONCLUSION

This work has discussed different aspects of financial stability and argues that the financial instability definition is the most relevant definition to be used in any operational financial stability framework. Further, there is a need for intensive work to measure financial stability and to develop instruments that can identify the building up of systemic risk threatening the financial system. For the prudential side, widening the range of macroprudential policy, applied only to the banking sector, to the non-banking sector (financial markets in particular) will enhance the resilience of the financial system. Finally, macroprudential tools are indeed effective instruments that contribute to the resilience of the financial system's financial stability, although they can't alone achieve that task. Therefore, there is a need to develop a holistic framework that mixes macroeconomic policies (monetary policy and fiscal policy) and prudential policies (macroprudential policy and microprudential policy) to maintain and safeguard financial stability.

REFERENCES

1. Issing, O.: Monetary and financial stability: is there a trade-off? *BIS Papers* **18**, 16–23 (2003)
2. Oosterloo, S., de Haan, J., et al.: A survey of institutional frameworks for financial stability. Tech. rep., Netherlands Central Bank, Research Department (2003).
3. Mishkin, F.S. (1990). Asymmetric information and financial crises: A historical perspective. Financial Markets and Financial Crises. University of Chicago Press, Chicago pp. 69–108 (1991)
4. Mishkin, F.S.: Global financial instability: Framework, events, issues. *Journal of Economic Perspectives* **13**(4), 3–20 (1999).
5. Schinasi, G.J.: Defining financial stability, International Monetary Fund, WP/04/187 (2004).
6. Allen, W.A., Wood, G.: Defining and achieving financial stability. *Journal of Financial Stability* **2**(2), 152–172 (2006).
7. Gadanecz, B., Jayaram, K.: Measures of financial stability-a review. *Irving Fisher Committee Bulletin* **31**(1), 365–383 (2008).
8. Padoa-Schioppa, T.: Central banks and financial stability: Exploring the land in between. *The Transformation of the European Financial System* **25**, 269–310 (2003).
9. Haldane, A., Saporta, V., Hall, S., Tanaka, M.: Financial stability and macroeconomic models. *Bank of England Financial Stability Review* **16**, 80–88 (2004).
10. Borio, C., Drehemann, M.: Toward an operational framework for financial stability:'fuzzy'measurement and its consequences. *Series on Central Banking, Analysis, and Economic Policies*, no. 15 (2010). doi: 10.2139/ssrn.1458294.
11. Elsinger, H., Lehar, A., Summer, M.: Risk assessment for banking systems. *Management Science* **52**(9), 1301–1314 (2006).
12. De Bandt, O., Hartmann, P.: Systemic risk: A survey. Available at SSRN 258430 (2000).
13. Chant, J., Lai, A., Illing, M., Daniel, F.: Essays on financial stability. Tech. rep., Bank of Canada (2003).
14. Houben, A.G., Kakes, J., Schinasi, G.J.: Toward a framework for safeguarding financial stability. IMF Working Papers **2004**(101) (2004)
15. Aldasoro, I., Borio, C.E., Drehmann, M.: Early warning indicators of banking crises: expanding the family. *BIS Quarterly Review*, March (2018). https://ssrn.com/abstract=3139160
16. Bunn, P., Cunningham, A., Drehmann, M.: Stress testing as a tool for assessing systemic risks. *Bank of England Financial Stability Review* **18**(116-26), 19–21 (2005).
17. Hannoun, H.: Towards a global financial stability framework. In: *Speech at the 45th SEACEN Governors' Conference*, Siem Reap province, Cambodia. pp. 26–27 (2010).
18. Adrian, M.T., He, M.D., Liang, N., Natalucci, M.F.M.: A monitoring framework for global financial stability. *International Monetary Fund* (2019)

19. Federal Reserve Bank of Kansas City: Rethinking Stabilization Policy: A Symposium. Federal Reserve Bank of Kansas City (2002)
20. Borio, C.E., White, W.R.: Whither monetary and financial stability? The implications of evolving policy regimes, No 147, BIS Working Papers, Bank for International Settlements (2004).
21. Roubini, N.: Why central banks should burst bubbles. *International Finance* **9**(1), 87–107 (2006).
22. Woodford, M.: Inflation targeting and financial stability. Tech. rep., National Bureau of Economic Research (2012).
23. Caruana, J.: The challenge of taking macroprudential decisions: who will press which button (s)? *Macroprudential Regulatory Policies: The New Road to Financial Stability*, pp. 19–28 (2012). https://www.worldscientific.com/doi/pdf/10.1142/9789814360678_0003.
24. Gambacorta, L., Murcia, A.: The impact of macroprudential policies and their interaction with monetary policy: An empirical analysis using credit registry data (2017)
25. Upper, C.: Using counterfactual simulations to assess the danger of contagion in interbank markets, BIS Working Papers, No 234 (2007).
26. Claessens, S., Ghosh, S.R., Mihet, R.: Macro-prudential policies to mitigate financial system vulnerabilities. *Journal of International Money and Finance* **39**, 153–185 (2013).
27. Cerutti, M.E., Claessens, M.S., Laeven, M.L.: The use and effectiveness of macro- prudential policies: new evidence. International Monetary Fund (2015).

15 The Behavior of Firms Regarding the Introduction of Financing by Profit- and Loss-Sharing Products in Morocco

Safa Ougoujil and Sidi Mohamed Rigar

15.1 INTRODUCTION

After more than four decades since their birth, Islamic banks have positioned themselves as institutions that play an important role in resource mobilization and allocation and are actively integrated into the economies of the respective countries where they operate [1]. Although Islamic banks offer many of the same products as conventional banks, the two entities differ conceptually. One of the main differences is that conventional banks earn their money by charging interest and fees for services. In contrast, Islamic banks earn money by sharing profits and losses, trading, renting, charging fees for services rendered, and using other sharia-compliant exchange contracts.

Islamic finance is seen as a compartment of ethical and socially responsible investment on the one hand and as a financial innovation on the other. This is so because it promotes financial inclusion, fair risk sharing, social justice, and equitable distribution of wealth [2]. A survey conducted in 2016 by the General Confederation of Moroccan Businesses revealed, based on a sample of 600 companies from various sectors, that unfair economic competition was leaders' main concern, followed by access to finance. The access of participatory banks to the financial scene has thus offered companies additional financing options, especially with financing difficulties representing the main obstacles to their growth. Therefore, Islamic financial institutions can play a leading role in Morocco by proposing new approaches to financing using the principle of profit and loss sharing (PLS), including mainly mudaraba and musharaka.

Conceptually, Islamic banking is based on the principle of risk-taking as a necessary condition for profit-making. However, it is a modern form of banking in which financial intermediation is based on several legal and judicial principles of Islamic law that emphasize PLS methods instead of financing based on the interest rate variable. Therefore, there is a partnership between the depositors and the Islamic bank and between the investor entrepreneurs and the Islamic bank.

PLS financing has become one of the unique features of Islamic banks because of its contribution to increasing the community's economic activities. Many scholars consider PLS as the most distinct interest-based financial instruments and the closest to the spirit of Islamic finance [3,4]. Indeed, PLS is characterized by risk sharing between the entrepreneur and the financier, which encourages entrepreneurial activities. In other words, the profit or loss on the trade of goods or the billing of services is shared between the financier and the beneficiary instead of transferring all the

DOI: 10.1201/9781032667478-17

risks to the borrower. In addition, PLS financing methods do not require collateral, making it easier to access funds. Furthermore, financial institutions provide technical support and efficiency through PLS contracts that combine the interests of both parties [5].

With the genesis of participatory banks in Morocco, the importance of this research lies in the fact that it may reveal some facets of the acceptance of these new modes of financing by firms as alternative solutions. In this context, the central problem to which we try to provide some answers is to examine the intention of Moroccan companies to adopt PLS products and to discover the factors that can influence it. The study is based on the decomposed theory of planned behavior. It uses the PLS method as the main data analysis tool, with the data collected through a questionnaire sent to Moroccan firms. Accordingly, the study has the following specific objectives:

- Examine the factors that motivate Moroccan firms to adopt PLS financing;
- Unveil the factors that influence the attitude of Moroccan firms' managers to use these products;
- Detect the reference groups that influence the perception of Moroccan firms' managers on PLS financing;
- Explore the factors that affect the perceived behavioral control of Moroccan business leaders to adopt PLS financing products.

This chapter is organized as follows: Section 15.2 is a brief literature review, including an overview of the theoretical framework and research hypotheses. Section 15.3 provides insight into the methodology employed. Section 15.4 presents the study's results, highlighting the reliability of the model and the results of the hypothesis testing. The final section contains conclusions based on the results and some future research directions.

15.2   LITERATURE REVIEW

15.2.1   Profit- and Loss-Sharing General Concept

The principle of PLS is unanimously accepted as the basis for contractual transactions in Islamic finance. Two instruments based on this principle have been designed and are used by Islamic banks:

Moudaraba. It is a long-term contract that establishes a relationship of solidarity between the bank and its client [6]. This formula can be assimilated to private equity. It is a partnership contract between the owner of the capital and the entrepreneur (mudarib) who brings his expertise and knowledge. The profit is divided between the two parties according to a ratio defined when signing the contract. The financial loss lies with the owner of the capital; the loss of the manager is the opportunity cost of his labor force.

Moucharaka. It is a contract where the entrepreneur and the financier participate in the contribution of capital and the management of the business. Profits are divided according to predetermined ratios, while losses are borne according to the initial contribution of each. Profits generated are distributed according to the terms set out in the contract. As for the losses, they remain relative to the share of each partner in the capital [7]. In contrast to the mudaraba, the bank is both co-owner and co-responsible for project management, which is equivalent to the joint venture mode.

15.2.2   Theoretical Framework and Hypotheses

It is difficult to understand the problem of financing a company independently of the characteristics of its manager. In a firm, the responsibility for the financing decision lies primarily with the owner-manager. In this context, we have chosen to adopt the decomposed theory of planned behavior as a theoretical framework. The first version of the theory of planned behavior (TPB) was introduced by Fishbein and Ajzen [8] and is called the theory of reasoned action (TRA). According to this theory, an individual's behavior is predetermined by his behavioral intention, which is jointly determined by the attitude of this individual toward the action and by the subjective norms related to this behavior [8,9]. Based solely on an attitude and a normative component, this theory has often been criticized because it assumes that all human behavior is volitional and rational. In other words, the individual has total control over his behavior. Hence, TPB's appearance is considered an extension of TRA, where a third component called perceived behavioral control is added. It refers to the perception of individuals of the ease or difficulty of performing a given behavior. Thus, three independent factors determine an individual's behavioral intention: attitude toward the behavior, subjective norms, and perceived behavioral control.

However, Taylor and Todd [10] suggest that the three components mentioned above should be decomposed to better understand the relationships between belief structures and the antecedents of behavioral intention. This then gave rise to the decomposed theory of planned behavior, which extends the TPB. Therefore, the belief structure will be decomposed into five main dimensions: perceived cost, perceived utility, perceived risk, reputation, and compatibility with religious beliefs. On the other hand, the control belief structure is decomposed into two main constructs. The first is internal, i.e., self-efficacy, while the second is external, i.e., facilitating conditions. Finally, Taylor and Todd [10] consider that the decomposition of normative beliefs should be related to the divergence of opinions between referent groups. In particular, in cases where the referent groups are expected to have similar opinions, the decomposition will not make any additional contribution. Therefore, the normative belief structure will not be decomposed further but will be kept as it is.

Researchers in Muslim and non-Muslim countries have widely studied consumer attitudes toward Islamic financial products. However, few empirical studies are conducted to examine firms' perceptions and attitudes. Among these works are: Jalaluddin [11], Ahmad and Haron [12], Bengrich [13], Echatibi [14], El Ouafy [15], El Meskine and Chakir [16], and Boubker et al. [17].

Jalaluddin's study [11] investigates the attitudes of 385 Australian small businesses toward the PLS method of financing. The results revealed that the variables that positively influence the adoption of this method of financing could be grouped into five factors: business support, risk sharing, risk of default in the traditional system, profitability, and suitability of the PLS method.

Ahmad and Haron [12] studied the perceptions of 45 companies toward the Islamic banking system in Malaysia. Most respondents are aware of the existence of Islamic banks; however, they have low knowledge of their products. 75% of the participants agreed that there is a need for a marketing strategy by Islamic banks in Malaysia to promote their products and services.

Bengrich [13] and Echatibi [14], having conducted their surveys of Moroccan SMEs, have shown that the manager's training significantly impacts the degree of access to finance. That said, a high level of knowledge of Islamic finance in general and PLS products in particular allows Moroccan business leaders to have better control over the use of PLS financing.

El Ouafy [15] surveyed a sample of 158 Moroccan small business firms. The author concludes that the managers of these companies seem hesitant toward participatory financial products. She finds that religiosity is a significant determinant of intention. The cost of financing and the manager's profile affect small business firms' attitudes toward Islamic products. Finally, the manager's entourage has an indirect positive influence on his intentions through social norms.

According to a quantitative study targeting 68 companies in the Souss-Massa region, El Meskine and Chakir [16] show that 50% intend to open an account in a participatory bank. The study revealed that companies interested in moudaraba and moucharaka contracts represent 15.22% and 9.78%, respectively. The study results show that 51.5% of the respondents are ready to give up on participatory financing because they would not accept additional costs compared to traditional credits.

The PLS analysis of the study conducted by Boubker et al. [17], based on 149 microenterprises, confirms the direct and positive association between attitude toward Islamic financing, perceived behavioral control, subjective norms, and intention to adopt Islamic financing in

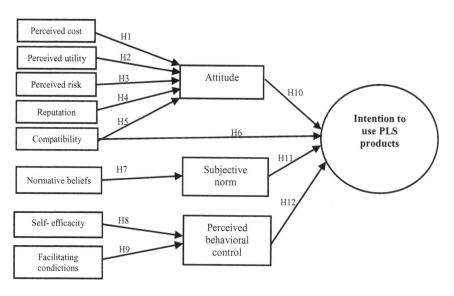

FIGURE 15.1   Research model.

business. These results are consistent with previous studies [18–20], concluding that attitude, behavioral control, and subjective norms significantly and directly influence the intention to adopt participatory financing.

The studies cited above provide a good starting point for studying and understanding corporate behavior toward participatory financial products. They will serve as a basis for developing our research model. Therefore, we propose dealing with PLS financing modes by developing a model that integrates religious, financial, social, contextual, and individual factors.

Hence, our research model summarizes the variables that influence the choice of participatory PLS products in the Moroccan context (Figure 15.1).

Based on the above model, the following hypotheses are formulated and will be tested in the following sections:

Perceived Cost. Cost is one of the main factors influencing firms' financial decisions [21]. Several studies assert the inclusion of financial cost in explaining behavioral intention [22,23]. Therefore, following these findings, we assume the following hypothesis:

H1. Perceived cost negatively and significantly influences the attitude of Moroccan firms' managers to use PLS financing.

Perceived Utility. Islamic finance offers profit- and loss-sharing contracts based on risk sharing between the bank and the company. Following the same logic, Islamic banks provide commercial support to the financed firms through PLS contracts to support the interests of both parties [5]. Therefore, we formulate the following hypothesis:

H2. Perceived utility positively and significantly influences the attitude of Moroccan firms' managers to use PLS financing.

Perceived Risk. PLS contracts may involve the bank's intervention in managing the beneficiary company. In using this mode of financing, the manager will have to face a number of constraints, including the risk of loss of control in terms of restrictions on the use of funds, mandatory disclosure of financial information, and auditing operations required by the financial partner. This study, therefore, hypothesizes the following:

H3. Perceived risk has a negative and significant influence on the attitude of Moroccan firms' managers to use PLS financing.

Reputation. Several studies have established a link between the reputation of Islamic banking and the attitude of customers [24,25]. They found that a bad reputation can lead to negative attitudes on the part of customers regarding their attitude toward using Islamic banking services. Therefore, this hypothesis was formulated as follows:

H4. Reputation negatively and significantly influences the attitude of Moroccan firms' managers to use PLS financing.

Compatibility with Religious Principles. Numerous studies on attitudes and intentions toward Islamic financial products have shown that religious people's values strongly influence their financial decisions [26–28]. We hypothesize that religious beliefs have a significant effect on firms' attitudes and intentions to use PLS financing:

H5. The compatibility of PLS financing with the religious beliefs of the firm's manager has a positive and significant effect on his attitude toward using this financing method.

H6. The compatibility of PLS financing with the religious beliefs of the firm's manager has a positive and significant effect on his intention to use this financing method.

Normative Beliefs. According to the TPB, the social norm is determined by the influence of relevant reference groups, including the owners' friends or family [29]. In addition, a financial decision must be made regarding the company's external financial consultants and internal advisors concerned with financing issues. Therefore, we make the following hypothesis:

H7. Normative beliefs have a positive and significant influence on subjective norms.

Self-Efficacity. This variable is measured by the degree of knowledge of the principles and products of Islamic banks, especially PLS products. Several studies have highlighted the positive correlation between the customer's knowledge of Islamic financial contracts and his willingness to accept them [15–30]. Therefore, this study raises the following hypothesis:

H8. Self-efficacy has a positive and significant influence on perceived behavioral control.

Facilitating Conditions. As noted in the study by Echchabi and Abd Aziz [26], this variable is firmly linked to the country's socio-political, financial, and economic state. In this research, we measure this variable regarding the presence of a well-established legal framework for the participatory finance segment in Morocco and the Moroccan government's promotion of Islamic finance. Therefore, we assume the following:

H9. Facilitating conditions have a positive and significant influence on perceived behavioral control.

Attitude. Several studies have asserted a positive and significant relationship between attitude and intention to adopt Islamic financing [24–26], hence the following hypothesis:

H10. Attitude has a positive and significant influence on the intention of Moroccan firms' managers to use PLS financing.

Subjective Norm. Subjective norm refers to the perceived social pressures that influence an individual's behavioral intention [31]. In the context of Islamic finance, previous studies show that subjective norms positively influence behavioral intention [23–26]. Thus, the study hypothesizes the following:

H11. Subjective norm has a positive and significant influence on the intention of Moroccan firms' managers to use PLS financing.

Perceived Behavioral Control. In this study, perceived behavioral control suggests that leaders will not use PLS funding methods if they feel they do not have behavioral control. Hence, the following hypothesis:

2 H12. Perceived behavioral control has a positive and significant influence on the intention of Moroccan firms' managers to use PLS financing.

15.3 RESEARCH METHODOLOGY

This study uses a quantitative approach, with data collected using a questionnaire from Moroccan business leaders to explore the factors that may influence their intentions to use PLS financing. The study covers not only Morocco's main cities but also other cities. It should be noted that Morocco is a panacea of a wide range of languages, cultures, races, etc., so care was taken to ensure that the sample studied reflected this variety in the Moroccan population. The questionnaire was written in French. This is since French is the country's second language and most Moroccans can speak it fluently. The questionnaire was tested in November 2020 with a dozen Moroccan business leaders. It revealed some misunderstandings, which prompted us to rephrase the questionnaire, simplify the language used, and eliminate some statements to reduce the length of the questionnaire. Thus, our questionnaire is broken down into three sections, the first of which deals with the general information of the companies solicited; the second deals with the behavior of the companies concerning PLS financing, based on the theory of decomposed planned behavior; and the last deals with the profile of the respondents. The items are measured using a five-point Likert scale (1=strongly disagree and 5=strongly agree).

The sampling technique used in this study was non-probability convenience sampling. This was a viable alternative due to the time constraints and the lack of a national database that we could use as a reference. Our survey lasted approximately five months, between December 2020 and April 2021. More than 1000 questionnaires were sent electronically, and 833 were received. After purifying our database, we discarded 56 questionnaires that were not in the scope of our study, resulting in a final sample of 777 companies. Structural equations will analyze the research model and the research hypotheses under the PLS approach using the SmartPLS 3.0 software. In practice, the implementation of the PLS method involves the evaluation of the measurement model and the structural model. First, the evaluation of the measurement model is done by assessing its convergent validity and discriminant validity. Second, evaluating the structural model requires checking a number of criteria, including coefficient of determination, effect size, predictive adequacy, model fit, and hypothesis testing.

Two out of 777 companies, 34% are small companies, and 53% are SMEs. In contrast, large companies with the highest annual turnover represent 13%. Most companies (38%) are located in Casablanca, as it is considered a Moroccan economic center. However, 14% are in Tangier, 11% in Marrakech, 10% in Rabat, 7% in Agadir, and 20% in other cities. The study sample includes companies from all sectors of activity, with a preponderance of the following three sectors: the service sector (30%), industry (22%), and trade (19%). Our sample's most dominant legal form is the limited liability company, with 49% of the companies surveyed. Our sample includes companies of all sizes. However, most companies (47%) employ between 10 and 200 employees. Thus, companies with a seniority of one to five years represent the majority of the sample, with a percentage of 43%. Men are in the majority and represent 71% of our sample, compared to 29% of women. Regarding age breakdown, the youngest respondents (under 20) and the oldest (over 60) represent only 1% of the population. On the other hand, the most represented age category (46%) is between 30 and 40 years old. Concerning the level of education, the majority (86%) of the respondents declared that they had a higher education degree.

15.4 RESULTS AND DISCUSSION

The evaluation of the measurement model shows strong convergent validity according to several criteria: Factor Loading, Cronbach's Alpha, Composite Reliability, and Average Variance Extracted (AVE). Thus, the external loading values of the indicators are above 0.7, except for the self-efficacy item (SE1), which has a slightly lower loading factor than the predetermined threshold (0.681).

However, we decided to keep it because, according to Chin [32], in the social sciences it is possible to keep an item whose loading factor threshold is between 0.5 and 0.7 provided that the other components (Cronbach's Alpha, Composite Reliability, and AVE) present satisfactory results, and this is the case for our study. All the values of Cronbach's Alpha and Composite Reliability are above 0.7, and the values of AVE are all above 0.5 (Table 15.1), which confirms their significant contribution to the model constructs.

Regarding discriminant validity, Table 15.2 reveals that the latent constructs have a much larger root mean square of variance extracted with the measures that form them than the other constructs. Therefore, the initial criterion for testing discriminant validity is met. However, Table 15.3 displays the second criterion used to check discriminant validity: Cross-Loading. It also turns out that this criterion is met, since the statistical contributions of the selected measures are more correlated with the latent constructs to which they belong than with the other latent constructs. Thus, considering the two criteria mentioned above, it turns out that the discriminant validity of the measurement scales is proven.

TABLE 15.1
Convergent Validity

Variables	Items	Factors Loading	Cronbach's Alpha	Composite Reliability	AVE
Intention to use PLS products	INT1	0.957	0.960	0.974	0.925
	INT2	0.972			
	INT3	0.957			
Attitude	ATD1	0.955	0.955	0.971	0.918
	ATD2	0.965			
	ATD3	0.955			
Subjective norm	SN1	0.935	0.928	0.954	0.874
	SN2	0.957			
	SN3	0.912			
Perceived behavioral control	PBC1	0.929	0.910	0.943	0.847
	PBC2	0.944			
	PBC3	0.887			
Perceived cost	PC1	0.919	0.935	0.954	0.837
	PC2	0.936			
	PC3	0.919			
	PC4	0.886			
Perceived utility	PU1	0.896	0.935	0.954	0.838
	PU2	0.935			
	PU3	0.928			
	PU4	0.902			
Perceived risk	PR1	0.886	0.927	0.948	0.820
	PR2	0.930			
	PR3	0.917			
	PR4	0.889			
Reputation	R1	0.865	0.919	0.940	0.759
	R2	0.899			
	R3	0.920			
	R4	0.901			
	R5	0.761			

(*Continued*)

TABLE 15.1 (*Continued*)
Convergent Validity

Variables	Items	Factors Loading	Cronbach's Alpha	Composite Reliability	AVE
Compatibility	C1	0.908	0.946	0.959	0.823
	C2	0.919			
	C3	0.918			
	C4	0.923			
	C5	0.866			
Normative beliefs	NB1	0.861	0.886	0.930	0.816
	NB2	0.938			
	NB3	0.909			
Self-efficacity	SE1	0.681	0.850	0.901	0.697
	SE2	0.903			
	SE3	0.885			
	SE4	0.851			
Facilitating conditions	FC1	0.941	0.954	0.967	0.880
	FC2	0.948			
	FC3	0.943			
	FC4	0.920			

TABLE 15.2
Fornell and Larcker Criterion

	ATD	SE	C	FC	PBC	PC	NB	INT	SN	PR	R	PU
ATD	0.958											
SE	0.412	0.835										
C	0.496	0.419	0.907									
FC	0.706	0.417	0.469	0.938								
PBC	0.898	0.413	0.515	0.701	0.920							
PC	−0.396	−0.536	−0.623	−0.403	−0.432	0.915						
NB	0.735	0.388	0.418	0.745	0.736	−0.362	0.903					
INT	0.869	0.434	0.557	0.696	0.851	−0.447	0.731	0.962				
SN	0.826	0.418	0.501	0.680	0.871	−0.421	0.733	0.858	0.935			
PR	0.355	0.497	0.558	0.361	0.381	−0.767	0.352	0.381	0.375	0.906		
R	−0.488	−0.453	−0.785	−0.452	−0.520	0.702	−0.420	−0.525	−0.493	−0.645	0.871	
PU	0.503	0.451	0.284	0.506	0.481	−0.242	0.523	0.461	0.457	0.248	−0.305	0.915

Given that the validity of the measurement model is assured, it is now time to assess the overall quality of the structural model. Therefore, the evaluation of the structural model requires the verification of the coefficient of determination R^2, the effect size f^2, the predictive relevance Q^2, and the goodness-of-fit (GoF) index.

According to our results (Table 15.4), the coefficient of determination R^2 shows a value greater than 0.67, which indicates that the explanatory variables attitude, subjective norms, and perceived behavioral control explain, on average, 83% of the variation in the intention to use PLS financing (the endogenous variable). Following Chin's recommendations [32], we can conclude that there is a strong association between the explanatory variables and the variable being explained. Regarding effect size f^2, perceived behavioral control has no direct and specific effect on intention. Similarly,

TABLE 15.3
Results of Discriminant Validity – Cross-Loading Criteria

	ATD	SE	C	FC	PBC	PC	NB	INT	SN	PR	R	PU
SE1	0.364	0.681	0.197	0.447	0.358	−0.268	0.377	0.362	0.340	0.247	−0.196	0.520
SE2	0.361	0.903	0.386	0.342	0.365	−0.474	0.319	0.387	0.365	0.448	−0.405	0.365
SE3	0.336	0.885	0.404	0.300	0.335	−0.521	0.299	0.358	0.353	0.465	−0.455	0.314
SE4	0.297	0.851	0.414	0.281	0.305	−0.530	0.285	0.327	0.323	0.504	−0.465	0.278
ATD1	0.955	0.385	0.446	0.670	0.853	−0.343	0.708	0.815	0.787	0.305	−0.442	0.494
ATD2	0.965	0.395	0.482	0.670	0.851	−0.387	0.694	0.837	0.789	0.344	−0.467	0.474
ATD3	0.955	0.404	0.498	0.689	0.878	−0.406	0.712	0.846	0.798	0.369	−0.493	0.480
PBC1	0.883	0.378	0.474	0.662	0.929	−0.395	0.703	0.822	0.799	0.339	−0.484	0.474
PBC2	0.848	0.405	0.490	0.665	0.944	−0.408	0.685	0.807	0.819	0.339	−0.486	0.457
PBC3	0.742	0.356	0.458	0.605	0.887	−0.389	0.642	0.717	0.790	0.377	−0.465	0.392
FC1	0.669	0.379	0.432	0.941	0.654	−0.352	0.710	0.662	0.644	0.303	−0.411	0.493
FC2	0.670	0.393	0.451	0.948	0.672	−0.361	0.703	0.665	0.643	0.320	−0.421	0.492
FC3	0.651	0.407	0.427	0.943	0.647	−0.401	0.687	0.641	0.638	0.355	−0.419	0.454
FC4	0.657	0.387	0.449	0.920	0.655	−0.401	0.694	0.643	0.625	0.379	−0.443	0.458
NB1	0.648	0.368	0.348	0.647	0.644	−0.304	0.861	0.654	0.663	0.283	−0.362	0.495
NB2	0.683	0.349	0.390	0.707	0.683	−0.327	0.938	0.677	0.682	0.323	−0.375	0.476
NB3	0.660	0.334	0.393	0.663	0.666	−0.351	0.909	0.647	0.639	0.348	−0.400	0.446
PC1	−0.387	−0.500	−0.579	−0.376	−0.414	0.919	−0.334	−0.417	−0.388	−0.686	0.661	−0.263
PC2	−0.377	−0.512	−0.601	−0.374	−0.404	0.936	−0.337	−0.433	−0.382	−0.714	0.656	−0.248
PC3	−0.342	−0.488	−0.558	−0.374	−0.380	0.919	−0.342	−0.392	−0.386	−0.714	0.625	−0.206
PC4	−0.337	−0.458	−0.537	−0.352	−0.379	0.886	−0.313	−0.393	−0.387	−0.695	0.624	−0.161
C1	0.467	0.412	0.908	0.464	0.474	−0.542	0.395	0.532	0.451	0.480	−0.706	0.292
C2	0.460	0.367	0.919	0.429	0.475	−0.527	0.376	0.527	0.474	0.475	−0.681	0.275
C3	0.462	0.386	0.918	0.431	0.482	−0.572	0.400	0.503	0.458	0.508	−0.709	0.257
C4	0.464	0.357	0.923	0.418	0.474	−0.588	0.364	0.516	0.459	0.526	−0.752	0.217
C5	0.391	0.378	0.866	0.381	0.428	−0.605	0.359	0.439	0.427	0.550	−0.716	0.244
INT1	0.837	0.413	0.512	0.672	0.824	−0.428	0.716	0.957	0.842	0.358	−0.482	0.439
INT2	0.824	0.422	0.544	0.671	0.810	−0.439	0.687	0.972	0.822	0.380	−0.510	0.444
INT3	0.847	0.418	0.551	0.666	0.822	−0.424	0.706	0.957	0.811	0.361	−0.524	0.447
SN1	0.794	0.387	0.483	0.649	0.846	−0.402	0.680	0.808	0.935	0.349	−0.476	0.440
SN2	0.793	0.396	0.471	0.632	0.827	−0.398	0.694	0.823	0.957	0.351	−0.457	0.424
SN3	0.728	0.388	0.450	0.625	0.770	−0.382	0.683	0.774	0.912	0.352	−0.450	0.417
R1	−0.438	−0.371	−0.675	−0.384	−0.463	0.526	−0.342	−0.453	−0.435	−0.480	0.865	−0.265
R2	−0.431	−0.380	−0.679	−0.385	−0.455	0.574	−0.393	−0.459	−0.445	−0.508	0.899	−0.273
R3	−0.445	−0.416	−0.735	−0.426	−0.463	0.644	−0.396	−0.485	−0.459	−0.613	0.920	−0.296
R4	−0.458	−0.420	−0.706	−0.410	−0.479	0.670	−0.383	−0.501	−0.447	−0.604	0.901	−0.283
R5	−0.344	−0.394	−0.619	−0.362	−0.399	0.662	−0.307	−0.378	−0.355	−0.624	0.761	−0.200
PR1	0.324	0.450	0.494	0.326	0.352	−0.651	0.331	0.327	0.354	0.886	−0.550	0.247
PR2	0.339	0.459	0.531	0.357	0.349	−0.718	0.314	0.358	0.327	0.930	−0.610	0.242
PR3	0.312	0.458	0.500	0.325	0.332	−0.724	0.327	0.353	0.331	0.917	−0.592	0.220
PR4	0.309	0.433	0.493	0.299	0.345	−0.684	0.304	0.343	0.346	0.889	−0.585	0.187
PU1	0.441	0.395	0.199	0.450	0.425	−0.165	0.464	0.388	0.394	0.147	−0.204	0.896
PU2	0.464	0.395	0.257	0.468	0.444	−0.225	0.476	0.431	0.411	0.211	−0.277	0.935
PU3	0.471	0.435	0.278	0.476	0.447	−0.243	0.479	0.437	0.431	0.266	−0.300	0.928
PU4	0.468	0.425	0.301	0.458	0.444	−0.251	0.496	0.431	0.435	0.279	−0.331	0.902

TABLE 15.4
Quality of the Overall Model

	R Square	f² Values	Q²	Goodness-of-Fit
Intention to use PLS products	0.828		0.763	*0.689*
Attitude	0.402	0.205	0.366	
Compatibility		0.055		
Perceived behavioral control	0.509	0.004	0.429	
Subjective norm	0.538	0.190	0.469	

religious compatibility has a weak effect (0.055) on intention. In contrast, attitude and subjective norms have a strong direct impact on intention. As for the predictive capacity of the structural model, the Q^2 values of the different variables are globally positive and greater than zero, which justifies that our research model has a good predictive capacity. Similarly, the GoF value (GoF=0.689) is sufficiently large to conclude that the overall model quality of our study is largely valid, given that the GoF value far exceeds the value of 0.36 as recommended by Wetzelset et al. (2009).

We are moving to the hypothesis test, which checks the significance of the correlation coefficients (path coefficients) and the t-values. From a statistical point of view, a hypothesis is significant according to the following thresholds: 0.05 if t-value>1.96; 0.01 if t-value>2.57; and 0.10 if t-value>1.645. Table 15.5 summarizes the hypothesis test results performed using the bootstrap technique.

The evaluation of these results leads us to conclude that among the 12 direct relationships tested, nine relationships show a highly significant link at the threshold of less than 0.01 with Student's t-values significantly higher than 2.57. These are the relationships reflected by hypotheses H2, H4, H5, H6, H7, H8, H9, H10, and H11. On the other hand, hypotheses H1, H3, and H12 are rejected because they are not significant at the threshold of less than 0.05 with Student's t-values greater than 1.96, nor at the threshold of less than 0.10 with Student's t-values greater than 1.645. In other words, leaders' attitude toward adopting PLS financing depends only on three variables: perceived utility, the reputation of participative banks, and the compatibility of PLS financing with leaders' religious principles. Furthermore, the intention to use PLS financing depends only on attitude and subjective norms, which reflect Ajzen's original model of the TRA.

Our model assumes several types of relationships between the dependent variable (intention) and the independent variables: direct and indirect relationships through the intermediation of the mediating variables (attitude, subjective norm, and perceived behavioral control). These variables help explain the process through which different types of beliefs influence intention. To test the significance of the mediating effects, we referred to the results of the indirect effects test produced by SmartPLS. The results show that attitude plays a significant mediating role between some exogenous variables, which are compatibility (t=4.328; p=0.000), perceived utility (t=7.345; p=0.000), reputation (t=2.720; p=0.007), and the endogenous variable, which is the intention to use PLS financing. Indeed, the mediation effect is negated in the relationships linking perceived cost (t=1.093; p=0.275) and perceived risk (t=0.428; p=0.669) to intention. We can find the explanation for these results by returning to the structural relationship tests, which show that these two variables have no significant effect on attitude. The test also indicates that the indirect impact of normative beliefs on intention to use PLS financing is significant (t=7.562; p=0.000). Thus, the subjective norm significantly mediates between normative beliefs and intention. Finally, perceived behavioral control has a non-significant mediating effect on both relationships: between facilitative conditions and intention (t=1.390; p=0.165), and between self-efficacy and intention (t=1.244; p=0.214). In contrast, the indirect effect of self-efficacy and facilitating conditions on intention to use PLS financing via perceived behavioral control was insignificant.

TABLE 15.5

Hypothesis Testing – PLS Approach

	Original Sample (O)	T Statistics	P Values	Decision
H1. PC -> ATD	−0.061	1.111	0.267	Rejected
H2. PU -> ATD	0.378	11.843	0.000**	Confirmed
H3. PR -> ATD	−0.019	0.433	0.665	Rejected
H4. R -> ATD	−0.153	2.829	0.005**	Confirmed
H5. C -> ATD	0.242	4.826	0.000**	Confirmed
H6. C -> INT	0.115	6.537	0.000**	Confirmed
H7. NB -> SN	0.733	34.810	0.000**	Confirmed
H8. SE -> PBC	0.146	4.931	0.000**	Confirmed
H9. FC -> PBC	0.640	22.720	0.000**	Confirmed
H10. ATD -> INT	0.438	10.189	0.000**	Confirmed
H11. SN -> INT	0.379	7.984	0.000**	Confirmed
H12. PBC -> INT	0.069	1.394	0.164	Rejected

The literature has allowed us to identify some factors that may motivate the attitude of Moroccan firms' managers to use PLS financing. Specifically, perceived cost, perceived utility, and perceived risk in terms of loss of control, reputation, and compatibility of the said products with the religious principles of the managers. For this, we assumed that cost has a significantly negative impact on the attitude of Moroccan firms toward PLS financing. The results obtained after the analysis allowed us to disprove this hypothesis. This result is consistent with several researchers' theoretical and empirical contributions [17–34]. This is equivalent to saying that the surveyed firms are indifferent to the perceived cost variable when considering the PLS principle of finance. In other words, this variable does not affect the acceptance or rejection of this type of finance.

However, the hypothesis regarding the impact of perceived risk on attitude is not significant. Our results are inconsistent with those of Jalaluddin [11], who showed that involvement in the firm's management is among the reasons that imply the rejection of PLS financing. On the other hand, we can conclude that the Moroccan firms in our sample are indifferent to the possibility of the bank interfering in their management decisions as financial partners. Because, in principle, Islamic banks function exactly like venture capitalists and partner with the firms they finance and share gains and losses. In terms of perceived utility, measured by risk sharing and technical, financial, and managerial support provided by the bank to the company. Our study found that it positively and significantly influences attitude, which is in line with the results of several studies [5–21,35]. The choice of PLS financing is also influenced by the reputation of participatory banks granting this product. According to our results, there is a significant relationship between the reputation and the attitude of Moroccan firms' managers toward contracting this kind of product. These results corroborate those of [17–20,36].

The results of our analysis reveal that the compatibility of PLS financing with the religious principles of the leader is a factor with a strong impact on the attitude of Moroccan firms. In addition, Echchabi and Abd.Aziz [26] and Ali and Puah [27] equate the phenomenon of Islamic finance with innovation and find a positive and significant relationship between the perceived attribute of innovation (compatibility) and the attitude toward Islamic finance. Our results also confirm the existence of a positive and highly significant relationship between normative beliefs and subjective norms, which is consistent with the results of studies conducted in Morocco with individuals [26] and firms [15]. However, our results reveal a positive and significant relationship between self-efficacy and the manager's perceived behavioral control over his ability to use this mode of financing. That is, a high level of knowledge of Islamic finance in general and PLS products in particular allows

Moroccan corporate executives to have better control over the use of PLS financing. Our results are in line with those of [15–30]. In addition to self-efficacy, control beliefs incorporate the executive's perceived facilitating conditions. Facilitating conditions were found to have a significant positive influence on perceived behavioral control, which aligns with the findings of [35].

Regarding the direct determinants of intention, which are attitude, subjective norm, and perceived control, we demonstrated that only attitude and subjective norm significantly impact intention. In contrast, perceived behavioral control does not influence managers' intentions, despite the considerable effect of self-efficacy and facilitating conditions on this variable. Our study's results align with the work of El Ouafy [15] and Rachidi and Abourrig [37]. According to these authors, the absence of the effect of the perception of control refers to the TRA model, whose explanatory capacity is considered sufficient. In addition to the variables included in the TPB model, we were interested in the role of the religious factor as a direct determinant of intention. However, hypothesis testing reveals that this factor has a positive and significant influence on intention, consistent with the results of [15–37].

15.5   CONCLUSION AND FUTURE RESEARCH DIRECTIONS

The main objective of this exploratory study was to examine the willingness of Moroccan firms to adopt PLS financing and to identify the variables that may influence managers' adoption of these financing methods, based on the theory of decomposed planned behavior. The results showed that perceived utility, reputation, and compatibility with religious principles significantly impact the attitude toward PLS financial products. Similarly, normative beliefs were found to have a positive and significant influence on the subjective norm, with particular reference to family and internal and external company advisors as the main referent groups. In addition, facilitative conditions and self-efficacy had a positive and significant influence on perceived behavioral control. Finally, attitude and subjective norms significantly impact Moroccan firms' intention to adopt PLS financing.

In terms of contribution to the body of knowledge, the study explored the prospects of financing PLS products in Morocco's emerging participatory financial context. The results generated by this study have important implications for policymakers, researchers, financial institutions, corporate executives, and financial managers, which should be considered in developing policies needed to improve the participatory banking industry in Morocco.

Our study has a number of limitations that should be considered in future research. First, our proposed model determines the use of PLS products for business financing based solely on demand, so another questionnaire with financial institutions is highly desirable. In addition, this research model on behavioral intention can be extended by testing other variables, mainly those related to the relationship between the company and the bank, such as trust, transparency, commitment, and morality.

REFERENCES

1. Čihák M., Hesse H. (2008) Islamic Banks and Financial Stability: An Empirical Analysis. IMF Working paper Monetary and Capital Markets Department WP/ 08/ 16.
2. ElMassah S., Abou-El-Sood H. (2021) Selection of Islamic banking in a multicultural context: The role of gender and religion. *All Works*. 4672. https://zuscholars.zu.ac.ae/works/4672.
3. Siddiqi M.N. (1988) Islamic banking: theory and practice. In Mohammad, A. (Ed.), *Banking in South East Asia*, Institute of Southeast Asian Studies, Singapore, pp. 34–67.
4. Khan M.F. (1992) *PLS System: Firms' Behavior and Taxation*, IRTI, Jeddah.
5. Khan T. (1995) Demand for and supply of mark-up and PLS funds in Islamic banking: Some alternative explanations. *Islamic Economic Studies*, 3(1), 39–78.
6. Ruimy M. (2018) La Finance Islamique, Cheval De troie De la finance Moderne? *Regards croisés sur l'économie*, 22(1), 37.
7. Doligez F., et al. (2012) *Experiences De Microfinance Au Senega.l*, Karthala, Paris, p. 97.

8. Fishbein M., Ajzen I. (1975) *Belief, Attitude, Intention and Behaviour: An Introduction to Theory and Research.* Addison-Wesley, San Francisco, CA.
9. Ajzen I., Fishbein M. (1980) *Understanding Attitudes and Predicting Social Behavior.* Prentice-Hall, Englewood Cliffs, NJ.
10. Taylor S., Todd P. (1995) Decomposition and crossover effects in the theory of planned behaviour: A study of consumer adoption intentions. *International Journal of Research in Marketing*, 12, 137–55.
11. Jalaluddin C. (1999) Attitudes of small business firms and financial institutions towards the profit/ loss sharing method of finance, PhD dissertation, University of Wollongong, Wollongong.
12. Ahmad N., Haron S. (2002) Perceptions of Malaysian corporate customers towards Islamic banking products and services. *International Journal of Islamic Finance Services*, 3(1), 13–29.
13. Bengrich M. (2006) Contribution à l'étude du comportement financier des petites et moyennes entreprises marocaines. PhD dissertation, Cadi Ayyad University, Morocco.
14. Echatibi M. (2010) Financement bancaire des PME au Maroc. PhD dissertation, Mohamed V-Agdal University, Morocco.
15. El Ouafy S. (2016) Contribution à l'étude des déterminants de l'intention d'utiliser les produits financiers islamiques: Cas des très petites entreprises marocaines", PhD dissertation, Ibn Zohr University, Morocco.
16. El Meskine L., Chakir A. (2017) La perception des produits participatifs islamiques à base PPP par les entreprises marocaines. *Cas de la région Souss Massa. Recherches et Applications en Finance Islamique (RAFI), [S.l.],* 1(2), 147–164.
17. Boubker O., Douayri K., Ouajdouni A. (2021) Factors affecting intention to adopt Islamic financing: Evidence from Morocco, *MethodsX*, 8, 101523.
18. Shahab A., Zahra A. (2018) Adoption of Islamic banking in Pakistan an empirical investigation. *Cogent Business & Management*, 5(1), 1548050, doi: 10.1080/23311975.2018.1548050.
19. Jaffar M.A., Musa R. (2014) Determinants of attitude towards islamic financing among halal-certified micro and SMEs: A preliminary investigation. *Procedia - Social and Behavioral Sciences*, 130, 135–144.
20. Jaffar M.A., Musa R. (2016) Determinants of attitude and intention towards Islamic financing adoption among non-users, *Procedia Economics and Finance*, 37, 227–233.
21. Gait A., Worthington A. (2009) An empirical survey of individual customer, business firms and financial institution attitudes towards Islamic methods of finance. *International Journal of Social Economics*, 35(11), 783–808. https://doi.org/10.1108/03068290810905423.
22. Ramayah T., Oh S.M., Omar A. (2008) Behavioral determinants of online banking adoption: Some evidence from a multicultural society. *Journal on Management*, 2(3), 29–37. https://doi.org/10.26634/jmgt.2.3.335.
23. Amin H., Abdul Rahman A.R., Sondoh S.L., Chooi Hwa A.M. (2011) Determinants of customers' intention to use islamic personal financing. *Journal of Islamic Accounting And Business Research*, 2(1), 22–42. https://doi.org/10.1108/17590811111129490.
24. Maryam S.Z., Ahmad A., Aslam N., Farooq S. (2021) Reputation and cost benefits for attitude and adoption intention among potential customers using theory of planned behavior: An empirical evidence from Pakistan. *Journal of Islamic Marketing*, 13(10), 2090–2107.
25. Elhajjar S. (2022) An investigation of consumers' negative attitudes towards banks. *Corporate Reputation Review*, 25, 1–10.
26. Echchabi A, Abd Aziz H (2012) Empirical investigation of customers' perception and adoption towards Islamic banking services in Morocco. *Oman Chapter of Arabian Journal of Business and Management Review*, 1(10), 849–858. https://doi.org/10.12816/0002190.
27. Ali M., Raza S.A., Puah C.H. (2017) Factors affecting to select Islamic credit cards in Pakistan: The TRA model. *Journal of Islamic Marketing*, 8(3), 330–344. https://doi.org/10.1108/ JIMA-06-2015-0043.
28. Wahyuni S. (2012) Moslem community behavior in the conduct of Islamic bank: The moderation role of knowledge and pricing. *Procedia-Social and Behavioral Sciences*, 57(10), 290–298. https://doi.org/10.1016/j.sbspro.2012.09.1188.
29. Matthews C., Vasudevan D., Barton S., Apana R. (1994) Capital structure decision making in privately held firms: beyond the finance paradigm. *Family Business Review*, 7(4), 349–367.
30. Beiginia A.R., Besheli A.S., Suluklu M.E., Ahmadi M. (2011) Mobile banking adoption based on the decomposed theory of planned behavior. *European Journal Economics Finance and Administration Sciences*, 1450–2275.
31. Ajzen I. (1991) The theory of planned behavior. *Organizational Behavior and Human Decision Processes*, 50(2), 179–211.

32. Chin W.W. (1998) The partial least squares approach to structural equation modeling. *Modern Methods for Business Research*, 295(2), 295–336.

33. Metwally M. (1996) Attitudes of Muslims towards Islamic banks in a dual-banking system. *American Journal of Islamic Finance*, 6, 11–17.

34. Al-Sultan W. (1999) Financial Characteristics of Interest-Free Banks and Conventional Banks, University of Wollongong: Wollongong, Australia.

35. Mahdzan N.S., Zainudin R.S., Au S.F. (2017) The adoption of Islamic banking services in Malaysia. *Journal of Islamic Marketing*, 8, 496–512.

36. Hoque M.N., Rahman M.K., Said J., Begum F., Hossain M. (2022) What factors influence customer attitudes and mindsets towards the use of services and products of islamic banksin Bangladesh? *Sustainability*, 14, 4703. https://doi.org/10.3390/su14084703.

37. Abourrig A., Rachidi L. (2016) Prédiction de l'acceptation des banques islamiques: Une extension de la théorie de l'action raisonnée. *Revue Marocaine de Recherche en Management et Marke*

16 Influence of Financial Innovation on Business Performance

Evidence from the SME Food Industry in Malaysia

Shreen Almas Mohamed Buhary and Hussen Nasir

16.1 INTRODUCTION

Financial innovation (FI) represents financial products and services in financial economics and identifies negative impacts on financial institutions. However, to cope with current rapid changes in the market, FI combines numerous financial institutions, financial products and services, and financial system procedures to support economic activity (Qamruzzaman et al., 2021). Innovation enhances the creation of goods and services in the financial markets and industry. In addition, innovation was used to describe advancements in technological solutions that resulted in unique combinations of productive means, produced above-average rates of return, and enhanced the dynamic development of the entire economy (Targalski, 2006). The "Theory of Innovation" by Mutlu and Er (2003) was referenced by Schumpeter because he believed that innovation significantly fosters growth and is one of the key factors driving a capitalist economy. According to several studies, innovation is essential to the company and economic development (Szirmai et al., 2011).

FI refers to innovations that exist in financial markets. Innovation studies have taken place not only in the real sector but also in the financial sector. During the 1980s, FI drew increased interest from the academic community, particularly concerning technical advancement and financial liberalization. Positive effects on FI were noted during the most recent global financial crisis. However, modern FI was criticized since it was believed to be to blame for the disaster (Michalopoulos et al., 2009).

Perspectives from the community on the food industry are required to create innovative technologies or ideas at the cutting edge of the industry as enterprises begin to operate in the global business market. According to Eniola and Entebang (2015), FI plays an increasingly major role in contemporary organizations by promoting the growth of businesses and contributing to the overall expansion of the economy. According to the research conducted by Syed et al. (2016), the critical variable that affects company success is innovation, rather than other variables. This finding shows that innovation may be used to provide a strong predictor of business performance. The rapid advancement of ideas, inventions, information, and strategies significantly impacts current small and medium-sized enterprises' (SMEs) performance, leading to an FI system rather than a traditional financial system.

The growth and development of SMEs depend heavily on their capacity to obtain funding for ongoing operations and commercial growth. The application of FI has created importance in business fields due to enhancing business performances that depend more on workers than physical assets. However, a prior study by Mehta and Brahmbhatt (2020) discovered that employees from small businesses were not aware of E-Finance, Trade Cards, Innovative Financing Structures & Instruments, and New Proxies for Credit Information, which can smooth the daily operation process and business performance. They were not adaptable enough to deal with this new financing, claim Mehta and Brahmbhatt (2020). Workers must firmly believe that implementing FI techniques in their interactions would help them become more financially competitive (Mehta and Brahmbhatt, 2020). The rapid development of ideas, innovations, strategies, and technology significantly impacts how well SMEs are currently performing, which has resulted in changes in the FI system from traditional financial. Therefore, this study was intended to increase awareness of the importance of FI in the SME food industry.

16.2 LITERATURE REVIEW

16.2.1 Financial Innovation

In essence, FI plays an increasingly significant role in the current organization by assisting companies in expanding and providing a competitive edge to the economy (Eniola and Entebang, 2015). This includes helping companies deal with demand-driven forces and the competitive pressure that comes from financial markets. The innovation aims to boost a company's competitiveness and add shareholder value. Besides, creating new ways to make profits and minimizing the impacts of taxes and regulations were two important factors in FI. It has been fresh prospective earnings that provide incentives for innovation and can be obtained, for instance, through cost reduction or technological advancements. (Shreen Almas, 2022). Moreover, improvements in computer and communication technologies have benefited financial markets. This alternative benefit undoubtedly promotes saving and investing and provides tools for risk management, among other things.

Information on their methods, operations, and finances may be captured and gathered using FI. FI can manage inherent risks and move money across time and location while facilitating financial transactions and services through payment systems (Mehta and Brahmbhatt, 2020). Moreover, a previous group of studies focused on FI. They define FI as creating new financial services, products, and procedures. They claimed that FI successfully developed and popularized new financial institutions, markets, and technologies. Most of FI was located in payment systems and financial instruments for lending and borrowing money, such as mobile banking, online banking, online payments, e-insurance, and other electronic banking and finance services.

FI have existed with technological advancements from the beginning; thus, they are not a recent occurrence. It is well recognized that technological and FI are connected, and they develop together over time (Michalopoulos et al., 2009). On the one hand, FI provides a method to finance innovative technological ventures when traditional funding sources are unavailable. Due to high investment risk, technological and economic advancements have increased the complexity of business processes and introduced new types of risk, which has forced the financial system and financial markets to adapt to the changes and modernize to meet the unique demands of corporate entities and the problems of the modern world. This leads to the conclusion that without FI, national-state wealth would decline, as would technical and economic development. At the same time, the application of the FI would be constrained without the demand arising from the technical progress.

Innovation is a sustained process involving planning, developing, testing, and developing processes. Three categories – institutional innovation, product innovation, and process innovation – were created from the general definition of FI. Institutional innovation is developing new categories of financial institutions, like direct banks and electronic trading platforms. Product innovation is the introduction of new products, such as credit, deposits, insurance, leasing, and hire-purchase. In contrast, process innovation introduces new ways to conduct financial transactions, such as online and telephone banking. New financial products are also being developed to improve the efficiency of the financial system (Dabrowski, 2017).

Regulations, economic policies, market conditions, and scientific and technological advancements all contributed to the development of FI. The purpose of FI is to generate revenue for the innovator and to set up conditions for a business's competitive edge. Finance professionals frequently use FI because it helps to raise profitability, promote financial development and growth, boost efficiency, and reduce or transfer corporate risk. It presents new challenges and risks and creates chances for social and economic advancement.

However, not all economies can access SME financing equally (Hewa Wellalage, Locke, & Samujh, 2020). The literature review showed a lack of knowledge avout accessing FI in the SME food industry. Very few studies on financial inclusion were found in wealthy nations like the United States, Hong Kong, etc. Still there was very little research on financial inclusion in emerging countries like Malaysia.

FI plays a crucial role in maintaining the survival of SME firms and helps increase their performance level. Therefore, for this reason, a study on FI is significant in finding out the importance of FI in determining SMEs' business performance.

16.2.2 SME Food Industry

This study highlights the fast-growing food market in Malaysia as one of the nation's primary income sources. The food sector was estimated to be worth RM109.96 billion in 2018 and was expanding at a 7.6% annual pace. The predicted Gross National Income (GNI) in 2019 increased from RM1424.27 billion in 2018, with an estimated GNI per capita of RM46,094.66, to RM1503.08 billion in 2019, with an estimated GNI per capita of RM44,010.10 (based on the Food Industry Report, 2020). SME business owners have recently been working to establish and investigate alternatives for marketing their food products in Malaysian supermarkets or hypermarkets by creating high-quality products (Abdul et al., 2017). Furthermore, SMEs in the food industry have opened the way for greater interest in market growth and product development (Azmi et al., 2019). As a result, it allows SMEs to improve their efforts to maintain the industry's rapid growth locally and globally (Yahya and Mokhtar, 2020).

16.2.3 Financial Innovation and Business Performance of the SME Food Industry

FI is often related to business and financial development growth. Researchers also indicate that FI is a vital determinant and influences SMEs' business performance. According to Eniola and Entebang (2015), FI plays an increasingly important role in the current organization by helping firms grow and contribute to the economy. The purpose of the study, which was carried out in Nigeria, was to raise awareness about the sources of financing for SME firms, conceptualize their financing difficulties, identify their root causes, and find creative ways to improve financing through crowdsourcing. Creating a regulatory environment to support SME growth and advancement was also necessary. Usually, successful firms continue to earn more returns on their innovation and achieve higher profit margins. Previous studies on innovativeness concluded that innovativeness is important in improving business performance by minimizing risks and achieving strategic objectives. Besides, the time factor is also important in evaluating FI and SMEs' business performance.

The ability of SMEs to use FI is crucial to gaining a competitive edge in the market. Companies must develop unique capabilities and resources that are hard to copy, cannot be simply replaced by other capabilities or resources, and safeguard businesses from companies that deploy or use them to survive in a competitive and demanding world. Hence, there are four reasons FI has been around for as long as it has: (i) macroeconomic condition volatility, (ii) regulatory and firm-based restrictions, (iii) technological developments, and (iv) competitiveness and market structure. To assist businesses in performing better, FI plays a crucial role in the survival of SME performance.

Global innovation has considerably expanded as a result of the COVID-19 epidemic. This transformation has already been noticed in regions where the virus first appeared (such as China), where businesses like Huawei have increased their R&D efforts (Davey, 2020). The pace and scope of the global food industry have recently changed due to innovation, which also expands applications for food sustainability and security while maintaining food safety. Advanced solutions to ensure 24/7 order taking are expected to grow, whereas companies must promote their values, brand, and quality commitments (Askew et al., 2020). Meanwhile, e-commerce utilizes mobile apps for shopping purposes, helping smallholders and producers find different customers in a small city (Askew et al., 2020) to help companies sustain this pandemic. The continuous changes internally and externally in the organization lead the firm to attain a competitive advantage and achieve better performance as a strategic response.

FI is capable of producing new financial products, services, and processes. However, FI is still not well explored in business performance, even though it is considered the heart of entrepreneurship. The public does not understand the relationship between the independent variable (FI) and the dependent variable (business performance of the SMEs in the food industry). In connection with this, documents or articles on FI need to be developed to obtain more thorough results and determine the moderating impacts on financial risks and business hazards. Since innovation is a drive to obtain higher performance in a company, more research should emphasize the significant impact of FI on the business performance of SMEs.

16.3 METHODOLOGY

The main objective of this study is not only to find out the importance of FI but also to examine the influence of FI on business performance in the SME food industry in Malaysia. Individual customers served as the unit of analysis in the current study. The most significant person to supply relevant details about one's ideas, attitudes, and intentions is the person because this is within the individual's control, according to a previous study by Ajzen and Fishbein (1980). The top management of food companies provided this data as they represent a range of roles, knowledge, experiences, and abilities in intellectual capital and affect the enhancement of business performance.

This investigation utilized a quantitative research methodology and collected data from 181 participants. A total of 200 questionnaires were delivered to the senior management of Malaysian food firms; however, only 181 (or 90.5% of those who responded) were willing to submit specific replies. There was an incomplete response to one out of every 19 questions (9.5%). The kind of sampling used in this investigation was a stratified random sample. This method involves dividing the population into subgroups that are easier to handle. The questionnaire is the primary data source used to create replies from respondents. Use the content validity and the Cronbach alpha test to assure the device's dependability and to guarantee the instrument's validity. Both of these tests can be found here. The FI and business performance indicators of the SME food industry are measured using five points: the Likert scale consists of 1=Strongly Disagree, 2=Disagree, 3=Neutral, 4=Agree, and 5=Strongly Agree. This study uses SPSS (statistical package for the social sciences) to evaluate hypotheses. Table 16.1 includes the ethnicity, age, position, and years of experience of the 181 respondents who were able to be analyzed based on the number of questionnaires that were possible to analyze and satisfied the criterion. There were 49 male respondents and 132 female respondents. Table 16.1 provides a brief synopsis of the respondent's background information.

TABLE 16.1
Profile of Respondents (N = 181)

Variable	Category	Frequency	Percentage (%)
Gender	Male	49	27.1
	Female	132	72.9
Race	Malay	131	72.4
	Chinese	10	5.5
	Indian	33	18.2
	Others	7	3.9
Age	20–30 years old	57	31.5
	30–40 years old	50	27.6
	40–50 years old	43	23.8
	50–60 years old	10	5.5
	60–70 years old	21	11.6
Position	Director/Founder	55	30.4
	Manager	35	19.3
	Assistant Manager	10	5.5
	Head of Department	81	44.8
Years of	Less than 10 years	25	13.8
Experience	11–15 years	56	30.9
	16–20 years	52	28.7
	21 years and above	48	26.5

16.4 FINDINGS

16.4.1 RELIABILITY ANALYSIS

The reliability test is normally employed to assess the consistency and stability of the variable. The reliability analysis computes various commonly used measures of scale reliability and provides information on the relationships between individual items in the scales. Cronbach's Alpha measures internal consistency between items on a scale. The closer Cronbach's Alpha is to 1, the higher the internal consistency and reliability. In general, reliability values less than 0.60 are considered to be poor, those in the 0.70 range are acceptable, and those over 0.80 are considered good (Sekaran and Bougie, 2010).

The Cronbach's Alpha for each variable in this study is displayed in Table 16.2 below. The independent variable, FI, is represented by five components. The Cronbach's Alpha value is 0.848. The value is acceptable and was considered able to deliver the meaning of FI. Measuring the business performance of the SME food industry is the dependent variable in this study. 15 items represent this variable. Three dimensions for measuring the performance of the SME food industry obtained good and trustworthy values due to the score results of 0.90, and above, which are the rate of new product development of 0.851, customer satisfaction of 0.881, and customer retention of 0.901. Altogether, the Cronbach's Alpha value for variables in this study was approximately acceptable and was positively correlated to one another. It concludes that all items in the questionnaire have good internal consistency and stability.

16.4.2 MEAN AND STANDARD DEVIATION

Each study variable's mean and standard deviation were used in a descriptive analysis. The analysis can provide an understanding of how respondents view each variable. The results of the

TABLE 16.2
Reliability Analysis

No. of Items	Variable	Cronbach's Alpha
	Independent Variable:	
5	Financial Innovation	.848
	Dependent Variable:	
5	Rate of New Product Development	.851
5	Customer Satisfaction	.881
5	Customer Retention	.901

TABLE 16.3
Mean, Standard Deviation, and Total Variance of the Study Variables

Variables	Minimum	Maximum	Mean	Std. Deviation	Variance
Financial Innovation	3.00	5.00	4.31	0.60	0.37
Business Performance	1.60	5.00	4.23	0.61	0.38

study's variables, assessing the performance of the SME food sector and FI, are summarized in Table 16.3, which includes their minimum, maximum, mean, standard deviation, and total variance values.

For FI, a high mean score was observed (4.31). In light of this, respondents concurred that the food business had successfully implemented beneficial FI. According to the mean score of 4.23 for business performance in the food industry, the respondents agreed that the dimensions of the rate of new product creation, customer satisfaction, and customer retention should be used as the metrics to assess business success in the SME food sector.

16.4.3 REGRESSION

The major purpose of this test is to analyze whether FI will influence the business performance of the SME food sector. A multiple regression analysis was done. Table 16.4 accurately presents the regression analysis findings for the impact of FI on the performance of the SME food industry. Because the R-value for the interaction term (FI as an independent variable) is relatively high at 0.811, it is evident that this term affects the linkages between the independent variables and the dependent variable, which is company performance. Moreover, the significant value for the independent variable, FI, is 0.000 ($p_{0.05}$), demonstrating a high significance level. In addition, the Beta value indicates a good value. The study model has shown sufficient predictive power regarding the variance explained (R^2) of the endogenous variables. The R^2 and F-value of the entire model, which includes the interaction effect, indicate that 66% and 344,435, respectively, of the variance of the dependent variable are explained. Together, the results below provide evidence in favor of the hypothesis. As a result, there is a strong correlation between FI and the business performance of the SME food industry. Table 16.6 displays the results of the b1 t-test. The value of the tcount coefficient was calculated from the results of the t-test using SPSS-based computer processing version 24, and it was found to be 3,587, while the value of the tcount coefficient for FI was 18,559. It demonstrates that the performance of the food business in Malaysia is inversely correlated with how well FI impacts an organization (Table 16.5).

TABLE 16.4
Model Summary

Model	R	R Square	Adjusted R Square	Std. Error of the Estimate
1	.811[a]	.658	.656	.35941

[a] Predictors: (Constant), FI.

TABLE 16.5
ANOVA[a]

Model	Sum of Squares	df	Mean Square	F	Sig.
Regression	44.492	1	44.492	344.435	.000[b]
Residual	23.122	179	.129		
Total	67.614	180			

[a] Dependent Variable: BP.
[b] Predictors: (Constant), FI.

TABLE 16.6
Value of *t* Count[a]

Model	Unstandardized Coefficients		Standardized Coefficients	*t*	Sig.
	B	Std. Error	Beta		
(Constant)	.691	.193	.811	3.587	.000
FI	.822	.044		18.559	.000

[a] Dependent Variable: BP.

16.5 CONCLUSION

FI has been cited as a key factor in numerous studies. FI was tested as an important variable in the respective studies conducted by Syed et al. (2016). Few research studies have used FI to explain the relationship between intellectual capital and business performance in the SME food industry. Compared to developing countries like Malaysia, there has been very little research on FI in industrialized nations like the United States, Hong Kong, and others.

Based on the findings of this study, it can be concluded that the model accurately captures the FI benefits for the SME food business, which may be regarded as a methodological advancement for future research. The Cronbach's Alpha statistics for the FI variable are 0.848, generally indicating strong reliability (Sekaran and Bougie, 2010). Also, because the score results were above 0.80, three performance measurement variables for the SME food business obtained good and trustworthy values. These factors include the rate at which new products are created (0.851), the level of consumer happiness (0.881), and the percentage of customers that remain loyal (0.901). The strongest correlation between FI and company success in the SME food industry reveals the highest link ($r=.83$, $p=.01$). This suggests that FI and business performance are significantly and favorably connected.

In addition, a multiple regression analysis was carried out to investigate FI's impact on the economic performance of small and medium-sized food businesses. Because the R-value for the interaction term (FI as an independent variable) is relatively high at 0.81, it is clear that this term significantly impacts the linkages between the independent variables and the dependent variable, business performance. In conclusion, FI will influence the business performance of the SME food industry.

Although specific FIs may benefit a single player while harming others, they may also have unfavorable side effects on the financial system. So, effective use of a particular FI necessitates an in-depth understanding of how it works and carefully examining its consequences.

This study has explored business performance in the Malaysian food industry, providing a glimpse of the impact of FIs on the financial system. However, this topic's complex nature suggests that further theoretical and empirical research is needed. To enhance the consistency of the results, the study could be expanded to include other sectors. Additionally, including other drivers to measure business performance could improve our understanding of the various factors that impact business performance. Further research in these areas could provide valuable insights into the role of FIs in the broader economy.

REFERENCES

Abdul, M., Rahman, R. A., Kamarulzaman, N. H., Ibrahim, S., Rahman, M. M., & Uddin, M. J. (2017). Small medium-sized enterprises (SMEs) to hypermarkets: Critical quality aspects in delivering food and beverage products. *International Journal of Economics & Management*, 11, 777–793.

Ajzen, I., & Fishbein, M. (1980). *Understanding Attitudes and Predicting Social Behaviour*. Hoboken, NJ: Prentice-Hall.

Askew, B., Brown, A., Christy, M., Gomez, A., Gooch, M., Kubicki, J. M., ... & Slutskaya, S. A. (2020). Georgia libraries respond to Covid-19 pandemic. *Georgia Library Quarterly*, 57(3), 10.

Azmi, F. R., Abdullah, A., Musa, H., & Mahmood, W. H. W. (2019). Perception of food manufacturers towards adoption of halal food supply chain in Malaysia: Exploratory factor analysis. *Journal of Islamic Marketing*. doi: 10.1108/JIMA-12-2018-0236.

Dabrowski, M. (2017). Potential impact of financial innovation on financial services and monetary policy. Case Research Paper, (488).

Davey, G. (2020). The China-US blame game: Claims-making about the origin of a new virus. *Social Anthropology*, 28, 250–251.

Eniola, A. A., & Entebang, H. (2015). SME firm performance-financial innovation and challenges. *Procedia-Social and Behavioral Sciences*, 195, 334–342.

Food Industry Report, 2020: Prepared by Flanders Investment & Trade, Malaysia Office. C/O TAPiO Management Advisory Sdn. Bhd.

Hewa Wellalage, N., Locke, S., & Samujh, H. (2020). Firm bribery and credit access: Evidence from Indian SMEs. *Small Business Economics*, 55(1), 283–304.

Mehta, N., & Brahmbhatt, M. (2020). Financial Innovation: The Trend-Changing Facet of SME Competitiveness.

Michalopoulos, S., Leaven, L., Levine, R. (2009). Financial Innovation and Endogenous Growth. National Bureau of Economic Research, Working Paper 15356, Cambridge, September, pp. 1–33.

Mutlu, B., & Er, A. (2003). Design Innovation. In *5th European Academy of Design Conference*, Barcelona.

Qamruzzaman, M., Mehta, A. M., Khalid, R., Serfraz, A., & Saleem, H. (2021). Symmetric and asymmetric effects of financial innovation and FDI on exchange rate volatility: Evidence from South Asian Countries. *The Journal of Asian Finance, Economics, and Business*, 8(1), 23–36.

Sekaran, U. & Bougie, R., (2010). *Research Methods for Business: A Skill-Building Approach*. (5th Ed.) UK: John Wiley & Sons.

Syed, A. M., Riaz, Z., & Waheed, A. (2016). Innovation, firm performance and riskiness: evidence from the leading worldwide innovative firms. *International Journal of Innovation Management*, 20(07), 1650066.

Szirmai, A., Naudé, W., & Goedhuys, M. (Eds.). (2011). *Entrepreneurship, Innovation, and Economic Development*. Oxford: Oxford University Press.

Targalski, J. (2006). Innowacyjność-przyczyna i skutek przedsiębiorczości. *Zeszyty; Naukowe/Akademia Ekonomiczna w Krakowie*, 730, 5–10. https://www.flandersinvestmentandtrade.com/export/sites/trade/files/market_studies/FB%20Industry%20Report.pdf.

Yahya, S. B., & Mokhtar, M. B. (2020). Small medium enterprises in food and beverage industry behavioural intention towards halal and fourth industrial revolution technology. *Jurnal Penyelidikan Sains Sosial (Jossr)*, 3(8), 16–21.

Section 3

Fintech Ecosystems, Collaboration, and Analysis

17 The Role of Collaborative Skills and Knowledge Sharing in the Emergence of Fintech Ecosystems
Case of Casablanca Finance City

Smyej Oumaima and Si Mohammed Ben Massou

17.1 INTRODUCTION

Lately, innovation has become crucial to business success in all industries, especially technology-related ones. Firms are increasingly aware of the importance of innovation to survive and evolve. However, competition has become so high that companies tend to adopt new strategies that can help them cooperate with their competitors. It's about being part of new inter-organizational entities that are based on the theory of the death of competition (Moore, 1993). Those are what we call innovation ecosystems. It refers to an interconnected network of actors (firms, universities, incubators, labs, state, etc.) who share knowledge, resources, and skills to generate common value by developing new products and services (Nambisan and Baron, 2013). They are typified by three fundamental characteristics: mutual goals, linked dependencies, and shared capabilities (Adner and Kapoor, 2010; Iansiti and Roy, 2004; Teece, 2009).

Being part of innovation communities comes down to the inability of companies to follow the fast innovation flow in an individual way. Not only should they go further and deeper in their field of expertise, but they also have to make use of knowledge and technologies from other disciplines. Besides, the complexity of innovation processes requires more than one actor's input.

Given that ecosystem communities connect various actors from different sectors and fields, the solution to these challenges remains to go beyond companies' boundaries and competition and to promote collaborative innovation. In this respect, many organizations have decided to participate in inter-organizational networks, whether by integrating the existing ones or by fostering the emergence of new ones (Asheim and Coenen, 2005; Tallman et al., 2004).

Despite the numerous benefits provided by innovation ecosystems, little is known about how they come into existence. While several studies have highlighted the importance of knowledge sharing and common skills in structured networks, the literature remains underdeveloped on the subject. This chapter addresses this gap by focusing on the factors that favor the advent of innovation ecosystems in general and Fintech ecosystems in particular. We pay special attention to "knowledge sharing" and "collaborative skills" as the main factors in both behavioral finance and technological innovation.

Casablanca Finance City (CFC) serves as an excellent example of a rapidly growing Fintech ecosystem. Recognized as Africa's leading financial center and a partner of major international financial centers, CFC has successfully built a robust community of members comprising financial companies, regional headquarters of multinationals, service providers, and holding companies.

DOI: 10.1201/9781032667478-20

This study meticulously considers the following research question: What are the emergence factors of the CFC Fintech ecosystem? Answering this question holds at least two significant interests. The first is theoretical, as it contributes to deepening knowledge about innovation ecosystems. The second is related to management, as it guides firms' boards in recommending strategies within communities. To address this issue, we will first present a brief literature review on Fintech ecosystems. The second part will be devoted to the methodology, while the third part will focus on presenting the results and discussion.

17.2 LITERATURE REVIEW

By reviewing the literature, we will clarify the foundations of the term ecosystem in management (Section 17.2.1), ranging from business ecosystems through innovation ecosystems to Fintech ecosystem. Then we will present the different phases through which an ecosystem community goes (Section 17.2.2) at that level. We will focus on the emergence phase, which is the fundamental stage of our research. Finally, we will make use of the literature in order to highlight the behavior of financial actors when it comes to information sharing and collaborative skills training (Section 17.2.3).

17.2.1 THE WORLD OF ECOSYSTEMS IN MANAGEMENT

The concept of ecosystems has recently gained attention as a form of inter-organizational collaboration that connects actors from diverse sectors to create value through collaborative innovation. Initially borrowed from the field of biology, the interest behind this appropriation is to emulate the natural environment and its efficient functioning. Moore (1993) introduced the term "business ecosystem", which has since been expanded to include other types of ecosystems, such as knowledge ecosystems, innovation ecosystems, and business ecosystems (Gomes et al., 2018). Knowledge ecosystems rely on researchers, institutes, and innovators to generate new knowledge, while business ecosystems consist of large companies dedicated to creating customer value. Innovation ecosystems combine both knowledge exploration and exploitation for value co-creation, involving large companies, innovators, researchers, and labs.

Overall, the study of ecosystems and their various forms represents a significant contribution to the field of inter-organizational collaboration. However, further research is needed to examine the specific factors that drive the success of these ecosystems and to explore their impact on innovation and value creation. Such research should adopt rigorous empirical methods that go beyond the descriptive level and employ statistical tests to confirm the hypotheses formulated. By doing so, we can enhance our understanding of how ecosystems can be effectively managed to generate value and drive innovation.

Due to their innovative nature and pecuniary interests, Fintech ecosystems are characterized as innovation ecosystems. Furthermore, the absence of sectoral boundaries between technology and finance aligns perfectly with the ecosystem community's logic, which necessitates crossing sectoral boundaries to earn the title of an ecosystem.

Financial technology manifests itself when technology is used to provide innovative products and services to the financial sector (Muthukannan et al., 2020). This new combination constitutes an emergent market that is highly influencing the traditional business models where consumers' experiences and early users' feedback are integrated into the innovation process. Henceforth, they are part of the ecosystem's actors, and they contribute to the improvement of financial services (Muthukannan et al., 2020).

In this regard, Fintech ecosystems can be identified as inter-organizational entities characterized by heterogeneous, dynamic, and complex networks of actors that cooperate with each other to deliver an innovative array of financial products and services. Lee and Shin (2018) distinguish Fintech ecosystems from other ecosystems by five major components: start-ups, technology firms, government, customers, and traditional financial institutions like banks. The significant impact of Fintech companies on the financial sector has forced exclusively financial firms to forge partnerships with innovative companies within Fintech ecosystems. Consequently, Fintech ecosystems have become important organizations that not only attract stakeholders' interest but also the attention of researchers and academics.

Therefore, exploring the emergence of ecosystems enables us to assist managers in making strategic decisions to build or integrate into an ecosystem. Moreover, studying the factors that facilitate the emergence of ecosystems is crucial, as numerous ecosystems tend to fail before their advent. The following section focuses on the various stages of an ecosystem's life cycle, from its nascent stage to either its decline or renewal.

17.2.2 INNOVATION ECOSYSTEMS' LIFE CYCLE

Like any other organizational entity, managed ecosystems have their own life cycle. Moore (1993, 1996) has developed an S-curve life cycle consisting of four phases: birth, expansion, authority, and renewal. The birth period is characterized by a strategic watch that allows seizing new opportunities through the establishment of value chains and the creation of value for customers. In the expansion phase, the created value will be captured by business ideas, and the incoming concept will be offered to a larger market. The authority stage is the one where components and processes are stable. Thus, leaders tend to encourage partnerships and welcome the advent of fresh contributors. Thenceforward, when relationships between actors have matured, ideas are developed, and innovation is highlighted, a whole new ecosystem emerges.

Quite a few authors have been interested in the birth phase of an ecosystem. This mainly comes down to the role played by this chapter in the growth of the entity, which can either lead to its development or its decline (Attour and Barbaroux, 2016). Furthermore, the emergence phase of ecosystems is critical and contains determining factors that have a significant impact on the entire life cycle. This phase serves as a learning period and an opportunity-taking stage (Attour and Barbaroux, 2014). During this phase, actors are trained under pressure, and collaborative skills are built through collaboration, innovation, and the creation of offers (Gulati et al., 2000). Then, opportunities might be taken when transformed into a real-value proposition that was initially confronted by many others. Therefore, an ecosystem's birth results from the development of a value proposition, which stems from learning through experimentation and seizing opportunities. The significant momentum during the birth stage leads us to inquire about the factors that reinforce it. The following section emphasizes factors specific to the field of behavioral finance, which undoubtedly impact community relationships and network structuring.

17.2.3 BEHAVIORAL FINANCE IN THE LIGHT OF THE LITERATURE

Whether in finance or any other managerial discipline, human beings cannot be completely robotized. Their actions are usually influenced by human nature, which often causes decision-making to be biased by emotions, psychology, beliefs, and opportunism. Especially in a social setting where the community is strongly present and societal connections are abundant and frequent. Behavioral finance relies on the psyche of investors and its impact on financial decision-making. This theory assumes that individuals have emotions that can affect their choices. These choices might be inefficient and irrational and, as a result, lead to disasters.

After the Nobel Prize in economics was awarded to the founders of behavioral finance theory in 2002, the study of finance through psychology became so interesting that brokers tried to predict stock prices based on stakeholders' behavior. Henceforth, the presence of psychological biases when making financial decisions is no longer a doubt. Psychological biases can be due to mood (Wright and Bower, 1992), emotion (Baumeister, 1999), or the momentum effect, i.e., the herding behavior of traders (Jegadeesh and Titman, 1993).

Behavioral finance does not only arise from psychological bias. It also occurs when financial actors demonstrate dishonest behavior in order to gain more or achieve a desired situation during conflicts of interest. Information withholding (Akerlof, 1970), leader entrenchment (Paquerot, 1996), and agency theory (Jensen and Meckling, 1976) are some examples of opportunistic behavior.

Considering that ecosystem communities are based on collaboration and sharing for innovation purposes, the opportunistic behavior of financial actors may hinder an ecosystem's birth. In fact, networks require sharing, while financial actors often prefer to withhold information. The question arises: how do financial actors manage their knowledge and skills within a nascent Fintech-ecosystem community? The contradiction between the requirements of ecosystems and the preferences of opportunistic actors raises questions about how Fintech ecosystems come into existence.

17.3 METHODOLOGY

This chapter tries to identify factors enabling the emergence of CFC Fintech ecosystem in general and its management of knowledge and skills sharing in particular. We have decided to proceed according to the qualitative approach in order to explore, as much as possible, the birth conditions of these complex organizations.

Thanks to the constructivist paradigm (Gavard-Perret et al., 2008), we were able to co-construct the research problem with practitioners. Then, we highlighted the relationships between emergence factors according to a model that promotes the understanding of complex processes such as the emergence of an ecosystem.

Qualitative analysis is the ideal approach when it comes to exploring a little-known area (Miles and Huberman, 2003). It allows better analysis and understanding, especially when dealing with complex organisms with highly developed network links. The research method used is an inductive approach to data collection through in-depth interviews. The use of this mode of collection is explained by our desire to extract as much data as possible. It is a research on the content (Schreier, 2012) whose objective is to discover and understand the studied object through its partition into sub-elements, which will subsequently be described distinctly but also together.

During data collection, three elements contributed to the progress of the work. To begin with, we relied on documentation (articles, theses, press, website, etc.) to determine both the actors belonging to CFC and likely to respond to our study and to prepare the interview guide in themes according to the birth factors of Fintech ecosystems proposed by the literature.

Afterward, we moved on to an in-depth description of the interviews by giving enormous importance to the vocabulary and language used by the interviewees. This is because the field of study remains little known. Lastly, we combined the results of the documentation and those of the interviews in order to improve the internal validity of the research.

In the next sections, we will first explain how we proceeded to select our sample, and then we will carry on by identifying our data collection method (Section 17.3.1). As a final point, we will conclude with the analysis of the collected data (Section 17.3.2).

17.3.1 SAMPLE SELECTION AND DATA COLLECTION

By reviewing both the literature on Fintech ecosystems and publications about CFC, we were able to select various experts who were likely to provide insights on specific entities. The main criterion for potential interviewees was "being part of CFC's Fintech ecosystem." The sample size was determined using the saturation method, a qualitative technique where the researcher aims to collect the maximum number of responses until no new information emerges from additional data (Bloor and Wood, 2006). We were concerned about the actors' willingness to participate. In this regard, we sent 30 emails and made 15 phone calls; only 5 CFC members responded positively to our request. Two interviews were conducted face-to-face, and three were carried out via video call. We chose

TABLE 17.1

Sample's Description

Verbatim	Actor	Date	Data Collection	Duration
E1	Interbank payment center	05/09/2022	Face to face	1 hour 50 min
E2	Central bank	06/09/2022	Face to face	2 hours 05 min
E3	Telecommunication and computer service company	07/09/2022	Video call	1 hour 30 min
E4	Fuel company	10/09/2022	Video call	2 hours
E5	Insurance company	11/09/2022	Video call	1 hour 45 min

to conduct semi-structured interviews that lasted a total of 9 hours and 10 minutes. To encourage respondents to speak and open up, we made sure to be attentive listeners. Our interview guide had six themes: Fintech-ecosystem representation; actors' outside in; actors' inside out; barriers to collaboration and their solutions; knowledge sharing; and collaborative skills.

CFC, also known as CFC, is an African financial hub located in Morocco. It's recognized as the financial leader in Africa (GFCI, 2021)[1] and a partner of more than 15 other financial centers in the world. CFC has more than 200 members, including government, industry, start-ups, customers, tech vendors, universities and research institutions, investors and incubators, accelerators, and innovation labs. Its main goal remains to afford its community an attractive value proposition by promoting the business deployment of actors' activities. On the whole, we've accomplished 5 semi-structured interviews with 1 representative of an interbank payment center, 1 representative of the central bank, 1 representative of an insurance company, 1 representative of a fuel company, and 1 representative of a provider of telecommunications and computer services. The interviews were transcribed verbatim, and 3 of them were recorded with the actor's permission. This study allowed us, among other things, to better understand our field of research by taking a step back from the discourse, practices, and perceptions of these actors. It also limited misunderstandings about Fintech ecosystems, the discovery, and in-depth analysis of the constraints and conditions of the emergence of CFC. Table 17.1 presents the data from the sample of our study.

17.3.2 DATA PROCESSING AND ANALYSIS

Contents analysis is a scientific approach generally used for recent issues that have weak theoretical foundations and modest empirical materials (Schreier, 2012). It consists of four fundamental steps: First, we identify relevant ideas, and then we classify them according to predefined themes. Finally, we end up coding and counting them so as to reach an inference (Thiétart, 2007). The classification into themes is made through their categorization according to a grid (Table 17.2). This process facilitates the coding and the analysis, whether it's horizontal or vertical. It might be accomplished manually or using computer software. In our case, we decided to perform it manually since the data was not too voluminous.

The first theme was dedicated to understanding the actor's vision of reticular entities. In other words, we've intended to figure out how CFC members perceive the perfect Fintech ecosystem and how they would describe it.

In the second theme, we've tried to comprehend the reasons why an actor would decide to integrate a Fintech ecosystem. Actually, the conditions of the emergence of Fintech ecosystems are, in one way or another, related to motives that encourage actors to be part of these communities. This purely inductive interest also has theoretical origins, namely the outside, which, as Attour and Ayerbe (2015) point out, refers to enriching the company's knowledge base through access to technologies from third parties in the same ecosystem.

TABLE 17.2
Theme Reading Grid

Themes	Description
Fintech ecosystems' representation	Fintech ecosystems' definition
	• Ideal Fintech ecosystems' description
Outside in	• The main reasons for joining Casablanca Finance City
	• The resources sought from other actors in the ecosystem
Inside out	• The resources that the interviewed organization provides within Fintech ecosystems
Barriers to partnerships and their solutions	• Risks and barriers associated with partnerships within the financial ecosystem
	• Solutions to obstacles that block collaboration during exchanges
Knowledge sharing	• Characteristics of the shared knowledge in Fintech ecosystems
	• The role played by knowledge sharing in Fintech ecosystems
	• How information is shared within Fintech ecosystems
Collaborative skills	• The role played by collaborative skills in Fintech ecosystems

Unlike the outside in, the inside out (Attour and Ayerbe, 2015) is known as actors' contribution to their ecosystem, i.e., what would an actor offer to other actors in the same ecosystem? In this respect, the third theme is devoted to the inside out. In reality, integrating into an ecosystem is not that simple. Potential members must demonstrate their ability and willingness to contribute in order to be admitted to the community and subsequently receive and benefit from external resources. In other words, membership in ecosystem communities must be earned through collaborative innovation.

While some factors facilitate the advent of Fintech ecosystems, others impede their development. Furthermore, Attour and Barbaroux (2016) discuss the disappearance of certain ecosystems during the first stage of their life cycle, even before they grow. In this regard, the emergence of obstacles in CFC constitutes the fourth theme of our analytical framework. The objective is to identify hindrances to partnerships and, at the same time, discuss potential solutions to address these challenges.

In recent years, the financial and digital fields have increasingly converged, giving rise to significant technological innovations that are based on knowledge sharing. However, the sensitivity of financial information necessitates special treatment, especially in a heterogeneous ecosystem environment. The fifth theme of our framework is devoted to the role played by knowledge sharing in Fintech ecosystems. Specifically, we aim to determine the characteristics of shared knowledge in Fintech ecosystems, as well as the perspectives and methods involved.

The pooling of skills can be considered a strategic resource, giving actors access to new skills that are likely essential for their development. Consequently, the final theme concerns skill management in CFC, which serves as an important factor in the birth of Fintech ecosystems.

17.4 HYBRID EXPLORATION OUTCOMES

Recognizing that internal and external validity are important in research, we decided to rely on multiple methods and strategies (Mathison, 1988). This approach aims to use a combination of several procedures to reinforce the results obtained. For this reason, we began by exploring the literature and documents related to the study, followed by an empirical investigation using a qualitative approach.

17.4.1 THEORETICAL EXPLORATION RESULTS

In the documentary analysis, we examined a collection of articles and theses to identify general ideas and establish connections between them. Article selection was made from various sources (Aims, Cairn, TIM Review, Springer, etc.), using the term "Fintech ecosystem" as a search criterion. Two imperatives helped refine our selection: relevance to the field of management and the effective presence of the notion in the text. A general analysis of the publications was conducted, focusing on passages that referred to ecosystem emergence factors. The exploration results are as follows (Table 17.3).

17.4.2 EMPIRICAL EXPLORATION RESULTS

In addition to emergence factors taken from the literature, the in-depth interviews showed new CFC-specific factors (Table 17.4).

From different respondents' points of view, we were able to conduct a detailed horizontal analysis of each theme of the reading grid. The results of our study revealed that: the representation of Fintech ecosystems remains unclear for community actors (Section 17.4.1.1), the main motivation for integrating CFC is financial gain, and the capabilities and skills of each member constitute their main contribution to the community (Section 17.4.1.2). Disagreements, poor visibility, and

TABLE 17.3

Literature Ecosystem's Emergence Factors

Emergence Factor	Explanation	Source
Government strategic vision in response to technological trends	The emergence of Fintech ecosystems begins with the government establishing strategic visions to promote digital governance and technology-enabled jobs, and subsequently executing the plan by improving conditions and fostering partnerships within the community. When interconnections between members are strengthened, symbiotic synergies are developed, and operational efficiency is promoted.	Muthukannan et al. (2020)
Technologies and platforms	Platforms play a significant role in shaping the relationships between entities and structuring ecosystems. They consist of multiple interconnected modules that are accessible through interfaces. Platforms enable users to access shared resources and facilitate collaborative innovation within the ecosystem.	Daidj (2011)
Knowledge sharing and close interactions	To achieve project outcomes, agents within communities need to engage in close interactions, share knowledge and collaborate. This type of learning process reinforces emerging structures such as Fintech ecosystems.	Attour and Lazaric (2018)
Copyright management	Copyright management reduces the risk of conflicts thanks to legal anticipation of collaboration outcomes depending on the contribution of each actor.	Attour and Ayerbe (2015)
Absence of a central authority	The emergence of Fintech ecosystems is not controlled by a central authority or a global controller. Instead, it is related to the coherent actions of multiple agents. The government can play a critical role in promoting the community, but it cannot control its real emergence.	Muthukannan et al. (2020)
Collaborative skills	Experimentation within a community of agents can encourage individuals to identify their unique skills and expertise, as well as develop new collective skills through collaborative innovation, ultimately contributing to the achievement of project goals.	Loilier and Malherbe (2012), Attour and Barbaroux (2016)

TABLE 17.4

Interview's CFC Emergence Factors

Emergence Factors	Explanation
Collaborative financial engineering	The connection of various experts in the fields of finance, technology, institutional regulation, and start-ups enables the achievement of financial innovation and engineering, which is the key to the emergence of Fintech ecosystems.
Actor's complementarities	Complementarity between services, products, activities, and resources in knowledge and materials is a necessary factor for collaboration and innovation.
Trust	In order to collaborate, actors need to trust and be trusted. In the absence of trust, actors tend to conceal valuable information and knowledge that could lead to significant innovations. Various solutions are suggested to overcome opportunistic behaviors, such as contracts, regulatory agents, and copyright management.

opportunistic behavior are the main obstacles to forming alliances (Section 17.4.1.3). Knowledge sharing and collaborative skills are fundamental for the birth of Fintech ecosystems (Section 17.4.1.4).

17.4.1.1 Fintech Ecosystem from the Perception of CFC Actors

Ecosystems are communities of heterogeneous and multisectoral actors that gravitate around a keystone because it provides them with value through skill co-evolution (Moore, 1993). Fintech ecosystems are a particular case of ecosystems that link the financial and technological sectors, encouraging the collaboration of five specific actors: start-ups, technology firms, government, customers, and traditional financial institutions (Lee and Shin, 2018).

All of the interviewees shared the idea that CFC is associated with collaboration between companies in order to generate value through collaborative financial engineering. They came close to the definitions proposed by the literature but did not manage to differentiate it from other collaborative entities such as clusters, hubs, or networks, which are limited to sectoral interactions, unlike ecosystems that have cross-sectoral boundaries.

17.4.1.2 Casablanca Finance City Coupled Process

The coupled process is an open innovation approach that includes both outside in and inside out. Outside in is when an actor integrates new knowledge into his organization (Spithoven et al., 2010) and uses it to improve a product or a service. Inside out is when he shares and values his resources and knowledge with his environment for innovation purposes (Attour and Ayerbe, 2015). This section highlights both inside out and outside in since what an actor is looking for constitutes what another actor is willing to give. Firstly, the interviewees agreed that the main reason behind integrating CFC is pecuniary. It might occur directly: value capture, fiscal advantages, doing business, or indirectly: notoriety, data collection, and regular meetings allowing the presentation of investment opportunities.…. Second, two of the five actors agreed that the exchange of skills and capabilities is essential in ecosystems, especially when actors have big experience in their field. Finally, complementarity is fundamental in a sharing community. It allows us to make a global offer.

17.4.1.3 Fintech-Ecosystem Emergence Obstacles

A significant number of inter-organizational forms disappear even before their inception (Attour and Barboux, 2016). This reality obliges us to study obstacles that put an end to possible collaborations. The in-depth interviews shed light on various obstacles, namely disagreements related to diversity, poor visibility about innovation, and opportunistic behaviors of certain actors. Given that Fintech ecosystems go beyond sectoral boundaries, they include various heterogeneous actors. Managing diversity becomes very difficult. Moreover, innovation is unknown and not guaranteed, which means that actors are not certain about the investment of being part of the ecosystem, especially since CFC requires fees. Finally, certain opportunistic behaviors might appear during collaborative exchanges, such as not respecting confidentiality, voluntarily maximizing the outside in while minimizing the inside out, or seeking individual value capture of a collaboratively created value.

17.4.1.4 Knowledge Sharing and Collaborative Skills

Knowledge and skills are crucial resources for boosting innovation. Sharing them between regulators and Fintech companies might improve regulatory awareness of consumer habits and desires, contributing to the building of regulatory systems that help create consumer trust in Fintech ecosystems (Mention, 2019). However, Fintech-ecosystem actors are often reluctant to share sensitive and valuable knowledge and skills due to concerns about opportunistic behavior arising from information asymmetry. To limit the risks of conflicts, actors suggest the management of copyrights and intellectual property and the legal anticipation of collaboration outcomes.

When knowledge-sharing problems are resolved and actors collaborate in a trustworthy environment, they begin a collaborative process by giving and receiving different types of information and skills. Interviewees acknowledged three major characteristics of shared knowledge: volume, velocity, and variety. The amount of data is important because it is used to detect trends and predict financial behavior. Velocity is crucial since traders need to make quick decisions, especially when it comes to market stock. Data variety, such as images, text, videos, and sources, is also important because it allows users to make connections and improve understanding.

Skills are important due to their contribution to the Fintech innovation process. In most cases, traditional financial institutions (e.g., banks, insurance companies) are stiff and poorly digitized, if at all in tune with innovation. Their connection with technological operators will allow them to acquire innovative skills that they can use in the financial field. In summary, knowledge sharing and skill allocation are essential factors for innovation and, in turn, for Fintech-ecosystem emergence.

17.5   DISCUSSION

This chapter makes a twofold contribution. Firstly, it highlights the importance of inter-organizational partnerships for the development of the economy and the nation. Secondly, it enriches the theoretical framework and fills the gap that exists between research on ecosystems and Fintech. In this regard, we have identified that Fintech ecosystems are communities that link together start-ups, technology firms, government, customers, and traditional financial institutions to create advantages such as fiscal benefits and administrative facilities and to enhance innovation in the financial field through collaboration, opportunities, and complementarities. Furthermore, we have demonstrated that Fintech ecosystems' birth is favored by knowledge sharing in the community and skill allocation between members. Different barriers might limit the emergence of such ecosystems, such as opportunistic behaviors, lack of trust, and disagreement. However, these barriers can be overcome through intellectual property management, diversity management by the keystone, contracts, and other solutions.

17.6   CONCLUSION

In conclusion, our qualitative analysis has shed light on various factors that contribute to the emergence of CFC's Fintech ecosystem. These factors include value capture and financial revenues, exchange of expertise through knowledge and skill sharing, and complementarity between actors from different sectors. However, diversity management remains a challenge in terms of establishing trust among actors. The use of contracts and copyright management can help overcome this challenge and enable collaboration and innovation within the ecosystem.

Our research provides valuable insights into the important role of inter-organizational partnerships in the development of both the economy and the nation. Moreover, our study fills the gap between research on ecosystems and Fintech, and contributes to the theoretical framework of Fintech ecosystems. However, we acknowledge that our qualitative approach requires refinement through statistical tests to confirm our hypotheses.

We believe that our findings offer important implications for financial firms, which can leverage partnerships with Fintech start-ups to co-create strategies and offer value-added services. Furthermore, knowledge and skills sharing within Fintech ecosystems can lead to innovative products and services such as customer segmentation, fraud detection, real-time stock market insights, and credit reassessment.

Future research could build on our work by conducting a quantitative empirical study to validate our findings and provide further insights into Fintech ecosystems, not only in Casablanca but also in other regions. Such research can contribute to the development of more effective strategies for ecosystem emergence and sustainable growth.

NOTE

1   The Global Financial Centres Index.

REFERENCES

Adner, R. & Kapoor, R. (2010, mars). Value creation in innovation ecosystems : how the structure of technological interdependence affects firm performance in new technology generations. *Strategic Management Journal*, 31(3), 306–333. https://doi.org/10.1002/smj.821.

Akerlof, G. A. (1970, août). The market for "Lemons": Quality uncertainty and the market mechanism. *The Quarterly Journal of Economics*, 84(3), 488. https://doi.org/10.2307/1879431.

Asheim, B. T. & Coenen, L. (2005, octobre). Knowledge bases and regional innovation systems: Comparing Nordic clusters. *Research Policy*, 34(8), 1173–1190. https://doi.org/10.1016/j.respol.2005.03.013.

Attour, A. & Ayerbe, C. (2015). Le management amont et aval des droits de propriété intellectuelle au sein des écosystèmes-plateformes naissants. *Systèmes d'information & management*, 20(3), 47. https://doi.org/10.3917/sim.153.0047.

Attour, A. & Barbaroux, P. (2014, mars 8). Naissance des écosystèmes d'affaires: une articulation des compétences intra et inter organisationnelles. *Gestion*, 33(4), 59–76. https://doi.org/10.3917/g2000.333.0059.

Attour, A. & Lazaric, N. (2018, 13 avril). From knowledge to business ecosystems: emergence of an entrepreneurial activity during knowledge replication. *Small Business Economics*, 54(2), 575–587. https://doi.org/10.1007/s11187-018-0035-3.

Baumeister, R. F. (1999, 1 novembre). *Self in Social Psychology: Key Readings (Key Readings in Social Psychology)* (1re éd.). Psychology Press, Philadelphia, PA.

Daidj, N. (2011, 1 décembre). Les écosystèmes d'affaires: une nouvelle forme d'organisation en réseau? *Management & Avenir*, 46(6), 105–130. https://doi.org/10.3917/mav.046.0105.

Gavard-Perret, M. L., Jolibert, A., Haon, C. & Gotteland, D. (2008, 1 mars). Methodologie de la recherche: reussir son memoire ou sa these en sciences de gestion. CHRETIEN.

Gomes, L. A. D. V., Facin, A. L. F., Salerno, M. S. & Ikenami, R. K. (2018, novembre). Unpacking the innovation ecosystem construct: Evolution, gaps and trends. *Technological Forecasting and Social Change*, 136, 30–48. https://doi.org/10.1016/j.techfore.2016.11.009.

Iansiti, M. & Levien, R. (2004, août 1). *The Keystone Advantage: What the New Dynamics of Business Ecosystems Mean for Strategy, Innovation, and Sustainability (Illustrated)*. Harvard Business Review Press, Boston, MA.

Jegadeesh, N. & Titman, S. (1993, mars). Returns to buying winners and selling losers: Implications for stock market efficiency. *The Journal of Finance*, 48(1), 65–91. https://doi.org/10.1111/j.1540-6261.1993.tb04702.x.

Jensen, M. C. & Meckling, W. H. (1976, octobre). Theory of the firm: Managerial behavior, agency costs and ownership structure. *Journal of Financial Economics*, 3(4), 305–360. https://doi.org/10.1016/0304-405x(76)90026-x.

Lee, I. & Shin, Y. J. (2018, janvier). Fintech: Ecosystem, business models, investment decisions, and challenges. *Business Horizons*, 61(1), 35–46. https://doi.org/10.1016/j.bushor.2017.09.003.

Loilier, T. & Malherbe, M. (2012, 28 mars). Le développement des compétences écosystémiques. Le cas de l'ESA émergent des services mobiles sans contact. *Revue française de gestion*, 38(222), 89–105. https://doi.org/10.3166/rfg.222.89-105.

Mathison, S. (1988, mars). Why Triangulate? *Educational Researcher*, 17(2), 13–17. https://doi.org/10.3102/0013189x017002013.

Mention, A. L. (2019, 26 juin). The future of fintech. *Research-Technology Management*, 62(4), 59–63. https://doi.org/10.1080/08956308.2019.1613123.

Miles, M. B. & Huberman, A. M. (2003). Analyse des données qualitatives. Dans M.B. (2e éd.). De boeck superieur.

Moore, J. F. (1993, mai). Predators and prey: A new ecology of competition. *Harvard Business Review*, 71, 75–86.

Moore, J. F. (1996, 4 avril). *The Death of Competition: Leadership and Strategy in the Age of Business Ecosystems*. Harperbusiness, New York.

Muthukannan, P., Tan, B., Gozman, D. & Johnson, L. (2020, décembre). The emergence of a Fintech Ecosystem: A case study of the Vizag Fintech Valley in India. *Information & Management*, 57(8), 103385. https://doi.org/10.1016/j.im.2020.103385.

Nambisan, S. & Baron, R. A. (2013, septembre). Entrepreneurship in innovation ecosystems: Entrepreneurs' self-regulatory processes and their implications for new venture success. *Entrepreneurship Theory and Practice*, 37(5), 1071–1097. https://doi.org/10.1111/j.1540-6520.2012.00519.x.

Gulati, R., Nohria, N. & Zaheer, A. (2000, mars). Strategic networks. *Strategic Management Journal*, 21(3), 203–215. https://doi.org/10.12691/jbms-9-4-3.

Paquerot, M. (1996). L'enracinement des dirigeants et ses effets. *Revue française de gestion*, 111, 212–225.

Schreier, M. (2012, 31 mai). Qualitative content analysis in practice. *Scientific Study of Literature*, 3(1), 165–168. https://doi.org/10.1075/ssol.3.1.15aaf.

Spithoven, A., Clarysse, B. & Knockaert, M. (2010, février). Building absorptive capacity to organise inbound open innovation in traditional industries. *Technovation*, 30(2), 130–141. https://doi.org/10.1016/j.technovation.2009.08.004.

Tallman, S., Jenkins, M., Henry, N. & Pinch, S. (2004, avril). Knowledge, clusters, and competitive advantage. *Academy of Management Review*, 29(2), 258–271. https://doi.org/10.5465/amr.2004.12736089.

Teece, D. J. (2009, 22 janvier). Dynamic capabilities and strategic management: Organizing for innovation and growth. *By David J. Teece. R& D Management*, 41(2), 217–218. https://doi.org/10.1111/j.1467-9310.2011.00638.x.

Wright, W. F. & Bower, G. H. (1992, juillet). Mood effects on subjective probability assessment. *Organizational Behavior and Human Decision Processes*, 52(2), 276–291. https://doi.org/10.1016/0749-5978(92)90039-a.

18 Toward Characterization of the Fintech Ecosystem
A Systematic Literature Review

Ali Mwase, Ernest Ketcha Ngassam, and Singh Shawren

18.1 INTRODUCTION

Technological advancements in areas like mobile and the internet and their extensive global usage have resulted in rapid changes and investments in the financial industry. Financial technology (Fintech) has emerged as a result [1,2]. Fintech is viewed as a gateway to expanding business opportunities in the financial and banking sectors [2]. Fintech is the new marriage of financial services with information technology, and it is a highly essential method for increasing the quality of service delivery in today's financial institutions [3]. It provides customers with more automated, transparent, and user-friendly goods and services [4]. Fintech further facilitates low-cost, faster, convenient, secure, and multi-channel accessibility to payments. Fintech includes not only the digitization and datafication of global financial markets but also the technological transformation of finance through digital financial services. It also includes the emergence of new Fintech startups around the world and, most recently, the emergence of large technology platforms ("BigTechs") involved in finance, such as Facebook, Apple, Amazon, Alibaba [5], Tencent in China, Mercado Credito in Argentina, Paytm in India, and Amazon Lending in the USA [6].

Global Fintech investments demonstrate phenomenal growth. According to [7], a report by Business Research Company projects that the value of global Fintech investment will rise to $309.98 billion by 2022 at a rate of 24.8% per year. In 2018, it was estimated to be worth $127.66 billion. Globally, Fintech offers several advantages for social, economic, and financial stability. Fintech's underlying technologies, for instance, Distributed Ledger Technologies and Artificial Intelligence (AI), enhance regulatory compliance, fraud detection, credit risk management, and collateral management [8]. Fintech in developing nations makes existing services more convenient and builds new infrastructure to enable greater financial inclusion for millions of people in the real economy [9]. Fintech facilitates the consumption of financial services by developing new applications for delivery, such as making payments, saving money, borrowing money, managing risk, and getting financial advice [10]. Fintech can be leveraged to promote economic and sustainable development in several ways, for example, by unlocking the gains of technological advances and accelerating the digitalization of economies. Moreover, KPMG [11] asserts that companies in the financial services industry should integrate Fintech innovation across the entire organization as a mainstream activity and further build a Fintech strategy to capitalize on Fintech opportunities to the fullest extent possible.

The literature on Fintech has recently received significant attention among academics and policymakers. However, studies characterizing the Fintech ecosystem are rare. Thus, this chapter aims to explore this gap by systematically reviewing the body of knowledge and debates relevant to Fintech published in the literature to characterize the Fintech ecosystem.

DOI: 10.1201/9781032667478-21

The remaining part of this chapter is structured as follows: We define Fintech and discuss its advantages and main drivers in Section 18.2 below on related work. The methodology used for the current research is covered in Section 18.3, and Section 18.4 presents the findings of our study that ultimately led to the characterization of the Fintech ecosystem. Finally, Section 18.5 presents the study's conclusion.

18.2 RELATED WORKS

18.2.1 Fintech Defined

Numerous researchers have offered various definitions of Fintech. In [12], Fintech is defined as new financial services or products offered through technology. Fintech, according to [13], is a method for developing efficient financial services using IT technologies. An industry called Fintech uses mobile-based IT to increase the financial system's effectiveness [14]. Contrarily, Fintech describes technological developments implemented or planned to automate and enhance banking and financial services [15]. A non-financial corporation using cutting-edge technology to offer services like remittance, payment, settlement, and investment without collaborating with a financial institution is known as Fintech in the industry. In terms of financial services, it is an innovative financial service that provides unique financial services using cutting-edge technologies like mobile, social media, and IoT [14].

Fintech is further defined as using mobile devices and other digital platforms to access a bank account, receive transaction alerts, and receive debit and credit alerts via pop-up messages, SMS, or other forms of communication [3]. After analyzing the above definitions and concepts given by different authors of Fintech, this paper defines Fintech as the intersection of financial services and technology to create new products, business models, and increased efficiency in service delivery in the financial industry.

18.2.2 Benefits of Fintech

Fintech is an innovative reference for businesses that want to rethink their business models or suggest new businesses [16]. With the help of the Fintech infrastructure, it is possible to produce digitalized data that can be used to design and tailor new financial services for the economically disadvantaged. An example is M-Shwari, a mobile payment platform in Kenya that uses M-Pesa's digital data and mobile money infrastructure to determine credit-scoring decisions [17].

Additionally, Fintech facilitates low-cost, faster, more convenient, secure, and multi-channel accessibility to payments. Fintech further lowers SME funding gaps, reduces costs and delays in cross-border remittance markets, and improves efficiencies and transparency in government operations that help reduce corruption [8]. Co-creation, personalization, and time optimization in more decentralized networks are other benefits that Fintech offers to consumers [18].

18.2.3 Stimulators of Fintech

According to [19], the following seven factors are the main forces behind Fintech: (1) a supportive policy and regulatory environment; (2) altered consumer behavior; (3) growing digital and mobile services; (4) an accelerating rate of change; (5) trust levels on the decline; (6) easier entry for digital disruptors; and (7) accessible, attractive profit pools.

In another survey by [20], it is mentioned that the factors driving Fintech include the high cost of traditional finance, the unmet demand for financial services like basic banking, payment methods, and money transfer services, a supportive regulatory environment, and other macroeconomic factors. Furthermore, demographics are another key factor boosting the Fintech industry because younger generations are more likely to trust and use Fintech services [20].

18.3 METHODOLOGY

This chapter is based on a systematic literature review. A systematic review was used because it has a more rigorous and rule-driven search process than a regular literature review. This helps to eliminate bias in selection or publication. Secondly, the transparent criteria used in this review type are emphasized [21]. These criteria are used to abstract data from studies and evaluate the quality of the evidence supporting those studies. A systematic literature review consists of six steps: choosing a topic, searching for relevant literature, developing arguments, surveying relevant literature, critiquing the literature, and writing the literature review [22].

The main research study objective served as the basis for the first stage, and steps two through six were carried out iteratively while conducting backward searches on pertinent references. Reference [23] describes a backward search as looking at earlier content cited in the articles that the keyword search turned up. Therefore, we used various combinations of the keywords in the introduction to search for articles in reputable international journals such as Scopus, IEEExplore, Elsevier, ProQuest, Science Direct, Emerald Insight, and ResearchGate Index. The articles selected for review were published in the last 11 years, from 2011 to 2021. However, older articles were selected once they had relevant findings for answering the research objective. First, 271 articles that matched the keyword combinations were chosen. After being initially screened by reading all of their abstracts, 105 publications were then disregarded because they bore no direct relation to the subject of the study. After a more thorough review to determine whether the remaining 166 papers were more precisely relevant to the research question, 78 articles were eliminated, and 88 were left. In certain cases, we additionally examined the articles' conclusions and introductions to more clearly determine how applicable they are to this study. To analyze the data collected, a three-column table containing the different papers found was designed. The columns were the publication year, journal, and methodology. Finally, we formed themes to characterize the Fintech ecosystem based on the commonalities presented in the results section below.

18.4 RESULTS

18.4.1 ARTICLE DISTRIBUTION BY JOURNAL

The review shows that out of the 88 papers included in the study, most were from ResearchGate (27%), 18% from Emerald Insight, 15% from Science Direct, 14% from Elsevier, and 11% from IEEExplore. ProQuest and Scopus had the minority article representation at 9% and 6%, respectively (Figure 18.1).

18.4.2 ARTICLE DISTRIBUTION BY YEAR OF PUBLICATION

The findings reveal that articles included in this study were mostly written between 2015 and 2021, with the highest percentage written between 2018 and 2020. This increasing trend implies an increase in studies in the area of Fintech (Figure 18.2).

18.4.3 ARTICLE DISTRIBUTION BY METHODOLOGY

Most articles reviewed followed a quantitative design, qualitative design science, or mixed methods design methodology. Figure 18.3 shows that most of the articles reviewed used the quantitative design (48%), 23% used a qualitative design, 18% used mixed methods, and only 11% used the design science methodology.

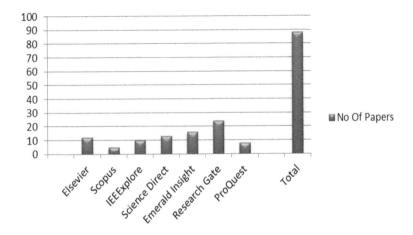

FIGURE 18.1 Article distribution by journal.

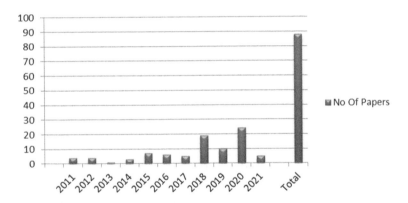

FIGURE 18.2 Distribution of articles by year of publication.

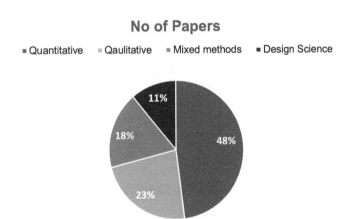

FIGURE 18.3 Article distribution by methodology.

18.4.4 TYPES OF FINTECH

Concerning the types of Fintech, researchers such as [24] assert that the types of Fintech include T-commerce, mobile banks, mPOS-acquiring, P2P-crediting, Bitcoins, and e-wallets.

P2P lending is a widely used technique defined as peer-to-peer lending from one person to another, implying that only two parties are involved in the loan issuing process, with no middlemen such as banks or credit agencies [25]. Online peer-to-peer lending services include Zopa (Zone of Possible Agreement) in the United Kingdom, Prosper, and Lending Club in the United States [25].

An e-wallet, or electronic wallet, is an online payment system that allows you to pay for goods or services [25]. E-Wallets, which can be open or closed wallets, are used by both retailers and consumers as a cashless alternative because they speed up the order processing and delivery of goods [26]. Webmoney Transfer, Yandex Money, QIWI, RBK Money, and PayPal are examples [25]. Other examples include Alipay, GrabPay, Paytm, Pay It, Boost, Visa Checkout, Google Pay, and Masterpass [26]. Bitcoins are a peer-to-peer electronic cash system based on distributed ledger technology [26]. Bitcoins are exchanged directly between clients without using any middlemen [24]. Litecoin, Ethereum, Ripple, and Dash are examples of bitcoin alternatives [26].

Customized POS terminals are used to process credit card transactions. These terminals are referred to as mPOS-acquiring or the mobile point of sale. Through the use of mPOS terminals, businesses can increase their profitability. The top mPOS companies include Square, SumUp, iZettle, and mPowa [24]. T-commerce is another name for tablet commerce. At the dawn of the mobile era in the economy, experts from Silicon Valley created this term. The global market for smartphones and tablets is expanding steadily, creating the perfect environment for T-commerce growth [24]. Smart Fintech is another new breed of financial technology primarily driven by data science and AI methodologies. Smart Fintech alters finance and economies by enabling intelligent, automated, whole-of-business, and personalized financial and economic businesses, services, and systems [27].

18.4.5 CHARACTERISTICS OF FINTECH

According to Katarzyna [28], the DIPLOMA approach comprises seven Fintech features that can be used to create new value in the financial technology business [28]. These characteristics are digitization, innovation, pricing, learning, openness, modernity, and agility. In a similar spirit, [29] posits that the six most important disruptive elements of Fintech are: (1) innovation, (2) disintermediation, (3) convergence, (4) low costs and low entry barriers, (5) borderless platforms, and (6) democratization of financial and investment opportunities.

18.4.6 FINTECH BUSINESS MODELS

Payment, wealth management, crowdfunding, lending, capital markets, and insurance services are among the six Fintech business models adopted by many Fintech companies [30]. Accordingly, a study by [31] proposes nine Fintech business models: online lending or online peer-to-peer lending or P2P lending, crowdfunding or crowd investing, payment terminals, mobile point of sale, cryptocurrency or digital money, robo-advisors, personal finance management, e-banking, and InsurTech [31].

18.4.7 FINTECH APPLICATIONS CATEGORIES AND OPERATIONAL BUSINESS PROCESSES

Fintech industry applications focus on clients and their processes [32]. These applications link consumers' finances to Fintech, thus providing easy access for consumers to their finances. Platforms for crowdsourcing, blockchain and cryptocurrencies, mobile payments, insurance, robot advisory, equity trading apps, budgeting apps, and Fintech stocks are examples of such applications [33].

Others include the InsurTech and Regtech platforms [34]. References [35] argue that Fintech service platforms enhance the accessibility and availability of resources and the transferability of resources in in-service systems.

Crowdfunding platforms like Kickstarter and GoFundMe allow entrepreneurs and early-stage enterprises to raise money from the global market without regard to local restrictions [14]. Quick Pay and Union Pay are two examples of payment apps, along with PayPal (an online payment service), Alipay (Alibaba Groups payment platform), ApplePay (contactless mobile payments), GoCardless (recurring direct debit), Google Wallet (peer-to-peer payments service), M-Pesa (mobile phone-based money transfer), Payoneer (online money transfer and e-commerce payment services), Samsung Pay (contactless mobile payments), and Stripe (an online platform) for currency transfer and exchange [36]. Smava, a crowdlending platform, offers loan comparisons for conventional banks and similar types of information on their online marketplace [14]. Another Fintech application that is now very popular is the use of mobile payment gateways. These gateways allow customers to conduct ordinary banking operations via their mobile devices without visiting a bank physically [14]. Coinbase, a broker and bitcoin exchange, is another. Coinbase is a cryptocurrency exchange that enables users to trade various cryptocurrencies, digital assets, and fiat currencies with other users [37].

InsurTech is another application related to a new insurance model that offers a secure and customized service, streamlining the insurance process and making it easier to file claims and manage policies online [14]. Platforms that support Regtech applications of Fintech make it possible to quickly and efficiently automate the compliance process for regulating financial organizations [14]. For instance, the brokerage app Robinhood provides a zero-transaction-cost trading platform geared toward retail investors. They generate income by providing incumbents with real-time order flow and attempting to extract time value from funds that aren't already invested [37].

18.4.7.1 Fintech Applications Operational Business Processes

It is posited that payments, advisory services, financing, and compliance are the four major categories of operational business processes for Fintech applications [16]. Investment advice, insurance service, customer support, asset management consultation, and management decision-making fall under advisory services. For instance, the Robo-advisor service, which uses cutting-edge technologies like AI, big data, and machine learning, offers clients individualized recommendations more efficiently and allows the recommendations to be updated following real-time data [16].

In addition, the Cambridge Centre for Alternative Finance report [38] postulates that Fintechs provide services and products under segments of payments and remittances, lending and credit risk assessment, savings, insurance, and digital identification. Reference [2] proposes six broad categories of Fintech services: borrowing, insurance, money transfers, payments, savings, and investments.

Reference [39] points out that Fintechs can provide payment services using blockchain and cryptocurrency. For instance, Tencent QQ Coins (Q Bi) are used in China [32], and SureRemit is used in Nigeria [40]. Consumers can also use online foreign exchange and international remittances, Swish, a peer-to-peer mobile payment system, and BankID, a digital identification app. Both of these services were developed by traditional banks and are included in the Fintech category of payments and remittances [41]. In China, WeChat Pay (Tencent) and Alibaba's Alipay Fintech innovations are used [32]. Vodafone introduced M-Pesa in 2007 for use in developing countries like Kenya in East Africa [40].

Reference [39] highlights that crowdsourcing and crowdfunding are channels through which financing and loan services can be advanced to Fintech users. For example, in China, Alibaba Microfinance Company established the WEbank platform [32]. Other examples are Avant.com, which allows customers to take out an unsecured personal loan and customize their payment plan online. Moreover, numerous Fintech applications have been adopted for savings. Online investment

TABLE 18.1

Fintech Applications Categories and Operational Business Processes

Author	Fintech Applications Category
Lee and Shin [30]	Wealth management, crowdsourcing, Payment, lending, capital markets, and Insurance services
Mehrotra and Menon [33]	Blockchain and cryptocurrency, equity trading apps, mobile payments, Robo-advisory, crowdfunding platforms, budgeting apps, insurance, and Fintech stocks
Ren et al. [14]	InsurTech, Regtech, and mobile payment gateways
Knewtson and Rosenbaum [37]	Cryptocurrency
Author	Operational business processes
Leong and Sung [16]	Payment, advisory service, financing, and compliance
The Cambridge Centre for Alternative Finance report [38]	Savings, insurance, digital identification, payments, remittances, lending, and credit risk assessment
Gulamhuseinwala, Bull, and Lewis [2]	Borrowing, payments, money transfers, insurance, savings, and investments

advice and investments, equity or rewards crowdfunding, online budgeting and financial planning, online stockbroking, and online spread betting are examples of peer-to-peer (marketplace) platforms for investments [2]. Fintech uses big data solutions, primarily for risk control, especially in the insurance industry [36]. For instance, telematics-enhanced car insurance or aggregators for health premiums are designed to reduce costs [2]. Zhong and Alihealth, Internet insurance companies, established WeSure, an insurance platform in China, in 2016 [32]. Table 18.1 summarizes the Fintech application categories and the operational business processes.

18.4.8 FINTECH ARCHITECTURE

There is a rapid growth of Fintech services and applications globally, and so does their utilization by businesses and consumers. However, these applications are developed on different architectures [32]. According to [42], Fintech applications or systems are built on monolithic or microservice architectures.

In a similar study by [43], the Fog-Oriented Banking Architecture was presented and described as a flexible, effective, secure, and configurable architecture for gathering and analyzing data produced by mobile devices, applications, and the bank's local equipment. With this idea, it is possible to construct Fog nodes using hardware with minimal processing power and network or storage connectivity. The Fog-Oriented Banking Architecture's recommendation and categorization system ensures that the items a bank recommends are tailored to the individual. Additionally, it presents a chance to enhance the user experience in the bank's physical channels, enabling clients to solve their problems quickly and successfully while also improving the capacity of office managers to resolve issues [43].

Another intriguing architecture is the cloud-centric IoT architecture for intelligent IoT in the banking-Fintech sector [44]. There are three layers to this architecture (physical, service, and application layers). Layer 1: The physical layer is made up of objects with embedded sensors, RFID tags, or barcode tags that collect data about the environment and send it via the internet. The information gathered is subsequently sent to the service layer. Layer 2: The service layer is a middleware that interfaces the physical and application layers and incorporates the cloud. This layer protects clients from complex operations: (1) storage services, (2) computing capability services, and (3) the cloud layer provides, and AI services. Layer 3: The Application Layer is an interface representing the uppermost layer in the hierarchy. This layer provides specialized services and a range of user tasks triggered by the service layer beneath. It combines the components and delivers abstract-level intelligent applications like sending bad debt warning signals and presenting offers aware of a user's location and activities [44]. Accordingly, while there is now a scarcity of research concerning architecture, it is likely to grow in the future.

18.4.9 FINTECH ECOSYSTEM AND STAKEHOLDERS

An ecosystem is a particular economic community upheld by a base of collaborative organizations and people [28]. For customers who are also ecosystem members, this economic community produces valuable goods and services [28]. In another study, [45] defines an ecosystem as a group of interdependent actors working together to generate complementary assets. Fintech ecosystems, according to [28], are crucial for nurturing the technological innovation necessary to improve the effectiveness of financial markets and systems, resulting in a better overall customer experience.

A study by [46] argues that several stakeholders or role players collaborate to create a holistic Fintech ecosystem and encourage the financial industry's transition and upgrade. These stakeholders have a direct or indirect influence on the Fintech ecosystem, assisting in detecting industry key areas for cooperation and places of contact with entrepreneurs and determining how risks can be distributed across stakeholders [5].

In the Fintech ecosystem, collaboration between financial institutions, entrepreneurs, and governments is important [47]. Therefore, the Fintech ecosystem comprises interconnected and ever-evolving technology to improve product, process, and management performance [1]. In addition, heterogeneous, non-linear, dynamic, and complex networks of agents that interact with one another to provide various financial products and services to end customers define the Fintech ecosystem [1]. According to [41], five main components make up the Fintech ecosystem: Fintech startups (offering services in the areas of payments and transfers, money management, lending and financing, securities trading, insurance, etc.); technology developers (offering services in the areas of big data analytics and AI, blockchain and cryptocurrencies, cloud computing, social networks, etc.); clients (individuals and legal entities); traditional financial institutions (traditional banks, insurance companies, brokerage firms, and venture capitalists); and government organizations (financial regulators and legislative bodies) [41].

In a similar study, the five primary Fintech ecosystem attributes are the demand attribute, the talent attribute, the solutions attribute, the capital attribute, and the policy attribute [48]. To comprehend the structure of a Fintech ecosystem, it is advised that these five fundamental features be connected to the connections between the subsystems and stakeholders [48]. Furthermore, [49] states that B2B, B2C, or B2B plus B2C relationships can be used to categorize consumer relationships in Fintech startups. B2B Fintechs are growing, while B2C Fintechs are increasingly focusing on the Millennial generation, seeking better and more innovative experiences through digital channels, according to a global analysis [23,14].

In their study, [5] asserts that Fintech firms are in charge of creating novel business models that benefit consumers. By providing funding (financial institutions and venture capital firms), direction and advice (incubators/accelerators, legal counsel, and consultancy firms), research (academic institutions), and regulatory guidance (international knowledge partners), all other actors contribute to the ecosystem's expansion. Financial institutions, entrepreneurs, and governments are all included in the Fintech ecosystem, according to [28]. According to a related study by [50], a Fintech ecosystem primarily consists of the following: new technologies and tools that enable innovations; telecom and technology companies that create infrastructure for distribution; startups that create innovative business models; government and regulators that set the rules of the game; financial institutions that work with startups; customers and users that benefit from innovations; investors, incubation centers, and accelerators that provide both financial aid and workspace for innovators [50]. It is further explained that the Fintech ecosystem's characteristics include talent, demand, capital, and policy and regulation, according to further explanation [51].

A study by [52] discovered three stages for tracing the Fintech ecosystem's progress; specifically, (1) the significant industry maturity stage, (2) the symbiosis stage, and (3) industry resilience, as well as the very visible role of new entrants who take over and restructure the sector as incumbents fear being replaced. Through the first collaboration with incumbents, the major industry maturity level opens a channel for technological breakthroughs [52]. More radical technologies like cryptocurrency and blockchain are introduced in the symbiosis stage. Because there are now no prior lock-in investments, existing incumbent organizations are working to cohabit with new businesses that are starting to gain market share.

Additionally, more venture capital is entering the market to help young companies hire fresh talent [52]. The third stage, called industry resilience, is characterized by the significant restructuring that new entrants bring about while incumbents fret about being replaced. At this point, the industry is changing, with new initiatives having more influence than long-standing companies. AI greatly facilitates efficiency and atomization gains. Unique and valuable banking services and products are disappearing while new client needs are continuously expanding [52]. The Fintech ecosystem is presented in Figure 18.4, and Table 18.2 summarizes the Fintech ecosystem stakeholders.

It's important to remember that the Fintech ecosystem, like any other successful market system, is supported by a variety of meso-level structures and institutions that are essential to the ecosystem's success as a whole [53]. As a result, Fintech startups specializing in Payments & Transfers, Capital Management, Insurance, Securities & Trading, and Lending & Financing provide B2B,

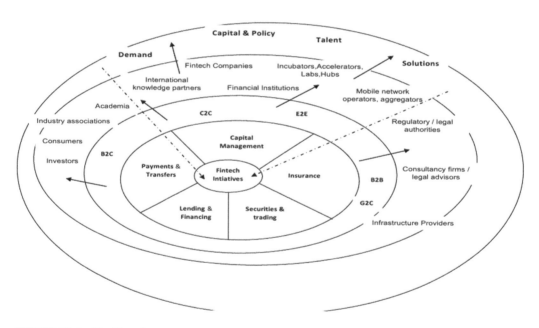

FIGURE 18.4   The Fintech ecosystem.

TABLE 18.2
Fintech Ecosystem Stakeholders

Author	Stakeholders
Soloviev [41]	Governmental agencies, clients, traditional financial institutions, Fintech startups, and technology developers
Katarzyna [28]	Governments, entrepreneurs, and financial institutions
Yazici [50]	Telecom and technology organizations, startups, the government and regulators, financial institutions, new technologies, end users, investors, incubators, and accelerators
UNCDF [51]	Capital, policy, demand, talent, and regulation
Santosdiaz [54]	Aggregators, telecoms, banks, and providers of mobile and digital wallets

B2C, C2C, and E2E services to the stakeholders who form and share their environment as market players (on the supply and demand sides). Furthermore, stakeholders create a conducive environment (demand, talent, solutions, capital, and policy) that determines the type of Fintech Initiatives at the heart of the ecosystem.

18.4.10   A Comparison of the Fintech Ecosystem in Developed and Developing Countries

Fintech appears as a cutting-edge startup that enables bankers and traders to re-evaluate the financial model and, consequently, economical solutions [55]. Fintech strengthens financial decentralization while fostering technological innovation, which stimulates the growth of the digital economy [56]. The development of the Fintech sector depends on a stable, symbiotic ecosystem [57]. Fintech startups, financial institutions, regulators, governments, investors, and talent institutions interact in over 30 different Fintech ecosystems [58]. In this review, we examine both the developed and developing world's Fintech ecosystems. The nations chosen for this comparison are determined by the availability of data with a long enough time series. We do grasp the success factors for the Fintech ecosystems as discussed in the paragraphs below:

In developed nations, the best Fintech ecosystem is found in the United States, followed by Singapore and the United Kingdom. An edge that is challenging to overcome has been made possible by the availability of talent, healthy competition, enabling regulations, and financial stability [59]. Additionally, it is argued that over 134 Fintech companies are based in the United Arab Emirates (UAE), with the payment/e-wallet, blockchain/cryptocurrency, and InsurTech sectors being the most developed. Regulatory changes have aided the blockchain industry's growth. Yallacompare Bayzat, a business-to-business (B2B) Tech Company, and Sehteq, the third-largest provider of health insurance plans in the UAE and the provider of a usage-based health insurance platform [60], are just a few startups in the InsurTech space that have found success. The third-largest global Fintech ecosystem in the world is located in India. This success is built on four essential elements: a significant future demand, a supportive legal environment, a high level of tech skill availability, and a solid investment [59]. The Fintech industry includes financial institutions, credit and insurance companies, investment tech neobanks, and banking infrastructure. According to estimates, 28% of the Fintech ecosystem is made up of investing technology, 27% of it is made up of payment technology, and 16% is made up of lending technology. The Fintech sector in India is made up of these three divisions to a degree of about 70%. Only 9% of Fintech businesses provide banking infrastructure; the rest are spread across industries, including insurance and neobanks [61].

The three most advanced African Fintech ecosystems in developing economies in Africa are found in South Africa, Nigeria, and Kenya [60]. Even though there are various levels of diffusion in various countries, these levels are primarily influenced by consumer needs, financial and technological infrastructure, level of development, regulatory framework, and the amount of capital available for the required investment. East Africa outpaces Southern, Western, and Central Africa regarding Fintech investment inflows [62]. One of the fastest-growing Fintech ecosystems in Africa is found in Ghana, where progressive digitalization regulations and significant mobile money use have contributed to rapid growth [60]. The present players in Ghana's Fintech ecosystem work in various industries, including wealth tech, InsurTech, payments and remittance, blockchain, e-commerce, business administration, investTech, lending and finance, personal finance, and cybersecurity.

Nigeria owns Africa's largest Fintech ecosystem [60], mainly due to increasing smartphone penetration, the surge in demand for affordable payments, quick loans, flexible savings and investments, financial inclusion, and cashless payments [63]. Fintech began with payments but swiftly spread to other industries like lending, asset management, investment, and InsurTech. In South Africa, the three most significant segments are payments (30%), business-to-business (B2B), tech support (20%), and lending (12%) [60]. Around 200 Fintech businesses are thought to be based in South Africa, offering a diverse range of services from banking and wealth tech to InsurTech and cryptocurrencies. The emergence of e-commerce and the quick development of the Internet are the key reasons why payments and remittances are currently the sectors with the quickest rate of growth. The majority of the market comprises third-party payment products and payment service providers (PSPs), which include companies like Yoco and Zoona specializing in money transfers, electronic voucher payments, and agent payments [60]. Yoco offers a card reader and app that enable users to convert smartphones into payment terminals. Loans are another rapidly expanding Fintech. In Ethiopia, the Fintech ecosystem is still tiny; however, favorable market factors, including high mobile phone penetration, a sizable unbanked population, and the government's supportive posture toward the digital economy, are positioning Ethiopia's Fintech industry for rapid expansion [60].

Many startups that are pursuing international trends are based in Kenya. For instance, Umba offers a complete digital banking app. An InsurTech business called Turaco offers individuals micro-insurance solutions. Pezesha runs a digital platform where banks, microfinance organizations, and retail lenders are paired with micro and SMEs to get loans [60]. The mobile money revolution started by M-Pesa, the SIM-based mobile banking service introduced by telco giant Safaricom in 2007, has fueled the development of Fintech in Kenya. M-Pesa combines the mobile infrastructure of Safaricom with an agency model to enable users to deposit, withdraw, transfer, pay for products and services, and access credit and savings all through a mobile device. The emergence of mobile money and M-Pesa has been attributed to promoting economic growth and improving financial inclusion, which has increased from 26% in 2006 to 82.9% today, the highest rate in sub-Saharan Africa [60].

A study by the United Nations Capital Development Fund found that Rwanda has more than 40 Fintech companies, with the most developed sectors being payments/remittances (17 companies) and lending/financing (14 companies) [64]. The existence of policies and regulations, as well as foreign investors, are major contributors to the expansion of this ecosystem [64]. In Uganda, the majority (47%) of the Fintech ecosystem is made up of payments, followed by bank infrastructure (23%), and investments and savings (16%), lending (7%), insurance (5%), and markets (2%) [54]. The growth of Uganda's Fintech ecosystem is attributed to enabling regulations, venture capital and private equity, associations and facilitators, accelerators, hubs, and incubators [65] (Table 18.3).

TABLE 18.3
Success Factors for the Fintech Ecosystems in Developed and Developing Economies

Country	Success Factors
United States United Kingdom Singapore	Talent availability, healthy competition, supportive laws, and financial stability
United Arab Emirates	Regulatory changes
India	Strong prospective demand, a robust investment, Favorable regulatory conditions, and a large pool of tech talent
South Africa	A third of the country's population now uses smartphones, and many people are underbanked or without a bank account, the E-commerce boom, and the Internet's quick growth
Nigeria	Increasing smartphone penetration, financial inclusion, cashless payments, the surge in demand for affordable payments, quick loans, flexible savings, and investments
Kenya	Technological infrastructure, technology hubs, internet penetration, Partnerships, Mobile money revolution
Ghana	Progressive digitalization policies and Strong mobile money adoption
Ethiopia	Government's supportive approach toward the digital economy and Favorable market conditions, such as high mobile phone penetration and a sizable unbanked population
Rwanda	Regulations, foreign investors, and Policies
Uganda	Venture capital and private equity, associations, and facilitators, Availability of regulators, accelerators, hubs, and incubators

18.5 CONCLUSIONS

Fintech is anticipated to play a larger role in the financial sector. This study conducts a systematic review and classifies the Fintech ecosystem. This classification contributes to improving education about Fintech, and the selection of articles and journal databases can provide a reference for Fintech research to obtain quality literature. Research Gate provided most of the publications evaluated, followed by Emerald Insight, Science Direct, and Elsevier journals. Furthermore, according to the findings, the biggest percentage of Fintech articles were written between 2018 and 2020. Moreover, it was discovered that the quantitative design was utilized in most of the studies reviewed, with mixed techniques employed in the smallest number of articles.

Given the benefits of Fintech ecosystems, we recommend that governments play a more significant role in developing the ecosystems in developing countries. Additionally, there is a need for concerted coordination among various stakeholders to make the ecosystems work. Future research may focus on Fintech architectures.

REFERENCES

1. Muthukannan, P., Tan, B., Gozman, D., Johnson, L.: The emergence of a Fintech ecosystem: A case study of the Vizag Fintech Valley in India. *Information & Management*, 57(8): p. 103385 (2020).
2. Gulamhuseinwala, I., Bull, T., Lewis, S.: FinTech is gaining traction and young, high-income users are the early adopters. *The Journal of Financial Perspectives: FinTech* (2015). https://ssrn.com/abstract=3083976.
3. Arner, D.W., Barberis, J.N., Buckley, R.P.: The Evolution of Fintech: A New Post-Crisis Paradigm? (2015). doi: 10.2139/ssrn.2676553.
4. Dorfleitner, G., Hornuf, L., Schmitt, M., Weber, M.: *FinTech in Germany*. Springer International Publishing AG, Cham (2017).
5. Alliance for Financial Inclusion (AFI): Creating Enabling Fintech Ecosystems: The Role of Regulators. Special Report (2020).
6. Huang, Y., Zhang, L., Li, Z., Qiu, H., Sun, T., Wang, X.: Fintech Credit Risk Assessment for SMEs: Evidence from China. IMF Working. (2020).
7. Yovev, A.: Microservice Architecture for Fintech Software (Doctoral dissertation) (2020).
8. Lukonga, I.: Fintech, inclusive growth, and cyber risks: A focus on the MENAP and CCA regions. IMF Working Paper (2018).
9. Koffi, H.W.S.: The fintech revolution: An opportunity for the West African financial sector. *Open Journal of Applied Sciences*, 6(11): pp.771–782 (2016).
10. Singh, S., Sahni, M.M. and Kovid, R.K..: What drives FinTech adoption? A multi-method evaluation using an adapted technology acceptance model. *Management Decision*, 58: pp. 1675–1697 (2020).
11. KPMG: Forging the future: How financial institutions are embracing Fintech to evolve and grow. KPMG International global Fintech survey (2017).
12. Lee, D., Kuo, C., Teo, G.: Emergence of fintech and the LASIC principles. *Journal of Financial Perspectives*, 3(3): pp. 1–26 (2015).
13. Jung-Oh, P., Byung-Wook, J.: A study on authentication method for secure payment in fintech environment. *The Journal of the Institute of Internet, Broadcasting, and Communication (IIBC)*, 15(4): pp. 25–31 (2015).
14. Yonghee, K., Young-Ju, P., Jeongil, C., Jiyoung, Y.: An Empirical Study on the Adoption of "Fintech" Service: Focused on Mobile Payment Services (2015).
15. Noor, U., Anwar, Z., Amjad, T., Choo, K.K.R.: A machine learning-based Fintech cyber threat attribution framework using high-level indicators of compromise. *Future Generation Computer Systems*, 96: pp. 227–242 (2019).
16. Leong, K., Sung, A.: Fintech (Financial Technology): What is it and how to use technologies to create business value in a Fintech way? *International Journal of Innovation, Management, and Technology*, 9(2): pp. 74–78 (2018).
17. Cortina Lorente, J.J. and Schmukler, S.L.: The FinTech revolution: A threat to global banking? World Bank Research and Policy Briefs, (125038). (2018). https://ssrn.com/abstract=3255725.
18. Felländer, A., Siri, S. and Teigland, R.: The three phases of FinTech. In: *The Rise and Development of FinTech* (pp. 154–167). Routledge, Milton Park. (2018).

19. Dang Thi, N.A.: FinTech ecosystem in Vietnam. Turku School of Economics Master's Thesis in Global Innovation Management (2018).

20. Frost, J.: The economic forces driving fintech adoption across countries. The technological revolution in financial services: How banks, Fintech, and customers win together, 70 (2020).

21. Jesson, J.K.., Lacey, F.M.: How to do (or not to do) a critical literature review. *Pharmacy Education*, 6(2): pp. 139–148 (2006).

22. Machi, L.A., McEvoy, B.T.: The literature review: Six steps to success (2016).

23. Vom Brocke, J., Simons, A., Niehaves, B., Riemer, K., Plattfaut, R., Cleven, A. (2009). Reconstructing the giant: On the importance of rigor in documenting the literature search process. In *ECIS* (Vol. 9, pp. 2206–2217).

24. Ryabova, A.V.: Emerging FinTech market: Types and features of new financial technologies. *Journal of Economics and Social Sciences*, 7: p.4. (2015).

25. Kalmykova, E. and Ryabova, A.: FinTech market development perspectives. In *SHS Web of Conferences* (Vol. 28, p. 01051). EDP Sciences (2016).

26. Kalra, D.: Overriding Fintech. In *2019 International Conference on Digitization (ICD)* (pp. 254–259). IEEE (2019).

27. Cao, L., Yang, Q. and Yu, P.S.: Data science and AI in FinTech: An overview (2021).

28. Katarzyna, B.: Impact of digital transformation on value creation in fintech services: An innovative approach. *Journal of Promotion Management*, 25(5): pp. 631–639 (2019).

29. Brummer, C. and Gorfine, D.: FinTech: Building a 21st-century regulator's toolkit. Milken Institute, 5. (2014).

30. Lee, I., Shin, Y.J.: Fintech: Ecosystem, business models, investment decisions, and challenges. *Business Horizons*, 61: pp. 35–46 (2018).

31. Liu, J., Li, X., Wang, S.: What have we learned from 10 years of fintech research? A scientometric analysis. *Technological Forecasting and Social Change*, 155: p.120022 (2020).

32. Zhang-Zhang, Y., Rohlfer, S., Rajasekera, J.: An eco-systematic view of cross-sector FinTech: The case of Alibaba and Tencent. *Sustainability*, 12(21): p. 8907 (2020).

33. Mehrotra, A., Menon, S.: Second round of fintech-trends and challenges. In *2021 2nd International Conference on Computation, Automation and Knowledge Management (ICCAKM)* (pp. 243–248). IEEE (2021, January).

34. Ren, X., Aujla, G.S., Jindal, A., Batth, R.S., Zhang, P.: Adaptive recovery mechanism for SDN controllers in Edge-Cloud supported FinTech applications. *IEEE Internet of Things Journal*, 10: pp. 2112–2120 (2021).

35. Breidbach, C.F., Ranjan, S.: How do fintech service platforms facilitate value co-creation? An analysis of twitter data. In: *ICIS* (2017).

36. Broby, D., Karkkainen, T.: FINTECH in Scotland: Building a digital future for the financial sector. The Future of Fintech Supported by International Financial Services District (IFSD) The Technology Innovation Centre, Glasgow Date: 2nd September. (2016). doi: 10.2139/ssrn.2839696.

37. Knewtson, H.S., Rosenbaum, Z.A.: Toward understanding FinTech and its industry. *Managerial Finance*, 46: pp. 1043–1060 (2020).

38. Cambridge Centre for Alternative Finance: FINTECH IN UGANDA: Implications For Regulation. Financial sector deepening (2018).

39. Alt, R., Beck, R., Smits, M.T.: FinTech and the transformation of the financial industry. *Electronic Markets*, 28(3): pp. 235–243 (2018).

40. Yermack, D.: FinTech in Sub-Saharan Africa: What Has Worked Well, and What Hasn't (No. w25007). National Bureau of Economic Research (2018).

41. Teigland, R., Siri, S., Larsson, A., Puertas, A.M. and Bogusz, C.I.: Introduction: Fintech and shifting financial system institutions. In *The Rise and Development of FinTech* (pp. 1–18). Routledge, Milton Park (2018).

42. Anam, F.: Future of Fintech-The Enterprise Architecture. Accessed on 22-08-2020 from https://www.rapidvaluesolutions.com/whitepapers/future-of-Fintech/.(2020).

43. Hernández-Nieves, E., Hernández, G., Gil-González, A.B., Rodríguez-González, S. and Corchado, J.M.: Fog computing architecture for a personalized recommendation of banking products. *Expert Systems with Applications*, 140: p. 112900. (2020).

44. Oualid, A., Y. Maleh, L. Moumoun: Federated learning techniques applied to credit risk management: A systematic literature review. *EDPACS*, pp. 1–15 (2023). doi: 10.1080/07366981.2023.2241647.

45. Grabher, G. and König, J.: Disruption, embedded: A Polanyian framing of the platform economy. *Sociologica*, 14(1): pp. 95–118 (2020).

46. Bu, Y., Li, H. and Wu, X.: Effective regulations of FinTech innovations: The case of China. *Economics of Innovation and New Technology*, 31: pp. 1–19 (2021).
47. Soloviev, V.: Fintech ecosystem in Russia. In *2018 Eleventh International Conference Management of Large-Scale System Development (MLSD)* (pp. 1–5). IEEE. (2018, October).
48. Ebrary.net: Ecosystems. Accessed on 12th-7-2021. Available at https://ebrary.net/79588/business_finance/ecosystems (2021).
49. Pollari, I.R.A., The Pulse of Fintech.: Biannual global analysis of investment in Fintech, KPMG. (2018).
50. Yazici, S.: The analysis of fintech ecosystem in Turkey. *Journal of Business Economics and Finance*, 8(4): pp. 188–197 (2019).
51. United Nations: The Sustainable Development Goals Report. New York (2019).
52. Palmié, M., Wincent, J., Parida, V., Caglar, U.: The evolution of the financial technology ecosystem: An introduction and agenda for future research on disruptive innovations in ecosystems. *Technological Forecasting and Social Change*, 151: p.119779 (2020).
53. Fsdafrica: Zimbabwe fintech ecosystem study Report, March 2020 (2020).
54. Santosdiaz, R.: The Fintech Ecosystem of Uganda (2022). https://thefintechtimes.com/the-fintech-ecosystem-of-uganda/.
55. Ponnusamy, V., Rafique, K., & Zaman, N. (Eds.): *Employing Recent Technologies for Improved Digital Governance*. IGI Global, Hershey, PA (2019).
56. Chen, X., Teng, L., Chen, W.: How does FinTech affect the development of the digital economy? Evidence from China. *The North American Journal of Economics and Finance*, 61: p. 101697 (2022).
57. Albarrak, M.S., Alokley, S.A.: FinTech: Ecosystem, opportunities and challenges in Saudi Arabia. *Journal of Risk and Financial Management*, 14(10), p. 460 (2021).
58. OXEPR: Building FinTech Ecosystems: Emerging Trends & Policy Implications: Insights from the 5th Annual Oxford Entrepreneurship Policy Roundtable.University of Oxford: Said Business School (OXEPR) (2019).
59. Stężycki, P.: 10 Rising Fintech Ecosystems Shaking up the Order of the World's Best Countries to Invest In. Available at https://www.netguru.com/blog/best-country-for-fintech-business. Accessed on 20-6-2022 (2021).
60. Fintechnews Middle East: UAE Fintech Report and Map 2021: Fintech is Booming in Dubai and Abu Dhabi, July 2021. Abu Dhabi, Dubai (2021).
61. The Hans India: India, Home of the 3rd Largest Fintech Ecosystem in the World the Hans India Hans News Service (2022). https://www.thehansindia.com/technology/tech-news/india-home-of-the-3rd-largest-fintech-ecosystem-in-the-world-726039.
62. Rizopoulos, E., Murinde, V.: FinTech: An inclusive pathway to economic recovery INSIGHTS & IDEAS.Commentary. African Center for Economic Transformation (2021).
63. Rufus, Z.: Top players transforming the Nigerian FinTech ecosystem. https://marketingedge.com.ng/top-players-transforming-the-nigerian-fintech-ecosystem/.
64. UNCDF: The Fintech Landscape in Rwanda: Results of a UNCDF Study to Identify Its Current State, Challenges, and Opportunities for Growth, November 2019 (2020).
65. FITSPA: Study on the state of Uganda's Fintech Industry (2022).

19 Analyzing the Impact of the Fear of COVID-19 on Stock Market Returns Using Twitter Text Mining

Ayoub Razouk, Youness Madani, and Fatima Touhami

19.1 INTRODUCTION

Today, people's risk-taking choices and decisions are influenced by their current moods and emotions [1]. People in a positive mood can make more advantageous choices and better judgments than individuals in a negative mood. These facts affect the prices of various assets. Several studies have found that external factors can capture emotions and are linked with stock results. Reference [2] examined the link between weather-induced emotions and stock return auto-correlation, References [3] and [4] investigated the association between air pollution and people's emotions, as well as the relationship between people's moods and stock returns, and References [5] and [6] studied the Twitter emotion states and the stock market. Reference [7] investigated how investor mood affected the Vietnamese stock market by using international football match scores as exogenous shocks to investor mood. The purpose of our study is to examine how a retail investor's emotions about the COVID-19 outbreak influence the stock market by using Twitter sentiment analysis.

The latest COVID-19 epidemic began in Wuhan in December 2019 and has aggressively expanded across continents. On March 11, 2020, the World Health Organization classified the COVID-19 disease as a worldwide pandemic. Corona transmission has grown rapidly since then. After a year of the pandemic, the number of affected persons was more significant than 634 million, with more than 6.5 million deaths. The exponential increase in corona infections prompted the government to implement various preventative measures to flatten the COVID-19 curve. The measures include closing borders, canceling international fights, prohibiting mass meetings, and closing academic institutions, amusement parks, gyms, restaurants, and pubs. The restrictions involve closing borders, canceling international flights, prohibiting mass meetings, and closing academic institutions, amusement parks, and restaurants. As the COVID-19 epidemic continued, variations of the virus appeared, and more are predicted as the virus evolves. WHO has established an integrated strategy for examining and controlling SARS-CoV-2 variants of concern and their effect on public healthcare systems, including social measures, vaccinations, diagnostics, therapies, etc. The COVID-19 epidemic caused fear and anxiety about the virus's impact on health and survival. Fear about the loss of jobs and uncertainty through economic downturns has placed investors in a negative state of mind. This anxiety and fear increased with the emergence of new variants that were more contagious and deadly.

In this chapter, we use sentiment analysis to analyze the tweets related to COVID-19. The collected tweets are analyzed to classify them as positive, negative, or neutral. This chapter captures the fear emotion caused by the COVID-19 epidemic and its influence on market returns by collecting data based on tweets containing COVID-19 terms and phrases. We create a time series of the sentiment in the tweets daily from January 2020 to September 2021. The following stock market

DOI: 10.1201/9781032667478-22

indexes were assessed to evaluate the influence of COVID-19 fear on stock returns: the S&P 500 index, Dow Jones index, Russell 1000 ETF, S&P 500 ETF, and Nasdaq 100 ETF.

For the collection of tweets to analyze, we have based it on the SnScrape Python library, which allows us to extract historical tweets using keywords and a period (in our case, from January 2020 to September 2021). After the collection, it is the step of text preprocessing to prepare the tweet for classification and to delete unuseful data. For the classification of each tweet, we used CrystalFeel to extract new features from the tweet and find the sentimental and emotional degrees.

Analyzing the emotions and sentiments of people using intelligent algorithms in real time helps decision-makers make optimal decisions. In this chapter, our method gives the specialists in stock return an idea of how people feel during this pandemic, how it affects the stock return indexes, and how they should react to make the best decisions.

The rest of this chapter is structured as follows: Section 19.2 explores the literature on the COVID-19 epidemic. Section 19.3 presents the data and methodology used. Section 19.4 summarizes the empirical findings, and Section 19.5 concludes this chapter with some perspectives.

19.2　LITERATURE REVIEW

Different pandemics have an impact on the stock market's performance. Several significant pandemics have been identified in previous research as impacting such returns. For example, researchers in [8] and [9] investigated the impact of the SARS pandemic on Taiwan's stock market. They found a link between the disease outbreak and hotel, tourism, wholesale, and retail stock returns. On the other hand, during the epidemic, Taiwan's biotechnology industry showed a strong, positive link with the SARS crisis. Reference [8] was using an event study approach to examine the SARS outbreak's impact on Taiwanese hotel stocks' efficiency. It found seven publicly traded hotel companies had their income, and stock prices declined significantly during the SARS outbreak. But according to [10], except for China and Vietnam, there is no evidence that SARS negatively influenced the stock markets in the nations impacted.

The infectious disease poses various economic and societal risks, reflected in stock prices. According to [11], the number of human avian influenza A (H7N9) cases reported daily correlates significantly with the Shanghai Composite Index's stock price. Also, the Ebola virus outbreak had a significant impact on US stock markets. Reference [12] indicates that the most significant impact of Ebola was linked to US companies operating in West Africa and the US. Ebola has a greater impact on the stock results of smaller companies than on the bigger ones. They also found that the biotechnology, healthcare supply, food and beverage, food and beverage, and pharmaceutical businesses were all favorably linked to the Ebola outbreak. In contrast, other industries were strongly linked to the outbreak in the opposite direction.

With COVID-19, various studies have documented the negative effects of the COVID-19 epidemic on financial markets worldwide. For example, [13] examined the impact of the pandemic on stock market performance in 64 countries and discovered a negative correlation between the number of confirmed cases and stock returns. It's worth noting that stock markets have responded more aggressively to growing confirmed cases than to increasing death rates. And to investigate the relationship between the spread of COVID-19 and volatility shocks, oil price, geopolitical risks, the stock market, and economic policy uncertainty in the US, [14] used a wavelet-based approach and daily COVID-19 observations (i.e., the number of cases in the country). They demonstrate that the COVID-19 pandemic has a much more massive impact on geopolitical concerns than economic uncertainty in the United States. Reference [15] confirms COVID-19's adverse effects on the stock markets of the 10 nations with the highest number of confirmed cases in March 2020, as well as the stock markets of Korea, Singapore, and Japan [16]. Also, the pandemic has a spreading influence on Asian, European, and American countries.

Reference [17] is attempting to study the stock market response by using the GARCH and EGARCH models to compare this pandemic with the financial crisis of 2008 on the important stock markets of the countries of the BRICS (Brazil, Russia, India, China, and South Africa). The results of the models GARCH and EGARCH show that the 2008 financial crisis generated increased volatility in stocks for India and Russia. In contrast, the stock markets of China, Brazil, and South Africa were more volatile during the COVID-19 epidemic.

However, the consequences of the COVID-19 pandemic on stocks differed from sector to industry [18]. The worst-affected industries on stock exchanges due to COVID-19 were the gas, petroleum, textiles, automotive, transport, machinery, and hospitality industries.

Many researchers have considered the influence of the COVID-19 lockdown on stock market performance. For illustration, [19] evaluated the impact on stock market indices of 45 nations from social isolation and COVID-19 lockdowns. They discovered an inverse link between the lockdown and the performance of global stock markets [20]. When analyzing the impacts of COVID-19 and its lockdown on the US financial markets, it is revealed that the lockdown caused a decrease in market liquidity and stability [21].

According to their research conducted during COVID-19, investors reduced the amount of leverage they used while simultaneously increasing the number of weekly trades they executed. Reference [22] constructed a global fear index for use with COVID-19. They created the fear index by analyzing information on reported cases and deaths. They concluded that the fear index is a very good predictor of stock returns in countries that are members of the OECD. Reference [23] expands the study done in [22] by using the Google search volume index to calculate the fear index; they find a strong negative relationship between the stock return and the COVID-19 fear index. Reference [23] also confirms that this fear index is negatively correlated with the stock return.

19.3 DATA AND METHODOLOGY

Our work is based on data collected from Twitter. An important step consists of collecting relevant tweets, preparing the tweets for classification, and analyzing each tweet to extract sentiments and emotions. The following subsections give more details.

19.3.1 DATA COLLECTION

This subsection describes how we construct our database of work and how we extract useful information from tweets. For optimally collecting old tweets from Twitter, we proposed a new method based on Snscrape, a scraping tool for social networking services. It scrapes users, user profiles, hashtags, searches, threads, and list posts and returns the discovered items without using Twitter's API.

To collect tweets, Snscrape uses keywords to retrieve only the target tweets (related to a specific domain). In this work, we used keywords such as "Covid19", "Covid", "Coronavirus", "StayHome", "LockdownNow", "corona", "Wuhan", and "wuhan virus". Also, all the collected tweets were published from January 2020 to September 2021.

Each collected tweet has several attributes we can retrieve automatically from Twitter using Snscrape. In this work, we collect the following features:

- **tweet_ID**: A unique value that every published tweet on Twitter has.
- **tweet_created_at**: The UTC when a tweet was created.
- **text**: The textual content of the tweet.
- **retweet_count**: Number of times a tweet has been retweeted.
- **favorite_count**: Number of times a tweet has been liked.
- **hashtag_text**: Name of the hashtags, minus the leading '#' character.
- **user_ID**: The unique identifier for the user who created the tweet.
- **location**: The user-defined location in an account's profile.
- **followers_count**: Number of followers account (the account of the tweet owner) currently has.

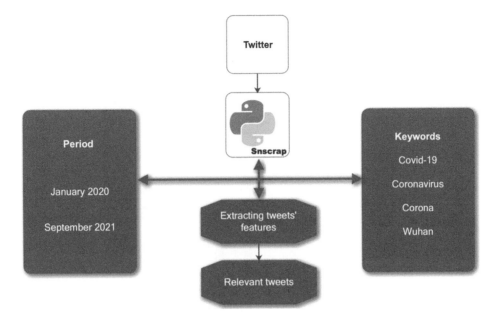

FIGURE 19.1   Data collection steps.

- **friends_count**: Number of users account (the account of the tweet owner) is following.
- **statuses_count**: Number of tweets (including retweets) issued by the user.
- **user_created_at**: The UTC date-time that the user account was created on Twitter.

Figure 19.1 shows how we extracted COVID-19 tweets from January 2020 to September 2021.

After all these steps, we collected more than 120 million Twitter posts from more than 20 million unique users.

19.3.2   TWEETS PREPROCESSING

For analyzing tweets and extracting useful information, one needs to apply some text preprocessing methods to the tweets' text to prepare them for classification and to remove useless information.

In this work, we apply the following text preprocessing methods:

- **Removing the Hashtags**
- **Removing web links**
- **Removing special characters**
- **Substitute the multiple spaces with single spaces**
- **Removing all the single characters**
- **Removing the Twitter handlers**
- **Removing numbers**
- **Remove characters** that have a length of less than 2
- **Tokenization:** Split each tweet into individual words
- **Removing stopwords:** deleting any word that does not emphasize emotions, such as the preposition, are, is, among the articles (a, an, the), etc.
- **Stemming and lemmatization:** reduce inflectional forms and sometimes derivationally related word forms to a common base form. We use this preprocessing method because its use has shown very positive results in the classification of social network data (Figure 19.2).

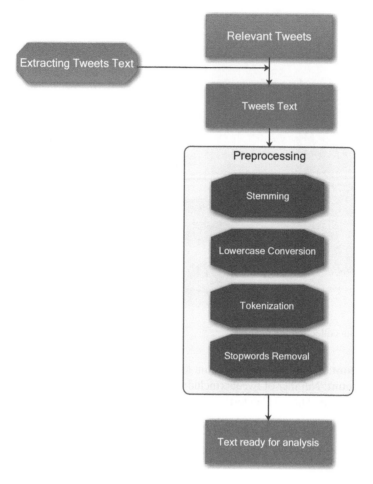

FIGURE 19.2   Text preprocessing.

19.3.3   Data Analysis

After collecting the pertinent tweets and preparing them by applying the necessary text preprocessing methods, it's time to analyze them to extract the necessary information.

Our analysis aims to extract sentiment and emotion from each tweet. In this work, the sentiment can have three possible values (classes): neutral, negative, and positive, and the same for the emotion with five classes: no specific emotion, happiness, anger, fear, and sadness. And to find these classes, we need to calculate for each tweet five new features, which are: valence_intensity, fear_intensity, anger_intensity, happiness_intensity (or joy_intensity), and sadness_intensity.

We calculated these new attributes based on CrystalFeel [24] in 2018. CrystalFeel is a set of emotion analysis algorithms based on machine learning designed to analyze the emotional intensity features present in natural language. CrystalFeel is built on a strong foundation of theoretical concepts and empirical knowledge from the field of emotion science. In addition, innovative computational algorithms have been developed for its development. These algorithms incorporate an in-house curated emotion intensity lexicon, recent natural language processing advances, and affective computing techniques. CrystalFeel will assign a number between 0 and 1 to each tweet based on the text of the tweet in order to represent the new characteristics.

Figure 19.3 shows an application example of CrystalFeel to find the sentiment and emotion of the text: "I am happy".

Quantitative Outputs				
Fear Intensity	Anger Intensity	Joy Intensity	Sadness Intensity	Valence Intensity
0.19	0.163	**0.712**	0.216	0.776
Qualitative Outputs				
Sentiment: Very Positive			Emotion: Joy	

FIGURE 19.3   Application example.

After applying CrystalFeel, each tweet has the following attributes:

- **tweet_ID**
- **tweet_created_at**
- **text**
- **retweet_count**
- **favorite_count**
- **hashtag_text**
- **user_ID**
- **location**
- **followers_count**
- **friends_count**
- **statuses_count**
- **user_created_at**
- **Valence_intensity**
- **Fear_intensity**
- **Anger_intensity**
- **Happiness_intensity**
- **Sadness_intensity**

For finding each tweet's sentiment and emotion, we based it on the following CrystalFeel algorithms that use the new five attributes to find the equivalent class (Figure 19.4).

Algorithm 1 Finding the Tweet's Sentiment

Ensure: a tweet (Initialize the sentiment category in a neutral or mixed class)

Require: tweet's sentiment

    valenceintensity <− CrystalFeel(tweet's text)

    sentiment <− "neutral or mixed"

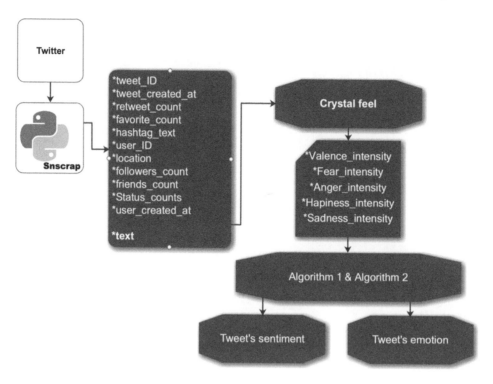

FIGURE 19.4 Steps for detecting sentiment and emotion.

```
if valenceintensity<=0.30 then

    sentiment <- "very negative"

                    else if valenceintensity < 0.48 then

    sentiment <- "negative"

else if valenceintensity > 0.70 then

    sentiment "very positive"

else if valenceintensity > 0.52 then

        sentiment <- "positive"

 End if
```

Algorithm 2 Finding the Tweet's Emotion

```
Ensure: a tweet(Initialize the sentiment category in a "no specific
emotion" class)
Require: tweet's emotion
valencejntensity <- CrystalFeel(tweet's text)
```

```
fearjntensity <- CrystalFeel(tweet's text)

angerjntensity <- CrystalFeel(tweet's text)

sadnessintensity <-  CrystalFeel(tweet's text)

    emotion <- "no specific emotion"

if valencejntensity<=0.52 then

    emotion <- "happiness"

else if valenceintensity < 0.48 then

    emotion <- "anger"

else if (fearjntensity > angerintensity) and (fearjntensity >=

sadnessintensity) then

    emotion <- "fear"

else if (sadnessintensity > angerintensity) and (sadnessjntensity

> fearjntensity) then

    emotion <- "sadness"

end if
```

19.3.4 COVID FEAR AND STOCK RETURN

The S&P 500 index, Dow Jones index, Russell 1000 ETF, S&P 500 ETF, and Nasdaq 100 ETF are regarded as measures of the influence of fear caused by the corona epidemic on stock market returns. From January 2020 to September 2021, the indexes' daily closing prices were sourced from Yahoo Finance. The descriptive statistics are presented in Table 19.1.

Note that negative, positive, and neutral tweets represent the number of tweets with negative, positive, and neutral sentiment by day.

In addition, we examine how COVID-19 sentiment has changed over time. Figure 19.5 shows the number of negative, positive, and neutral tweets from January 2020 to September 2021.

TABLE 19.1
Descriptive Statistics

Variables	N	Mean	Median	Std. Deviation	Minimum	Maximum
S&P 500	403	0.000348	0.000729	0.007819	−0.055439	0.038949
Dow jones	403	0.000223	0.000534	0.008270	−0.060114	0.046749
Russell 1000 ETF	403	0.000365	0.000708	0.007808	−0.055053	0.039337
Nasdaq 100 ETF	403	−2.95E-05	0.000386	0.006156	−0.046867	0.035196
Negative tweets	403	56866.35	56381.00	28387.99	358.0000	177042.0
Positive tweets	403	25562.23	24562.00	12950.72	72.00000	65182.00
Neutral tweets	403	15366.82	14797.00	8196.941	63.00000	46272.00

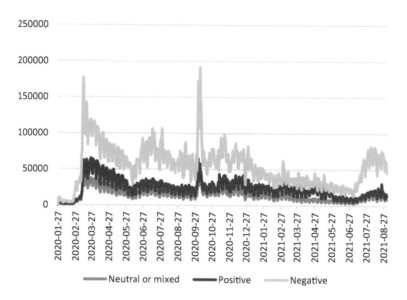

FIGURE 19.5   Sentiment evolution over time.

The following model is used to assess the link between *negative sentiment related to COVID-19 and* stock market returns:

$$R_t = \beta_0 + \beta_1 \text{CovidNegSent}_t + \sum_n \lambda_n \text{Control}_{i,t} + \varepsilon_t \qquad (19.1)$$

where:

- R_t denotes the return on day t with $R_t = \ln(\text{Price}_t / \text{Price}_{t-1})$.
- CovidNegSent_t denotes the *negative sentiment related to COVID-19, and it* is calculated as follows:

$$\text{CovidNegSent}_t = \ln(\text{NumberNegatifTweets}_t / \text{NumberNegatifTweets}_{t-1})$$

- The control variables included in the model lag asset returns by up to five lags: changes in the volatility index (VIX), changes in the economic policy uncertainty index, changes in the ADS business condition index, and changes in the number of new cases. We used weekday dummies in the regressions to treat seasonality.

Also, we used model (2) to examine the relationship between the emotion of fear and stock market returns:

$$R_t = \beta_0 + \beta_1 \text{CovidFear}_t + \sum_n \lambda_n \text{Control}_{i,t} + \varepsilon_t \qquad (19.2)$$

With CovidFear_t denotes the emotion of fear from COVID, it is calculated as follows:

$$\text{CovidFear}_t = \ln(\text{NumberFearTweets}_t / \text{NumberFearTweets}_{t-1})$$

19.4   RESULTS AND DISCUSSION

Table 19.2 displays the estimated impact of the negative sentiment related to COVID-19, proxied by negative tweets, on aggregate market returns (Dow Jones and S&P 500 returns). After controlling

TABLE 19.2

Impact of Negative Sentiment Related to COVID-19 on Stock Return Indexes

	SP500	Dow Jones	Russel 1000 ETF	NASDAQ 100 ET
CovidNegSent	−0.022776[c]	−0.025118[c]	−0.022510[c]	−0.016226[c]
R(t−1)	−0.119018[c]	−0.105295[c]	−0.103346[c]	−0.253838[c]
R(t−2)	0.191134[c]	0.207708[c]	0.200998[c]	0.084993[b]
R(t−3)	0.011814	0.001189	0.009809	0.095778[a]
R(t−4)	−0.093142[b]	−0.106602[c]	−0.097293[c]	0.020543
R(t−5)	0.088974[b]	0.093654[b]	0.091895[b]	0.038084
EPU	−0.000809	0.000676	0.000882	0.001033
ADS	−0.001495	−0.001589	−0.001240	0.003882[b]
VIX	−0.137150[c]	−0.132473[c]	−0.137358[c]	−0.101720[c]
Adjusted R^2	0.572492[c]	0.503952[c]	0.570317[c]	0.538650[c]

[c, b,] and [a] denotes 1%, 5% and 10% significance level, respectively.

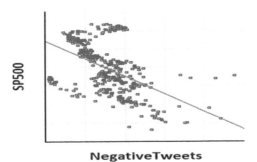

FIGURE 19.6 The effect of negative tweets on the SP 500 index.

for lagged returns and contemporaneous VIX, EPU, and ADS, the COVID-19 fear coefficient is determined to be negative and significant. The negative sentiment generated by COVID-19 is negatively associated with index returns. An increase in negative sentiment is associated with a decrease in Dow Jones and S&P 500 returns.

We also look at how COVID-19 sentiment affects highly liquid exchange-traded funds. We analyze two highly liquid ETFs: the Russell 1000 ETF and the Nasdaq 100 ETF. The findings are equivalent to the S&P 500 and Dow Jones market indexes.

These findings are consistent with the behavioral school of thought, which holds that investor pessimism may influence stock returns [25] (Figure 19.6).

In addition, we examine how fear emotion affects stock return indexes. Table 19.3 displays the estimated impact of the fear of COVID-19. The findings in Table 19.3 are comparable to the negative sentiment. The fear emotion evoked by COVID-19 shows a negative relationship with index returns. The COVID-19 fear coefficient is determined to be negative and significant. An increase in COVID-19 fear is associated with decreased stock return indexes.

Figure 19.7 shows the negative effect of the fear emotion on the SP 500 index.

We also noticed that the coefficients of Table 19.2 relating to the impact of the CovidNegSent (Covid-19 Negative Sentiment) on the stock return indexes are significantly high compared to those of Table 19.3 relating to the impact of the CovidFear variable on the same stock return indexes. This difference can be explained by the fact that the negative sentiment includes all the negative emotions, namely anger, fear, and sadness, which makes the impact more obvious.

TABLE 19.3

Covid Fear Impact on Stock Return Indexes

	SP500	Dow Jones	Russel 1000 ETF	Nasdaq 100 ETF
CovidFear	−0.018506[c]	−0.019265[c]	−0.018794[c]	−0.013990[c]
R(t−1)	−0.115633[c]	−0.101103[b]	−0.099908[c]	−0.251460[c]
R(t−2)	0.191924[c]	0.209439[c]	0.201391[c]	0.083410[b]
R(t−3)	0.010734	−0.000363	0.008708	0.094616[b]
R(t−4)	−0.093096[b]	−0.105973[c]	−0.097386[c]	0.020948
R(t−5)	0.092015[b]	0.097437[b]	0.095088[c]	0.040854
EPU	0.000983	0.000865	0.001071	0.001177
ADS	−0.001542	−0.001651	−0.001290	−0.003921[b]
VIX	−0.137639[c]	−0.133729[c]	−0.137656[c]	−0.101764[c]
Adjusted R^2	0.572189[c]	0.501791[c]	0.570422[c]	0.539171[c]

[a,] [b,] and [c] denotes 10%, 5%, and 1% significance level, respectively.

FIGURE 19.7   The effect of the fear emotion on the SP 500 index.

19.5 CONCLUSION

During the COVID-19 pandemic, people published a lot of content on social media that is full of negative sentiments and fear toward the coronavirus. All that affected many domains negatively. Many researchers try to develop new approaches to analyze social network data automatically using machine learning and artificial intelligence methods. Among these proposed approaches in the literature, we find a lot of techniques based on the domain of social network analysis.

In this chapter, we have proposed a new model based on a sentiment analysis approach to extract and analyze Twitter data published from January 2020 to September 2021 and related to COVID-19. To extract only COVID-19 tweets, we used the SnScrape Python library and keywords such as "Covid19", "Covid", "Coronavirus", "StayHome", "LockdownNow", "corona", "wuhan", and "wuhan virus". Using the CrystalFeel algorithms, we extracted from each tweet its sentiment and its emotion.

The following stock market returns were used to examine the influence of COVID-19 fear on stock returns: the S&P 500 index, Dow Jones index, Russell 1000 ETF, S&P 500 ETF, and Nasdaq 100 ETF. Experimental results show that tweets' negative sentiment and fear emotions negatively affected stock return indexes.

REFERENCES

1. Mann L. (1992) Stress, affect, and risk taking. In: J. F. Yates (Ed.) *Risk-Taking Behavior.* John Wiley & Sons, Oxford, England, pp 202–230.
2. Khanthavit A. (2020). Weather-induced moods and stock-return autocorrelation. *Zagreb Int Rev Econ Bus* 23:19–33. https://doi.org/10.2478/zireb-2020-0002.
3. Wu Q., Hao Y., Lu J. (2018) Air pollution, stock returns, and trading activities in China. *Pac-Basin Finance J* 51:342–365. https://doi.org/10.1016/j.pacfin.2018.08.018.
4. Lepori G.M. (2016) Air pollution and stock returns: Evidence from a natural experiment. *J Empir Finance* 35:25–42. https://doi.org/10.1016/j.jempfin.2015.10.008.
5. Bollen J., Mao H., Zeng X. (2011) Twitter mood predicts the stock market. *J Comput Sci* 2:1–8. https://doi.org/10.1016/j.jocs.2010.12.007.
6. Nofer M., Hinz O. (2015) Using twitter to predict the stock market: where is the mood effect? *Bus Inf Syst Eng* 57:229–242. https://doi.org/10.1007/s12599-015-0390-4.
7. Truong Q.-T., Tran Q.-N., Bakry W., et al. (2021) Football sentiment and stock market returns: Evidence from a frontier market. *J Behav Exp Finance* 30:100472. https://doi.org/10.1016/j.jbef.2021.100472.
8. Chen M.-H., Jang S. (Shawn), Kim W.G. (2007) The impact of the SARS outbreak on Taiwanese hotel stock performance: An event-study approach. *Int J Hosp Manag* 26:200–212. https://doi.org/10.1016/j.ijhm.2005.11.004.
9. Chen C.-D., Chen C.-C., Tang W.-W., Huang B.-Y. (2009) The positive and negative impacts of the sars outbreak: A case of the Taiwan industries. *J Dev Areas* 43:281–293.
10. Nippani * S., Washer K.M. (2004) SARS: A non-event for affected countries' stock markets? *Appl Financ Econ* 14:1105–1110. https://doi.org/10.1080/0960310042000310579.
11. Sun W. (2017) H7N9 not only endanger human health but also hit stock marketing. *Adv Dis Control Prev* 2:1. https://doi.org/10.25196/adcp201711.
12. Ichev R., Marinč M. (2018) Stock prices and geographic proximity of information: Evidence from the Ebola outbreak. *Int Rev Financ Anal* 56:153–166. https://doi.org/10.1016/j.irfa.2017.12.004.
13. Ashraf B.N. (2020) Stock markets' reaction to COVID-19: Cases or fatalities? *Res Int Bus Finance* 54:101249. https://doi.org/10.1016/j.ribaf.2020.101249.
14. Sharif A., Aloui C., Yarovaya L. (2020) COVID-19 pandemic, oil prices, stock market, geopolitical risk and policy uncertainty nexus in the US economy: Fresh evidence from the wavelet-based approach. *Int Rev Financ Anal* 70:101496. https://doi.org/10.1016/j.irfa.2020.101496.
15. Zhang D., Hu M., Ji Q. (2020) Financial markets under the global pandemic of COVID-19. *Finance Res Lett* 36:101528. https://doi.org/10.1016/j.frl.2020.101528.
16. He Q., Liu J., Wang S., Yu J. (2020) The impact of COVID-19 on stock markets. *Econ Polit Stud* 8:275–288. https://doi.org/10.1080/20954816.2020.1757570.
17. Kumar S., Kaur J., Tabash M.I., et al. (2021) Response of stock market during covid-19 and 2008 financial crisis: A comparative evidence from brics nations. *Singap Econ Rev* 1–24. https://doi.org/10.1142/S0217590821500387.
18. Schoenfeld J. (2020) The invisible business risk of the COVID-19 pandemic. Tuck School of Business Working Paper (3567249).
19. Eleftherioua K., Patsoulis P. (2020) COVID-19 Lockdown Intensity and Stock Market Returns: A Spatial Econometrics Approach. 2020 Univ. Libr. Munich Munich.
20. Baig A.S., Butt H.A., Haroon O., Rizvi S.A.R. (2021) Deaths, panic, lockdowns and US equity markets: The case of COVID-19 pandemic. *Finance Res Lett* 38:101701. https://doi.org/10.1016/j.frl.2020.101701.
21. Ortmann R., Pelster M., Wengerek S.T. (2020) COVID-19 and investor behavior. *Finance Res Lett* 37:101717. https://doi.org/10.1016/j.frl.2020.101717.
22. Salisu A.A., Akanni L., Raheem I. (2020) The COVID-19 global fear index and the predictability of commodity price returns. *J Behav Exp Finance* 27:100383. https://doi.org/10.1016/j.jbef.2020.100383.
23. Subramaniam S., Chakraborty M. (2021) COVID-19 fear index: Does it matter for stock market returns? *Rev Behav Finance* 13:40–50. https://doi.org/10.1108/RBF-08-2020-0215.
24. Gupta R.K., Yang Y. (2018) CrystalFeel at SemEval-2018 Task 1: Understanding and Detecting Emotion Intensity using Affective Lexicons. In: *Proceedings of the 12th International Workshop on Semantic Evaluation.* Association for Computational Linguistics, New Orleans, Louisiana, pp. 256–263.
25. Zouaoui M., Nouyrigat G., Beer F. (2011) How does investor sentiment affect stock market crises? Evidence from panel data. *Financ Rev* 46:723–747. https://doi.org/10.1111/j.1540-6288.2011.00318.x.

20 Behavior Analysis of Lenders in P2P Lending Platforms
Identifying Cognitive Effort by Response Time Method

Benhmama Asmaa, Sabiri Brahim, and Melliani Hamza

20.1 INTRODUCTION

The normative economic theory relies on cost-benefit analysis to evaluate, choose, and adopt economic decisions (Hausman and McPherson, 2008). Based on the principle of rationality, the individual always opts for decisions that generate a benefit greater than the costs involved. The development of behavioral economics (Kahneman, 2011; Tversky and Kahneman, 1979) and then neuroeconomics (Camerer et al., 2004; Chorvat and McCabe, 2005; Glimcher and Fehr, 2013) has called into question the reliability of the principle of individual rationality. That said, cost-benefit analysis often does not coincide with certain behaviors adopted in everyday life. The emergence of these disciplinary fields has opened up new avenues of research that can provide more realistic answers about the actions of economic agents.

The structural change that economic science has welcomed has made it possible to mobilize new academic approaches capable of implementing different analytical tools to explore new angles in terms of knowledge. The financial field has benefited from this methodological support in recent years. Personal finance consists of studying individuals' behavior concerning their financial situation. The neoclassical approach qualifies this behavior as rational. Still, with the development of behavioral sciences and the emergence of behavioral economics, the rationality of individuals in their financial choices is not always verifiable. At this level, new research avenues have been adopted to explore further, more realistic explanations of individual behavior.

Neuroeconomics embodies a new field of research specialized in the functioning of cognitive and neural mechanisms in economic decision-making (Rangel et al., 2008; Rustichini, 2009). The work in neuroeconomics is multidisciplinary: Psychology, Economics, Neuroscience, and Computer Science, and it addresses economic issues at different levels: Biological, Cognitive, and Behavioral. Neuroeconomics has experienced a boom in recent years, especially with the technological development that has facilitated the application of the sophisticated tools proposed by the discipline of neuroscience.

Debt financing represents a topic that has been heavily exploited in work in experimental and behavioral economics and neuroeconomics (Cornée et al., 2012; Fehr and Zehnder, 2009; Oukarfi and Sabiri, 2022). With technological development, we have witnessed a new form of debt related to the financial technology context, namely Fintech Credit.

This work proposes an analytical framework that is part of behavioral economics and neuroeconomics to identify the lender's behavior on platforms dedicated to fintech credit. It is research that emphasizes the psychological aspect of economic science, mobilizing the time spent to answer as

DOI: 10.1201/9781032667478-23

a criterion of behavioral evaluation of the cognitive effort of the lender. For this purpose, the topic will be divided into the following categories: Section 20.1 treats social lending as a framework we have mobilized to analyze the lender-borrower relationship. Section 20.2 presents the hypotheses we wish to test. Section 20.3 highlights the results we have obtained. Section 20.4 is devoted to the discussion, and finally, there is a last section for the conclusion.

20.2 SOCIAL LENDING: PEER-TO-PEER LENDING

Social lending, or peer-to-peer lending, presents an alternative to the activity of debt financing with banks in which there is a direct financial exchange between two individuals (lender and borrower) without the intervention of a bank or a traditional financing organization (Chen et al., 2009). This is indeed a new mode of debt financing that belongs to a family of online financing modes, or fintech credit. According to Claessens et al. (2018), fintech embodies any credit activity that passes through an online electronic platform that commercial banks or credit institutions do not operate. Fintech credit activity has been booming recently, with financial revenues estimated at 92 billion USD in 2018 (Deloitte, 2020).

Social lending adopts a simple structure that brings together two individuals, the lender and the borrower. The novelty that gives rise to social lending is the platform that brings these two agents together. It is an electronic platform that offers the lender an investment opportunity by lending his funds to a borrower (Figure 20.1). And for the borrower, immediate access to funds to finance their purchases. Estimates According to Price Waterhouse Coopers (2015), the volume of social lending transactions will reach 150 billion USD in 2025.

20.2.1 SOCIAL LENDING AND ALTRUISM

Altruism enhances social lending. Berentsen and Markheim (2021) studied the behavior of the altruistic investor (lender) in a social lending process. They found that lenders with a strong preference for altruistic behavior are willing to finance projects even if they may result in future losses.

Altruism is associated with subjective behavior related to the individual's willingness to lend to the lender. Let's say an individual (i) is faced with choosing to grant a loan. The degree of altruism in the process is presented in the following equation:

$$\text{AltruisticLending}_i = \alpha_i \text{Benefit} - \beta_i \text{Cost} \qquad (20.1)$$

The lender often adopts a cost-benefit analysis. However, contrary to the normative economics approach, other factors directly impact the process. αi is the coefficient of anticipated social and psychological benefits ($-1 < \alpha_i < 1$). The more the coefficient tends toward 1, the more the social and psychological benefits reinforce the lender's financial benefit. β_i is the cognitive cost generated by the individual (i). The costs take two forms: psychological costs (risk aversion, etc.) and cognitive costs (cognitive effort) with $-1 < \beta_i < 1$.

$$\text{SelfishLending}_i = -\alpha_i \text{Benefit} + \beta_i \text{Cost} \qquad (20.2)$$

When the costs exceed the benefits, including the social and psychological benefits, the lender refuses to grant the loan to the borrower.

Lender   P2P Platform   Borrower

FIGURE 20.1 Simplified diagram of the lender-borrower.

20.3 HYPOTHESIS

The subject proposes a new vision based on cognitive and behavioral analysis to study the lender's actions. For this, we will rely on two hypotheses we wish to verify: the first hypothesis aims to mobilize the time-to-response method to determine the lender's behavior, and **H1: the time spent making a decision impacts the lender's behavior**. The second hypothesis is interested in evaluating the altruistic behavior of an individual and his or her perceived cognitive effort, **H2: the cognitive effort pushes the lender to adopt an altruistic behavior**.

20.4 EXPERIMENTAL METHOD

20.4.1 PARTICIPANTS

To test our hypotheses, we recruited graduate students randomly assigned to play the roles of lender ($n = 19$) and borrower ($n = 16$). The experiment was conducted in the economics department of the Faculty of Legal, Economic, and Social Sciences, Ain Sebaa, at Hassan II University of Casablanca.

20.4.2 PROCEDURE

We created a simple computer program that embodies a P2P platform. The technical structure applied is based on an internet network that links computers together. All the data collected from the subjects' computers are stored in an online cloud system. The mobilized platform has been built only for this experiment, programmed by the computer language JavaScript. The experiment is presented in vitro, with the subject who plays the role of the lender participating first in the game. He sits in front of the computer and then follows the scheme of the experiment. Each player plays five consecutive times (Figure 20.2).

The subject should choose between three cards. His decision provides him with a gain convertible into monetary gain. The subject receives a proposal to lend to the subject (2), who plays the role of the borrower. If he accepts, the subject should determine the amount he will give to the borrower. Similarly, he receives a return on loan of 1.5 on the initial amount.

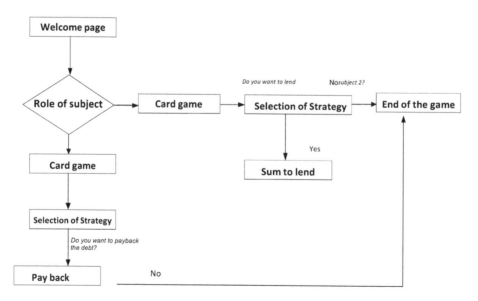

FIGURE 20.2 Schematic of the experiment.

20.5   RESULTS

The game's structure allows the lender to choose between three cards whose amounts are randomly ranked: $\pi_p = \{100, 200, 300\}$.

Figure 20.3 presents the total sum of monetary gains obtained during the five sessions. The gains vary in [3700 MAD, 4400 MAD]. The total sum of loans increased between the first and fourth sessions, from 1010 MAD to 1390 MAD, and decreased by 55 MAD in the fifth session (Figure 20.4).

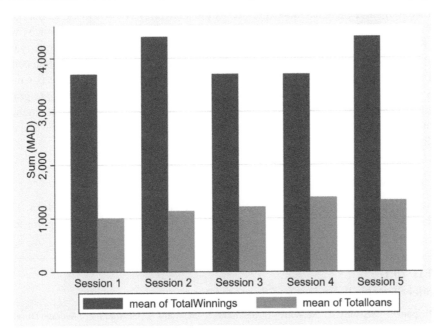

FIGURE 20.3   Graphical presentation of gains and loans transferred.

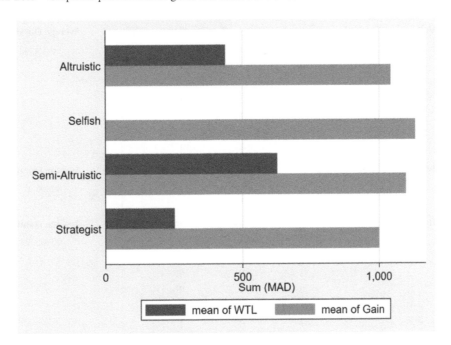

FIGURE 20.4   Graphical presentation of DMP averages, GM.

The average earnings for the four lender types are almost similar, in contrast to the average willingness to lend, which has changed depending on the lender's profile. We see that the semi-altruistic lender offers more than the one presented. Table 20.1 illustrates the lender strategic behaviors and response times.

20.5.1 RESPONSE TIME

We identified four strategic behaviors expressed by the lender: (1) The Altruistic lender: he adopted the "lend" strategy during all sessions. The average duration attributed to the choices was determined to be 5 min, 45 sec. (2) The Selfish lender: he adopted the "do not lend" strategy during all sessions. The average duration is 6 min, 30 sec. (3) The Strategist lender: he adopted a combination of the mixed strategies "lend, do not lend" during all sessions. The average time is 6 min, 7 sec. (iv) The Semi-Altruist Lender: adopted altruistic behavior with the first acceptance to grant a loan, which created a successive series of acceptances that were not always obvious to his financial situation and motivated him to adopt the "don't lend" situation to limit his lending. The average duration is 5 min, 22 sec.

20.5.2 ECONOMETRIC MODELING

The altruistic behavior of the lender is identifiable by his willingness to increase a loan. This unobserved variable associated with a binominal parameter embodies this behavior by assigning acceptance

TABLE 20.1
Lender Strategic Behaviors and Response Times

Participant	Session1	Session2	Session3	Session4	Session5	Average Duration
1	0	0	0	0	0	00 :06 :03
2	0	0	0	0	0	
4	0	0	0	0	0	

Participant	Session1	Session2	Session3	Session4	Session5	Average Duration
3	1	1	1	1	1	00 :05 :45
6	1	1	1	1	1	
7	1	1	1	1	1	
9	1	1	1	1	1	
11	1	1	1	1	1	
15	1	1	1	1	1	
18	1	1	1	1	1	

Participant	Session1	Session2	Session3	Session4	Session5	Average Duration
16	0	1	1	1	0	00 :06 :07
17	0	1	1	1	1	

Participant	Session1	Session2	Session3	Session4	Session5	Average Duration
5	1	0	1	0	1	00 :05 :22
8	1	0	1	1	0	
10	1	1	0	0	1	
12	0	0	1	0	1	
13	1	1	0	0	1	
14	0	1	0	1	0	
19	0	0	1	1	0	

the value (1) and refusal the value (0). It is an unobserved variable related to a binominal parameter that embodies this behavior by assigning to acceptance the value (1) and to refusal the value (0).

$$\begin{cases} \text{Choice}_{\text{Lender}} = \text{Accepttolend, when} \text{Altruism}_{\text{Lender}} = 1 \\ \text{Choice}_{\text{Lender}} = \text{Rejecttolend, when} \text{Altruism}_{\text{Lender}} = 0 \end{cases}$$

The random-effects logistic regression model is presented as the following equation:

$$P\left(\text{Altruism}_{\text{lender}} = 1 \mid Y_i\right) = \frac{\exp\left(Y_i^*\right)}{1 - \exp\left(Y^*\right)} = \beta X_i + \delta_i$$

with Y_i^* is the latent binary dependent variable $Y_i^* \approx Y_i = 1$, β is the vector of parameters to be estimated, X_i is the vector of independent variables, and δ_i is the error term modeled as an independent random-effect variable. Table 20.2 presents the results of random-effect logistic regression analysis.

The results show the absence of statistical significance between the choice of accepting or refusing to grant the loan to the borrower and the earnings generated. We also note that the running of the game several times did not impact the lender's altruistic behavior. The time taken to choose is statistically significant for whether or not to grant a loan ($p<0.1$). We can explain these results through the cognitive effort applied by the lender to implement the optimal strategy in a situation characterized by information asymmetry. We find that the cognitive process is costly in terms of the energy consumed by the brain. This pushes the individual to adopt a decision as quickly as possible. Altruistic behaviors remain one of the strategies that will allow the subject to overcome the cognitive load while preserving a good image at an ethical level. This explains why altruistic subjects' choice duration is lower than that of egoistic subjects. The choice made by the lender is not associated solely with the borrower's strategy, but there is a cognitive explanation that clarifies altruistic and selfish behavior.

TABLE 20.2
Results of Random-Effect Logistic Regression Analysis

	Random-Effect Logistic Regression		
	All Sessions		
Variables	**Coef.**	**CI 95%**	
Gain	0.003	−0.003	0.010
	(−0.003)		
Session	0.165	−0.216	0.548
	(0.195)		
Duration	0.006*	−0.001	0.014
	(0.003)		
Constant	−2.447	−5.550	0.656
	(−1.583)		
Wald Chi2	4.91		
Log Likelihood	−51.78		
Observations	95		

*$p<0,1$; **$p<0,05$; ***$p<0,01$

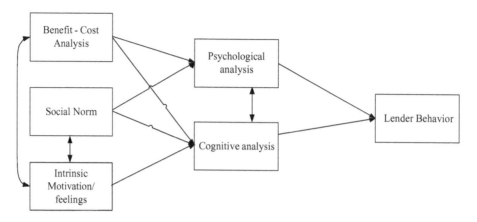

FIGURE 20.5   Model of lender behavior factors and processes.

20.6   DISCUSSION

The general model that frames the behavioral and cognitive processes of the lender is presented as follows (Figure 20.5).

The lender's behavior is based on two complex analytical modes: the psychological analysis fed by heuristics and biases, and the cognitive analysis based on the mental calculation process linked to several factors. Indeed, cognitive analysis relies on an effort that becomes difficult without information on the lender's environment. Three factors impact both modes: (1) the economic benefit-cost analysis; (2) the social norms and the impact of society on the lender's behavior in front of the debt situation; and (3) the intrinsic motivation or fear that pushes the lender to react positively or negatively. These factors impact analytical abilities insofar as the interaction between two or more factors makes analysis more difficult, which explains the long time the lender takes to adopt a decision. Based on our model, we find that social norms impact altruistic behavior more than that of the selfish. $\alpha_{iAltruism} > \alpha_{iSelfish}$, on the other hand, $\beta_{iAltruism} < \beta_{iSelfish}$ psychological and cognitive costs are often more intense in the analysis of the selfish than the altruistic.

20.6.1   BEHAVIORAL ASPECT

Selfish behavior confirms the neoclassical economic hypothesis that considers the individual a selfish agent. This behavior leads his employer to decline any request for the loan. The altruistic behavior of the individual is always to agree to give loans to borrowers. Strategist behavior is associated with individuals who adopt mixed strategies, accepting at times and declining at others. It is indeed a Tit for Tat strategy: the lender adopts an N strategy based on the borrower's $N-1$ strategy (Axelrod Hamilton, 1984). Semi-altruistic behavior embodies altruistic behavior for a specific period of time. Indeed, its behavior is associated with the Grim Trigger strategy: after a series of altruistic strategies (N_{A1}, N_{A2}, ..., N_{An}), the lender refuses to continue lending to the borrower (Gillman and Housman, 2019).

20.6.2   COGNITIVE ASPECT

The lender's behavior can be explained by intense cognitive effort. The experimental design carried out does not provide any relevant information to the lender about the identity or creditworthiness of the borrower; this creates a situation of ambiguity, which pushes him to make a decision as quickly as possible. During a financial decision, most people prefer to avoid cognitively difficult tasks when given a choice, a phenomenon called avoidance demand (Embrey et al., 2022; Kool et al., 2010; Shenhav et al., 2017). It is a natural behavior expressed by the brain to avoid high energy consumption in a cognitive process by adopting less costly alternatives on a neural scale (Kahneman, 1973). The lender, confronted by the situation of ambiguity, proceeds to exhibit similar behavior by adopting altruistic behavior, for two reasons: first, to avoid the high cognitive effort applied in the analysis of the situation, and second, to leave a favorable social image in front of his entourage. Altruistic behavior is very sensitive to the environment, so any change in an environmental factor (space, experience conditions, the information communicated, etc.) impacts the cognitive effort of the lender.

20.7 CONCLUSION

The study aimed to investigate the behavior of lenders when issuing a loan. The use of an experiment showed several behavioral patterns. Individuals expressed high distrust by adopting selfishness as the dominant strategy. Others adopted a contradictory strategy by adopting altruism as the dominant strategy. In contrast, two modes leaned toward mixed strategies: periodically stable (semi-altruistic) and unstable (strategist).

Similarly, the choice of strategy adoption was related to the time spent making the decision. This was statistically verifiable using random-effects logistic regression. The study showed that social norms significantly impacted altruistic lenders, while cognitive effort was more intense in analyzing selfish lenders. This leads us to admit that the analysis based on the psychological and cognitive aspect s is more intense than the rational reasoning based on cost-benefit.

REFERENCES

Axelrod Hamilton, 1984. *The Evolution of Cooperation.* Basic Books, New York.

Berentsen, A., Markheim, M., 2021. Peer-to-peer lending and financial inclusion with altruistic investors. *International Review of Finance* 21, 1407–1418.

Camerer, C.F., Loewenstein, G., Prelec, D., 2004. Neuroeconomics: Why economics needs brains. *Scandinavian Journal of Economics* 106, 555–579.

Chen, N., Ghosh, A., Lambert, N., 2009. Social lending. In: *Proceedings of the 10th ACM Conference on Electronic Commerce*, pp. 335–344.

Chorvat, T., McCabe, K., 2005. Neuroeconomics and rationality. *Chicago-Kent Law Review* 80, 1235.

Claessens, S., Frost, J., Turner, G., Zhu, F., 2018. Fintech credit markets around the world: Size, drivers and policy issues. *BIS Quarterly Review.* https://ssrn.com/abstract=3288096.

Cornée, S., Masclet, D., Thenet, G., 2012. Credit relationships: Evidence from experiments with real bankers. *Journal of Money, Credit and Banking* 44, 957–980.

Deloitte, 2020. Fintech | On the brink of further disruption. The Netherlands.

Embrey, J.R., Donkin, C., Newell, B., 2022. Is All Mental Effort Equal? The Role of Cognitive Demand-Type on Effort Avoidance, *Cognition* 236, 105440.

Fehr, E., Zehnder, C., 2009. Reputation and credit market formation: How relational incentives and legal contract enforcement interact, IZA Discussion Paper No. 4351.

Gillman, R., Housman, D., 2019. *Game Theory: A Modeling Approach*, CRC Press. ed. Taylor & Francis Group, UK.

Glimcher, P.W., Fehr, E., 2013. *Neuroeconomics: Decision Making and the Brain.* Academic Press, Cambridge, MA.

Hausman, D.M., McPherson, M.S., 2008. The philosophical foundations of mainstream normative economics. In: *The Philosophy of Economics: An Anthology*, 226–250. doi: 10.1017/CBO9780511819025.017.

Kahneman, D., 1973. Attention and effort. Citeseer.

Kahneman, D., 2011. *Thinking, Fast and Slow.* Farrar, Straus and Giroux, New York.

Kahneman, D., Tversky, A., 1979. Prospect theory: An analysis of decision under risk. *Econometrica* 47, 263–291. https://doi.org/10.2307/1914185.

Kool, W., McGuire, J.T., Rosen, Z.B., Botvinick, M.M., 2010. Decision making and the avoidance of cognitive demand. *Journal of Experimental Psychology: General* 139, 665.

Oukarfi, S., Sabiri, B., 2022. Analysis of lender trust in a risk situation. In: *Economic and Social Development: Book of Proceedings*, 86–96.

Pricewaterhousecoopers, 2015. Peer pressure: How peer-to-peer lending platforms are transforming the consumer lending industry. https://doi.org/10.4324/9781351163729-4.

Rangel, A., Camerer, C., Montague, P.R., 2008. A framework for studying the neurobiology of value-based decision making. *Nature Reviews Neuroscience* 9, 545–556.

Rustichini, A., 2009. Neuroeconomics: Formal models of decision making and cognitive neuroscience. In: *Neuroeconomics*. Elsevier, pp. 33–46. https://doi.org/10.1016/B978-0-12-374176-9.00004-X.

Shenhav, A., Musslick, S., Lieder, F., Kool, W., Griffiths, T.L., Cohen, J.D., Botvinick, M.M., 2017. Toward a rational and mechanistic account of mental effort. *Annual Review of Neuroscience* 40, 99–124.

Tversky, A., Kahneman, D., 1974. Judgment under uncertainty: Heuristics and biases. *Science* (1979) 185, 1124–1131

Section 4

Blockchain, Security, and Sustainability

21 Conceiving a Blockchain-Based Upstream Supply Chain Management System Enhancing Innovation and Sustainability

Ahmed El Maalmi, Kaoutar Jenoui,
and Laila El Abbadi

21.1 INTRODUCTION

For the first time, blockchain was initially introduced as an innovative solution for cash transfers by Satoshi Nakamoto in his paper "Bitcoin: A Peer-to-Peer Electronic Cash System" [1]. The aim is to eliminate the mediation role of the trusted third party in the exchange normally insured by banking institutions. It solves the problem of establishing trust in a distributed system. This groundbreaking technology is the foundation of all cryptocurrencies and has wide applications in the traditional financial system. The blockchain concept can be assimilated into a continuously growing list of logs known as secured and cryptographically connected blocks.

The blockchain network nodes check and verify the blocks before they are added. For example, handling a single piece of data necessitates thousands of instances, and each one requires a significant amount of work and time. Access to information in the blockchain is associated with higher quality in various blockchain systems. Here are a few qualities that distinguish blockchain from other technologies [2]. On the blockchain system, data are immutable and tamper-resistant. Their validation involves several nodes in a decentralized network. Scientists are listing three forms of blockchain: public or unauthorized, private or permitted, and consortium blockchain. Each one has specific characteristics because of the uniqueness of the network's geographic area [1–3].

Blockchain technology extends more and more to new domains, especially supply chain management. The main challenges facing innovation and sustainability in supply chain management are information access, traceability issues, heaviness of supply chain processes in terms of procedures, information flow, and long validation circuits. Companies are deploying several tools and systems for managing their supply chain activities. The literature includes some relevant models for enhancing innovation in supply chain systems [4–6]. Traditional supply chain management, even with the new models and strategies, responds to crucial needs and serves a broad goal but falls short of full compliance and still implies several limits, such as the heaviness and complications of the procedures, waste of time, and weak flexibility for reverse transactions and flow. Redesigning supply chain management based on Industry 4.0 technologies such as blockchain induces an important potential for empowering innovation and sustainability in the field and improving their performance in traceability, information sharing, access management, process automation, and flexibility [7–10].

DOI: 10.1201/9781032667478-25

This chapter tends to develop a blockchain-based supply chain model, ensuring, at the same time, consideration of the main constructs revealed for enhancing innovation and sustainability. Meanwhile, the blockchain system is serving the benefits of blockchain technology to improve transparency, traceability, information sharing, information control access, suppliers' product and service evaluation, process execution speed, and automation of transactions. The rest of the paper is organized as follows: Section 21.1 introduces the main steps in blockchain history and related technologies. In Section 21.2, the method is deployed to fulfill the paper's objectives. Section 21.3 is dedicated to describing the blockchain-based supply chain system innovation model; it formulates system interaction rules, writes and deploys smart contracts, and builds a supply chain system. Section 21.4 summarizes this work.

21.2 LITERATURE REVIEW

A simple blockchain system is basically founded on a peer-to-peer network and a consensus, ensuring the validity and security of the system [1]. Transactions in this network are validated according to the consensus protocols and stored in distributed ledgers in an immutable way across the network [1]. Note that any alteration of data made to the ledger is reflected and copied to all participants in seconds or minutes. The concept of a distributed ledger was initiated in 1976 by Helman et al. [11].

In 1990, Stuart Haber and Scott Stornetta proposed, in their paper, computationally practical procedures for digital time-stamping documents [12]. This stamping protects the data (document) from any attempt either to back-date or to forward-date, even with the collusion of a time-stamping service [12]. It ensures the complete privacy of the documents and requires no record-keeping by the time-stamping service [12]. Two years later, D. Bayer revealed in his work some improvements in the efficiency and reliability of existing time-stamping services [13,14]. On the same line, H. Massias et al. proposed an improved design of time-stamping services with a minimum trust requirement [15]. During this time, a wonderful advancement was made regarding an essential idea: electronic cash (or digital currency). It was discovered using a proposed model by David Chaum and his colleagues [16]. Adam Back presented the idea of hashcash in 1997 [17], which at the time promised a solution to the problem of unwanted emails known as spam. Because of this, Wei Dai came up with the idea of using a peer-to-peer network to create a new form of currency that he named "b-money" [18]. Satoshi Nakamoto, widely regarded as the person who invented blockchain technology, is credited with publishing a paper on bitcoin [1] based on earlier ideas and developments. The most challenging problem that Nakamoto had to solve was making a safe and direct online payment from one source to another that did not depend on a third-party source and was instead based on the idea of cryptography [19]. The article written by Nakamoto offered a solution to the problem of double spending by proposing that digital money could not be replicated. Hence, no one could use it more than once [19].

The issue of double spending was addressed in the study by proposing the implementation of a public ledger that would allow for the tracking and verification of an electronic currency's transaction history [1,18,19]. This would help to eliminate the possibility of a coin being used more than once. An open-source application that could be used to implement the bitcoin system was made available a few months after the first bitcoin was created and the bitcoin network was established. Since then, additional Bitcoins have been mined and sold on the market, and a sizable community has sprung up around them to provide assistance and fix different problems in the bitcoin code. Today, even nation-states (like El Salvador and the Central African Republic) and private businesses use this cryptocurrency (Tesla, Microsoft, Overstock). There are currently hundreds of various cryptocurrencies available on the market, including Ethereum, Ripple XRP, Tether, Litecoin, NEO, and others. Bitcoins continue to command a predominant market share and have established themselves as the most widely used cryptocurrency. It was able to attract the attention of the users owing to its capacity to keep its users unified, but it was due to its transparency that it became famous. The Ethereum platform, which enabled the blockchain to interact with loans and connections, was released in 2015 [19]. It was built on software known as a smart contract, which was responsible for ensuring the execution of action between the two parties. The widespread adoption of Ethereum's technology may be attributed to the platform's capacity to provide an environment that is simultaneously quicker, safer, and more productive [18,19].

Blockchain use is expanding to new application domains, such as healthcare. It was an example of patient-centric health records [20] and staff credential verification [21]. It is also used for security enhancement purposes in IoT systems [22]. Government services also benefit from blockchain to manage registries by promoting transparency and reducing fraud [23].

In supply chain management, the blockchain can serve in order fulfillment, supplier relationship management, manufacturing flow management, and demand management [24]. Manoshi Das Turjo et al. [8] advance smart supply chain management using blockchain and smart contracts. The system focuses on transaction deployment steps and their security strengthening based on access management and immutable ledger use. Jing Li [1,2] and Yafei Song [25] proposed a supply chain model that guarantees transaction deployment security using blockchain technology. This system involves two platforms dedicated, respectively, to information management and trade management. It allows suppliers to upload and acquire product information. Meanwhile, it is ensuring the management of orders, from creation to delivery. This solves the issues of asymmetric information, weak traceability, and common collaboration efficiency issues.

In this paper, we are proposing a blockchain-based supply chain management system that enhances sustainability and innovation. The new model has continuous interaction with a decision-making system. This is considering the following relevant constructs: competition capability based on cost, quality, capacity, and delivery; information management mainly for information sharing flexibility, information access control, information availability, integrity, and security; business continuity management through policy and objectives, risk approach, change management, business continuity plan, knowledge management, and awareness; supply chain integration granted by supplier integration and customer integration; supply chain orientation having as subconstruct attitude and intention. The company can use the decision-making system to deploy a multi-sourcing strategy for its strategic raw materials and consumables [6,26,27]. The material requirement plan (MRP) is considered an input to the decision-making system and a factual evaluation per material/consumable per supplier. The output of this system is a matrix distribution of material/consumable quantities that maximizes innovation and sustainability for the enterprise. The new model ensures the functionality of supplier account creation, product information upload and update by the suppliers, products and suppliers' information download, order initiation, validation, shipment, and delivery. It allows, moreover, order evaluation according to decision-making system criteria.

21.3 METHODS

This section delves into the approach and methods used to achieve the aims of this paper (Figure 21.1). It includes a literature review of blockchain technology history. It sorts out the available blockchain-based supply chain management systems in the literature and the purposes that are solved. Blockchain-based supply chain models were analyzed in terms of structure and functionalities. The analysis of these systems' limitations points out an interesting gap in our research. In parallel, we explain the decision-making system for strategic materials and consumable multi-sourcing that promote innovation and sustainability in the supply chain systems. The outputs of this process are taken into consideration while developing a blockchain platform that grants distributed and decentralized information ledgers and covers the weaknesses of traditional supply chain systems, The next step is dedicated to conceiving a supply chain management system that is built with blockchain

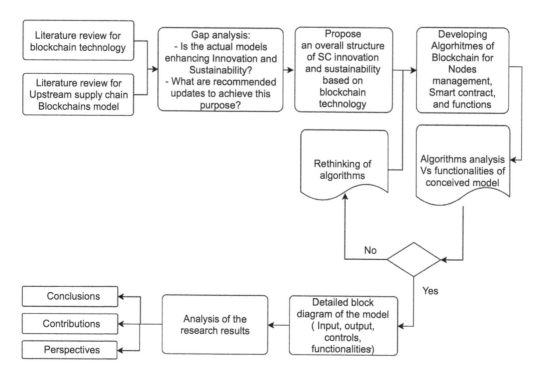

FIGURE 21.1 Method structure for developing a blockchain-based upstream supply chain management system enhancing innovation and sustainability.

technology and operates in collaboration with the decision-making system. In this step, the main processes of the model are analyzed and schematized for algorithm definition. The algorithms of blockchain for node management, smart contracts, and functions are analyzed in comparison with the objectives and functionalities required for our system. Each gap or limitation induces the correction and rethinking of the algorithms in an iterative process. Then a detailed blockchain of the system is proposed. The paper ends with research perspectives on this supply chain model and lists some system limitations.

21.4 DEVELOPING A BLOCKCHAIN-BASED SUPPLY CHAIN SYSTEM

21.4.1 A CONCEPTUAL FRAMEWORK FOR INNOVATIVE AND ENVIRONMENTALLY RESPONSIBLE SUPPLY CHAIN MANAGEMENT BASED ON BLOCKCHAIN

Regarding information sharing, the most important things to consider are accessibility, dependability, and safety. However, the supply chain system has stringent criteria for the real-time nature of the trading process. The capability of exchanging information can present an excellent opportunity for businesses to strengthen the collaboration between themselves and their suppliers. The primary selection factors can be used as the basis for multi-objective decision-making that can be implemented following the enterprise emphasis. Increasing innovation and sustainability within an organization's supply chain systems is one of the primary problems businesses face. The capability of businesses to compete in their respective markets is a critical component, and other factors such as business continuity management, supply chain integration, and supply chain orientation play crucial roles in developing innovative and sustainable supply chains [8,20,21].

To meet the new difficulties posed by the amount of data and the complexity of the interactions between organizations, it is essential to tackle the issues of asymmetric information, poor traceability, and inefficient cooperation, particularly in the conventional centralized supply chain. Blockchain technology is being pitched as the long-term answer to the problem of traditional centralized supply chains. This article aims to provide a blockchain system that combines aspects of supply chain innovation and sustainability with blockchain technology.

The blockchain concept concerns two platforms: the trade chain platform (related to the physical supply chain: materials, resources, money, equipment, and product flow) and the information chain platform (purchasing orders, technical and financial offers, documents, etc.). The schema in Figure 21.2 illustrates the two platforms and the related interactions inside each one and between both. It considers the upstream supply chain of the enterprise, where the main/critical materials are presented with P1, …, and PN. Those products are provided by the M suppliers shown by Supplier 1, …, and Supplier M.

The trading functions of the supply chain system are managed and supervised by the trade chain platform. The blockchain-based trading alliance chain, where smart trade contracts are placed, is the primary player in this network. The platform for the supply chain fulfills two primary functions. To begin, it accomplishes a trading method that is both efficient and convenient. It manages the trading process between the businesses that make up a supply chain. During this process,

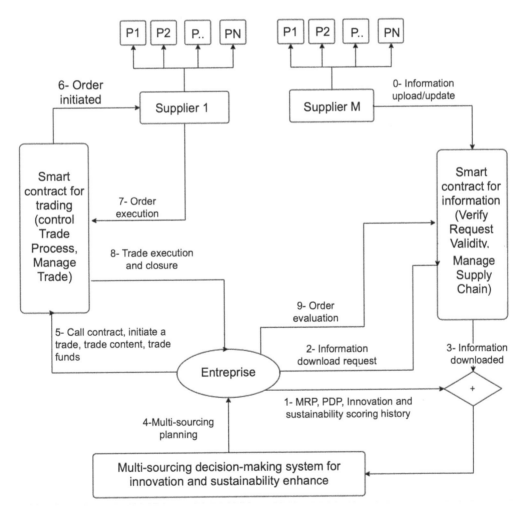

FIGURE 21.2 The overall schema of the blockchain-based supply chain system enhances innovation and sustainability, where PDP refers to production planning, and MRP refers to material requirements planning.

a smart trade contract will verify a transaction's legitimacy, completion, and other information. This will reduce the number of steps required for manual verification while improving the transaction's transparency and traceability, completion status, and other information. This information will then be made available to enhance the transaction's transparency and traceability. When the trade chain platform carries out a transaction, it logs the moment of its commencement, the parties involved in the transaction, the state of the transaction's completion, and other information. It then makes this data accessible to supply chain businesses.

The management of information is the platform's responsibility for the information chain. The information alliance chain, built on the blockchain and where smart information contracts are deployed, is the primary player in this network. The information chain platform maintains essential data on various goods and providers involved in the supply chain. Here, businesses can acquire information about their supply chains and exchange information about the products they sell. The bare minimum of essential information has to cover competition capabilities, business continuity management, supply chain integration, and supply integration, in addition to the sub-constructs associated with each. This is essential to meet the requirements of the multicriteria selection method that promotes innovation and sustainability. The beginning of commercial transactions will be achieved as a result of this therapy. A product cycle management system is another application that may be run on the information chain platform.

In detail, a complete transaction process includes:

Step 0: suppliers must update any changes in product information, such as product specification, availability, stocks, and delivery time.

Step 1: MRP, PDP, Innovation, and sustainability scoring history. This is an important stage where the enterprise consolidates production planning based on customer ordering and forecasting. The MRP is then performed according to the enterprise production and storage capacity. In parallel, the enterprise proceeds with the combination Material-Supplier evaluation based on history, new samples, and technical specifications if any changes are due. This evaluation includes the relevant constructs of the sustainable supply chain innovation model.

Step 2: information download request. The request concerns the combination Material-Supplier (M-S), and it is sent to the smart contract. The system analyzes the permissions requested by the user, and the results are added to the information portion of the smart contract. The substance of the interactions is [IdEse, InfoEse, InfPdt (Idsup, IdPdt, QtyPdt, InfoPdt)], where IdEse Id is the enterprise's address, InfoEse is the information related to the enterprise (Name, description.), InfPdt is the information related to the product or material, where Idsup is the address of the supplier, IdPdt is the identifier of the product, QtyPdt is the quantity to acquire, and InfoPdt is the information related to the product. Suppose the content of the request and the enterprise address matching are both verified by the smart information contract, which also grants access to the information for the enterprise. In that case, the corresponding operation will be performed, and the operation record will be uploaded to the information alliance chain. If the content of the request and the enterprise address matching are not verified, then the request will be denied.

Step 3: the enterprise acquires related information from all-combination materials-suppliers.

Step 4: those data are deployed in the multi-sourcing decision-making system to get the optimal allocation of quantity Q between suppliers for each product. Where Q= $\sum$Qpdtij and Qij is the optimal quantity for the product j from the supplier i.

Step 5: for each Qij, the enterprise must initiate a transaction, which includes [Idord, Stord IdEse, InfoEse, InfPdtj (Idsupi, IdPdtj, QtyPdtj, FQij, TQij, InfoPdtj)], where Idord is the order identifying, Stord is the status of the order, IdEse Id is the enterprise's address, InfoEse is the information related to the enterprise (Name, description), InfPdtj is the information related to the product or material j, Idsupi is the address of the supplier i,

IdPdtj is the identification of the product, QtyPdtj is the quantity to acquire for the j from the supplier j, FQij the found associated to this order TQij is the target delivery time, and InfoPdt is the information related to the product j delivered by the supplier i.

Step 6: upon receiving the request, the smart contract verifies the transaction content. After the verification, the transaction's content is sent via the smart trade contract to the supplier's account, and the order status is updated. If the requested content doesn't match the smart contract rules, an information message is sent to the enterprise. This transaction includes, by nature, a timestamp and the sender's address, which is the smart contract.

Step 7: in this stage, the order status is changed to order shipped, at the time of shipping. The order status change is performed with the supplier i with the correct access for this update. This information is time-stamped and shared with the enterprise.

Stage 8: when received, the enterprise must update the order status with material and service evaluation.

Different orders can be managed following the same steps defined here before, according to the needs of the enterprise.

21.4.2 THE DETAILED PROCESS OF THE BLOCKCHAIN-BASED SYSTEM

The core enterprise will have a special currency for the blockchain. Nodes are deployed according to precise rules to avoid abuse of normal system operation. Nodes are mandatory for permission verification for information requests or transaction initiation. The blockchain system includes the core enterprise node with the highest permissions and ensures coordination with system nodes. It also includes upstream supply chain enterprises.

The blockchain system is based on a distributed architecture, which produces considerable pressure and heaviness. Defining some rules is mandatory to minimize this issue. Figure 21.3 is a diagram of the whole mechanism of this system. The main functionalities are ensured in this diagram. It allows supplier account creation, which can be done only with the enterprise based on the supplier wallet. The supplier information can be updated by the supplier if needed. It allows product information to be uploaded by the suppliers. It can include product identification, technical description, available stock, and delivery time. Each supplier can also update the information according to any potential changes.

Conversely, based on a specific and explicit information request, the enterprise can download information per product and supplier to proceed to the multicriteria evaluation and the multi-sourcing strategy using the decision-making system. Products and suppliers' information was downloaded. Most of all, the diagram explains the stages of order initiation, validation, shipment, delivery, and evaluation and clarifies the interaction between the enterprise, the suppliers, and the smart contract.

The deployment of this blockchain-based system remains an important step in the operability analysis. This requires developing the smart contract based on the block diagram using a dedicated coding language and deploying the smart contract on a blockchain network.

For further study, we propose the following steps that can be handled in a future paper:

Coding the smart contract, nodes of the system, and functions using blockchain language. Solidity program language can be an interesting language for this purpose.

Deploying the smart contract on an adequate network. Remix – Ethereum IDE, the main network of Ethereum can be used. A virtual machine can be used for pre-tests and finetuning of the system.

Creating the suppliers of the enterprise based on suppliers' addresses in the network.

Each supplier can create a list of products and upload relatives' information.

Downloading information on products per suppliers to proceed with the decision-making system.

Initiating orders according to the MRP and the output of the decision-making system.

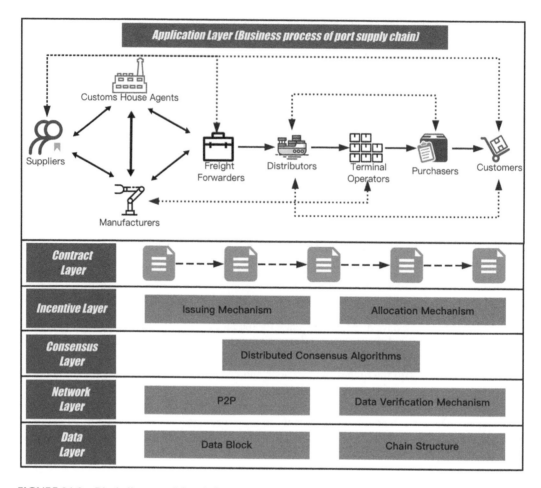

FIGURE 21.3 Block diagram of the whole system.

The orders' status can be checked easily with the blockchain-based system. Other functionalities can be explored.

The conceived blockchain-based system provides a relevant solution for traceability and centralization issues. Moreover, it improves procedure execution and eliminates waste from upstream supply chain processes. Transactions and information are saved and time-stamped. Access to information is fast and controlled by the system. Data are no longer centralized but are shared on the network. The MRP is defined using the muti-sourcing decision-making system and is based on production planning and downloaded data from the blockchain-based system. It is promoting suppliers and products with high scores of innovation and sustainability. The blockchain-based system ensures order status transparency between the enterprise and the suppliers. It allows service and product evaluation for each order. The evaluation score is saved and accessed in the network, and it can be an input in the next iteration of the MRP definition.

21.5   CONCLUSION AND FUTURE WORKS

Due to the permanent flow of innovative technologies and dynamic changes in supply chain system challenges, smart solutions such as blockchain technology were introduced to the field. In this work, we sought to alleviate some of the issues associated with the supply chain by proposing an innovation and sustainability system based on blockchain technology. One smart contract serves as the foundation for the whole system, which includes not only a trade chain platform but also an information chain platform. The first platform is devoted to trading functions based on the iterations of the multicriteria decision-making system, and the second platform grants easy and trustworthy information sharing. A smart contract for the two platforms was deployed mainly for suppliers' account management, product information management, supplier service, product evaluation, and product or service management. The developed blockchain-based system also considers access and role management for suppliers and the owner company. Indeed, blockchain transactions are highly beneficial in terms of traceability, transparency, execution speed, process simplification, and computational costs.

This study is limited to conceiving a blockchain-based upstream supply chain system. The performance in terms of transaction speed and security is not evaluated at this stage of work. The blockchain-based system was not tested in this work, and the operability test is still a must to prove that the system can meet the basic requirements of the supply chain. This operability test can first be done on a small scale based on a small volume of data and executed on a standard computer with limited treatment and storage capacity. With the large volume of data and considerable infrastructure, the system's operability can be ensured. The security of such a system is also a relevant indicator explored in the present work.

Performance evaluation through stress tests would be more significant in terms of operability. Indeed, the storage and processing expenses associated with blockchain transactions will benefit tremendously from their implementation. Increasing throughput can also be accomplished by implementing decentralized database systems. In addition, the architecture can accommodate the addition of tracking devices and more characters if used on a large scale. We may decide to store the original data using an off-chain architecture as the amount of data continues to expand, but the proof of existence may continue to be stored on the blockchain itself. These limits, in terms of scale level, performance, and security, might be a viable future study topic for the analysis that is now being conducted.

REFERENCES

1. Nakamoto, S. Bitcoin: A peer-to-peer electronic cash system. Decentralized Business Review, 2008, p. 21260. https://bitcoin.org/bitcoin.pdf.
2. Maleh, Y., Lakkineni, S., Tawalbeh, L. and AbdEl-Latif, A.A. Blockchain for cyber-physical systems: Challenges and applications. In *Advances in Blockchain Technology for Cyber Physical Systems* (2022): 11–59. https://doi.org/10.1007/978-3-030-93646-4_2.
3. Alfandi, O., Otoum, S., and Jaraweh, Y. Blockchain solution for IoT based critical infrastructures: byzantine fault tolerance. In: *Proceedings of the 2020 IEEE Network Operations and Management and Symposium*. Budapest, Hungary, April 2020. pp. 1–4.
4. Jenoui, K. and Abouabdellah, A. Implementation of a decision support system heuristic for selecting suppliers in the hospital sector. In: *2015 International Conference on Industrial Engineering and Systems Management (IESM)*. IEEE, 2015. pp. 625–632.
5. Jenoui, K. and Abouabdellah, A. Single or multiple sourcing strategy: a mathematical model for decision making in the hospital sector. In: *2016 11th International Conference on Intelligent Systems: Theories and Applications (SITA)*. IEEE, 2016. pp. 1–6.
6. Jenoui, K. and Abouabdellah, A. Determining a new multi-level criteria evaluation strategy for medicines suppliers. In: *2018 4th International Conference on Logistics Operations Management (GOL)*. IEEE, 2018. pp. 1–6.
7. Magazzeni, D., Mcburney, P. and Nash, W. Validation and verification of smart contracts: A research agenda. *Computer*, 2017, vol. 50, no. 9, pp. 50–57.
8. Turjo, M.D., Khan, M.M., Kaur, M., et al. Smart supply chain management using the blockchain and smart contract. *Scientific Programming*, 2021, vol. 2021. https://doi.org/10.1155/2021/6092792.
9. Xu, K. and Cong, H. A framework of sustainable supply chain management in Beijing environmental logistics. In: *2012, Fifth International Joint Conference on Computational Sciences and Optimization*. Kunming, Yunnan China, 2011. pp. 1263–1266.

10. El Maalmi, A., Jenoui, K., and El Abbadi, L. Innovative and sustainable supply chain model in industry 4.0 based on moroccan industrial field. In: *2021 IEEE International Conference on Industrial Engineering and Engineering Management (IEEM)*. IEEE, 2021. pp. 124–128.

11. Hellman, M., et al. New directions in cryptography. The Work of Whitfield Diffie and Martin Hellman, *IEEE Transactions on Information Theory*, 1976, vol. 22, no. 6, pp. 644–654.

12. Haber, S., Stornetta, W.S. How to time-stamp a digital document. In: Menezes, A. J., Vanstone, S. A. (eds.) *Advances in Cryptology-CRYPTO' 90. CRYPTO 1990*. Lecture Notes in Computer Science, vol 537. Springer, Berlin, Heidelberg, 1991. https://doi.org/10.1007/3-540-38424-3_32.

13. Bayer, D., Haber, S., Stornetta, W.S. Improving the efficiency and reliability of digital time-stamping. In: *Sequences II: Methods in Communication, Security and Computer Science*, 1993. pp. 329–334. https://doi.org/10.1007/978-1-4613-9323-8_24.

14. Soni, G., Kumar, S., Mahto, R.V., Mangla, S.K., Mittal, M.L., Lim, W.M. A decision-making framework for Industry 4.0 technology implementation: The case of FinTech and sustainable supply chain finance for SMEs. *Technological Forecasting and Social Change*, 2022, vol. 180, p. 121686. ISSN 0040-1625.

15. Massias, H., Avila, X.S., and Quisquater, J.J. Design of a secure timestamping service with minimal trust requirements. In *20th Symposium on Information Theory in the Benelux*. 1999.

16. Chaum, D., Fiat, A., and Naor, M. Untraceable electronic cash. In: *Conference on the Theory and Application of Cryptography*. Springer, New York, 1988. pp. 319–327. https://doi.org/10.1007/0-387-34799-2_25.

17. Back, A. A partial hash collision-based postage scheme. Retrieved December 1997, vol. 29, 2018.

18. Dai, W. b-money, 1998. URL https://www.weidai.com/bmoney.txt. (Last access: 08.04. 2019), 1998.

19. Sarmah, S. Understanding Blockchain Technology, 2018, vol. 8, pp.23–29. https://doi.org/10.5923/j.computer.20180802.02.

20. Prybutok, V.R., Sauser, B. Theoretical and practical applications of blockchain in healthcare information management. *Information and Management*, 2022, vol. 59, p. 103649.

21. Abbas, A., Alroobaea, R., Krichen, M., Rubaiee, S.; Vimal, S.; Almansour, F.M. Blockchain-assisted secured data management framework for health information analysis based on Internet of Medical Things. *Personal and Ubiquitous Computing*, 2021, pp. 1–14. https://doi.org/10.1007/s00779-021-01583-8.

22. Huo, R., Zeng, S., Wang, Z., Shang, J., Chen, W., Huang, T., Wang, S., Yu, F.R., Liu, Y. A comprehensive survey on blockchain in industrial internet of things: Motivations, research progresses, and future challenges. *IEEE Communications Surveys and Tutorials*, 2022, vol. 24, pp. 88–122.

23. Verma, S., Sheel, A. Blockchain for government organizations: Past, present and future. *Journal of Global Operations and Strategic Sourcing*, 2022, vol. 15, pp. 406–430.

24. Yousuf, S., Svetinovic, D. Blockchain technology in supply chain management: preliminary study. In: *2019 Sixth International Conference on Internet of Things: Systems, Management and Security (IOTSMS)*. IEEE, October 2019. pp. 537–538.

25. E- Li, J., Song, Y. Design of supply chain system based on blockchain technology. *Applied Sciences*, 2021, vol. 11, p. 9744. https://doi.org/ 10.3390/app11209744.

26. Jenoui, K., Abouabdellah, A. System of multisourcing supplier's selection and evaluation in the hospital sector integrating the criteria: Total cost Gap time Risk performance. *ARPN Journal of Engineering and Applied Sciences*, 2015, vol. 11, no 17, pp. 10433–10437.

27. Jenoui, K. and Abouabdellah, A. Proposal of an evaluation system for monitoring suppliers and controlling risks in the hospital sector. *International Journal of Supply Chain Management*, 2017, vol. 6, no 4, p. 157–166.

22 Green Finance in the Moroccan Mining Sector
Cases of Sustainable Development and CSR

Insaf El Atillah and Mohamed Azeroual

22.1 INTRODUCTION

Morocco is a mining country. Indeed, the mining sector in the Kingdom is one of the pillars of the national economy. "The importance of this sector in Morocco is perceptible beyond the investments it drains (11.9 billion DH in 2020), its share in national exports (more than 21% in value in 2020), and its beneficial impact on regional development, without omitting its positive impact on the dynamics in the transport sector and on the port activity," said Ms. Benali, Minister of Energy Transition and Sustainable Development as part of its participation in the 6th edition of the International Exhibition of Mining Senegal (SIM Senegal 2021) which took place from November 2–4 in Dakar, under the theme "Promotion and development of local content, a lever for optimizing socio-economic benefits in the extractive sector" [1,2]. However, the mining sector scored 49 points in the Natural Resources Governance Index 2021 [1]. This places it in the group of countries with a "weak" performance. This leaves a large gap in the development of the mining sector in Morocco, which will have to address challenges affecting its solvency and sustainability. The mining activity goes through several stages, each with particular environmental effects: the prospecting of the deposits and their exploration, the setting up of the mines and their preparation and exploitation, and the treatment of the minerals obtained to extract commercial products. This generates different types of waste that can pollute the ground, water, air, fauna, or flora. In addition, the extraction of mines can have a negative impact on the environment due to several factors, such as dust, noise, and vibrations resulting from blasting and drilling. Socially speaking, the mining activity is sometimes the reason behind the relocation of the population.

Moreover, implementing a mine can also cause health problems for the surrounding population. This is why integrating a green finance approach is essential to protect the environment and society. Especially as it has strong links with Islamic finance since Morocco is an Islamic country. Indeed, Green Sukuk is considered 'innovative' because it is the first instrument of finance in the world that considers environmental, economic, and Islamic values [3]. In addition, green finance and Islamic finance have achieved a growing trend worldwide. And this, over the past two decades [4]. The integration of green finance translates into adopting the principles of sustainable development and social responsibility.

At this constant, it is legitimate to ask: Is the Moroccan mining sector an extractive activity toward a green economy?

The objective of this work is to shed light on the actions taken by the Kingdom in the mining sector to achieve a green economy. In this sense, we will discuss Morocco's actions in the green finance framework to achieve sustainable development. In the same context, we administered a questionnaire to study the behavior of Moroccan mining companies toward adopting sustainable development and corporate social responsibility (CSR) practices. The chapter is organized into three main sections. The first section focuses on sustainable development and CSR in the mining sector. This section provides an overview of the challenges faced by the mining industry regarding sustainable development and the role of CSR in addressing these challenges.

The second section of this chapter presents sustainable development and CSR in the mining sector in Morocco. Section 22.3 discusses the survey methodology. Section 22.4 presents the results of a survey on the adoption of sustainable development practices by mining companies in Morocco. This section provides insights into the current state of sustainability practices in the Moroccan mining industry and identifies areas for improvement.

The chapter's final section concludes by summarizing the survey findings and discussing their implications for the Moroccan mining industry. This section provides recommendations for how mining companies in Morocco can adopt more sustainable practices and promote CSR.

22.2 SUSTAINABLE DEVELOPMENT AND CSR IN THE MINING SECTOR

22.2.1 The Importance of the Mining Sector in Morocco

The mining sector plays a crucial role in the Moroccan economy. The country is rich in minerals, including phosphates, lead, zinc, cobalt, and copper, among others. These minerals are extracted by mining companies and used in various industries, such as agriculture, construction, and manufacturing. The mining industry is a major source of employment and contributes significantly to the country's GDP.

However, the extractive activity of the mining sector can have adverse effects on the environment, such as deforestation, soil degradation, and water pollution. To mitigate these negative impacts, the Moroccan government and mining companies are adopting sustainable development practices and prioritizing green finance. This is in line with the global trend of promoting a green economy and reducing the carbon footprint of industries.

The Moroccan government has enacted legislation, such as Law No. 33-13, which requires mining companies to consider the environment in their activities. The "Plan Maroc Mines 2021–2030" also prioritizes sustainable development, making it a key aspect of the mining sector. The mining companies, for their part, are committed to implementing sustainable development practices, such as reducing their carbon footprint, conserving natural resources, and protecting the environment. Figure 22.1 describes the Moroccan mining sector.

22.2.1.1 Definition of Green Finance

Traditional finance directs savings to the most profitable projects without considering the environmental aspects of the investments made. On the other hand, green finance projects do not harm the environment or allow the development of a sustainable economy [5].

22.2.2 Definition of Sustainable Development

Sustainable development can be defined as "development that meets the needs of the present without compromising the ability of future generations to meet their own needs" (Brundtland, 1987). Usually, sustainable development consists of balancing these three pillars: (1) The environment: preserving the diversity of species as well as natural and energy resources; (2) The social: satisfying the needs in education, health, housing, prevention of exclusion, employment, and intergenerational equity; and (3) The economic (or economy): creating wealth and improving material living conditions. We can add a fourth pillar: governance, which allows coordination between the three pillars mentioned. This last one is often mistakenly forgotten [6].

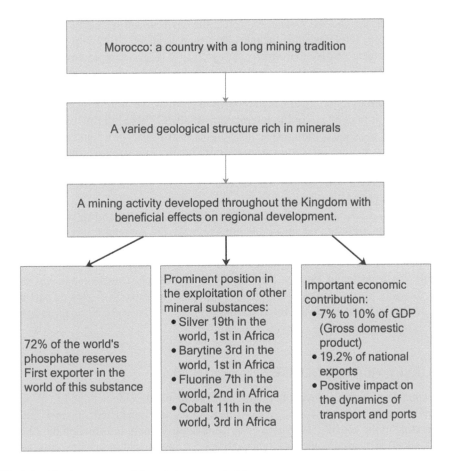

FIGURE 22.1 The importance of the mining sector in Morocco.

22.2.3 DEFINITION OF CSR

CSR refers to the voluntary practice by companies of incorporating environmental and social considerations into their business operations and interactions with stakeholders. Companies demonstrate CSR when they exceed the minimum legal requirements and obligations imposed by collective agreements to fulfill the needs of society. Business leaders are seen as servants of society and should not ignore accepted societal values or prioritize their values over those of society [7]. CSR helps companies, regardless of size, to contribute to reconciling social, environmental, and economic ambitions in cooperation with their stakeholders. This is why CSR has become an increasingly important theme in the debate on globalization, competitiveness, and sustainable development [8].

22.2.4 REGULATORY TEXTS, POLICIES, AND STRATEGIES

The old Moroccan mining legislation, the Dahir (a Decree Law promulgated by the King of Morocco) of April 16, 1951 [9], does not include provisions that obligate mining permit holders to take necessary measures to protect the environment. This has led to many mining sites being abandoned without rehabilitation when operations cease.

However, the current mining industry in Morocco is based on Law No. 33-13, which came into force in 2016 and places emphasis on environmental protection. This law includes articles such as 52, 56, 57, 59, 60, and 62, which consider the concepts of the environment, safety, protection of human resources, and conservation of deposits.

Morocco also has a rich legal arsenal to protect the environment from any potential nuisances resulting from mining activities. This includes laws such as law n° 11-03 [10] relating to the protection and development of the environment, law n° 12-03 [11] relating to environmental impact studies, law n°13-03 [12] relating to air pollution, and the framework law n°99-12 relating to the National Charter of the Environment and Sustainable Development [13].

The public authorities have implemented reforms to promote the mining sector and ensure favorable conditions for it to compete in the international market. These reforms include modernizing the regulatory and legislative framework, aligning administrative procedures, reducing the risk of discretionary power from the administration, introducing a prospecting authorization process, regulating small-scale mining, extending provisions of mining regulations to maritime areas and environmental protection, establishing the National Program of Geology, and reforming the institutional framework [14].

The Kingdom has adopted the Morocco Mines Plan 2021–2030 [15,16], which builds on the achievements of the 2013–2025 strategy and aims to make the national mining sector a driver of sustainable and responsible development. This plan emphasizes investment in geological knowledge, transparent and equitable exploitation of mineral resources, and the sharing of expertise. Mrs. Benali, the Minister of Energy Transition and Sustainable Development, has also shared important thoughts for developing a sustainable extractive industry in Africa as part of her participation in the sixth session of the International Mining Exhibition in Senegal on November 2, 2021 [1,2].

22.3   SURVEY METHODOLOGY

The study aimed to assess the level of adoption of sustainable development and CSR practices by mining companies in Morocco. To gather the data, a survey was conducted among 43 mining companies operating in the country. The survey was designed to capture information on the size of the companies, adoption of sustainable development practices, investment in sustainability, certification for environmental protection, and awareness and adoption of CSR practices.

22.3.1   Data Collection

The data for this study was collected using a self-administered questionnaire. The questionnaire was designed to capture information on the size of the companies, duration of adoption of sustainable development practices, investment in sustainability, certification for environmental protection, and awareness and adoption of CSR practices. The questionnaire was distributed to 43 mining companies operating in Morocco and was collected through electronic mail. The data was then analyzed to identify trends and patterns.

22.3.1.1   Data Analysis

The data collected through the survey was analyzed using descriptive statistics. The data was presented in the form of tables, charts, and graphs to provide an overview of the key findings. The data was then interpreted to identify trends and patterns and to draw conclusions about the level of adoption of sustainable development and CSR practices by mining companies in Morocco.

22.3.1.2   Sample Size

The sample size for this study was 43 mining companies operating in Morocco. The sample was selected using a convenience sampling technique, where the companies were selected based on their willingness to participate in the study. The sample size was considered sufficient for the purpose of this study, as it allowed for a representative overview of the level of adoption of sustainable development and CSR practices in the sector.

22.3.1.3 Validity and Reliability of the Data

The validity of the data was ensured by designing the questionnaire in a way that captured accurate and relevant information. The reliability of the data was ensured by conducting a pilot study before distributing the questionnaire to the final sample. The pilot study allowed for any errors or inconsistencies in the questionnaire to be corrected, ensuring the reliability of the data collected.

The study used a survey methodology to collect data from 43 mining companies operating in Morocco. The data was analyzed using descriptive statistics to provide an overview of the key findings and draw conclusions about the level of adoption of sustainable development and CSR practices in the sector [17]. The sample size was considered sufficient, and the validity and reliability of the data were ensured through appropriate measures.

22.4 SURVEY ON THE ADOPTION OF SUSTAINABLE DEVELOPMENT PRACTICES BY MINING COMPANIES IN MOROCCO

22.4.1 SURVEY RESULTS

On February 24, 2021, we administered a questionnaire on Google Forms to study the behavior of mining companies in Morocco toward adopting sustainable development.

From a sample of 43 Moroccan mining companies, we obtained the following results (see Figure 22.2).

51.2% of the responding companies are small (over half). 26% are medium, and 23% are significant (Figure 22.3).

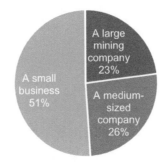

FIGURE 22.2 Nature of the company. (Survey; $n=43$.)

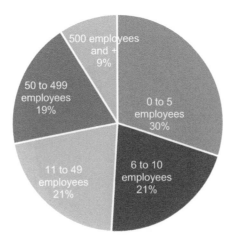

FIGURE 22.3 Number of employees. (Survey; $n=43$.)

Among the respondents, companies with 0–5 employees are the most numerous, with a percentage of 30% of the sample. For companies with 6–10 employees, 11–49 employees represented the same percentage: 21%. Followed by companies with 50–499 employees: 19%. Finally, companies with 500 or more employees represented 9% of the sample (Figure 22.4).

Almost half of the companies surveyed are beginners (their ages are between 0 and 5 years). In contrast, 28% of the sample was in operation for at least 11 years (which is a critical period). Companies aged between six and ten years represented 23% of the sample (Figure 22.5).

70% of responding companies have already carried out sustainable development actions, compared to only 30% who have not (Figure 22.6).

Lack of funding is the obstacle for most respondents, with a percentage of 38% of the sample, followed by the need for competent resources (30%) and lack of time (23% of the sample, Figure 22.7).

The company's internal policy represents the factor of adopting sustainable development practices for the majority of the responding companies (43%). Followed by the requirements of the state, which is the second dominant factor with a percentage of 20% of the sample. For international standards, 16.6% of the sample chose it as a factor (Figure 22.8).

33.33% of responding companies consulted with investors, shareholders, and employees to carry out sustainability projects. 13% consulted suppliers and the community. At the same time, 6.66% represents the percentage of companies that consulted with customers as a stakeholder (Figure 22.9).

51% of the responding companies have adopted sustainable development actions for 0–2 years. 28% of the sample have adopted them for a period between three and six years. A percentage of 7% for those who adopted it for durations between 7 and 10 years, 11 and 20 years, 21 years, and more (Figure 22.10).

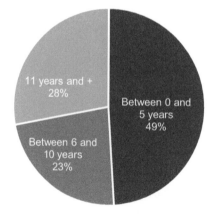

FIGURE 22.4   The company's age. (Survey; $n=43$.)

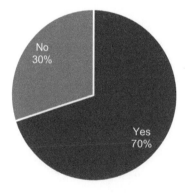

FIGURE 22.5   The realization of sustainable development actions. (Survey; $n=43$.)

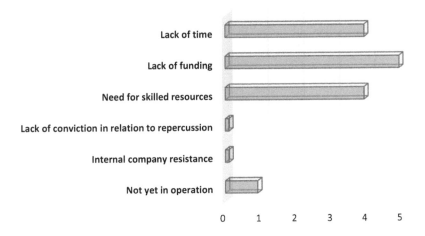

FIGURE 22.6 Brakes for companies without sustainable development actions. (Survey; $n=43$.)

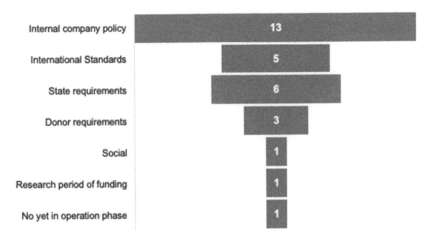

FIGURE 22.7 Factors for companies that have carried out sustainable development. (Survey; $n=43$.)

41.9% of the sample devotes 0%–2% of the annual investment budget to investing in sustainability. 32.6% of the sample dedicates an annual investment budget ranging from 3% to 5%. While 2.3% of the sample dedicates a budget between 26% and 50% (Figure 2.11).

Most responding companies have not been certified by a third party in terms of environmental protection (81% of the sample), compared to 19% that have been certified. Some are certified by ISO 9001/2000, others by SGS, the general surveillance company.

Have you ever heard of CSR? (Figure 22.12).

77% of responding companies have heard of CSR, while 23% have not.

70% of responding companies adopt practices that consider CSR, compared to 30% that do not. Some companies have been adopting CSR since 2000, others for 12 years. There are those who have adopted it for only two months (Figure 22.13).

79% of the responding companies have mining permits, while 21% do not.

There are companies that have more than 700 mining permits; others have more than 200. And some have less than 5.

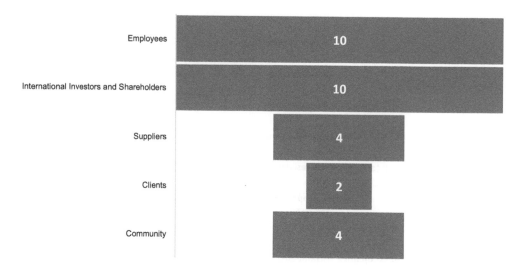

FIGURE 22.8   Stakeholders consulted to carry out sustainable development projects. (Survey; $n=43$.)

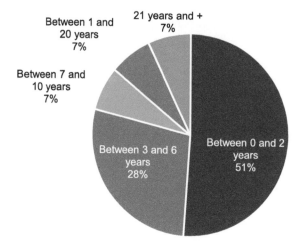

FIGURE 22.9   Duration of adoption of sustainable development actions. (Survey; $n=43$.)

22.4.2   Questionnaire Synthesis

The survey of 43 Moroccan mining companies provides valuable insights into the level of adoption of sustainable development and CSR practices in the sector [18]. The following is a synthesis of the key findings from the survey, with in-depth analysis and elaboration.

Size of Companies: The majority of companies operating in the mining sector in Morocco are small, with 51.2% of the surveyed companies falling into this category. Moreover, 49% of the companies surveyed are start-ups, operating in the sector for 0–5 years. On the other hand, 28% of the sample has been operating in the sector for 11 years or more, indicating a significant presence of established companies in the sector.

Adoption of Sustainable Development: The survey results show that although most mining companies are beginners in the sector, they have adopted sustainable development actions. This highlights the growing orientation of the Moroccan mining sector toward sustainable development and the recognition of its importance.

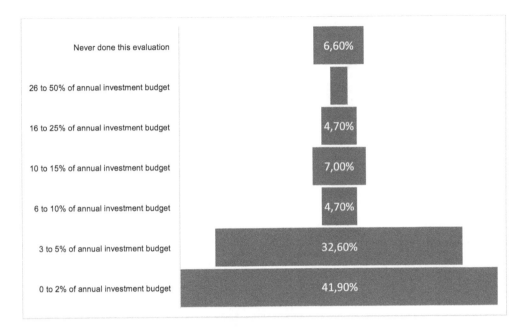

FIGURE 22.10 The rate of the annual investment budget devoted to investing in sustainable development (Survey; $n=43$.)

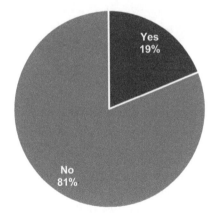

FIGURE 22.11 Certification, by a third party, in environmental protection. (Survey; $n = 43$.)

Timing of Adoption: The dominant majority of companies that have adopted sustainable development actions started implementing them in the last 0–2 years, with 51% of the responding companies. This indicates a recent increase in the interest of mining companies toward sustainability. The main driver behind the adoption of sustainable development practices, according to 43% of the responding companies, is internal company policy. On the other hand, the lack of financing represents the main obstacle for companies that have not yet adopted sustainable development practices, with 38% of the sample citing this as a reason.

Investment in Sustainability: 41.9% of the surveyed companies allocate 0%–2% of their annual capital budget toward investing in sustainability, which is a modest share. This can be attributed to the recent direction toward sustainability for most mining companies,

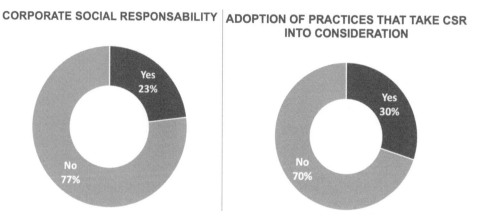

FIGURE 22.12   Corporate social responsibility (CSR), and the adoption of practices that consider it. (Survey; *n*=43.)

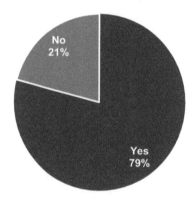

FIGURE 22.13   Possession of mining permits. (Survey; *n*=43.)

which are still in their early stages and small in size. On the other hand, 32.6% of the sample allocates an annual investment budget of 3%–5%, while 2.3% devotes a budget of 26%–50%, which is a considerable share.

Certification for Environmental Protection: 19% of the responding companies have been certified by a third party for environmental protection, which is a significant percentage. This indicates that a considerable portion of the mining companies in Morocco are taking proactive steps toward implementing sustainable development practices.

Adoption of CSR: 70% of the responding companies adopt practices that consider CSR, while 30% do not. This shows that most mining companies are aware of the importance of CSR and are taking steps toward incorporating it into their business practices.

Mining Licenses: 79% of the responding companies have mining licenses, while a small minority do not. This could be because these companies are still new to the sector or have not yet started their operations, or are operating as subcontractors.

The survey provides valuable insights into the level of adoption of sustainable development and CSR practices in the Moroccan mining sector. It highlights the recent orientation of the sector toward sustainability and the recognition of its importance [19]. The results also indicate the presence of both small and established companies in the sector, with most companies adopting sustainable development practices in recent years. The results also show the modest investment in sustainability

TABLE 22.1

Summary of the Survey Results

The adoption of sustainable development practices by mining companies in Morocco

The realization of the actions of sustainable development?	Yes	**70%** of the sample
	No	**30%** of the sample
Duration of adoption of sustainable development actions?		**51%** of the responding companies have adopted sustainable development actions for **0 and 2 years**.
Certification, by a third party, in the field of environmental protection?	Yes	**19%** of the sample
	No	**81%** of the sample
Corporate social responsibility (CSR)	➡	**77%** of responding companies have heard of CSR **70%** of them adopt practices that take CSR into consideration

Source: Survey; *n*=43.

by most companies and the significance of certification for environmental protection [8,20]. Most responding companies are aware of the importance of CSR and are taking steps toward incorporating it into their business practices (Table 22.1).

22.5 CONCLUSION

The conclusion of the research on adopting sustainable development and CSR practices by Moroccan mining companies highlights the key role played by the mining sector in the Moroccan economy. The negative impact of the extractive activities of the sector on the environment has prompted the government and mining companies to adopt sustainable development practices to achieve a green economy [21,22]. This is in line with the trend toward green finance.

The Moroccan mining legislation, Law No. 33-13 [23], emphasizes the protection of the environment in the activities of mining companies, and the "Plan Maroc Mines 2021–2030" prioritizes sustainable development. Most mining companies surveyed have adopted sustainable development and CSR practices, despite being relatively new to the sector. This suggests that the mining sector is taking its initial steps toward a green economy [24].

REFERENCES

1. Site du Maroc.ma.
2. Site du ministère de la Transition énergétique et du Développement durable: www.mem.gov.ma.
3. Liu, F., & Lai, K. (2021). Ecologies of green finance: Green sukuk and development of green Islamic finance in Malaysia. *Environment and Planning A: Economy and Space*, 53(8), 1896–1914.
4. Sekreter, A. (2017). Green finance and islamic finance. *International Journal of Social Sciences & Educational Studies*, 4(3). doi: 10.23918/ijsses.v4i3p115.
5. Ansidei, J., Leandri, N. (2021). Introduction. Dans: Julie Ansidei éd., La finance verte. Paris: La Découverte (pp. 3–7).
6. Securities Commission Malaysia (2019). *Islamic Green Finance Development, Ecosystem and Prospects*. Kuala Lumpur: World Bank Group Global Knowledge and Research Hub.
7. Bowen, H. R. (1953). *Social Responsibilities of the Businessman*. New York: Harper.
8. Jounot, A. (2010). 100 questions pour comprendre et agir- RSE et développement durable. AFNOR.
9. Dahir du 9 rejeb 1370 (16 avril 1951) portant règlement minier, au Maroc.
10. Loi n° 11-03 relative à la protection et à la mise en valeur de l'environnement.
11. Loi n° 12-03 relative aux études d'impact sur l'environnement.
12. Loi n°13-03 relative à la lutte contre la pollution de l'air.
13. Loi-cadre n°99-12 portant Charte Nationale de l'Environnement et du Développement durable.

14. Plan Maroc Mines 2021-2030.
15. Babi, K. (2011). Perceptions du développement minier durable par les acteurs locaux, gouvernementaux et industriels au Maroc. Université du Québec à Chicoutimi.
16. Plan National de cartographie géologique 2021-2030.
17. Shneider, L. (2013). Le developemet durable territorial. AFNOR.
18. El Amine, S., Amine, N.B. (2021). Développement durable un levier de compétitivité des entreprises marocaines: Cas de l'OCP. *Revue Française d'Economie et de Gestion*, 2(5), 220–241.
19. Figuière, C., Rocca, M. (2012). Gouvernance: Mode de coordination innovant? Six propositions dans le champ du développement durable. *Innovations*, 3, 169–190.
20. Hamiti, D., Bouzadi-daoud, S. (2021). Etude du concept du développement durable. مجلة ابن خلدون للإبداع والتنمية, 3(2), 133–147.
21. Kasbaoui, T., Nechad, A., El Yamani, K. (2018). Responsabilité sociale des entreprises au Maroc: Etat des lieux et nouveaux enjeux. Revue Marocaine de la Prospective en Sciences de Gestion, (1).
22. La responsabilité sociale des entreprises au Maroc -Vue par des chercheurs en sciences de gestion, cordonné Jacques IGALENS et Farid CHAOUKI (2017). Editions MPE.
23. Loi N° n°33-13 relative aux mines.
24. Note de présentation du projet de loi n°33-13 relative aux mines.

23 Cloud Data Integrity Auditing and Deduplication Using an Optimized Method Based on Blockchain and MAS

Mohamed El Ghazouani, Abdelouafi Ikidid,
Charafeddine Ait Zaouiat, Layla Aziz,
Yassine El Khanboubi, Moulay Ahmed El Kiram,
and Latifa Er-Rajy

23.1 INTRODUCTION

Many companies store massive amounts of data locally, which creates several issues, including processing, security, and the necessity to invest in hardware and human resources. The key is to outsource the data to the cloud.

For clients, cloud computing provides access to many technologies while reducing barriers to entry, including technical expertise and cost. The cloud services market is generally categorized into three main service models: infrastructure, platforms, and software. According to business needs and security worries, cloud clients can opt for private, public, or hybrid cloud deployment patterns.

Given the huge technical and economic benefits of cloud computing and the various challenges threatening cloud adoption, our research work is set in the context of this technology. Once data is stored on cloud storage servers, the cloud client loses control. While this technology provides many benefits, it also poses security concerns, particularly those relating to data integrity, which is arguably one of the most sensitive aspects of any system. To maintain the integrity of the externalized data, the data owner must activate audit processes.

Furthermore, owing to the evolution of the immense amount of data, much of which is redundant, storage efficiency is another essential requirement that must be guaranteed via, for example, deduplication technology.

To this end, we have proposed a model to implement a new blockchain-based architecture that ensures data integrity verification and deduplication in the cloud environment.

Nevertheless, implementing this technique in the cloud is complicated since large amounts of data must be deduplicated rapidly. To address this concern, we propose a new architecture that introduces a multi-agent system (MAS) where multiple intelligent agents collaborate to establish a flexible system for server-side deduplication.

More specifically, in this research, we will explore the deduplication technique and data integrity as two of the most crucial aspects of cloud adoption. Our primary concern is to rely on the existing and most relevant protocols that ensure data integrity and deduplication to propose a new and efficient architecture that addresses the cited issues and is suitable for CC environments.

DOI: 10.1201/9781032667478-27

This chapter's organization is as follows: Section 23.2 presents the related works. The preliminaries and concepts used in our proposal are discussed in Section 23.3. Section 23.4 gives a comprehensive depiction of the proposed system. Section 23.5 provides a performance and results analysis. Lastly, a conclusion is offered in Section 23.6.

23.2   RELATED WORKS

In recent years, deduplication and data integrity issues have gained significant attention, and a set of approaches and protocols have been proposed. Ateniese et al. [1] have presented a provable data possession (PDP) protocol that allows the user to check the accuracy of the externalized data without retrieving it. Juels and Kaliski [2] proposed a new PDP variant: Proof of Retrievability.

Thus, support for data dynamics is also of crucial relevance to auditing schemes in the cloud data storage model. Erway et al. [3] presented an efficient protocol for dynamic provable data possession. Wang et al. [4] presented a public dynamic auditing system based on Merkle Hash Tree (MHT), which can simultaneously handle audit delegations on requests from different owners. Liu [5] extended MHT to rank-based MHT (R-MHT) with auditable and efficient, accurate updates. Zhu et al. [6] presented an auditing system known as public auditing based on an index hash table (IHT-PA).

Tian et al. [7] presented a dynamic hash table-based protocol that allows both public and dynamic auditing. Yu et al. [8] invented an identity-based auditing scheme to verify the integrity of externalized data. Lee et al. [9] presented a novel system for verifying the integrity of remotely stored data.

Li et al. [10] presented a storage protocol with deduplication that allows keyword search by using the converged encryption technique to encrypt data before outsourcing. Miao et al. [11] presented a secure deduplication protocol assisted by multiple servers in cloud computing. Zhang et al. [12] proposed a content-defined asymmetric extremum segmentation algorithm for data deduplication. El Khanboubi et al. [13] presented a new data deletion protocol for a blockchain-based deduplication system in the cloud. Jiang et al. [14] suggested a secure fuzzy deduplication system for cloud storage. These latter systems primarily considered data deduplication but did not address data integrity.

23.3   PRELIMINARIES AND CONCEPTS USED IN OUR
PROPOSAL BLOCKCHAIN: DEFINITION AND TYPES

Blockchain was introduced in 2008, along with the cryptocurrency Bitcoin [15]. It has quickly caught the attention of the industry and the financial communities. As a secured and decentralized infrastructure, it is broadly recognized as an effective solution to the issues of centralization and security when registering, tracking, managing, supervising, and sharing not just financial transactions but also all other items, including contracts, deeds and titles, medical procedures, birth and death certificates, etc.

The blockchain is a publicly distributed database containing every activity or transaction distributed to all the participants, where the transaction data is permanently recorded (Figure 23.1).

Blockchain gained strength from a crypto protocol that permits all participants to agree on the ledger's status, thus ensuring its security.

The blockchain is a collection of blocks generated and organized in a linked list. Distinct network nodes, called miners, are constantly handling new transactions. Once a transaction is included at the end of this linked list, nobody can change or delete the added block.

The cloud is composed of a cluster of virtual machines. It may be applied successfully to the data structure of the blockchain. In our system, the blockchain is defined as a database that maintains all the file information in block form. It is used as a database for logging that provides the function of auditing the integrity of the data.

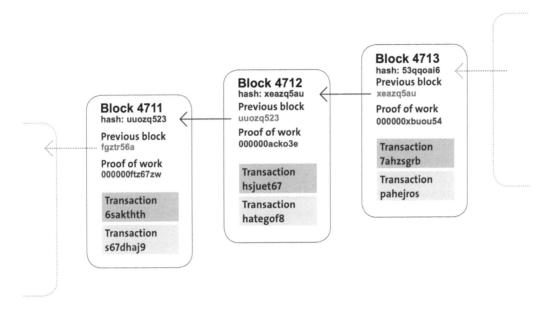

FIGURE 23.1   Structure of the blockchain.

Blockchain has three main types: Private, Public, and Consortium Blockchain [16].

The type of blockchain we have decided to implement in our model is a private blockchain. We use the blockchain database to store metadata about the original file, the MHT root, the leaves, and hashes where access to the database is restricted and requires approval.

Many fields and industries already use blockchain or plan to do so in the coming years. These sectors include insurance, banking, logistics, energy, real estate, aeronautics, health, education, etc. The report [17] details the fields of use of blockchain.

23.3.1   Merkle Hash Tree (MHT)

The MHT is an algorithm created in 1979 by Ralph Merkle [18]. MHT aims to divide the transactions into chronologically ordered blocks and timestamp each block with its corresponding cryptographic hash. The MHT, in bitcoins, consists of one tree per block, which includes all transactions related to the block. Blocks are organized in an orderly chain named blockchain, in which every block refers to the previous one; this is a smart strategy developed by Nakamoto to ensure that entire blocks cannot be deleted or replaced.

Critical parameters of blocks (such as the Merkle root) are gathered in a block header to facilitate blockchain auditing. Hence, ensuring the immutability of transactions is equivalent to ensuring the immutability of block headers. The immutability of transactions is obtained thanks to the properties of hash functions.

23.3.2   Decentralizing Cloud Data via Blockchain

The authors in [19] highlight the need for a decentralized architecture to make storage, retrieval, and file sharing sustainable on a decentralized network. In addition, the authors in [20] demonstrate the effectiveness of using a database based on blockchain to guarantee data integrity in the CC [21].

Like data integrity, storage efficiency is critical when storing user data in CC. In this chapter, we propose an elegant system that maintains data integrity and provides data deduplication via blockchain databases transparently and securely.

23.3.3 DEDUPLICATION PROCESS

Transferring and storing identical data many times in the same storage space is time-consuming and resource-intensive. When a backup solution stores unique data, the data owner can reduce storage space and network load. With deduplication technology, the data owner can realize these savings. Based on the study performed by [22], only 25% of the data can be unique.

The deduplication technique reduces storage costs and network bandwidth by removing duplicate chunks of data when backing up and transferring data to storage servers. Data deduplication is a better approach to ensuring the effectiveness of data storage. This technique can be carried out at either the file or chunk level.

During deduplication, files are divided into file blocks. The uniqueness of these file blocks is checked using different techniques and technologies. Unique files (or parts of files) are sent to the cloud storage, while duplicates are ignored. Deduplication is performed by calculating the hash value of each file or file block. Two different types of deduplication techniques exist: client-side deduplication and server-side deduplication. The first type removes redundant data and retains only one copy of each file or file block before transferring data to the storage server. The second type removes redundant data and retains only a single copy of each file or file block. After that, the data is transferred from the local disk to the CSP server.

Chunking is a complex technique and a key aspect of the deduplication workflow. This can be achieved using a variety of algorithms [23].

23.3.4 MULTI-AGENT SYSTEM

MAS is a set of intelligent agents using resources and knowledge and interacting with their environment to solve a problem or accomplish a goal [24]. The intelligent agent perceives the evolving environmental conditions, acts to influence the environmental conditions, and reasons to interpret the perceptions. It also solves problems, draws inferences, and determines actions [25].

Agents enjoy the following features: Autonomy, Reactivity, Proactivity, and Social [26]. Recently, many researchers have proposed cloud-associated MASs employing knowledge bases to manage data storage [27].

The main challenges that introduce complexity into cloud computing are managing cloud resources and communication and accounting for each user's resource and/or service usage [28]. MAS is widely used to overcome this complexity.

23.4 THE PROPOSED SYSTEM

The proposed system provides data integrity auditing and deduplication using blockchain technology. We call our proposed system B-DID for blockchain-based data integrity and deduplication. Therefore, we are looking to attain the following targets in our suggested system: server-side deduplication, the confidentiality of data, data integrity auditing, and batch auditing.

23.4.1 STORAGE PHASE

Whenever a cloud user attempts to upload a file, the CSP verifies the existence of some file blocks or the whole file. Our method utilizes the blockchain, where each block includes the user ID (U_{id}), file *ID* (F_{id}), the preceding block hash, version number ν, timestamp t, number of file blocks N, and Merkle root. Figure 23.2 illustrates the architecture of our proposal.

The fixed-size chunking module and the MD5 hash algorithm manipulate the deduplication process. The Merkle root and the file-block hashes are calculated and compared to those stored in a local Merkle root database (MRDB) and a hash database (HDB), generated in the preceding operations to detect duplicate files or chunks of files.

The deduplication process is composed of two cases:

Case 1: If the generated Merkle root n_0 does not resemble any root (stored in the MRDB), the new root is stored locally in the MRDB along with the user's U_{id} for use in subsequent storage operations. After that, the generated hashes are compared with those stored in the HDB. There are two possibilities: If there are no identical file-block hashes, the hashes of all calculated file blocks will be stored in the HDB, and all file blocks will be externalized to the cloud storage servers. Otherwise, if there are identical hashes, the identical hashes will be ignored, and the other file-block hashes will be stored in the HDB. Then, only the unduplicated chunks will be stored in the cloud storage servers, which reduces the storage space usage. A new block is then created in the blockchain.

Case 2: If the generated Merkle root n_0 is identical to a root that has already been stored in the MRDB, and the current U_{id} already exists in the list of user *IDs* linked to that n_0. The file owner is informed that it has already stored the same file in a previous storage operation. Otherwise, if the current U_{id} does not exist in the list of user *IDs* linked to this root, the new U_{id} is added to this list, and a new block will be created in the blockchain. In this case, storing the data in the cloud storage servers is unnecessary, reducing storage space usage.

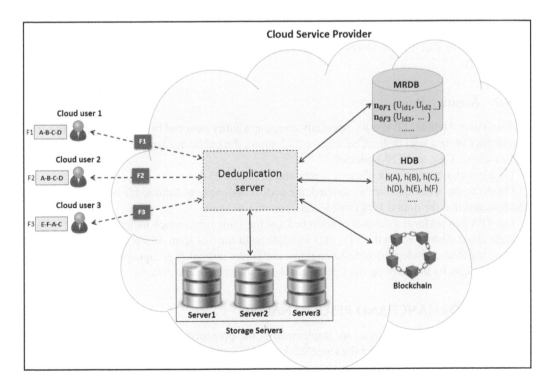

FIGURE 23.2 The proposed system architecture.

The Merkle root is a crucial element based on all the underlying file-block hashes. Consequently, it ensures optimal and secure data content checking.

23.4.2 MAS Applied to Ensure the Deduplication Process

The B-DID model, proposed in the previous subsection, exploits the server-side deduplication technique to identify and delete duplicate data before storing it on the CSP servers. The deduplication process involves a set of tasks that can be run in parallel, especially when thousands of users are simultaneously able to store massive amounts of data. Given this observation and through this subsection, we propose a reliable architecture in which we introduce the use of a MAS to handle server-side deduplication, which will allow reducing the volume of data stored on CSPs' servers.

This new system is motivated by the management of the deduplication process using a set of intelligent agents with a distribution, collaboration, and communication mode that we called BMAS-DID for blockchain and MAS applied for data integrity and deduplication.

We perform data deduplication on the CSP side, wherein a number of autonomous intelligent agents are modeled, and each one has several tasks to perform.

- **Interface agent**: Its role is to receive the file and transmit it to the MA so other agents can use it.
- **Mediator agent**: Manages the communication between the different agents.
- **Analysis agent**: Checks whether the Merkle root and the client ID exist in the MRDB to identify duplicate files. It also performs block-level deduplication by matching the calculated hashes to those stored in the HDB to distinguish duplicate file blocks.
- **Control Agent**: Calculates the Merkle root and forwards it to the AA for comparison. It also stores the file-block hashes in the HDB and the Merkle root with the corresponding user ID in the MRDB.
- **Data agent**: Builds a new block in the blockchain, which includes the corresponding metadata of the file. Then it stores the file blocks corresponding to a file on the CSP storage servers.

23.4.3 Auditing Phase

A Third-Party Auditor (TPA) is an externally competent entity assigned by the cloud client to verify the integrity of data held by the CSP. Figure 23.3 shows the public auditing system where the three entities (client, CSP, and TPA) interact.

The algorithm for the audit phase is mentioned in Figure 23.4.

Thanks to the technique that we applied, the auditor will have no knowledge of the client's data, which means that the data is kept confidential from the auditors.

The TPA can perform a number of audit tasks at the same time, which includes coming in from multiple cloud clients. When the TPA gets multiple audit queries from different users for the same file, treating them as independent tasks could be inefficient instead of grouping them and running a single audit task by intercepting the CSP to check the integrity of the data.

23.5 PERFORMANCE AND RESULT ANALYSIS

We used the following techniques for implementing the proposed system: the Java language, the JADE development platform, and the OpenStack Cloud Platform.

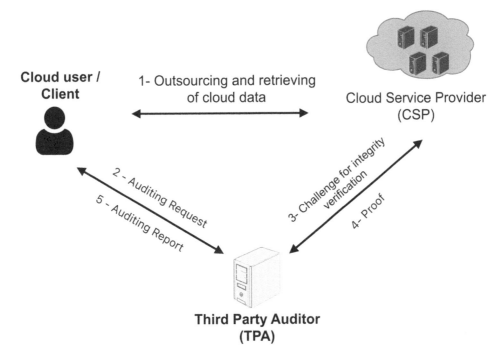

FIGURE 23.3   Participating entities in the public audit model.

1: **Begin**	
2:	*OWN* sends leaves and Merkle Root to *TPA*
3:	*TPA* computes the seed $r = h^P(n_0)$
4:	*TPA* derives leaf numbers in each P chunk as :
5:	$for\ j\ in\ 0..P - 1:\ l_j = G(r, j)$
6:	*TPA* sends leaf numbers $\{l_j\}$ to *CSP*
7:	*CSP* provides the appropriate sibling information to *TPA*
8:	*TPA* computes the new Merkle root n'_0
9:	checks if $n_0 = n'_0$
10:	if they match then the file is verified
11:	*TPA* sends the auditing result to *OWN*
12: **End**	

FIGURE 23.4   Algorithm of the audit phase.

23.5.1   IMPLEMENTATION AND EVALUATION OF THE B-DID MODEL

The application is based on the fixed-size chunking module algorithm, where it accepts a file and a fixed file-block size as input, and then calculates the fixed size of file blocks from the beginning of the file and stores them in a temporary folder. We used the MD5 hash algorithm, which generates a unique 32-character string. The main objective is to decide whether to upload the whole file, some parts of the file, or nothing at all.

In the demonstration, we used large files to demonstrate the benefits of the deduplication solution. Figure 23.5 shows a graph of the computation time versus the size of the input files in MB.

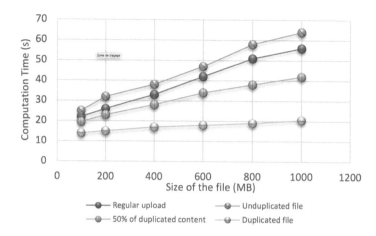

FIGURE 23.5   Comparison of calculated time in storage operations for files of different sizes.

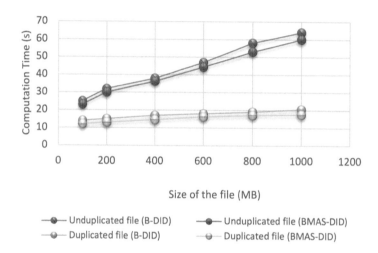

FIGURE 23.6   A plot of computation time versus input file size for the B-DID and BMAS-DID models.

We notice that the computation time for storing a replicated file is less than that for the other scenarios. Each time there are replicas in the file, the execution time decreases, which shows that the volume of storage on the servers is reduced, thanks to the deduplication technique.

23.5.2   IMPLEMENTATION AND EVALUATION OF THE **BMAS-DID** MODEL

The BMAS-DID architecture consists of a set of agents that communicate via messages using the ACL language. Figure 23.6 shows a plot of computation time versus input file size for the two proposed models (B-DID and BMAS-DID), specifically for duplicate and non-duplicate files.

We notice that the computation time for storing a duplicated or unduplicated file for the BMAS-DID model is lower compared to the other scenarios implementing the B-DID model, so the use of MAS via the distribution of the different tasks to a set of intelligent agents saves the execution time of the deduplication process.

Therefore, MAS is well suited to manage the deduplication process in cloud environments effectively.

23.6 CONCLUSION

In this chapter, we proposed a blockchain-based system to ensure data integrity and deduplication via the B-DID model. After that, we compared this model with the one integrating multi-agent systems (BMAS-DID). Then, we stated the importance of integrating MAS in the deduplication process, mentioning that both models ensure data integrity auditing.

To show the reliability of our models, we presented the implementation scenarios with the different implemented components and discussed the obtained results. The results demonstrated that our models are realistic and valuable for ensuring deduplication and data integrity in a complex and dynamic environment like cloud computing. The added value of our models over others is that our proposals allow for both data deduplication and data integrity checking.

Using blockchain in our solution is especially suitable for situations where historical data is very important, such as justice, real estate agents, medical records storage, or tax collection.

REFERENCES

1. Ateniese, G., Burns, R., Curtmola, R., Herring, J., Kissner, L., Peterson, Z., Song, D.: Provable data possession at untrusted stores. *Proc. 14th ACM Conf. Comput. Commun. Secur. – CCS'07*, p. 598 (2007). https://doi.org/10.1145/1315245.1315318.
2. Juels, A.., Kaliski Jr., B.S.: Pors: Proofs of retrievability for large files. *Proc. ACM Conf. Comput. Commun. Secur.*, pp.584–597 (2007). https://doi.org/10.1145/1315245.1315317.
3. Erway, C.C., Küpçü, A., Papamanthou, C., Tamassia, R.: Dynamic provable data possession. *ACM Trans. Inf. Syst. Secur.* 17, 1–29 (2015). https://doi.org/10.1145/2699909.
4. Wang, Q., Wang, C., Ren, K., Lou, W., Li, J.: Enabling public auditability and data dynamics for storage security in cloud computing. *IEEE Trans. Parallel Distrib. Syst.* 22, 847–859 (2011). https://doi.org/10.1109/TPDS.2010.183.
5. Liu, C., Chen, J., Yang, L.T., Zhang, X., Yang, C., Ranjan, R., Rao, K.: Authorized public auditing of dynamic big data storage on cloud with efficient verifiable fine-grained updates. *IEEE Trans. Parallel Distrib. Syst.* 25, 2234–2244 (2014). https://doi.org/10.1109/TPDS.2013.191.
6. Zhu, Y., Ahn, G.J., Hu, H., Yau, S.S., An, H.G., Hu, C.J.: Dynamic audit services for outsourced storages in clouds. *IEEE Trans. Serv. Comput.* 6, 227–238 (2013). https://doi.org/10.1109/TSC.2011.51.
7. Tian, H., Chen, Y., Chang, C.C., Jiang, H., Huang, Y., Chen, Y., Liu, J.: Dynamic-hash-table based public auditing for secure cloud storage. *IEEE Trans. Serv. Comput.* 10, 701–714 (2017). https://doi.org/10.1109/TSC.2015.2512589.
8. Yu, Y., Xue, L., Au, M.H., Susilo, W., Ni, J., Zhang, Y., Vasilakos, A. V., Shen, J.: Cloud data integrity checking with an identity-based auditing mechanism from RSA. *Futur. Gener. Comput. Syst.* 62, 85–91 (2016). https://doi.org/10.1016/j.future.2016.02.003.
9. Lee, K.M., Lee, K.M., Lee, S.H.: Remote data integrity check for remotely acquired and stored stream data. *J. Supercomput.* 74, 1182–1201 (2018). https://doi.org/10.1007/s11227-017-2117-4.
10. Li, J., Chen, X., Xhafa, F., Barolli, L.: Secure deduplication storage systems supporting keyword search. *J. Comput. Syst. Sci.* 1, 1–10 (2015). https://doi.org/10.1016/j.jcss.2014.12.026.
11. Miao, M., Wang, J., Li, H., Chen, X.: Secure multi-server-aided data deduplication in cloud computing. *Pervasive Mob. Comput.* 24, 129–137 (2015). https://doi.org/10.1016/j.pmcj.2015.03.002.
12. Zhang, Y., Feng, D., Jiang, H., Xia, W., Fu, M., Huang, F., Zhou, Y.: A fast asymmetric extremum content defined chunking algorithm for data deduplication in backup storage systems. *IEEE Trans. Comput.* 66, 199–211 (2017). https://doi.org/10.1109/TC.2016.2595565.
13. Khanboubi, Y. El, Hanoune, M., Ghazouani, M. El: A new data deletion scheme for a blockchain-based de-duplication system in the cloud. *Int. J. Commun. Networks Inf. Secur.* 13, 331–339 (2021). https://doi.org/10.54039/ijcnis.v13i2.4975.
14. Jiang, T., Yuan, X., Chen, Y., Cheng, K., Wang, L., Chen, X., Ma, J.: FuzzyDedup: Secure fuzzy deduplication for cloud storage. *IEEE Trans. Dependable Secur. Comput.* (2022). https://doi.org/10.1109/TDSC.2022.3185313.
15. Nakamoto, S.: Bitcoin: A peer-to-peer electronic cash system. www.Bitcoin.Org. 9 (2008). https://doi.org/10.1007/s10838-008-9062-0.

16. Maleh, Y., Lakkineni, S., Tawalbeh, L. and AbdEl-Latif, A.A.: Blockchain for cyber-physical systems: Challenges and applications. *Advances in Blockchain Technology for Cyber Physical Systems* (2022): 11–59. https://doi.org/10.1007/978-3-030-93646-4_2.

17. Du, C., Mission, P.A.R.L.A., Commune, I.: RAPPORT D' INFORMATION, pp. 1–137 (2018).

18. Merkle, R.C.: Information Systems Laboratory By (1979).

19. Zikratov, I., Kuzmin, A., Akimenko, V., Niculichev, V., Yalansky, L.: Ensuring data integrity using blockchain technology, In *2017 20th Conference of Open Innovations Association (FRUCT)* (pp. 534–539). IEEE (2017).

20. Gaetani, E., Aniello, L., Baldoni, R., Lombardi, F., Margheri, A., Sassone, V.: Blockchain-based database to ensure data integrity in cloud computing environments. *CEUR Workshop Proc.* 1816, pp. 146–155 (2017).

21. El Ghazouani, M., My, A., Kiram, E., Er-rajy, L., Khanboubi, Y. El: Efficient Method Based on Blockchain Ensuring Data Integrity Auditing with Deduplication in Cloud, 1–7 (2020). https://doi.org/10.9781/ijimai.2020.08.001.

22. Digital, T., Decade, U.: - I V I E W The Digital Universe Decade - Are You Ready? 2009, 1–16 (2020).

23. Venish, A., Siva Sankar, K.: Study of Chunking Algorithm in Data Deduplication. *Proc. Int. Conf. Soft Comput. Syst. Adv. Intell. Syst. Comput.*, vol. 398, pp. 319–329 (2016). https://doi.org/10.1007/978-81-322-2674-1.

24. El Ghazouani, M., El Kiram, M.A., Er-Rajy, L.: Blockchain & multi-agent system: A new promising approach for cloud data integrity auditing with deduplication. *Int. J. Commun. Networks Inf. Secur.* 11, 175–184 (2019).

25. Ikidid, A., El Fazziki, A., Sadgal, M.: A fuzzy logic supported multi-agent system for urban traffic and priority link control. *J. Univers. Comput. Sci.* 27, 1026–1045 (2021). https://doi.org/10.3897/jucs.69750.

26. Wooldridge, M., Jennings, N.R.: Intelligent agents: Theory and practice. *Knowl. Eng. Rev.* 10, 115–152 (1995).

27. Ikidid, A., Fazziki, A. El, Sadgal, M.: A multi-agent framework for dynamic traffic management considering priority link. *Int. J. Commun. Networks Inf. Secur.* 13, 324–330 (2021). https://doi.org/10.54039/ijcnis.v13i2.4977.

28. Bajo, J., De la Prieta, F., Corchado, J.M., Rodríguez, S.: A low-level resource allocation in an agent-based Cloud Computing platform. *Appl. Soft Comput. J.* 48, 716–728 (2016). https://doi.org/10.1016/j.asoc.2016.05.056.

24 Securing Crowdfunding Platforms with Blockchain to Boost the Real Estate Sector

Hibatou Allah Boulsane, Karim Afdel, and Salma El Hajjami

24.1 INTRODUCTION

The Moroccan economy is largely based on the real estate industry, and the country and its residents place great importance on real estate purchases. In Morocco, there are various types of real estate: All structures erected with the intention of housing people are considered to be residential real estate, including apartments, villas, buildings, etc. "business real estate" refers to a broad category of structures not used for habitation but for professional, commercial, or industrial activity (office floors, commercial premises, business premises, and warehouses). Leisure and tourist structures, such as hotels, are included in the category of tourism real estate. Several Moroccan real estate market issues include the inability to track transactions between real estate promoters, real estate funds, investors, and clients. The hiring of an intermediary (a human or a virtual entity) and the high cost of the recruiting procedure (real estate agency fees) further increase the length and cost of transactions. The biggest issue is that most Moroccans do not have enough money to invest in real estate. The real estate industry has significantly benefited from Fintech (financial technology). References [1,2] consider that, with its unique taxonomy, Fintech refers to a wide range of company or organizational activities that use information technology (IT) applications to enhance the quality of their services. The development of Fintech features cutting-edge technologies in numerous fields, such as mobile networks, big data, etc. It has been made possible by the ongoing rise of investment.

We might think of three main stages in the evolution of financial services in Fintech: Fintech 1.0 is the stage of digitalizing existing things. Platforms that mix many service kinds, such as those provided by applications (WeChat, Paytm), are referred to as Fintech 2.0. Fintech 3.0, which is based on the incorporation of blockchain technology, is the final category. Blockchain technology [3–9] is a collection of decentralized databases and systems that offer users precise, reliable, traceable, and secure services for minimal transaction costs and without human interference. The blockchain system uses P2P networks, cryptographic technology, proof-of-work, and proof-of-stake consensus models to provide infrastructure for decentralized applications. Crowdfunding [10–13] is a type of financial technology that enables many people, individuals, and businesses to contribute to funding a single project. The innovative and inventive phenomenon of crowdfunding in entrepreneurial financing allows project owners to ask for cash from investors. To determine whether potential customers are interested in the crowdfunding campaign's offer, it may also be utilized as a market test.

This study aims to demonstrate the effects of blockchain-based crowdfunding on real estate developments. A crowdfunding platform can lower the cost of financing real estate projects and make it easier for many Moroccan individuals who struggle with housing financing to purchase a home. We present a blockchain-based real estate management paradigm in this context to enable a more dependable and transparent real estate sector. Then, we analyze our model's technical and practical contributions to this research work.

DOI: 10.1201/9781032667478-28

The rest of the work is presented as follows: Section 24.2 comprises related work. Section 24.3 explains our funding approach and the planned application's architecture. Section 24.4 summarizes the outcome. The last section is the conclusion.

24.2 RELATED WORK

Various methodologies pertaining to this topic are discussed in this section. The authors indicate in [14] that real estate transactions are costly and complicated due to the number of middlemen engaged in the process. These intermediaries include escrow firms, inspectors, appraisers' notaries, and others. In addition, blockchain technology enables an effective and dependable workflow, increases the transparency and visibility of all phases, and assists investors in obtaining a history of real estate transactions.

Creta et al. [15] discussed the potential applications of blockchain technology and crowdfunding in the real estate industry. This article covers a number of topics, one of which is the best approach to make the real estate market more appealing. They suggested using a technology called blockchain to bolster and revitalize this industry. They mentioned that the most significant benefit of using blockchain technology in the real estate sector through crowdfunding is the reduced need for a centralized registry, which will be replaced by a distributed registry of real estate using digital property titles. This was cited as the most important advantage of using blockchain technology.

According to Morena et al. [16], several nations, including the Netherlands, are building their first applications on blockchain technology to document lease contracts in the real estate industry. This project intends to make it possible to digitize building data, digitize ownership status, complete rental contracts, and unlock contractual information for third parties. Due to blockchain, all building information is gathered. The project also wants to make it feasible to digitize building data.

A decentralized real estate transaction application built on blockchain was proposed by Yang et al. [17]. The five levels of this platform are as follows: the web application layer serves as the platform's interface; the smart contract layer. The connected smart contract is activated whenever the page requests a query or the storage of data. The proof of elapsed time and the PBFT algorithm are included in the third tier, which is the network layer. The P2P protocol is the data transfer mechanism between the nodes at this layer. And last but not least, the data layer employs blocks as the primary method for storing data such as the date and time of a real estate transaction and the names of the buyers and sellers.

Because of the difficulties caused by the complexity and lack of security of the network in various industries, Kumar et al. [18] proposed a platform for buyers and sellers that is based on the Ethereum blockchain and that enables the use of smart contracts to store data related to land, users, and transactions between customers and other participants. This platform would allow smart contracts to store data on land, users, and transactions between customers and other participants.

Jadye et al. [19] established two distinct forms of smart contracts. The first type of smart contract is a generator, and it deploys the second type of smart contract, which records the whole campaign progression.

Nik Ahmed et al. [20] have shown how blockchain technology might improve consumer confidence in crowdfunding sites by decreasing the prevalence of fraudulent activity and asymmetric information. They employ smart contracts to ensure that all transactions are open and transparent and keep track of where and how money is spent.

Garcia et al. [21] discussed the roles of intermediaries' real estate in the European Union and investigated how blockchain technology might improve the safety of transactions while also lowering the prices and timeframes associated with these transactions. They got rid of the notary to get rid of the middleman, but now they have a problem authenticating the formal identification of parties. In addition, they should include a legal phrase to control the legality of the situation.

Tilbury et al. developed two different real estate conceptual models, one for South Africa and one for the international real estate transaction process, based on blockchain technology.

They emphasized five areas of use for blockchain in the real estate industry, including smart contracts, costs, speed, preserving records in an immutable manner, and transparency. On the other side, they highlighted that trust and support from stakeholders need to be created, that getting proper information translated into blockchain is a difficulty that has to be overcome, and that to implement blockchain technology internationally, a legal framework for transactions is essential.

Blockchain, which is the subject of all of these articles and is believed to have huge promise for the financial industry, particularly for crowdfunding, is the primary focus of this discussion. However, the most important subject that is not discussed in any of these publications is how to incorporate legal provisions into blockchain-based crowdfunding in such a way as to guarantee that customers will pay less for real estate and will receive accurate information from all participants in the system [22]. Implementing blockchain technology in the real estate industry to streamline work processes and, in particular, cut expenses, reduce the amount of time it takes to finish each transaction, and strengthen this system with legal provisions is the primary objective of this article. The proposed method calls for creating a real estate financing mechanism based on crowdsourcing and blockchain technology. Additionally, some level of transaction transparency, traceability, and security must be achieved through homomorphic encryption. To put this concept into action, we developed a web platform in the form of a Dapp (which stands for "decentralized application") that customers may use to buy or rent real estate even if they do not have the required amount of money.

24.3 PROPOSED MODEL

24.3.1 How Blockchain Works

The blockchain system was created to operate without a central authority (no bank or organization that controls transactions) where the transactions must be authenticated. Thanks to cryptographic keys, a data string identifies a user and gives him access to his account and wallet. Each member of the system has a digital identity called a digital signature. With this signature, the user can authenticate himself and unlock the transaction he wants to perform. After the transaction is unlocked, it must be approved and authorized before it can be added to a block. Figure 24.1 shows the steps to add a transaction to the blockchain.

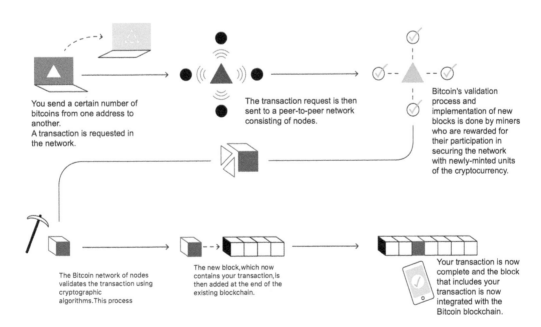

You send a certain number of bitcoins from one address to another.
A transaction is requested in the network.

The transaction request is then sent to a peer-to-peer network consisting of nodes.

Bitcoin's validation process and implementation of new blocks is done by miners who are rewarded for their participation in securing the network with newly-minted units of the cryptocurrency.

The Bitcoin network of nodes validates the transaction using cryptographic algorithms. This process

The new block, which now contains your transaction, is then added at the end of the existing blockchain.

Your transaction is now complete and the block that includes your transaction is now integrated with the Bitcoin blockchain.

FIGURE 24.1 Steps to add a transaction to the Blockchain.

In a public blockchain, the transaction is added after the consensus. The members connected to the network will have to verify the transactions with rewards (proof of work). The proof of work requires that the network members solve a complex mathematical problem to add a block to the chain. Solving this problem is called mining, and miners are paid for their work. But mining is difficult because the mathematical problem can only be solved by trial and error, and the chances of solving it are about 1 in 5.9 trillion. It requires considerable computing power. On top of that, the remuneration must be higher than the cost of computers and the electricity needed to run them.

24.3.2 ETHEREUM DAPP

A Dapp (decentralized application) can be a website or a simple application, but Dapps differ from traditional applications because they are built on a decentralized network. When creating smart contracts on the Ethereum blockchain, this code is a piece of Dapps' backend code, as shown in Figure 24.2.

The architecture of Ethereum Dapp is composed of three layers:

1. Client web browser: the frontend interface that allows users to engage with smart contracts, one another, and other users. To put the smart contracts into action. You must use the Metamask wallet with the WEB3.js framework to construct them.
2. The Web3 Provider is the component that acts as a go-between between the front end (web3js) and the back end (smart contract). It is necessary for the proper operation of the Dapp that these two components be connected by it.

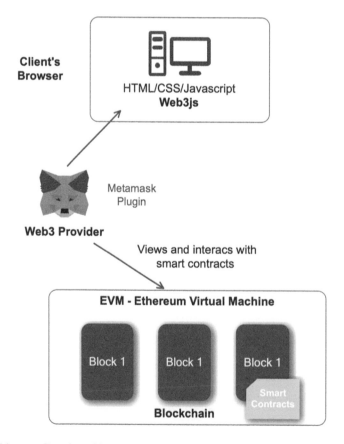

FIGURE 24.2 Ethereum Dapp's architecture.

3. Ethereum Virtual Machine: the Ethereum blockchain supplies a runtime to execute the code of the smart contract and then compile it in BYTECODE. This process is known as the Ethereum Virtual Machine. This latter is what Web3js employs to interface with smart contracts.

24.3.3 HOMOMORPHIC ENCRYPTION

Data can be encrypted using homomorphic encryption, which permits data manipulation without first decoding it. Enabling specific computations to be done on encrypted data without revealing the underlying data itself can be used to offer privacy in blockchain transactions.

Several studies have looked into the idea of homomorphic encryption and its uses. For instance, the first fully homomorphic encryption (FHE) system was published in 2009 by Gentry's seminal work, allowing arbitrary computations to be performed on encrypted data. Since then, several homomorphic encryption systems have been put out, such as the BGV, CKKS, and FHEW schemes.

FHE and partially homomorphic encryption are the two primary varieties of homomorphic encryption (PHE). While PHE only permits certain kinds of computations on encrypted data, FHE allows unrestricted computations to be performed on encrypted data. While FHE is more powerful than PHE, it also requires more computation.

It is also possible to share data securely between multiple parties using homomorphic encryption. Parties can work together on computations while keeping the underlying data private by encrypting data. When numerous parties need to compute a shared set of data but are unable to trust one another with the data, this can be helpful.

Blockchain and homomorphic encryption are two significant technologies that can offer safe and private data processing across various domains. Blockchain can provide a decentralized and tamper-proof ledger for storing data, while homomorphic encryption can be used to conduct computations on encrypted data. When these technologies are combined, decentralized ecosystems like blockchain-based applications may handle data securely and privately.

Ultimately, a strong, secure, and private data processing tool in decentralized situations can be created by fusing homomorphic encryption and blockchain. To make this technology more useful for a more extensive range of applications, ongoing research is concentrated on creating new strategies and optimizing existing ones.

24.3.4 THE PROPOSED MODEL ARCHITECTURE

Whether it be the investment fund, the final client, the investor, or the real estate promoter, our proposed system requires us to decide to adopt a precise and intelligent contractual relationship between all of the actors who are a part of the blockchain network by stating their position. This decision must be made to implement our system. Every transaction will employ smart contracts, from investor funding to customer purchases to subcontracting with real estate promoters. This includes all of the transactions. The conceptual model of the real estate financing process is depicted in Figure 24.3.

1. The investors decide on the most appropriate project for their money to be invested in, and they also select their participation level. The contract is immediately generated on the crowdfunding site and named "Modarabah [23]."
2. Once we have collected the appropriate cash for the investment, the fund will put out a call for offers on our website to entice clients interested in renting or purchasing a property in the area.
3. The purchaser is responsible for selecting the desired home from the crowdfunding site. After that, he will choose the kind of contract, such as a "Murabaha" [24], an "Ijara Montahiya bitamlik" [25], or a straightforward standard sales contract.

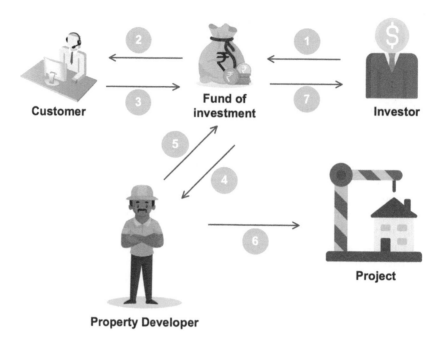

FIGURE 24.3   The proposed real estate finance process model.

4. After more than half of the customers have committed to purchasing, building will begin after the investment cash has been collected in full. The real estate investment fund will call for applications and provide prospective candidates with comprehensive information and specifics on the project.
5. The real estate promoter searches for projects compatible with his financial constraints, his projections of manageable expenditures, and the standing of his firm. An "Istisnaa" [26] will be established between the investment fund and the real estate promoter in the capacity of a subcontracting contract for the investment fund to initiate the house-building works. This contract will oversee the operation of the house-building works.
6. The promoter will turn the project over to the investment fund after it has achieved its final state. At that point, the sale and the transfer of ownership to the final clients will begin following the criteria that have been agreed upon.
7. After this sale transaction is completed, the generated profits will be transferred to the real estate fund. The real estate fund will then transfer the capital and the profits to the investors, per the return percentages discussed and agreed upon prior to the sale.

24.4   RESULTS AND DISCUSSION

24.4.1   IMPLEMENTATION OF THE MODEL

To implement this model as a Dapp, we created several smart contracts using the Vyper language. We used Web3.js, which is a Javascript library that allows interaction with the Ethereum blockchain, the Metamask browser extension that plays the role of a cryptocurrency wallet that connects to the Ethereum blockchain, the Ethereum Ganache development tool, which allows us to make a local simulation, and to test the deployed smart contracts. Finally, IPFS (Interplanetary File System) is a file system that allows the storage and follow-up of the versions in the tamps within a distributed network. We use Pallier as a homomorphic encryption scheme compatible with Ethereum to encrypt and manipulate data in a homomorphically encrypted form.

24.4.2   RESULT ANALYSIS

In this section, we will provide the results of a study that compares the costs of traditional financing with the costs of funding using blockchain technology (Table 24.1).

You want an apartment of $70\,m^2$ with an acquisition price of 800 000 MAD (Table 24.2).

240 monthly payments of 5233 MAD result in a total amount due of 1,256,005 MAD, indicating a profit margin of 456,005 MAD (Table 24.3).

Except for the fees associated with the real estate agent, acquisition costs are routinely incorporated into the transaction price. The technology behind blockchain does away with the need for intermediaries. We may arrive at this conclusion by contrasting the overall purchase cost when using credit with the total cost when utilizing blockchain. The proportion of gain for the consumer is determined to be 38% (Tables 24.4 and 24.5).

24.4.3   DISCUSSION

To guarantee the security, traceability, and transparency of the financial transactions that take place between the various system participants as well as to create a legal set of rules by applying contracts for rental and sale, this study will utilize blockchain technology for crowdfunding and add homomorphic encryption to enhance privacy in blockchain. This comparison analysis indicates that blockchain technology must unquestionably be incorporated into real estate financing to reduce housing finance expenses by doing away with all middleman fees and additional purchase charges.

TABLE 24.1

Benchmarking the Different Modes of Real Estate Financing

Payment Method	Different Charges
Real estate credit	The purchase price, Application fees, Guarantee-mortgage fees, Borrower's insurance, Additional loan, Expenses related to installation, Notary fees, and Agency fees.
Funding by "Murabaha"	The purchase price, administration fees, guarantee fees, installation fees, notary fees, and agency fees.
Own fund	The purchase price, notary fees, agency fees, and registration fees.
Online shopping blockchain	Purchase price.

TABLE 24.2

The Acquisition Cost Using the Real Estate Credit

Property Price	800 000 MAD
Notary fees	8400 MAD
File fees	800 MAD
Interest	414 687 MAD
Loan insurance	59 200 MAD
Total cost	1 283 087 MAD

TABLE 24.3

The Acquisition Cost Using the Murabaha Contract

Property price	800 000 MAD
Monthly payment	8400 MAD
Financing period	20 years
Profit margin	456 005 MAD

TABLE 24.4

The Acquisition Cost Using Our Funds

Property Price	16000 MAD
Registration fee	8400 MAD
Notary	8000 MAD
File fees	700 MAD
Agency fees	20 000 MAD
Total cost	844 700 MAD

TABLE 24.5

The Acquisition Cost Using Blockchain

Total cost	800 000 MAD

The key finding of our study is that legal contracts were added to the blockchain system based on crowdfunding to control the real estate transaction process and increase the system's security. All parties engaged in the lease or purchase must sign these contracts, which the blockchain network must validate.

24.5 CONCLUSION

Innovative digital financial technologies like blockchain and crowdfunding have recently emerged. These technologies point to new methods of designing financial procedures that do not involve intermediaries and increase the level of trust between various stakeholders by utilizing powerful cryptographic algorithms to guarantee the security and coherence of the financial system. In light of the foregoing, this chapter aims to provide a strategy based on blockchain technology that addresses problems relating to liquidity, security, financial disintermediation, and transparency. We contrasted the overall cost of acquiring credit with the total cost of utilizing the blockchain to evaluate the efficacy of the suggested approach. Due to the removal of any forms of intermediation and any additional purchase charges, we found that the cost of home finance had significantly decreased.

REFERENCES

1. Gai, K., Qiu, M., and Sun, X. (2018, févr.). A survey on FinTech. *J. Netw. Comput. Appl.*, 103, 262–273. doi: 10.1016/j.jnca.2017.10.011.
2. Bollaert, H., Lopez-de-Silanes, F., and Schwienbacher, A. (2021). Fintech and access to finance. *J. Corp. Finance*, 68, 101941.
3. Zheng, X. R. and Lu, Y. (2021, juin). Blockchain technology - recent research and future trend, *Enterp. Inf. Syst.*, 1–23. doi: 10.1080/17517575.2021.1939895.
4. Shrimali, B. and Patel, H.B. (2021, août). Blockchain state-of-the-art: architecture, use cases, consensus, challenges and opportunities, *J. King Saud Univ. - Comput. Inf. Sci.*, S131915782100207X. doi: 10.1016/j.jksuci.2021.08.005.
5. Hewa, T., Ylianttila, M., Liyanage, M. (2021). Survey on Blockchain based smart contracts: Applications, opportunities and challenges. *J. Netw. Comput. Appl.*, 177, 102857.
6. Rajasekaran, A.S., Azees, M., and Al-Turjman, F. (2022, août). A comprehensive survey on blockchain technology, *Sustain. Energy Technol. Assess.*, 52, 102039. doi: 10.1016/j.seta.2022.102039.
7. Aggarwal, S., Kumar, N. (2021). Blockchain 2.0: smart contracts. In *Advances in Computers* (Vol. 121, pp.301–322). Elsevier. https://doi.org/10.1016/bs.adcom.2020.08.015.
8. Cai, W., Wang, Z., Ernst, J. B., Hong, Z., Feng, C., Leung, V. C. (2018). Decentralized applications: The blockchain-empowered software system. *IEEE Access*, 6, 53019–53033.

9. Mohanta, B. K., Panda, S. S., Jena, D. (2018, July). An overview of smart contract and use cases in blockchain technology. In *2018 9th International Conference on Computing, Communication and Networking Technologies (ICCCNT)* (pp. 1–4). IEEE.

10. Cai, W., Polzin, F. and Stam, E. (2021, janv). Crowdfunding and social capital: A systematic review using a dynamic perspective. *Technol. Forecast. Soc. Change*, 162, 120412. doi: 10.1016/j.techfore.2020.120412.

11. Böckel, A., Hörisch, J., and Tenner, I. (2021, avr.). A systematic literature review of crowdfund- ing and sustainability: highlighting what really matters. *Manag. Rev. Q.*, 71(2), 433–453. doi: 10.1007/s11301-020-00189-3.

12. Alhammad, M. M., AlOthman, R., and Tan, C. (2021, déc.) Review of crowdfunding regulations across countries: A systematic review study. *J. Inf. Syst. Eng. Manag.*, 6(4), em0145. doi: 10.21601/jisem/11395.

13. Belleflamme, P., Omrani, N., Peitz, M. (2015). The economics of crowdfunding platforms. *Inf. Econ. Policy*, 33, 11–28.

14. Baiod, W., Light, J., and Mahanti, A. (2020). Blockchain technology and its applications across multiple domains: A survey, 43. doi: 10.58729/1941-6679.1482.

15. Creta, F. and Mazaj, J. (2021, avr.). Can fintech progress the real estate sector? The disruptive role of crowdfunding blockchain: A systematic literature review. *Eur. J. Islam. Finance*, 17, Art. no 17. doi: 10.13135/2421-2172/5323.

16. Morena, M., Truppi, T., Pavesi, A. S., Cia, G., Giannelli, J., and Tavoni, M. (2020, janv.). Blockchain and real estate: Dopo di Noi project. *Prop. Manag.*, 38(2), 273–295. doi: 10.1108/PM-01-2019-0005.

17. Yang, L. and Wang, J. (2020, avr.). Research on real estate transaction platform based on blockchain technology. *J.Phys. Conf. Ser.*, 1486(7), 072074. doi: 10.1088/1742-6596/1486/7/072074.

18. Kumar, P., Dhanush, G. A., Srivatsa, D., Nithin, A., and Sahisnu, S. (2019, June). A buyer and seller's protocol via utilization of smart contracts using blockchain technology. In *International Conference on Advanced Informatics for Computing Research* (pp. 464–474). Springer, Singapore.

19. Jadye, S., Chattopadhyay, S., Khodankar, Y., and Patil, D. N. (2021). Decentralized crowd- funding platform using ethereum blockchain technology. *Int. Res. J. Eng. Technol.*, 08(04), 6.

20. Nik Ahmad, N. A. and Syed Abdul Rahman, S. A. H. (2021, July). Applying ethereum smart contracts to blockchain-based crowdfunding system to increase trust and information symmetry. In *2021 7th International Conference on Computer Tech- nology Applications* (pp. 53–59).

21. Garcia-Teruel, R. M. (2020). Legal challenges and opportunities of blockchain technology in the real estate sector. *J. Prop. Plan. Environ. Law.* doi: 10.1108/JPPEL-07-2019-0039.

22. Tilbury, J. L., de la Rey, E., van der Schyff, K. (2019, August). Business process models of Blockchain and South African real estate transactions. In *2019 International Conference on Advances in Big Data, Computing and Data Communication Systems (icABCD)* (pp. 1–7). IEEE.

23. Rammal, H., Gulzar, G., Graeme, P. (2003). Mudaraba in Islamic finance: Principles and application (Doctoral dissertation, Franklin Publishing Company).

24. Miah, M. D., Suzuki, Y. (2020). Murabaha syndrome of Islamic banks: A paradox or product of the system? *J. Islam. Account. Bus. Res.* doi: 10.1108/JIABR-05-2018-0067.

25. Fares, L. I., Habachi, M. (2020). Contrat Ijara: cadre juridique et comptable marocain et rapprochement avec le crédit-bail. *Recherches et Applications en Finance Islamique (RAFI)*, 4(2), 230–251.

26. Al-Bashir, M., Al-Amine, M. (2001). Istisna and its application in Islamic Banking. *Arab LQ*, 16, 22.

25 Blockchain Technology
A Proposed Solution to Hike the Tax-to-GDP Ratio in Bangladesh

Prianka Ghosh Puja and Md Sadik Adnan

25.1 INTRODUCTION

Although Bangladesh is one of the fastest-growing economies in South Asia, according to the World Bank, the economic contribution of the portion as taxation to improve society's lifestyle is not on an uptrend. Instead, it is in the lower stage, which portrays a significant detachment among all countries in South Asia. The tax-to-GDP ratio represents the tax revenue compared to the comprehensive economic growth of a nation. In other words, the tax-to-GDP ratio can describe how well the government can mobilize its economic growth in terms of tax contributions.

Bangladesh, being addressed as a developing and emerging country, is historically vulnerable due to tax evasion and avoidance activities by taxpayers. Overall economic growth cannot determine the benefit to citizens. Moreover, society alone, the benefactor, can be assured that the government can arrange the utilization of the economic rise for society through a linear process from GDP to TAX.

This chapter identifies the problems behind Bangladesh's low tax-to-GDP ratio, which is the most improper and unorganized documentation process for direct tax and complicated indirect taxation. As a result, tax evasion, fraud, and a lack of transparency occurs. There is a knowledge gap and misunderstanding regarding the acceptable process for solving tax evasion and crime in Bangladesh. Due to this reason, there are considerable opportunities for tax fraud. As a result, the tax-to-GDP ratio declines in Bangladesh.

According to Kiabel and Nwokah (2009), every tax system suffers from the issue of tax evasion and avoidance. While in OECD countries, tax revenues are almost 35% of GDP, in the case of Bangladesh, they are only near 10%. While considering the OECD countries, we found that the tax-to-GDP ratio is even higher in Bangladesh compared to other Asian countries like India, Nepal, and Pakistan, which generate tax-to-GDP ratios of 16.8%, 23.1%, and 11.0%, respectively. According to data, the average tax-to-GDP ratio of developing countries is 15%, whereas this percentage is low, just 9.3%, for Bangladesh.

The taxation process in developing countries like Bangladesh is complex due to a lack of transparency and corruption; hence, the public is not interested in providing tax and even finds scope to commit fraud, resulting in a low tax-to-GDP ratio. Already, an alarming situation arises, and to enhance the interest of citizens in contributing to society, this analysis will help find a proper track that is transparent, cost-effective, and secure.

Various research has been done on blockchain and taxation. However, no constructive research has been done for Bangladesh on blockchain technology, the tax-to-GDP ratio, or direct and indirect taxation (Akeo, 2018). So, this chapter will fill the gap and provide information for further research on this topic.

The study is organized into five sections. Section 25.2 presents a comprehensive literature review of the existing research on individual investors and their investment behaviors, focusing on short-term, aggressive trading strategies. Section 25.3 outlines the methodology used in the study, including the research design, sample selection, and data collection and analysis techniques. Section

DOI: 10.1201/9781032667478-29

25.4 analyzes the reasons behind the reduced returns experienced by individual investors engaging in short-term, aggressive trading behaviors and identifies key factors contributing to these losses. Section 25.5 presents the study's conclusions, including potential solutions for mitigating the risks associated with short-term, aggressive trading strategies. The findings of this study will be of interest to individual investors, financial advisors, and policymakers.

25.2 LITERATURE REVIEW

Several studies on blockchain implementation in taxation and tax-to-GDP ratio analysis exist. Most are on relationship analysis of tax revenue and GDP, including analyzing developing countries' tax-to-GDP ratio data. However, no prior research has been found on blockchain technology to improve the tax-to-GDP ratio of developing countries like Bangladesh. Some critical literature on blockchain technology and the tax-to-GDP ratio is reviewed.

Yayman (2021) investigated the use of blockchain technology in taxation in his paper, specifically in Turkey, where he mentioned the effects of blockchain on Corporate Taxpayers, State Tax Management, and VAT. The paper added that the government could gain data from companies' periodic transactions in a more organized and easy way to maintain the direct tax. The government and the company get advantages from the indirect taxation process, where the procedure becomes more transparent and faster by providing real-time checks. In addition, blockchain will be a time-saving factor for the government and taxpayers and reduce tax disputes. However, the report said blockchain is not the only remedy for the taxation system. Rather, it is a potential facilitator to enhance transparency and reduce administrative burden.

Hima (2018) contemplated a question regarding the ability of blockchain technology to change the taxation process in her paper on the topic of blockchain in taxation. She found out that artificial intelligence can make the whole process more reliable by reducing the involvement of humans, and data manipulation will also be restricted. Hence, the entire process will be more authentic. While discussing the implementation of blockchain, the study says that blockchain is a comprehensive structure. Therefore, it is only possible when the corporate and government sectors enter the same interest group. Ryakhovsky et al. (2021) analyzed the hypothesis that blockchain can serve a digital structured tax administration where they use data from 33 experts such as employees of tax authorities and IT companies, and the result was in percentage. 75% consider blockchain as a source of smart contract holders, an easy path to transfer pricing, and a way boost economic efficiency.

Furthermore, 65% think blockchain will be a secure process, and 80% mark it as a transparent technology for tax administration. On the other hand, for VAT, blockchain gets 80% support for its real-time mode, and 75% think it will help to reduce duplicate records. And finally, the report confirmed the hypothesis and stated that FinTech tools could be applied to all stages of tax administration and make the process smoother.

Faccia and Mosteanu (2019) proposed blockchain to solve tax evasion and fraud. They mentioned that the main benefits of blockchain in taxation are human error reduction, hard manipulation of the system, automation techniques, cost savings, and efficiency. In addition, blockchain will help to reduce complicated tax obligations, provide proper tax planning, create positive collaboration with companies that would agree to fairness and acceptance below the 10% rate, provide an easy solution to generate corporate tax, and provide real-time verification, which will add value to downturn tax evasion.

Mansur, Yunus, and Nandi (2011) stated in their report that the weakness of tax administration and policy are two important reasons for tax suffering in Bangladesh, where modernization can help. The main reason behind this report was to analyze the sustainability of the long-term goal of Bangladesh, which is primarily dependent on tax revenue. The author considers that without ameliorating the taxation system, Bangladesh would never be able to reach FY15, where the GDP ratio is 12.4%, while the current situation is FY11, which is 10%.

Murshed and Saadat (2018) estimated the gap between in existing literature by modeling tax evasion in Bangladesh, India, Pakistan, Sri Lanka, and Nepal. The author uses macroeconomic data from 2001 to 2015. The author uses a regression model, fixed-effects panel estimation techniques, the Vector Error Correction Model approach, and Granger.

Causality tests were also considered to establish the findings. The report finds that political stability and public service positively impact the tax-to-GDP ratio, whereas per capita GDP and GDP growth are found inside a non-linear relationship. The paper would be helpful for proper taxation systems, designing public policies, improving tax administration, and other policy development for Bangladesh, India, Pakistan, Sri Lanka, and Nepal.

Most of the studies above show tax evasion as an essential issue of low tax collection, where blockchain can cause a revolution even though implementing new technology is primarily dependent on government policy.

25.3   METHODOLOGY

This chapter aims to provide a sustainable solution to reduce the tax gap in Bangladesh by using FinTech technology. To achieve this goal, the author has thoroughly analyzed qualitative data, including historical data and secondary sources from various papers and websites. The analysis is based on both fundamental and qualitative research methods.

The first step in the research process was to thoroughly review the existing literature on tax collection in Bangladesh, including government reports, academic papers, and news articles. This provided the author with a comprehensive understanding of the current state of tax collection in the country and the challenges faced by the government in this area.

The second step involved conducting a historical data analysis of Bangladesh's tax collection, focusing on the tax gap and the factors that contribute to it. This involved reviewing and analyzing government data, financial reports, and other relevant data sources to understand the trends and patterns in tax collection over time.

The final step in the research process was to conduct qualitative research to identify potential solutions for reducing the tax gap in Bangladesh. This involved conducting interviews with experts in the field of taxation and FinTech, as well as conducting online surveys and focus groups with stakeholders in the industry.

The data collected through these methods were then analyzed and synthesized to identify the key factors contributing to Bangladesh's tax gap and develop a comprehensive solution that leverages FinTech technology to reduce this gap and improve tax collection in the country.

25.4   REASONS AND SOLUTIONS

25.4.1   Reasons for Low Tax Collection in Bangladesh

The collection of taxes in Bangladesh has been facing numerous challenges, which can be attributed to several factors. Firstly, the central tax administration authority in Bangladesh is vested with the National Board of Revenue, which is affiliated with the Internal Resources Division of the Ministry of Finance. Despite the advancements in technology, the tax department has been lagging behind in leveraging technology to improve the tax collection process.

One of the major issues faced by tax administrators is information asymmetry, as the authority heavily relies on the information provided by taxpayers. This opens up the possibility for taxpayers to misrepresent or conceal their actual financial status to reduce their tax liability. On the other hand, taxpayers themselves may not have adequate knowledge of tax laws and procedures, leading to tax evasion. The lack of organized documentation makes it difficult for the tax administration to accurately assess the tax liability of taxpayers (Lederman, 2003).

Another contributing factor to low tax collection in Bangladesh is the existence of the informal economy. The informal economy refers to economic activities that are not documented and, hence, not subject to taxes. This lack of transparency in economic transactions results in tax evasion and corruption. In Bangladesh, the size of the informal economy is estimated to be around 30.2% of the country's GDP, which is equivalent to nearly $254 billion at PPP levels (World Economics, 2020). This not only reduces the size of the formal economy but also contributes to the growth of corruption.

Finally, the low level of enforcement in the taxation system in Bangladesh has also contributed to the low tax collection. People are not motivated to pay their taxes correctly if they believe that the chances of getting caught are low. Hence, proper tax legislation, coupled with the effectiveness of human resources, is necessary to improve tax collection in Bangladesh.

25.4.2 Solution for Low Tax Collection

The average tax-to-GDP ratio of developing countries is nearly 15%. To achieve the goal of 2041, a rapid change in the system is highly required. According to trade analysts, the Dhaka Tribune (2018) reported that the branch between GDP and tax ratio is 7.5%, the highest in the Asia-Pacific region. That was reported based on the "Economic and Social Survey of Asia and the Pacific, 2018". The current state of Bangladesh's tax-to-GDP ratio can only be improved by properly implementing technology in an organized way. A different tech solution can be blockchain, a distributed peer-to-peer digital ledger with blocks as a chain. A block is a space where one can lock the transactions securely, and the block is related to another by a cryptography hash function (Fortune Business Insights, 2021). Already, several countries have used blockchain to make their transactions more transparent and secure. Such countries are Malta, Estonia, Switzerland, the United Arab Emirates, Singapore, the UK, China, Japan, the USA, Sweden, and many more, and most of them use blockchain to continue their government services along with various sectors. Standard Chartered Bank was the first bank to execute Bangladesh's first blockchain trade transaction (Finextra, 2020).

Moreover, the first blockchain-based supply chain finance platform in Bangladesh was launched by IPDC Finance Limited. Prime Bank executed the Interbank blockchain Letter of Credit transaction (IPDC, 2019). In Bangladesh, blockchain technology can add value to building resilient infrastructure and establishing e-governance, which will help the Bangladesh government be more user-friendly toward its citizens in providing public services.

25.4.3.1 Blockchain

Blockchain technology is the updated technical version of the decentralized ledger, which keeps information regarding everyone's assets or transactions using cryptographic hashing (Figure 25.1).

As the figure shows, a hash attached one block to the previous one. In addition, we can easily say that a blockchain is a digital ledger where one block works as a page written with transactions and is connected by unique codes to work as a distributed ledger.

25.4.3.1.1 *Types of Blockchain*

1. A public blockchain is a transaction that is accessible to everyone. It is a non-restricted blockchain where anyone can observe the network but cannot modify it, such as in the bitcoin market. Open blockchains are available for anyone with an Internet connection.
2. A private blockchain works the opposite of a public blockchain. There are restricted entities who can control and access the data. Moreover, it is a restricted and permission-able closed network. Private blockchains are typically used by enterprises to solve business problems and support enterprise software.
3. Consortium Blockchain works like a private blockchain, but the significant difference is that more than one author can present it here. Two organizations can use consortium blockchain when they want to hold data. A consortium is a "semi-private" system with a controlled user group that operates in different organizations.

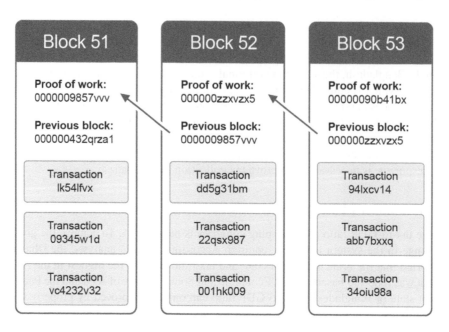

FIGURE 25.1 Blockchain flowchart.

25.4.3.1.2 Features of Blockchain

Nodes: A node is usually a device such as a computer, laptop, or server in a blockchain. Nodes save transaction history and have several functions, such as proof or rejection of a block of transactions. Anyone can create a node for a particular transaction system to keep track of the history of that specific blockchain transaction or various blockchains. However, running a node requires huge memory to keep the vast number of transactions in the blockchain. Namely, blockchain can draw breath on nodes.

The node can be of two types: one is the whole node, and the other is the light node. Full nodes refer to the preservation of entire data transactions. For example, one is keeping a full node of bitcoin until 2009. On the other hand, a light node means keeping a few replicas of ledger transactions. Another individual, as an illustration, has been keeping bitcoin transactions for the last three months. Full nodes require huge memory, whereas light nodes require less than full nodes.

Hash and Digital Signature: Cryptography algorithms are vital in blockchain technology. Hash works in a blockchain as the block's protector and intensifies the integrity the data. A hash is a digital signature that cannot be the same between blocks. The security code can be represented in front of the block owner only. Hash functions are widely used in Blockchain to protect the integrity and durability of the data stored in the distributed record. Moreover, we can say the security of Blockchain is dependent mainly on the hash function (Figure 25.2).

25.4.3.1.3 Work Process of Blockchain

Blockchain works as a secure connection that is transparent and has proper privacy. A transaction request occurs, taking a block position with a secure hash function. After this scenario, the block is checked by the nodes. After verification, the block is ready to add to the chain, and the transaction goes to its ultimate position. If the nodes disagree to accept the transaction as valid, then the block will go through the verification process.

On the other hand, blockchain in banks, corporations, or government authorities will work differently by following private or consortium blockchain. Here, a specific group of people or authorities will play the role of nodes. A consortium blockchain is a more used chain with more security and transparency for the bank (Figure 25.3).

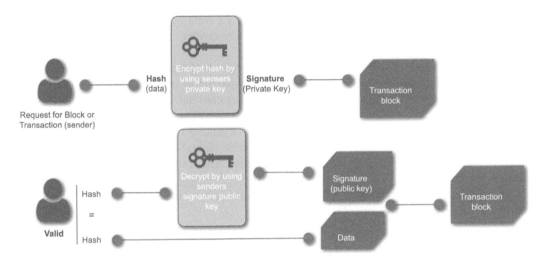

FIGURE 25.2 Blockchain hash function.

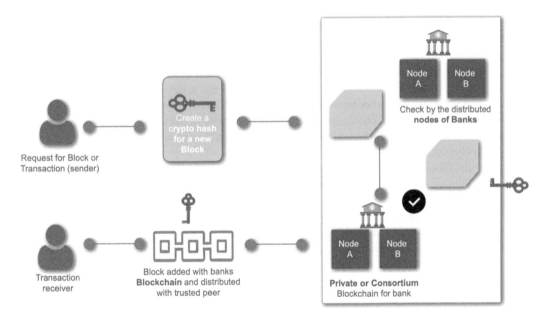

FIGURE 25.3 Private or consortium blockchain work process.

25.4.3.1.4 Market Share of Blockchain

Blockchain technology is one of the prominent technologies used in financial services, and the adoption rate of this technology is increasing daily. The global blockchain technology market size was USD 1.64 billion in 2017 and is projected to reach USD 21.07 billion by the end of 2025. According to Fortune business insights, the market size is expected to reach USD 69.04 billion by 2027 (The Daily Star, 2020).

There are so many players in the blockchain technology market. However, IBM Corporation is one of the most significant users of blockchain technology. North America holds the largest market share.

As per market analysis, the market size value of blockchain in 2020 is USD 3.84 billion, and the revenue forecast for 2025 is USD.

57.64 billion, along with a growth rate of CAGR is 69.4% from 2019 to 2025. Distributed ledger technology based on advanced analytics is one of the key drivers of market growth and is being adopted worldwide very quickly.

Suppose we have a look at the market growth of blockchain. In that case, we can consider –2017 as an introduction period for blockchain. When banks started to see the value of blockchain in their work processes and regulatory certainty drives, they considered blockchain for auditing and internal work.

After 2017, the consideration of blockchain in the work process increased gradually in the banking sector. Not only that, but many countries have adopted blockchain to make their governments more reliable and transparent, even in their voting systems. The realization of the benefits of blockchain is coming.

After COVID-19 this technology will take a high jump; after 2025, blockchain technology will entirely go into the maturity stage (Figure 25.4).

25.4.4 BLOCKCHAIN AND DIRECT TAX

Blockchain can enhance the documentation process and moreover, information systems. Direct tax mostly depends on correct information. There are several ways to commit fraud and tax crimes. According to National Board of Revenue data, Bangladesh has 10 million potential income taxpayers. However, only 3.8 million are registered in the system.

Moreover, tax returns and taxes are only filed by 1.5 million people. Hence, the main two problems of the taxation system in Bangladesh, which insist on tax evasion, are improper or unauthentic documentation and a lack of transparency. Blockchain can effectively provide solutions to those issues. This blockchain process will not only be helpful for the taxation system of Bangladesh but will also add value to every work process in a country, such as in banks, corporations, etc. The full document will be in the blockchain, and whenever an authority needs to use those documents, they can access them from the block with the approval of the authority who rectifies or approves the chain and plays the role of nodes.

In addition, blockchain will be helpful to know your customers, a well-known term in the banking sector. However, to serve the citizen, the government also needs to keep a database. A bank takes around 30–50 days to complete the KYC process to a perfect level just because this slow KYC process requires a high operational cost and has so many duplicate documentation requirements. Banks spend £40 million a year on KYC compliance on average, according to the Thomson Reuters Survey, and some banks spend up to £300 million annually on KYC, Anti-Money Laundering

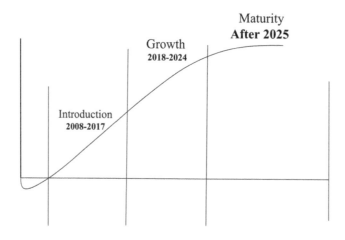

FIGURE 25.4   Market share of Blockchain.

("AML") checks, and Customer Due Diligence ("CDD"). Blockchain can be the perfect solution to all these issues because it has less documentation storage, but it will also be faster and more cost-friendly. Therefore, it is proven that blockchain could play a significant role in streamlining these KYC and AML processes. As per the Goldman Sachs Report, using blockchain technologies in the KYC process, the banking sector can achieve a 10% poll reduction, saving around $160 million in costs annually. Blockchain also helps reduce about 30%, or $420 million, of the budget for employee training. Overall, operation costs will save around 2.5 billion dollars. Table 25.1 shows the benefits of using Blockchain in the KYC process.

This is how a central database will go on, and every transaction will add to the noted block where or by whom the transaction will occur, and every authority will use their demanded data from that blockchain. This will drastically reduce corruption and fraud (Figure 25.5).

25.4.5 BLOCKCHAIN AND INDIRECT TAX

For indirect taxes such as VAT, GST, and Sales Taxes, blockchain can be a revolutionary solution. The tax collection process has a complex structure. Real-time data is the most significant and crucial point for tax collection. Therefore, it is hard to determine how, where, when, and what type of tax applies in any given situation, which is mandatory for every transaction to be correct. By making the whole system into a decentralized supply chain, blockchain can solve the problem. Chains of transactions and their tax liabilities are required to be found correctly. For a service or product, for instance, evaluate each point where the value is added.

Moreover, tracking the entire chain of transactions would make it much easier to secure precision and compliance. Using blockchain, from the tax administrator's side, speed, accuracy, and ease of collecting relevant tax data can be possible. On the other hand, from the business's perspective, higher rates of VAT recovery can be possible with less complexity.

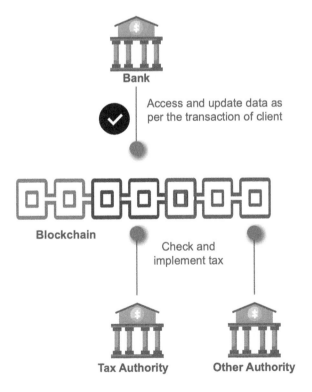

FIGURE 25.5 Blockchain in taxation.

TABLE 25.1

Benefits of Using Blockchain in the KYC Process

Factor	Issues with the Manual KYC Process	What Can Blockchain Do?
Manual processing	So many documents are required that have been performed in manual processes.	Digitalize process.
Information storage	Banks, companies, and governments all participants must store the common information.	All information is stored in the block, and anyone can get it based on their requirement.
Transparency	So many participants are involved here, such as applicants, companies, banks, government, etc., so transparency is highly required to enhance trust issues.	Blockchain can provide high transparency in the KYC process.
Trust	As mentioned before, many participants are involved, so there are trust issues.	This is a clear, open, and decentralized process, so it provides a high probability of maintaining trust problems.
Security	Less security is ensured.	High security by its working process.
Documentation	Paper-based work is required here. As a result, so much documentation happens for several participants.	All information is stored in a block, which is easy to find.

Tax authorities have invested heavily in data integration and analysis systems for digitalization to enhance tax collection and prevent tax fraud (Noor, 2020). Among indirect taxes, VAT is one of the crucial factors that contribute to government budgets. One of the critical VAT instruments is an invoice. A blockchain-based tax system can solve the problem by providing the service of validity checking, where VAT invoices need a digital fingerprint derived from the VAT blockchain consensus process. Through this procedure, the fingerprint will track that specific block. Same as browsers, a hand-held detector will attach to an authorized tax inspection program, and that will take to reach the entire commercial chain instantly for a valid postpaid item. For invoice-level data, each node must have instant access for both parties. Blockchain technology can create a safe business environment by providing DICE (Digital Invoice Customs Exchange) across boundaries. Table 25.2 shows Blockchain and Indirect Tax factors.

The integration of blockchain technology in the indirect tax system can bring numerous benefits, including increased transparency, improved efficiency, reduced errors, and enhanced security.

25.5   CONCLUSION

This study was conducted to provide a solution to enhance Bangladesh's low tax-to-GDP issue. In other words, this study examines the reasons and possible solutions behind the current taxation gap in Bangladesh, despite being one of the fastest-growing countries. A way out of this report is to fill the comprehension gap on whole economic growth and tax revenue to satisfy society and how the taxation system can be reestablished for errorless and transparent transactions. There has been a portion of a breach in the tax-to-GDP ratio of Bangladesh compared to other developing countries. Furthermore, this research not only analyzes historical data but also provides a practical solution to solve the tax evasion issue, which will help decode the inefficiency of the tax authority and the fraud of individuals or corporations in the direct and indirect tax areas. Bangladesh is facing massive tax evasion for a few reasons, where technology can rule the system conveniently by using blockchain. Nevertheless, it is a revolutionary step for developing countries like Bangladesh to rely on technology. For future studies, this report will provide an insight into how central blockchain implementation can make government services easily accessible and secure as well.

TABLE 25.2
Advantages of Using Blockchain in Indirect Taxation

Factor	Issues with the Current Indirect Tax System	What Can Blockchain Do?
Data Management	Maintaining records and data is complex and time-consuming.	Blockchain can provide a secure, tamper-proof, and transparent ledger of all transactions.
Tax Evasion	Companies can evade taxes by manipulating records and misreporting sales.	Blockchain can eliminate the possibility of fraud and manipulation, making tax evasion difficult.
Auditing	Auditing records is a difficult and time-consuming process.	Blockchain can provide a clear and auditable trail of all transactions, making the auditing process simpler and more efficient.
Transparency	The current indirect tax system lacks transparency.	Blockchain can provide a clear and transparent view of all transactions, enabling tax authorities to monitor compliance and enforce tax laws.
Efficiency	The indirect tax system is often slow and inefficient.	Blockchain can help streamline the indirect tax process, reducing the time and resources required to administer indirect taxes.

REFERENCES

Akeo (2018). Top 8 Countries Using Blockchain Technology around the World. Retrieve data from https://akeo.tech/blog/blockchain-and-dlt/top-8-countries-using-blockchain-technologyaround-the-world/.

The Daily Star (2020). SCB, bKash & Valyou launch Bangladesh's first blockchain remittance service. Retrieve data from https://www.thedailystar.net/bangladesh-first-blockchain-based-remittanceservice-launched-1958453.

Dhaka Tribune (2018). Bangladesh's tax gap to GDP ratio highest in Asia-Pacific. Retrieve From https://archive.dhakatribune.com/business/2018/06/28/bangladesh-s-tax-gap-to-gdp-ratiohighest-in-asia-pacific.

Fortune Business Insights (2021). Blockchain Technology Market. Retrieve from https://www.fortunebusinessinsights.com/press-release/blockchain-technology-market-9046.

Finextra (2020). Blockchain Use Cases for Banks In 2020. Retrieve from https://www.finextra.com/blogposting/17857/blockchain-use-cases-for-banks-in-2020.

Faccia, A., & Mosteanu, N. R. (2019). Tax evasion_information system and blockchain. *Journal of Information Systems & Operations Management*, 13(1), 65–74.

Globenewswire (2020a). Blockchain Market to Reach USD 69.04 Billion by 2027; High Demand for Blockchain-as-a-service to Aid Growth: Fortune Business Insights. Retrieve from https://www.globenewswire.com/news-release/2020/10/22/2112779/0/en/Blockchain-Marketto-Reach-USD-69-04-Billion-by-2027-High-Demand-for-Blockchain-as-a-service-to-AidGrowth-Fortune-Business-Insights.html.

Globenewswire (2020b). Blockchain Technology Market to Reach USD 21.07 Billion by 2025; Investments in Development of Research Centers to Aid Growth: Fortune Business Insights. Retrieve from https://www.globenewswire.com/news-release/2020/05/12/2032075/0/en/Blockchain Technology-Market-to-Reach-USD-21-07Billionby-2025-Investments-in-%20Development-of-Research-Centers-to-Aid-Growth-FortuneBusiness-Insights.html.

Hima, Z. (2018). Blockchain in taxation. In *Proceedings of FIKUSZ Symposium for Young Researchers* (pp. 168–177). Óbuda University Keleti Károly Faculty of Economics.

IPDC (2019). South East Asia's First Blockchain Based Digital Supply Chain Finance Platform. Retrieve from https://www.ipdcbd.com/home/orjon.

Kiabel, B. D., & Nwokah, N. G. (2009). Boosting revenue generation by state governments in Nigeria: The tax Consultants option revisited. *European Journal of Social Sciences*, 8(4), 532–539.

Lederman, L. (2003). The interplay between norms and enforcement in tax compliance. *Ohio St. LJ*, 64, 1453.

Mansur, A. H., Yunus, M., & Nandi, B. K. (2011, December). An evaluation of the tax system in Bangladesh. In *Conference on Linking Research to Policy: Growth and Development Issues in Bangladesh, Dhaka*, December (Vol. 20).

Murshed, M., & Saadat, S. Y. (2018). Modeling tax evasion across South Asia: Evidence from Bangladesh, India, Pakistan, Sri Lanka and Nepal. *Journal of Accounting, Finance and Economics*, 8(1), 15–32.

Noor, A. (2020). Blockchain: Emergence of a New Breed of Tax Professionals in the Digital VAT World. Retrieved from https://blockchain.news/opinion/blockchain-emergence-of-a-newbreed-of-tax-professionals-in-the-digital-vat-world.

Ryakhovsky, D. I., Ukhina, T., Duborkina, I. A., Kozhukhova, N. N., Goncharov, V. V., & Zharov, A. N. (2021). Applications of Blockchain in Taxation: New Administrative Opportunities. Management.

World Economics (2020). Bangladesh's Informal Economy Size. World Economics. Retrieve data from. https://worldeconomics.com/Informal-Economy/Bangladesh.aspx.

Yayman, D. (2021). Blockchain in taxation. *Journal of Accounting and Finance*, 21(4), 140–155.

26 Approach for Strengthening the Network Security Based on Boosting Algorithms
Performance Study

Sabrine Ennaji, Nabil El Akkad, and Khalid Haddouch

26.1 INTRODUCTION

The excessive use of the internet is becoming more and more revolutionary since the users all over the world are interconnected incessantly via the internet, either to benefit from certain services such as social networks, streaming services, storing their data in public storage servers with remote processing, or to benefit from location-based services, etc. Therefore, the technology revolution is constantly spreading, and we are increasingly drowning in the rampant growth of a massive amount of data containing people's sensitive information [1]. As is well known, the exponential growth of data volume results in the exponential growth of malicious activities that aim to breach the network system, device, or service, etc. Thus, these cyberattacks are considered a great challenge and present a high risk. The latter includes stealing private information, tracking and spying, or usurping identities, etc. To counteract all these risks, many techniques have been developed to ensure the security of applications on the internet. For instance, the distribution of data on several scattered servers and obfuscation or rerouting techniques that prevent tracking. There are also access control and cryptographic techniques that restrict access to authorized persons only. It is true that all these protection mechanisms have many advantages in preventing tracking, ensuring anonymity, and controlling the exchange of data. However, they usually provide only external protection, since some of them necessitate further security measures for strong protection [2]. For this reason, many research studies are well carried out to shed light on Intrusion Detection Systems (IDSs). They are known to be an appropriate solution for detecting external and internal cyberattacks [3]. Furthermore, they play an important role in examining and identifying different types of attacks across all the internet technologies:

Network-Based Intrusion Detection System (NIDS) – It monitors the traffic packet by packet at the network level to analyze the intrusion patterns upon which a warning is sent.

Host-Based Intrusion Detection System – It is a software package installed directly into the host. It examines the malicious activities in an individual host, unlike the NIDS that monitors the entire network.

Perimeter Intrusion Detection System – It is a sensor that identifies any threats on perimeter fences.

A Virtual Machine–Based Intrusion Detection System – It is applied remotely via a virtual machine, and combines the three types mentioned above (Figure 26.1).

Additionally, there are two different methods with which an IDS detects an attack: anomaly and signature-based methods. The anomaly-based IDS can examine and analyze any malicious activity. It is considered the most applicable type for such attacks [4]. The signature-based IDS necessitates

DOI: 10.1201/9781032667478-30

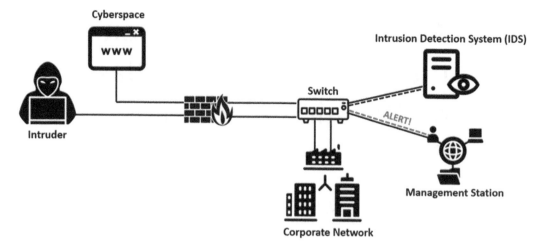

FIGURE 26.1   Intrusion detection system architecture.

a comparison of the signature and patterns with its database. Evidently, this method cannot detect unknown attacks since it cannot find any signatures [5]. Consequently, the anomaly-based IDS received more attention due to its ability to identify unknown and recent attacks [6]. However, the detection performance of this type of IDS still needs to be enhanced. Thus, it has been developed by many researchers in order to classify and detect unknown threats automatically. For instance, Masaaki Sato et al. relied on feature extraction from anomaly-based IDS alerts in order to handle this issue [7]. Whereas, Jungsug Song et al., proposed a technique for extracting unknown threats using data mining by applying the multiple training data that consist of IDS alerts [8].

In spite of the significant role of the IDSs in identifying different types of cyberattacks, they are unceasingly alert to false positives more than they are to real attacks [7,9]. Consequently, the performance of an IDS requires enhancement and updating. To address this void, this research study has been based on three machine learning algorithms, namely: Multilayer Perceptron (MLP), Random Forest (RF), and AdaBoost. It is well known that these learners have a powerful potential for prediction tasks without additional human effort, unlike the IDSs [9]. The purpose of this paper is to come up with new results for maximizing the predictive accuracy of an IDS model using the collaborative learning method. The latter consists of a combination of the upsides of multiple machine learning classifiers in order to build a robust model. Moreover, an analytical study is also provided to discuss the difference between using single and ensemble learning techniques. The experiments are evaluated within the NSL-KDD data set using the feature selection technique to boost the quality of the prediction process. Furthermore, this work reveals the weaknesses of each method used, which will lead to a new approach for the next work that can deal with these limitations.

The remaining chapter is organized as follows: Section 26.2 covers a literature study about the research done so far in the area of IDS based on single and collaborative learning methods. Section 26.3 presents the proposed hybridization modeling as well as the tools and dataset used. Section 26.4 discusses the results obtained and their analysis. Section 26.5 is concluded along with a general conclusion and future scope.

26.2   LITERATURE SURVEY

Much effort has been devoted to improving the intrusion detection system sector. Many researchers have integrated machine learning algorithms to make the IDSs more efficient and appropriate for the current security challenges. In this section, we have classified these efforts into two well-known learning methodologies adopted by different research studies:

26.2.1 Single-Learning Method

As far as is known, among the most efficient and widely used machine learning algorithms are Artificial Neural Networks [10]. Peng et al. approved the importance of the latter by proposing an approach based on MLP, LSTM, and other ANN algorithms [11]. The purpose of their contribution is to build an intelligent model that can identify the SQL injection attack within the HTTP traffic data in real time. Their obtained results show how an MLP classifier with three hidden layers performs well in achieving the main aim of their work. Additionally, their new approach is able to detect unknown attacks and can also handle the incomplete blacklist filtering issue. In another work, there is a new architecture of ANN based on IDSs, put forward by Shenfield et al. [12], which aims to identify the malicious activity and distinguish between a harmful and a benign threat within a complex network traffic. The authors came up with a satisfying result in terms of classification, accuracy, precision, and sensitivity over 1000 iterations of repeated 10-fold cross-validation, which proves the robustness of their detection system.

Naïve Bayes is also a popular classification algorithm known for its simplicity [13]. For instance, Panda et al. have developed this classifier and relied on it to build a robust IDS model. They have proposed a two-level classification approaches. The first level presents the base classifier using Discriminative Multinomial Naïve Bayes. In the second level, they used the nominal-to-binary supervised filtering with 10-fold cross-validation for experimental results over the NSL-KDD dataset, which reveals that the proposed approach works better than other ML algorithms such as Decision Tree, RF, and Support Vector Machine. In addition, Sravani et al. [14] confirmed and proved that Naïve Bayes and Neural Network algorithms are effective to classify and detect intrusions, with a focus on the abnormal ones. Their results proved that both of them decreased the false negatives and false positives. Besides, they increased the accuracy of their developed IDS.

RF is also an important classification algorithm, which has been enhanced in [15]. The authors attempted to detect four different types of attacks using the NSL-KDD dataset and evolved the accuracy of their proposed model, which was good at decreasing low alarms and increasing detection rates. Furthermore, many analytical and comparative studies have been carried out. They discuss the advantages and limitations of each classifier, which will result in the selection of the most performing algorithm. For example, Belavagi et al. built an enhanced IDS model using different steps [16]. They have been based on four ML classifiers, namely, RF, Logistic Regression, Support Vector Machine, and Gaussian Naïve Bayes. Their results proved that RF is better than other classifiers in terms of Accuracy, Precision, Recall, and F1-score. Besides, LR achieved better accuracy than SVM and GNB. In another comparative study, Pal et al. used two algorithms, RF and SVM, to improve the performance of an IDS. Their outcome showed that SVM reached the highest accuracy rate in detecting anomaly traffic within the KDD'99 dataset. However, RF takes less time to make predictions, and it has better precision compared to SVM [17].

As can be deduced, each machine learning algorithm has its weaknesses and strengths. There are several types of cyberattacks, including SQL injection, phishing, denial of service, man in the middle, eavesdropping attacks, etc. Hence, some classification algorithms can be performant in detecting a type of attack but cannot perform well on another type [9,18]. As a result, the single-learning method cannot always be considered a sufficient solution to provide strong security. For this reason, numerous research studies opt for collaborative learning.

26.2.2 Collaborative Learning Method

Many researchers agree that the collaborative learning method can achieve a high accuracy rate compared to the single-learning technique. Because it deals with statistical and computational problems by decreasing variance and bias [19]. Consequently, it provides better predictive results. Khonde et al. validated the efficiency of the ensemble method and applied it to boost the accuracy of an IDS in detecting normal and malicious activities [20]. They combined five classification algorithms: SVM, DT, RF, KNN, and ANN, using a novel method of feature selection. The latter had a significant capability for examining real-time traffic. Therefore, their method performs well and produces good results. Das et al. have also relied on the collaborative approach and grouped four different algorithms, namely J48, SVM, MLP, and KNN, in order to build a powerful IDS [21]. Their proposed method unfolds in two stages: Firstly, during the training step, each algorithm creates a model that is able to detect DDoS attacks and their recent unknown types. Secondly, the final model, which combines the trained algorithms, is evaluated in the test set using 10-fold cross-validation and feature selection techniques. Their experimental results showed the important potential of their model in detecting new types of DDoS attacks. Besides, it outperforms many other existing models. In another work, an enhanced IDS model was created in [22] using the fusion of the hybrid feature selection technique and a two-level classifier ensemble. The latter relies on two meta-learners (rotation forest and bagging) that are evaluated in two different datasets: UNSW-NB15 and NSL-KDD. The results obtained from their proposed model validated the importance of ensemble learning in maximizing the accuracy rate. In addition, the authors have considered the training time as a significant measurement that can impact the performance of the model. Thus, it should also be taken into consideration in order to improve the intrusion detection task.

To strengthen the ability of an IDS to provide strong security, Lian et al. have based their work on the stacking/blending technique, which is a two-level ensemble learning method. This necessitates training different single-learning algorithms on the training data. Then, it combines the outputs of each algorithm for the final prediction task over unseen data [23]. The authors used the decision tree-recursive feature elimination algorithm, which helps reduce the system size and the time required for training [24]. Therefore, their designed IDS achieved some excellent results in terms of detection rate accuracy for the classification of four types of attacks (Dos, Probe, U2R, and R2L). Moreover, it has a low false-positive rate. In [25], a new deep learning approach has been proposed using the AdaBoost Regression algorithm to improve the quality of cyberattack detection tasks and ensure strong network security in real time. The authors approved through their obtained results that AdaBoost performed effectively for the main aim of their contribution. This classifier boosts the classification accuracy by identifying the weak learners based on the decision tree concept with consideration of different levels.

26.3 PROPOSED HYBRIDIZATION APPROACH

In this section, the experimental setup, the intrusion detection dataset, and the proposed methodology used in this study are described in detail.

26.3.1 TECHNICAL TOOLS

The numerical modeling was performed using a Dell Inspiron 15 3000 Series Core i5 7th Gen. Therefore, as long as the latter does not have a high GPU, we have been based on Google Colab and connected our machine with it. Google Colab, short for Google Collaboratory [26], is a free cloud service hosted by Google that supports free GPUs and has a significant role in developing artificial intelligence and big data experiments. Furthermore, it comes with almost all the libraries needed for such domains. This environment has a fast network speed that makes data files easy to download. Moreover, every Google Colab notebook is stored in the cloud. The only problem faced during the use of this environment is that it does not have persistent storage. As a solution, there is Google Drive. Besides, it is necessary to be connected to a very strong internet connection and save every line written because Google Colab does not save the code automatically.

26.3.2 INTRUSION DETECTION DATASET

To the best of our knowledge, the creation of a powerful IDS that is able to make accurate predictions using several machine learning classifiers is required to use an effective dataset [27].

TABLE 26.1

Attack Classes in the NSL-KDD Dataset

Attack Class	Description	Attack Types
Denial of Service attack (DOS/DDoS)	It overcharges the service network with traffic, which results in the exhaustion of its resources and bandwidth [30].	Back, Land, Neptune, Pod, Smurf, Teardrop, Apache2, Udpstorm, and Processtable.
Probing (Probe)	A physical attack that joins several reports exploited by the attacker to access networks directly and extract confidential data [31].	Satan, Ipsweep, Nmap, Portsweep, and Mscan.
Remote to Local User (R2L)	It targets a network/system by sending a packet that can detect all the vulnerabilities the intruder could exploit [32].	Guess_Password, Ftp_write, Imap, Phf, Multihop, Warezmaster, Warezclient, Spy, Xlock, Xsnoop, Snmpguess, Snmpgetattack, Httptunnel, Sendmail.
User-To-Root (U2R)	By using a normal account, the attacker gets access to the victim's system. Thus, he obtains accessibilities that the victim has [33].	Buffer_overflow, Loadmodule, Rootkit, Perl, Sqlattack.

Based on many research studies about the available datasets we can select for our experiments, it appears that NSL-KDD is the most appropriate dataset for this analytical study. It is an advanced version of the KDD'99 Cup dataset, which includes a large amount of internet traffic records. The latter was a result of an International Knowledge Discovery and Data Mining Tools competition [28], which had the purpose of building a network intrusion model that can distinguish between a normal and abnormal connection. However, this old version of the NSL-KDD dataset causes some problems in the data preprocessing since it does not eliminate redundant and repeated records. Thus, it creates the issue of biasing the features and the classification algorithms. For this reason, the NSL-KDD dataset has been created. It has dealt with the problem of duplicated records; thus, it has the best reduction rate. Additionally, it includes a reasonable number of records in the training and test sets without the necessity of randomly selecting a subset of the dataset [29]. It concludes that 24 different types of attacks are categorized into four classes, as shown in Table 26.1.

26.3.3 Methodology

It is well known that real-world data usually contains redundant values, outliers, etc., which results in the complexity of analyzing the data and the degradation of the accuracy rate. As a matter of fact, data preprocessing is considered an important process. In this contribution, we have followed the process illustrated in the flow below:

Data preprocessing: This phase plays a vital role in removing the redundant entries. As is shown in Figure 26.2, it includes the distribution of each attack type to its attack class, which represents the target variable. Besides, to remove the mean and unit variance, a scaling of numerical attributes is applied in order to eliminate the mean and unit variance. Later on, the categorical attributes are encoded for data sampling and resampling techniques. The latter is useful for data reduction and can also deal with imbalanced class issues.

Feature selection: In order to reduce the computational cost of modeling and enhance the performance of the classification process, in this phase we have minimized the features of the dataset and kept 10 significant variables, that have a direct impact on the target variable (attack class), as is shown in Table 26.2.

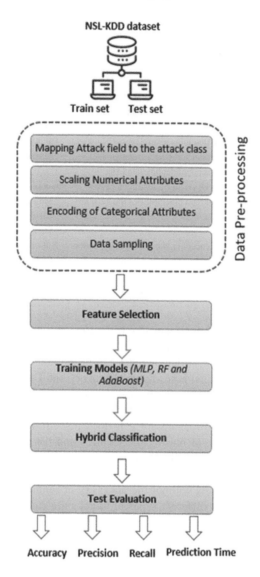

NSL-KDD dataset

Train set   Test set

Mapping Attack field to the attack class

Scaling Numerical Attributes

Encoding of Categorical Attributes

Data Sampling

Data Pre-processing

Feature Selection

Training Models (MLP, RF and AdaBoost)

Hybrid Classification

Test Evaluation

Accuracy   Precision   Recall   Prediction Time

FIGURE 26.2   Block diagram representing the proposed methodology.

Classification: Three important algorithms have been considered in this paper; Multilayer perceptron [34] and RF [35] that have been in conjunction with AdaBoost algorithm [36]. The latter has a vital role in maximizing the performance of the IDS model. It boosts the weaknesses of other algorithms, which results in the creation of a powerful hybrid model.

Evaluation: The performance of the hybrid model has been examined using the accuracy, the precision, the recall, the F1-score and the time of prediction for a binary classification (Table 26.3).

26.4   RESULTS AND DISCUSSION

In this proposed study, we approve the effectiveness of the hybridization method based on the collaboration of MLP and RF with the AdaBoost classifier. From the results obtained, it can be validated that this technique is positive for boosting the performance of an IDS model based on machine learning algorithms in detecting normal and abnormal activities within the network. As is indicated in the

TABLE 26.2

The 10 Significant Variables Selected in the NSL-KDD Dataset

Attribute No.	Name	Description	Feature Category
(5)	'src_bytes'	The sent data bytes were sent in single connection from the source to the destination.	Basic Feature
(6)	'dst_bytes'	The sent data bytes were sent in a single connection from the destination to the source.	Basic Feature
(12)	'logged_in'	It indicates the login status (1 or 0).	Content Feature
(23)	'count'	Connection number to the same destination host as the current connection in the 2 previous seconds.	Time-based Feature
(24)	'srv_count'	Connection number to the same service as the current connection in the 2 previous seconds	Time-based Feature
(33)	'dst_host_srv_count'	Connections that have the same port number	Connection-based Feature
(34)	'dst_host_diff_srv_rate'	Connections that were to the same service among the connections in 'dst_host_count'	Connection-based Feature
(36)	'dst_host_same_src_ port_rate'	Connections that were to the same source port among the connections in 'dst_host_srv_count'	Connection-based Feature
(38)	'dst_host_serror_rate'	Connections that have activated the flag s0/s1/s2/ s3 among the connections in 'dst_host_count'	Connection-based Feature
(3)	'service'	The used destination network service	Basic Feature

TABLE 26.3

Performance Measures

Measure	Definition	Formula
Accuracy	Percentage of the model's correctness	$\dfrac{TP+TN}{TP+TN+FP+FN}$
Precision	Percentage of positive predictions that are absolutely included in the positive class	$\dfrac{TP}{TP+FP}$
Recall	Percentage of the model's ability to identify true positives	$\dfrac{TP}{TP+FN}$
F1-Score	The weight average of the precision and recall measures	$2.\dfrac{precision.recall}{precision+recall}$

table below, our hybrid model outperforms other models based on single classifiers. It has achieved the highest accuracy rate of 99.76% in detecting a wide variety of attacks (Dos, Probe, R2L, and U2R), categorized in attack class, and also normal behaviors. Moreover, it returns very few false positives since it has a precision of 99.64%. Besides, it obtains a percentage of 99.80% in terms of false-negative identification. Considering the same methodology and dataset, the MLP and RF have reached an accuracy rate of 91.41% and 82.99%, respectively. Moreover, they are generating a great portion of false positives and have weak identification for true positives (Table 26.4).

Further, from the results provided in Figures 26.3 and 26.4, which present the time of computation and prediction required by each model for the detection task, it is evident that despite the highest accuracy reached by our model, it takes a lot of time to learn and make predictions. It is a computationally costly process since it requires more than 60 seconds to build the model and 289,400 milliseconds for the prediction time. However, the RF model is the less time-consuming classifier in this work, as it takes 16.57 seconds to be built and 4.59e-05 milliseconds to predict.

TABLE 26.4

The Performance of the Proposed Approach Compared to Single Classifiers Over the Test Set (%)

Classifiers	Accuracy	Precision	Recall	F1-Score
MLP	91.41	80.71	96.68	87.98
RF	82.99	76.71	82.85	79.67
Our hybrid model	99.76	99.64	99.80	99.72

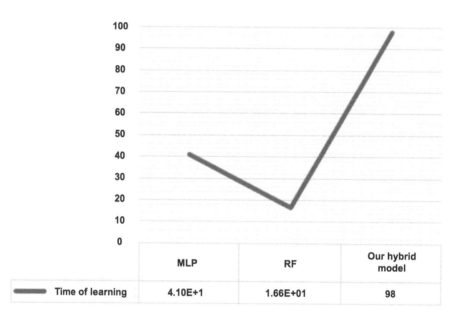

	MLP	RF	Our hybrid model
Time of learning	4.10E+1	1.66E+01	98

FIGURE 26.3 Time for the learning process in seconds.

It is known to be one of the fastest classifiers. MLP reached the second rank since it takes 41 seconds to learn and 1.00e-4 milliseconds for the prediction process.

To prove the effectiveness of our hybrid model, we compared it to some existing hybrid models, as Figure 26.5 depicts.

It is evident how this combination of MLP, RF, and AdaBoost algorithms comes up with a high-performance intelligent hybrid model for developing the task of intrusion detection within computer networks. It provides a better accuracy rate with fewer false alarms compared to other approaches, as the detection accuracy in [37–41] is reaching 99.3%, 99.4%, 98.77%, 90.41%, and 95.2%, respectively. On the contrary, our approach achieved 99.76%. However, the limitation of the proposed hybridization technique is the time required for the predictions.

26.5 CONCLUSION AND FUTURE WORK

An enhanced machine learning-based model has been proposed in this paper, in order to strengthen the performance of an NIDS for a binary classification (i.e., abnormal and normal behaviors). It is a hybridization approach based on the AdaBoost algorithm, which combines two other different machine learning classifiers, namely, MLP and RF, using the feature selection technique. The

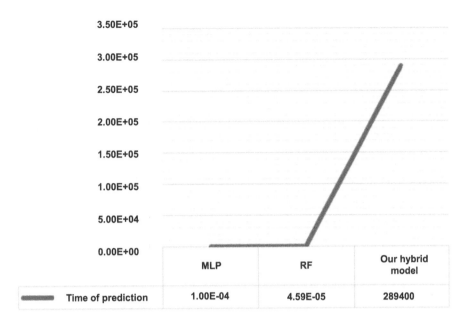

FIGURE 26.4 Time for the prediction process in milliseconds.

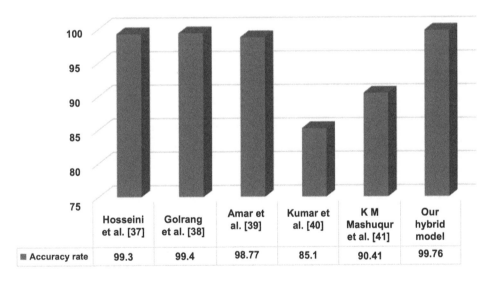

FIGURE 26.5 Comparison of the proposed hybrid model to another existing model (%).

performance of the latter was assessed on the networking NSL-KDD dataset, and the achieved outcomes approve the importance and efficiency of the designed hybrid model since it has maximized the accuracy of a single model from 91.41% and 82.99% to 99.76%. This chapter also takes into account an analytical study that compares hybridization and single-learning methods, considering the same methodology and dataset. Thus, a more accurate picture of the limitations of each method and algorithm used is provided. This will be helpful to propose novel techniques in order to handle these weaknesses, which include the time required by the model to be built and to make predictions. In addition, the presented model has been compared to other existing hybrid methodologies for network IDS, and it is shown that the proposed hybridization method is more efficient. In future work, the limitations of the presented cybersecurity modeling will be overcome in order to provide a robust IDS against the current threats, which are adversarial attacks. The latter can craft a sample by adding small perturbations, aiming to trick the model into getting wrong results and making incorrect predictions.

REFERENCES

1. Sadqi, Y., & Maleh, Y. (2022). A systematic review and taxonomy of web applications threats. *Information Security Journal: A Global Perspective*, 31(1), 1–27.
2. Ray, L. L. (2013). Training and testing anomaly-based neural network intrusion detection systems. *International Journal of Information Security Science*, 2(2), 57–63.
3. Can, O., & Sahingoz, O. K. (2015, May). A survey of intrusion detection systems in wireless sensor networks. In *2015 6th International Conference on Modeling, Simulation, and Applied Optimization (ICMSAO)* (pp. 1–6). IEEE.
4. Aydın, M. A., Zaim, A. H., & Ceylan, K. G. (2009). A hybrid intrusion detection system design for computer network security. *Computers & Electrical Engineering*, 35(3), 517–526.
5. Bushehr, I. (2014). A novel data mining method for malware detection. *Journal of Theoretical and Applied Information Technology*, 70(1), 1–9.
6. Brandon Lokesak (2008, December 4). A Comparison Between Signature Based and Anomaly Based Intrusion Detection Systems" (PPT). https://www.iup.edu.
7. Sato, M., Yamaki, H., & Takakura, H. (2012, July). Unknown attacks detection using feature extraction from anomaly-based ids alerts. In *2012 IEEE/IPSJ 12th International Symposium on Applications and the Internet* (pp. 273–277). IEEE.
8. Song, J., Ohba, H., Takakura, H., Okabe, Y., Ohira, K., & Kwon, Y. (2007, December). A comprehensive approach to detect unknown attacks via intrusion detection alerts. In *Annual Asian Computing Science Conference* (pp. 247–253). Springer, Berlin, Heidelberg.
9. Ennaji, S., Akkad, N. E., & Haddouch, K. (2021). A powerful ensemble learning approach for improving network intrusion detection system (NIDS). In *2021 Fifth International Conference On Intelligent Computing in Data Sciences (ICDS)*, pp. 1–6. doi: 10.1109/ICDS53782.2021.9626727.
10. Jahani, B., & Mohammadi, B. (2019). A comparison between the application of empirical and ANN methods for estimation of daily global solar radiation in Iran. *Theoretical and Applied Climatology*, 137(1-2), 1257–1269.
11. Tang, P., et al. (2020) Detection of SQL injection based on artificial neural network. *Knowledge-Based Systems*, 190, 105528.
12. Shenfield, A., Day, D., & Ayesh, A. (2018). Intelligent intrusion detection systems using artificial neural networks. *ICT Express*, 4, 95–99.
13. Panda, M., Abraham, A., & Patra, M. R. (2010, August). Discriminative multinomial naive bayes for network intrusion detection. In *2010 Sixth International Conference on Information Assurance and Security* (pp. 5–10). IEEE.
14. Sravani, K., & Srinivasu, P. (2014). Comparative study of machine learning algorithm for intrusion detection system. In *Proceedings of the International Conference on Frontiers of Intelligent Computing: Theory and Applications (FICTA) 2013* (pp. 189–196). Springer, Cham.
15. Farnaaz, N., & Jabbar, M. A. (2016). Random forest modeling for network intrusion detection system. *Procedia Computer Science*, 89(1), 213–217.
16. Belavagi, M. C., & Muniyal, B. (2016). Performance evaluation of supervised machine learning algorithms for intrusion detection. *Procedia Computer Science*, 89, 117–123.
17. Hasan, M. A. M., Nasser, M., Pal, B., & Ahmad, S. (2014). Support vector machine and random forest modeling for intrusion detection system (IDS). *Journal of Intelligent Learning Systems and Applications*, 6, 45–52.
18. Gao, X., et al. (2019). An adaptive ensemble machine learning model for intrusion detection. *IEEE Access*, 7, 82512–82521.
19. Abirami, M. S., Yash, U., & Singh, S. (2020). Building an ensemble learning based algorithm for improving intrusion detection system. In *Artificial Intelligence and Evolutionary Computations in Engineering Systems*. Springer, Singapore, pp. 635–649. doi: 10.1007/978-981-15-0199-9_55.
20. Khonde, S. R., & Ulagamuthalvi, V. (2019). Ensemble-based semi-supervised learning approach for a distributed intrusion detection system. *Journal of Cyber Security Technology*, 3(3): 163–188.

21. Das, S., et al. (2019). DDoS intrusion detection through machine learning ensemble. In *2019 IEEE 19th international conference on software Quality, Reliability and Security Companion (QRS-C)*. IEEE.

22. Tama, B. A., Comuzzi, M., & Rhee, K.-H. (2019). TSE-IDS: A two-stage classifier ensemble for intelligent anomaly-based intrusion detection system. *IEEE Access*, 7, 94497–94507.

23. Polikar, R. (2012). Ensemble learning. In *Ensemble Machine Learning*. Springer, Boston, MA, pp. 1–34. doi: 10.1007/978-1-4419-9326-7_1.

24. Lian, W., et al. (2020). An intrusion detection method based on decision tree-recursive feature elimination in ensemble learning. *Mathematical Problems in Engineering*, 2020, 1–15.

25. AlShahrani, B. M. M. (2021). Classification of cyber-attack using adaboost regression classifier and securing the network. *Turkish Journal of Computer and Mathematics Education (TURCOMAT)*, 12(10), 1215–1223.

26. Bisong, E. (2019). *Building Machine Learning and Deep Learning Models on Google Cloud Platform: A Comprehensive Guide for Beginners*. Apress, Berkeley, CA.

27. Dhanabal, L., & Shantharajah, S. P. (2015). A study on NSL-KDD dataset for intrusion detection system based on classification algorithms. *International Journal of Advanced Research in Computer and Communication Engineering*, 4(6), 446–452.

28. Meena, G., & Choudhary, R. R. (2017). A review paper on IDS classification using KDD 99 and NSL KDD dataset in WEKA. In *2017 International Conference on Computer, Communications and Electronics (Comptelix)*. IEEE.

29. Ravipati, R. D., & Abualkibash, M. (2019). Intrusion detection system classification using different machine learning algorithms on KDD-99 and NSL-KDD datasets-a review paper. *International Journal of Computer Science & Information Technology (IJCSIT)*, 11, 65–80.

30. Vissers, T., Somasundaram, T. S., Pieters, L., Govindarajan, K., & Hellinckx, P. (2014). DDoS defense system for web services in a cloud environment. *Future Generation Computer Systems*, 37, 37–45.

31. Wang, H., Forte, D., Tehranipoor, M. M., & Shi, Q. (2017). Probing attacks on integrated circuits: Challenges and research opportunities. *IEEE Design & Test*, 34(5), 63–71.

32. Alharbi, A., Alhaidari, S., & Zohdy, M. (2018). Denial-of-service, probing, user to root (U2R) & remote to user (R2L) attack detection using hidden markov models. *International Journal of Computer and Information Technology*, 7, 1–7.

33. Paliwal, S., & Gupta, R. (2012). Denial-of-service, probing & remote to user (R2L) attack detection using genetic algorithm. *International Journal of Computer Applications*, 60(19), 57–62.

34. Barapatre, P., et al. (2008). Training MLP neural network to reduce false alerts in IDS. In *2008 International Conference on Computing, Communication and Networking*. IEEE.

35. Resende, P. A. A., & Drummond, A. C. (2018). A survey of random forest based methods for intrusion detection systems. *ACM Computing Surveys (CSUR)*, 51(3), 1–36.

36. Schapire, R. E. (2013). Explaining adaboost. In *Empirical Inference*. Springer, Berlin, Heidelberg, 37–52. doi: 10.1007/978-3-642-41136-6_5.

37. Hosseini, S., & Zade. B. M. H. (2020). New hybrid method for attack detection using combination of evolutionary algorithms, SVM, and ANN. *Computer Networks*, 173, 107168.

38. Golrang, A., et al. (2020). A novel hybrid IDS based on modified NSGAII-ANN and random forest. *Electronics*, 9(4), 577.

39. Meryem, A., & Ouahidi, B. E. (2020). Hybrid intrusion detection system using machine learning. *Network Security*, 2020(5), 8–19.

40. Samriya, J. K., & Kumar, N. (2020). A novel intrusion detection system using hybrid clustering-optimization approach in cloud computing. *Materials Today: Proceedings*. doi: 10.1016/j.matpr.2020.09.614.

41. Mazumder, A. M. R., Kamruzzaman, N. M., Akter, N., Arbe, N., & Rahman, M. M. (2021, February). Network intrusion detection using hybrid machine learning model. In *2021 International Conference on Advances in Electrical, Computing, Communication and Sustainable Technologies (ICAECT)* (pp. 1–8). IEEE.

27 Smart Cities Technologies in the COVID-19 Context
A Bibliometric Analysis

Elafri Nedjwa, Boumali Badreldinne,
Lalmi Andallah, Sassi boudemagh Souad,
Yassine Maleh, Rose Bertrand, and Hemza Barkani

27.1 INTRODUCTION

Since the first appearance of the smart city as a research field in 1992. There are several definitions of a smart city. The concept is often identified by other terms such as digital city, smart city, etc. other terms such as digital city, smart city or virtual city. A study of smart city definitions, which examined 32 diverse definitions, resulted in the following comprehensive definition of a smart city [1]. Smart city is a concept that can be defined as follows:

A smart city is a system that enhances social and human capital by using and interacting with economic and natural resources through innovation and technological solutions to achieve sustainable development and a high quality of life effectively based on a multi-stakeholder and municipal partnership.

Since the 1990s, the concept of the "smart city" has been closely associated with most forms of technological innovation in the planning, development, operation, and management of cities [2].

Many researchers have conducted studies on urban planning, smart city governance, long-term development, energy technologies, internet of things (IoT), open challenges, smart city trends, components, and architecture. There are also many literature reviews on smart city-enabling technologies such as cloud computing, big data, etc. The literature approach is a useful tool for academics to develop their understanding of policymakers [2].

Bibliometric approaches are helpful in both identifying and quantifying cooperative patterns of research performance and the research patterns of authors, reviews, publications, countries, and institutes. On the other hand, they evaluate their contribution to specific topics [3].

The bibliometric methods can be implemented at the title and keyword list levels, in the publication abstracts, or possibly even in the overall citation file, to provide specific information on topics and themes. The co-occurrence of keywords would give an idea of the diversity of research topics and could identify the research area's multidisciplinary directions (domains/subdomains) [4].

27.2 BACKGROUND

27.2.1 SMART CITY

Understanding the concept of a smart city is essential to understanding its scope and content.

A practical definition of a smart city is yet to emerge, and various stakeholder definitions have been given from different points of view.

DOI: 10.1201/9781032667478-31

It is hard to standardize the definition, as the intelligence of a city can be as easy as a simple feature provided to a particular group of people.

It is important to note that a city's intelligence can be as complex as a function provided to a certain group of citizens or as complex as an entire administrative process representing the efforts to restructure a governmental process.

It can be as complex as an entire business process that represents the redesign efforts of a business process [5].

In recent years, smart cities have been the subject of significant and growing interest. The world aims to make the city smart due to the urgent need to improve human life. Smart cities are mainly based on the tremendous advancement of information and communication technology, IoT, big data, and cloud computing. These technologies are the key to realizing and developing smart cities [6].

27.3   METHOD

Our goal is to find publications that describe smart city technologies in COVID-19. We used Tranfield's systematic review technique [7], which looked at the advantages and drawbacks of expanding evidence-based research synthesis. The steps in a systematic literature review are as follows: (1) research design, (2) bibliometric data compilation, and (3) analysis and interpretation of results utilizing VOSviewer, a special program that facilitates data management and visualization. Figure 27.1 summarizes all of the stages.

27.4   RESULTS AND DISCUSSION

27.4.1   RESEARCH GROWTH: NUMBER OF PAPERS THAT STUDIED SMART CITIES TECHNOLOGIES IN COVID-19

Figure 27.2 shows the citations by year that these studies received. It can be seen from the increasing number of published studies on smart city technologies in COVID-19 from 2019 to 2022 on the Scopus database. This demonstrates two periods in the publication trajectory. Phase one comprises the period between 2019 and 2021, with the number of publications between 1 and 129 per year. Phase two comprises the period between 2021 and 2022, in which the average annual number of papers for this period is estimated to be between 129 and 34, with a high of 129 papers in 2019. Figure 27.2 shows the total number of publications (cumulative) and the total number of citations on smart city technologies in COVID-19 from 2019 to 2022.

27.4.2   SUBJECT AREA OF SMART CITIES TECHNOLOGIES IN COVID-19

This study provides a list of papers published on the Scopus database about smart city technologies through COVID-19, as summarized in Figure 27.3. This analysis was completed to determine the areas in which smart cities have been addressed. Overall, the distribution indicates that research is emerging in various fields ranging from computer science, social sciences, engineering, energy, environmental science, etc. Based on the papers shown, scientific papers on smart city technologies in COVID-19 are best presented from the perspectives of computer science (25.4%), social sciences (16.2%), engineering (12.6%), and energy (8.9%). We can see that there is some equity between computer science and social sciences, which is logical because it uses computer technologies such as artificial intelligence (AI), challenges, and urban discipline, and the smart city is based on its technological core, which is itself fueled by the advances made in the fields of computer science, which explains the high percentage of these areas.

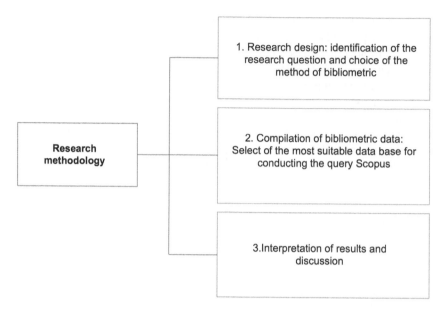

FIGURE 27.1   Research methodology.

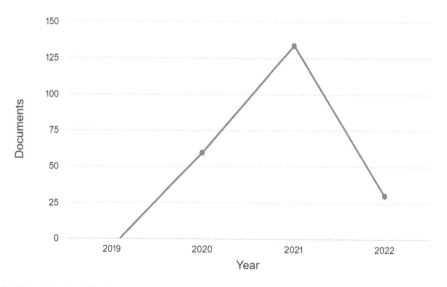

FIGURE 27.2   Total publications (accumulated) and total citations on smart city technologies in COVID-19.

27.4.3   GEOGRAPHICAL DISTRIBUTION OF SMART CITIES TECHNOLOGIES IN COVID-19

We note that the distribution of publications on smart cities is unevenly spread worldwide. At the top of the list is India with 21 publications, followed by Italy with 20 publications, then the United Kingdom with 19 publications, followed by China with 14 publications, etc. (Figure 27.4).

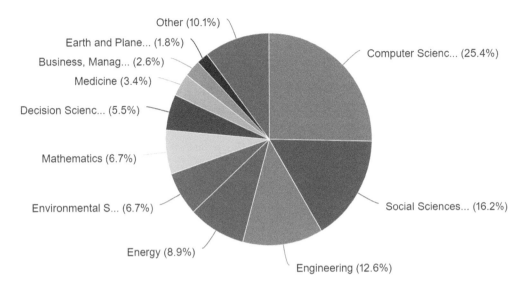

FIGURE 27.3 Subject area of smart cities technologies in COVID-19.

This indicates that these countries' smart city technologies in COVID-19 research have kept different attitudes in an open intellectual environment. For example, China has adopted a more collaborative attitude. Researchers in the field of smart cities were more willing to share their experience with China [8].

27.4.4 Citation Network Analysis

A network citation analysis was performed based on the Scopus data analysis to identify the links and research topics cited in the 225 articles analyzed. For the CAN analysis, the authors matched the keywords provided for each article using the VOS viewer software. Figure 27.5 shows the authors' papers for citations.

The most influential author was Kumitha R.K.R. for his paper: "Smart technologies for fighting pandemics: The techno- and human-driven approaches in controlling the virus transmission" [9]. He got 95 citations. He presented an introduction. This paper contributes to the literature by understanding the human-technology relationship and offering five practical observations for controlling virus transmissions during pandemics.

He was followed by Capolong for his article "COVID-19 and cities: from urban health strategies to the pandemic challenge" [10]. They tried to answer the following question: How can we redesign the concept of Public Health in relation to the built environment and contemporary cities? It was cited in 72 citations.

27.4.5 Keywords Analysis

First, we examined the keyword authorship in the articles and their occurrence, the thematic map, growths, trends, and thematic evolution [11]. Figure 27.6 represents a word cloud to show the 64 most dominant author keywords in the smart city articles. The size of the keywords displays the frequency of their occurrence. I was surprised to see the keyword "Smart cities" in the center of the graph, followed by "technologies" and "Covid", followed by the "pandemic", Smart technologies have been a hot topic in the last few years, particularly in areas such as urban planning and social sciences.

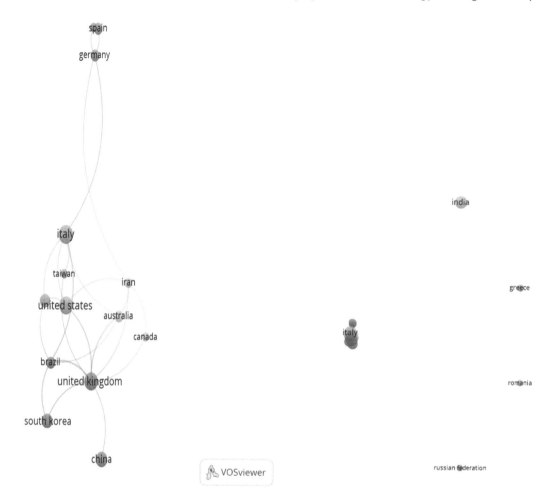

FIGURE 27.4   Geographical distribution of smart city technologies in COVID-19.

A topic map provides a visual representation of four different typologies of topics based on two aspects, namely centrality and density. Density is the strength of internal connections between all relevant keywords describing the search topic, while centrality is the strong connection to other topics, leveraging the keywords [12]. We have four clusters: "smart cities", "covid", "pandemic", and "technologies".

In the first ranking, the most frequently cited words were divided into several groups: smart cities, economic and social effects, health risks, sustainable development, etc.

The second ranking shows the keywords related to smart city technologies such as smart IoT, AI, deep learning, big data, Blockchain, automation, information and communication, etc.

The massive production of data in existing general-purpose networks and cities has opened the possibility for new technological trends that look for added value through the analysis of this digital data.

The IoT is introduced as a tool that supplies a series of special services that give low-level support to the applications offered to citizens. In Said et al. [13], IoT is described as a concept that allows objects to connect to people's environments via the Internet as if they were personal computers. What seems clear is that when talking about some of the things a smart city should offer, the concept of IoT plays a pertinent role, particularly in the information gathering from the environment via sensors and the performance of certain actions via actors.

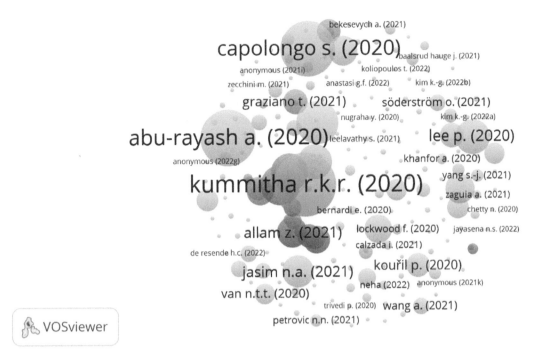

FIGURE 27.5   Citation network analysis (CAN).

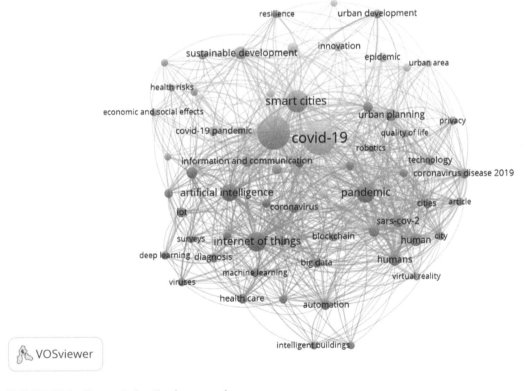

FIGURE 27.6   Keyword visualization network.

Researchers at the Computer Science and Artificial Intelligence Laboratory at the Massachusetts Institute of Technology have recently proposed a powerful new system known as the Data Science Machine [14], a program based on AI capable of identifying patterns in data relationships and predicting from those models better than many human experts and in a significantly shorter time. Smart cities can employ AI to visualize their impact on global warming, pollution levels, and the local environment. Using AI and machine learning in the fight against pollution and energy consumption enables authorities and cities to reach well-informed and environmentally friendly decisions.

27.5 CONCLUSION AND DISCUSSION

The smart city has been of interest to the authors lately. This study presents a bibliometric analysis of the journal's publications between 2019 and 2022. This work adopts two bibliometric methods: performance analysis and science mapping. The database used for this article is Scopus, with a total number of 225 publications. In this type of systematic literature review, our unit of review is the keywords that appear in the paper, which covers the title, abstract, keywords, and each paper's full paper. India, Italy, and the UK are the most productive countries in SC technologies in COVID-19. Computer science and social science are the most focused fields in this research theme. After analyzing the geographical distribution of smart city technologies in COVID-19, the best publications were ranked, and the total number of citations was analyzed. Then, the most common sources were visualized. Khumumma R.K.R. is the most cited author for his paper: "Smart technologies for fighting pandemics: The techno- and human-driven approaches in controlling the virus transmission". He got 95 citations. Then, smart city technologies in COVID-19 keywords are discussed. The most frequently cited words were divided into several groups: smart cities, technologies, COVID, and pandemic. In light of the results obtained and the interdependencies between smart cities and technologies in the COVID-19 context, we deduce the existence of numerous synergies between them. The most commonly used technologies are IoT, big data, and AI. In light of this study, it is mandatory to have the application of technologies in smart cites in the COVID-19 context for an interactive vision of the simulation. It would also be interesting to examine the barriers and contradictions that influence the application of technologies in smart cities in the case of crises such as COVID-19 or in the normal case, and it should focus on the maturity of cities and propose roadmaps or models. To do this, more thorough research should be conducted on a larger number of available documents. To this end, smart city technologies studies can also be found in other databases such as WoS, IEEE, etc. It will be very important to perform a bibliometric review of other databases to compare what is available in the literature. A further research perspective would be to further explore smart city technologies and involve the possibilities of improvement on a pillar such as sustainability, which is one of the most important perspectives, thus strengthening ecological smart cities as a performance factor.

REFERENCES

1. Parlina A., Murfi H., Ramli K. (2019) Smart city research in Indonesia: A bibliometric analysis. In: *2019 16th International Conference on Quality in Research (QIR): International Symposium on Electrical and Computer Engineering*. IEEE, pp. 1–5.
2. Guo Y.-M., Huang Z.-L., Guo J., et al. (2019) Bibliometric analysis on smart cities research. *Sustainability* 11:3606.
3. Nedjwa E., Bertrand R., Sassi Boudemagh S. (2022) Impacts of Industry 4.0 technologies on Lean management tools: A bibliometric analysis. *Int J Interact Des Manuf* 16:1–16.
4. Najwa E., Bertrand R., Yassine M., et al. (2022) Lean 4.0 tools and technologies to improve companies' maturity level: The COVID-19 context. *Procedia Comput Sci* 196:207–216.
5. Yin C., Xiong Z., Chen H., et al. (2015) A literature survey on smart cities. *Sci China Inf Sci* 58:1–18.

6. Yahya Y.A., Raed S., Talal A., et al. (2021) A review on smart cities technologies, challenges, and solution. *J Adv Comput Electron Eng* 6:1–7.

7. Xia B., Chan A.P.C. (2012) Measuring complexity for building projects: A Delphi study. *Eng Constr Archit Manag* 19:7–24.

8. Riva Sanseverino E., Riva Sanseverino R., Anello E. (2018) A cross-reading approach to smart city: A european perspective of chinese smart cities. *Smart Cities* 1:26–52.

9. Kummitha R.K.R. (2020) Smart technologies for fighting pandemics: The techno-and human-driven approaches in controlling the virus transmission. *Gov Inf Q* 37:101481.

10. Capolongo S., Rebecchi A., Buffoli M., et al. (2020) COVID-19 and cities: From urban health strategies to the pandemic challenge. A decalogue of public health opportunities. *Acta Bio Medica Atenei Parm* 91:13.

11. Wang L., Wei Y.-M., Brown M.A. (2017) Global transition to low-carbon electricity: A bibliometric analysis. *Appl Energy* 205:57–68.

12. Mumu J.R., Tahmid T., Azad M.A.K. (2021) Job satisfaction and intention to quit: A bibliometric review of work-family conflict and research agenda. *Appl Nurs Res* 59:151334.

13. Said O., Masud M. (2013) Towards internet of things: Survey and future vision. *Int J Comput Networks* 5:1–17.

14. Wu G., Talwar S., Johnsson K., et al. (2011) M2M: From mobile to embedded internet. *IEEE Commun Mag* 49:36–43.

Section 5

Fintech Innovations and Applications

28 Fintech Innovations for Supply Chain Resilience

*Asma Boujrouf, Sidi Mohamed Rigar,
and L'houssaine Mounaim*

28.1 INTRODUCTION

Crises such as COVID-19 and the Russian War are making considerable changes in the world, making access to financial services harder to reach. Fintech is believed to significantly help the most affected organizations by introducing new technology products and promoting sustainability and financial inclusion.

Fintech covers digital and technological business model innovations in the finance industry. Its innovations are providing new gateways to supply chains and how business is conducted in general (Philippon, 2017). Some authors focus on it as a catalyst for new businesses or business models.

Although Fintech innovations are considered opportunities for advancements in the financial system, their applications present many obstacles. Regulation issues, data protection issues, and an unfair competitive environment are among the challenges facing the application of these new yet powerful innovations.

In the scientific literature on Fintech, research claims that it is undergoing a revolution since it provides new technological innovations to supply chains, thus encouraging the diversification of financing sources and contributing to their financial performance (Chbaik et al., 2022). Furthermore, some preliminary indications indicate that Fintech innovations enhance supply chain resilience. Their positive impact on supply chain financial performance can influence supply chain preparedness, alertness, and agility, identified as the three dimensions of supply chain resilience (Li et al., 2017). Supply chain resilience refers to the capacity of the supply chain to recover from a breakdown. It can maintain a positive adjustment while undergoing a difficult situation, rebound, and emerge more resourceful (Vogus & Sutcliffe, 2007).

Therefore, applying Fintech innovations to the supply chain can drive change and the reconfiguration of supply chain management and breed new risks and challenges. This article attempts to address this issue by answering the following question: **How can the introduction of Fintech innovations impact supply chain resilience?**

Our article seeks to mobilize the existing literature to define concepts, elucidate applications of Fintech innovations to supply chain management, and discuss opportunities and challenges related to adopting this technology. Our study covers five sections. The first introduces the study. The second describes the methodology adopted to conduct the study. The third reviews the concepts of Fintech and supply chain resilience. The fourth examines the link between them, while the last provides conclusions.

28.2 RESEARCH METHODOLOGY

Our research is based solely on secondary data. We have examined papers on Fintech in different databases. We then tried to identify papers that addressed Fintech and its recent developments.

DOI: 10.1201/9781032667478-33

We also reviewed articles on supply chain resilience and supply chain financial performance. Finally, we reflect on the potential links between these concepts.

The article focused on scientific papers and reports. Databases such as Google Scholar, Scopus, and ScienceDirect were deployed with a keyword search using keywords such as 'Fintech Innovations' and 'Supply Chain Resilience'. Several papers were then considered to be examined. Thirty of them, which have a direct link to our main concepts, were selected. The critical feature was to obtain a wide range of relevance for the concepts of 'Fintech Innovations' and 'Supply Chain Resilience' and navigate their findings to find out the nature of the link between the two concepts. This was followed by an investigation of whether other concepts and variables, such as 'financial performance', were involved in the relationship between Fintech innovations and supply chain resilience that would eventually serve as mediating variables between them.

Based on the brief overview of Fintech and supply chain resilience, we concluded the link between them. We have illustrated our findings with a suggested framework where Fintech innovations are linked to each of the elements that form the supply chain resilience process. However, considering that empirical data adds more value to the research, one of the first limitations of the framework is that empirical studies were lacking in their elaboration.

28.3   FINTECH AND SUPPLY CHAIN RESILIENCE

In this section, we will provide a brief literature review of the two main concepts of our research: Fintech and supply chain resilience. We will introduce the emerging concept of Fintech and its recent advances. We will then continue by giving a glimpse of supply chain resilience (Christopher and Peck, 2004). We will not treat all aspects of it, as it is a large field with a lot of new information.

28.3.1   FINTECH

Due to the increasing number of crises today, companies are more challenged to find the appropriate funds to finance their activities. Firms had to turn to new alternatives as their main fund providers faced certain regulatory constraints. Thus, in this context, Fintech startups, firms, and innovations have risen to bridge this gap.

Given the fairly recent emergence of Fintech, the scientific literature on this field is not extensive. There is no consensus on the scope of the term Fintech, which remains relatively new (Varga, 2017).

Fintech is the contraction of the terms financial and technology. It refers to the innovation of technology-based financial services. It profoundly transforms financial products, payment methods, business models, market actors and structures, and even money (WBG, 2022). In general, its main objective is to use technology to improve the efficiency of financial systems.

On the one hand, Fintech is considered a company or a set of companies that provide the market with technological innovations to support financial transactions among businesses and consumers. On the other hand, some consider Fintech to be technological innovations brought to the market to improve financial services and encourage disintermediation (Findexable, 2020).

Fintech innovations are revolutionizing the financial sector. They are bringing up improvements to traditional financial methods in a way that fuzzes boundaries, promotes disintermediation, and democratizes access to financial services. Therefore, they are contributing to global financial inclusion.

Examples of innovations at the heart of Fintech today include cryptocurrencies and blockchain, new digital advisory and trading systems, artificial intelligence and machine learning, peer-to-peer lending, equity crowdfunding, and mobile payment systems. Based on the scientific literature, Fintech segments can be organized into four major categories: digital lending, payments, blockchain, and digital wealth management.

From another perspective, the World Economic Forum (Schwab, 2015) distinguishes between six areas of Fintech based on the type of offer. They are organized as follows:

- Payment methods, whether electronic or mobile currencies, include all decentralized and electronic payment methods that do not necessarily go through the banking system.
- Insurance includes new forms of insurance (decomposed insurance, connected insurance, etc.).
- Deposits and loans: These concern new platforms that innovate in the method of credit evaluation and the search for financial resources through the introduction of peer-to-peer lending; changing consumer preferences (mobile or virtual banking); alternative lending; and virtual technology.
- Crowdsourcing, such as crowdfunding.
- Investment management includes social trading, retail algorithmic trading, outsourcing processes, etc.
- Market supply concerns the availability of new, smarter, and faster machines on the market. It also concerns artificial intelligence, big data, and new platforms that improve connectivity between stakeholders, making the market more accessible, efficient, and with more cash flow.

For Findexable, the global data and analytics company for private market Fintech, which provides the Global Fintech Index and the Fintech Diversity Radar that organizations use to measure their growth and benchmark themselves with other competitors, Fintech involves ten major fields:

- Payments and transfers;
- Crypto; business and accounting;
- Analytics and risk management;
- Insurance;
- Infrastructure;
- Wealth;
- Capital markets; and
- Lending and market places.

Looking at these categories, it is apparent that Fintech competes with traditional financial institutions. It is understandable why these two ecosystem actors are generally positioned as competitors.

In addition, Fintech is using new technologies, such as blockchain, big data, artificial intelligence, cloud services, etc., to offer more efficient and user-oriented products and services (Varga, 2017). Its main drivers are twofold. They are identified as ubiquitous connectivity through mobile devices connected to the internet and communication networks, as well as low-cost computing and data storage.

Fintech innovations can, therefore, be useful in two ways. First, they will demonstrate to what extent technology can provide low-leverage solutions. Second, they are financed with far more equity than existing practices.

28.3.2 SUPPLY CHAIN RESILIENCE

Before addressing supply chain resilience, it is essential to examine the existing resilience literature from an organizational perspective. Then it is convenient to extend the concept to the supply chain level.

Resilience refers to the intrinsic capacity of a system to adjust its functioning before or after changes and disruptions so that it can continue to function even after a major incident or during ongoing stress (Hollnagel, 2006). From the same perspective. Vogus and Sutcliffe (2003, 2007) have defined it as 'the maintenance of positive adjustment under challenging conditions so that the organization emerges from those conditions strengthened and more resourceful.' The latter definition seems to sum up the whole process of being resilient and its main objective.

In addition, resilient organizations need resilient individuals, groups, and processes. The level of resilience of individuals and systems highly defines how resilient the organization will be. Resilience is thus embedded in the culture of the organization.

In times of disruption, sustainability and resilience are essential prerequisites for every trade and investment decision. To do this, improving the way risk is monitored and managed is essential. Therefore, it is difficult to address the concept of resilience without referring to the concept of risk. Organizational resilience is considered a risk management tool and is part of reactivity through adaptation and proactivity through anticipation and prevention. In all cases, it does not allow for the total and radical elimination of risk but aims to mitigate it through innovative methods (Dauphiné & Provitolo, 2007). Consequently, innovation is essential to achieving resilience.

From a supply chain perspective, resilience is considered the ability of a supply chain to recover from a failure and thrive. Melnyk et al. (2014) see it as a sequence of two major phases: resistance capacity and recovery capacity. While the former consists of facing and responding to an expected or an unexpected shock, the latter focuses on restating the normal or rebounding strongly.

Kamalahmadi and Parast (2016) have summarized the existing literature and developed a framework of three fundamental steps that form the supply chain resilience process: Anticipation, Resistance, and Recovery and responses. The first phase includes all the planning and preventive measures an organization needs to take proactively. The second phase suggests taking control of the organization's processes during the disturbance by ensuring the continuity of operations and absorbing the shock. Finally, the third phase encompasses the measures implemented to respond with agility to the disruption and restore a higher position. The organization's available resources and their use often generate a competitive advantage.

However, the growing number of activities, flows, and actors leads to increased risks within supply chains. Organizations are under pressure to manage their supply chain flows and the resulting risks, from increasing customer requirements to reducing costs, shortening lead time, and lowering inventory. In this sense, research has shed light on the risks incurred during the different stages of the supply chain. This step helps address each of the supply chain processes and improve them. Risk is an element that spreads across the whole chain; its impact generally affects the entire chain.

Tang and Nurmaya Musa (2011) define the concept of supply chain risk as any unpredictable event or disturbance that negatively impacts one or more components of the supply chain and thus its ability to achieve its performance objective both at the level of the companies involved and the service provided. Thus, the main risks in the supply chain are related to procurement, manufacturing, warehousing, and shipping. These types of risks often stem from the lack of information sharing between the different actors in the chain. This highlights how complex and important the management and control of information flows in supply chains are.

Supply chains generally face global and specific risks that directly affect operations. From a global perspective, some examples of macro risks that can hinder the organization's objectives are given below.

- Currency hedging risk of the local partner
- Payment defaults
- Political risk
- Intercultural risks

In addition, supply chains face not only macro risks but also micro risks. They are linked to its four parts: procurement, production, storage, and distribution. For example:

- Delivery delays
- The bullwhip effect
- Supply shortages
- Higher input prices
- Higher logistic costs, and more.

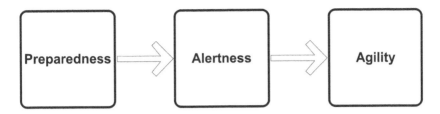

FIGURE 28.1   Supply chain resilience process.

The philosophy of supply chain management today emphasizes that all supply chains aim to maximize members' value for their companies and clients. It is based on continuous cooperation and external integration (Balan, 2008).

In summary, to ensure resilience in supply chains, organizations are suggested to invest more in supply chain management, risk prevention, and response practices that make the supply chain better prepared, more alert, and more agile in the face of disruptions. Li et al. (2017) summarized all this in a three-fold process for supply chain resilience (Figure 28.1).

28.4   FINTECH FOR SUPPLY CHAIN RESILIENCE

Supply chain financing issues have become critical since the financial crisis, making it more difficult to access bank financing. It also caused each supply chain member to delay payments and collect revenues as quickly as possible. The COVID-19 crisis, the Russian war, and subsequent crises deepened this matter and created more emerging financial issues through supply chains, forcing organizations to seek resilience to weather these new challenges and adapt (Gaugain et al., 2014).

Fintech has developed as a disruptive phenomenon that provides an alternative to the financial institutions blamed for the global financial crisis. The loss of trust in the sector's historical institutions proved to be the ideal context for the development and growth of Fintech (Varga, 2017). However, some applications of these technologies aim not simply to replace these trust mechanisms and machines but to establish trust where it is lacking.

Although crises tend to have a negative impact on processes, they remain an opportunity for organizations to invest more in new technologies. They have therefore accelerated the use of some innovations, including Fintech. According to the Financial Stability Board (Board, 2019), Fintech is a technologically enabled innovation in financial services that could result in new business models, applications, processes, or products with a material effect on financial institutions and the provision of financial services. Therefore, it is an opportunity to optimize financial flows within supply chains.

Otherwise, the concept of supply chain per se encompasses three types of flow: material, financial, and information flows. Fintech innovations have been developed to address the shortcomings of financial flows and improve them.

The problem of financing assets in the supply chain is made even more acute because these assets are difficult for banks to control. They are thus reluctant to finance them. Therefore, it is urgent to find innovative financing solutions to optimize the global chain (reduce the need for financing and its cost). These solutions can only be found through better collaboration between customers and suppliers.

While Fintech innovations have made it to the supply chain, its innovations are many. One of the applications of Fintech innovations is the payment system. Payment systems have been an early target of Fintech companies. Rysman and Schuh (2017) review the literature on retail payments and discuss three recent innovations: mobile payments, real-time payments, and digital currencies. These innovations will undoubtedly improve financial transactions in the supply chain, but they are unlikely to radically change the payment system. In particular, they are unlikely to reduce their reliance on short-term receivables.

Additionally, among the Fintech revolutions that changed the supply chain is blockchain. Blockchain is gaining in popularity due to its wide range of applications, from interbank transactions to supply chain management (Hussain et al., 2021). It refers to an extensive and secure information exchange platform that involves all stakeholders. It is a new concept and mechanism that may have disrupted the principle of transferring virtual currencies at first and has today turned the attention of actors in other sectors, such as finance and supply chain.

Blockchain is "a technology for storing and transmitting information that is transparent, secure, and operates without a central control body using an unforgeable digital database on which all exchanges made between its users" (Maleh et al., 2022). Its main promise is to create transparency, better coordinate actors, improve traceability, and reduce the number of intermediaries within the chain (Imeri, 2021).

On the financial level, the blockchain is considered a distributed digital register that decentralizes data sharing and ensures the secure storage and transmission of information without any control authority. It is believed that it would make it possible to secure financial exchanges throughout the supply chain (Hug, 2017; Kin et al., 2018).

In addition, blockchain is designed to improve traceability within the supply chain. Using blockchain as a traceability means appears to be an efficient and secure way to store financial information, disseminate it, and authenticate it (Galvez et al., 2018). It tends to decrease processing costs through a certain level of automation associated with consensus algorithms and improves the transparency of financial transactions (Ke & Tang, 2022). It contributes thus to enhancing supply chain alertness.

The benefits of integrating Fintech innovations into the supply chain make the aforementioned process more efficient. Its financial performance thus improves, allowing it to be better prepared for potential disruptions and have enough funds to stay alert and agile.

On another note, Fintech innovations are believed to be one of the key drivers of preparedness, alertness, and agility due to their ability to improve supply chain financial performance. Li et al. (2017) have revealed through their empirical study that supply chain resilience is based on three main dimensions: preparedness, alertness, and agility. While proactivity prepares the supply chain better for new changes, reactivity makes it more alert and agile in the face of disruptions. Therefore, supply chain resilience is mainly based on proactivity through preparation and reactivity through alertness and agility.

Therefore, how Fintech innovations affect supply chain resilience can be presented as follows (Figure 28.2).

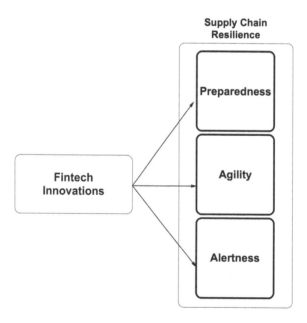

FIGURE 28.2 The link between Fintech innovations and supply chain resilience.

To demonstrate the effect of Fintech innovations on the supply chain resilience process, it is rec-ommended to address their effect in each phase. Empirical research is also required.

28.5   CONCLUSIONS

Although the concept of Fintech is new, it has captured the attention of researchers and experts in all fields. This new technology has become a solution with multiple uses. Therefore, this paper aims to provide some insight into the concept of Fintech and its application to supply chain management. It also seeks to address the impact of its innovations on supply chain resilience.

The Fintech concept refers to organizations or innovations made by organizations to improve the efficiency of the financial system. The scientific literature on Fintech indicates that its applica-tions may revolutionize the supply chain field. Many indicators show they can improve supply chain financial performance with supply chain resilience by making it more prepared, alert, and agile. However, the literature in this sense is still scarce and requires more in-depth research.

Although Fintech is considered an opportunity for the supply chain, it does not avoid the many threats hindering its application. For example, startups operating in the financial sector are not subject to the same regulations as banks. In the future, it is reasonable to imagine that regulatory exemptions will decrease, requiring an adaptation of the models of new financial actors (Laidroo & Avarmaa, 2020).

In addition, there has been a particular emphasis on data security and privacy management through digital tools in recent years. Therefore, Fintech will need to adapt to meet regulatory requirements and reassure users concerned about their data protection (Lee & Shin, 2018).

In addition, as Cochrane (2014) argues, some barriers are not technological, particularly with respect to accounting and taxation, to the extent that these transactions would generate capital gains. If regulators wish to reduce systemic reliance on short-term leverage, they must identify issues that often fall directly outside their traditional regulatory horizon. This presents the complex problem of unfair competition between new entrants and traditional actors. Ensuring a fair competitive envi-ronment is, in fact, a historical goal of regulation. From a microeconomic perspective, Darolles (2016) argues that regulators should indeed ensure a fair competitive environment. However, this reasoning hardly applies to many of the disruptions plaguing the finance industry.

In summary, there are indications that Fintech innovations can help build supply chain resilience. This paper briefly presents the concepts of Fintech innovations and supply chain resilience. We then tried to navigate their literature to determine how the former could affect the latter and what connections could be drawn. Finally, we relied on the findings to suggest a framework under which the link between the two concepts could be further examined. However, this link still needs to be thoroughly investigated. For example, consider how Fintech innovations affect each supply chain resilience phase. How can they help ensure preparedness? How do they operate in terms of alertness and agility? Contextualized empirical studies are also required.

REFERENCES

Balan, C. (2008, June). The effects of the lack of coordination within the supply chain. *The AMFITEATRU ECONOMIC Journal, Academy of Economic Studies - Bucharest, Romania,* 10(24), 26–40.

Board, F.S. (2019). Evaluation of the effects of financial regulatory reforms on small and medium-sized enterprise (SME) financing. *FSB, Basel.* https://www.fsb.org/2019/11/evaluation-ofthe-effects-of-financial-regulatory-reforms-on-small-and-medium-sized-enterprise-sme-financingoverview-of-responses-to-the-consultation.

Chbaik, N., Khiat, A., Bahnasse, A., and Ouajji, H. (2022, 1 janvier). The application of smart supply chain technologies in the moroccan logistics. *Procedia Computer Science, 12th International Conference on Emerging Ubiquitous Systems and Pervasive Networks / 11th International Conference on Current and Future Trends of Information and Communication Technologies in Healthcare,* 198 (pp. 578–83). https://doi.org/10.1016/j.procs.2021.12.289.

Christopher, M., and Peck, H. (2004, 1 janvier). Building the resilient supply chain. *The International Journal of Logistics Management,* 15(2): 1–14. https://doi.org/10.1108/09574090410700275.

Cochrane, J.H. (2014). Challenges for cost-benefit analysis of financial regulation. *The Journal of Legal Studies*, 43(S2), S63–S105.

Darolles, S. (2016). The rise of fintechs and their regulation. *Financial Stability Review*, (20), 85–92.

Dauphiné, A., and Provitolo, D. (2007). La résilience: Un concept pour la gestion des risques. *Annales de geographie*, 654(2), 115–25.

Findexable (2020). The Global Fintech Index 2020.

Galvez, J.F., Mejuto, J.C., and Simal-Gandara, J. (2018). Future challenges on the use of blockchain for food traceability analysis. *TrAC Trends in Analytical Chemistry*, 107, 222–32.

Gaugain, M., Viviani, J.L., and Vo, T.L.H. (2014). Collaborative finance between buyer and supplier as a means of competitiveness reinforcement inside supply chains. *Revue francaise de gestion*, 239(2), 107–20.

Hollnagel, E., Woods, D.D., and Leveson, N. (2006). *Resilience Engineering: Concepts and Precepts*. Ashgate Publishing, Ltd., Farnham.

Hug, M. (2017). Un nouvel outil numérique pour la fiabilisation des supply chains: la blockchain. *Réalités Industrielles*, 2017, 106.

Hussain, M., Javed, W., Hakeem, O., Yousafzai, A., Younas, A., Awan, M.J., Nobanee, H., and Zain, A.M. (2021, janvier). Blockchain-based IoT devices in supply chain management: A systematic literature review. *Sustainability* 13(24), 13646. https://doi.org/10.3390/su132413646.

Imeri, A. (2021). Using the Blockchain Technology for Trust Improvement of Processes in Logistics and Transportation (Doctoral dissertation, University of Luxembourg, Esch-sur-Alzette, Luxembourg).

Kamalahmadi, M., and Parast, M.M. (2016, 1 janvier). A review of the literature on the principles of enterprise and supply chain resilience: Major findings and directions for future research. *International Journal of Production Economics*, 171, 116–33. https://doi.org/10.1016/j.ijpe.2015.10.023.

Ke, J., and Tang, Z. (2022). Innovation of supply chain finance model based on blockchain technology. In *International Conference on Cognitive based Information Processing and Applications (CIPA 2021)* (pp. 680–6). Springer, Singapore.

Kin, B., Spoor, J., Verlinde, S., Macharis, C., and Van Woensel, T. (2018). Modelling alternative distribution set-ups for fragmented last mile transport: Towards more efficient and sustainable urban freight transport. *Case Studies on Transport Policy*, 6(1), 125–32.

Laidroo, L., and Avarmaa, M. (2020). The role of location in FinTech formation. *Entrepreneurship & Regional Development*, 32(7-8), 555–72.

Lee, I., and Shin, Y. J. (2018). Fintech: Ecosystem, business models, investment decisions, and challenges. *Business Horizons*, 61(1), 35–46.

Li, X., Wu, Q., Holsapple, C.W., and Goldsby, T. (2017, 1 janvier). An empirical examination of firm financial performance along dimensions of supply chain resilience. *Management Research Review*, 40(3), 254–69. https://doi.org/10.1108/MRR-02-2016-0030.

Maleh, Y., Lakkineni, S., Tawalbeh, L.A., & AbdEl-Latif, A.A. (2022). Blockchain for cyber-physical systems: Challenges and applications. *Advances in Blockchain Technology for Cyber Physical Systems*, 11–59.

Melnyk, S.A., Zobel, C.W., Macdonald, J.R., and Griffis, S.E. (2014). Making sense of transient responses in simulation studies. *International Journal of Production Research*, 52 (3): 617–32. https://doi.org/10.1080/00207543.2013.803626.

Philippon, T. (2017). L'opportunité de la FinTech. *Revue d'économie financière*, 127, 173–206. https://doi.org/10.3917/ecofi.127.0173.

Rysman, M., and Schuh, S. (2017). New innovations in payments. *Innovation Policy and the Economy*, 17(1), 27–48.

Schwab, K. (2015). World economic forum. *Global Competitiveness Report* (2014–2015).

Tang, O., and Nurmaya Musa, S. (2011, 1 September). Identifying risk issues and research advancements in supply chain risk management. *International Journal of Production Economics, Leading Edge of Inventory Research*, 133(1), 25–34. https://doi.org/10.1016/j.ijpe.2010.06.013.

Varga, D. (2017). Fintech, the new era of financial services. *Vezetéstudomány-Budapest Management Review*, 48(11), 22–32.

Vogus, T., and Sutcliffe, K. (2007). Organizational Resilience: Towards a Theory and Research Agenda, 3418-22. https://doi.org/10.1109/ICSMC.2007.4414160.

Vogus, T.J., & Sutcliffe, K.M. (2003). Organizing for resilience. *Positive Organizational Scholarship*, 94–110.

World Bank Group (2022). Fintech and the Future of Finance Overview Paper.

29 Strategic Tools for the Decision to Outsource Maintenance Activities in Moroccan Airports

Ahlam Boutahar, Mohamed Ben Ali, and Said Rifai

29.1 INTRODUCTION

The growth of outsourcing practices is the result of the economic climate. As mentioned in the early 1990s, it was discussed in academic business studies and operational training [1]. Moroccan companies seek new approaches to maintain and increase their competitiveness to cope with globalization, leading to increased market competition. They are gradually delegating some of their activities to specialized service providers.

Outsourcing can be defined as the act of entrusting an activity, usually performed internally, to an external service provider. According to [2], several firms have outsourced their non-core activities. The peculiarity of the current wave of outsourcing is that it affects an ever-increasing number of support activities in the value chain. This operation does not necessarily mean that the outsourced activity is less important for the company's performance, as mentioned in [3]. Many other functions across all industries have been actively outsourced, including information systems and technology, telecommunications, facility management and maintenance, food services, and management services. Among the forerunners of this new form of outsourcing are Eastman Kodak [4] and Continental Bank [5], among the first to outsource their entire IT operations. In the Moroccan context, this phenomenon is new. Indeed, many authors are working on the issue of outsourcing in all capacities. As for maintenance outsourcing, the given definition is quite like outsourcing in general. Thus, maintenance outsourcing can be defined as "the act of transferring some maintenance activities or the maintenance of specific equipment that was done internally by the company, as well as the decision-making responsibilities to external suppliers" [6].

The decision to outsource any significant function, such as maintenance, should not be taken lightly, and careful consideration of all critical issues is vital. The manager can only assume this decision in the long term if he has the necessary strategic elements [1]. The problems become much more numerous and complex concerning the resorting of the airport industry to outsource maintenance.

Many outsourcing approaches or strategies have been developed, and various tools can be used; the challenge is to integrate them in the right place at the right time to avoid dissatisfaction and miss the missing fits of outsourcing. Even if there is little research on outsourcing activities in the Moroccan airport field, there is abundant literature on the academic understanding of make-or-buy approaches, especially a rich literature on outsourcing synthesized by [7], who used transaction cost theory as the dominant paradigm [8]. It has been used in many empirical studies, as well as resource theory. This chapter is presented as follows: Section 29.1 gives a brief overview of outsourcing in general, followed by Section 29.2, which discusses outsourcing in Moroccan airport maintenance. Section 29.3 presents the research strategy for collecting and evaluating the literature. Section 29.4

DOI: 10.1201/9781032667478-34

aims to give the different tools that managers rely on to make a good decision on outsourcing maintenance in the airport industry; Section 29.5 is a discussion that summarizes the main tools. Section 29.6 deals with some limitations. Finally, Section 29.7 concludes the chapter.

29.2 RELATED LITERATURE

This section provides an account of the knowledge and ideas established in outsourcing, including a critical appraisal of the literature on maintenance outsourcing, strategy, and theories associated with theoretical tools that can be applied in the Moroccan civil aviation field. It includes reviewing current, relevant, and significant views about maintenance outsourcing and assessing previous research. This research reveals a lack of knowledge of the primary criteria for maintenance managers to outsource their activities and select service providers.

29.2.1 OVERVIEW OF OUTSOURCING

In the managerial literature, there are several definitions of outsourcing. We can cite those that encompass the same vision of the phenomenon in a broader sense: outsourcing transfers some activities to outside contractors to gain various benefits [9]. According to the "Outsourcing Barometer 2003 [10]: practices and trends in the outsourcing market in France," Ernst and Young define outsourcing as delegating certain functions on a contractual basis to service providers outside the company.

This definition represents a more general approach: Outsourcing is a management method based on delegating management of maintenance, accounting, IT, and administration.

According to [11], outsourcing is divided into four different levels, namely, strategic and traditional, resulting from the intersection of two criteria [12] (Table 29.1).

29.2.2 MAINTENANCE OUTSOURCING IN AIRPORTS

This section reviews the literature on maintenance outsourcing, including its reasons, benefits, and drivers. Organizations are becoming more conscious of the economic value of maintenance in the face of globalization's increasing market competition [13]. Decision-makers are deeply concerned with maintenance because it has experienced a significant transformation in its position within businesses and is a crucial strategic component for attaining business objectives [14]. Financially successful businesses must fulfill better performance standards while spending less money. They aim to minimize operating costs so they can better compete. Organizations can take advantage of appropriate expertise, the newest methods, and technologies by outsourcing maintenance. A CEGOS observatory study found that France is the second-largest European outsourcing market, with 82% of French enterprises outsourcing some aspect of their operations [15]. To cope with the tough competition that all players in the market are facing, there is no miracle solution except to be more efficient. One of the keys to achieving this goal is to focus on these strengths (20%) to sharpen and improve them and thus become even more efficient [16]. For this reason, several

TABLE 29.1
Levels of Outsourcing

Strategic Outsourcing (With the Transfer of Resources)	Strategic Outsourcing
Traditional Outsourcing (With the transfer of resources)	Traditional Outsourcing

TABLE 29.2

Drivers of Outsourcing in Moroccan Industries

Refocus on the core business	20%
Competition	20%
Cost efficiency	17%
Meeting international standard	14%
Customer requirement	11%
The quest for flexibility	9%
Need to innovate	9%

reasons (Table 29.2) push organizations to outsource activities to companies that perfectly master the associated know-how, such as maintenance, IT, gardening, reception, and cleaning.

Like most businesses, airports are also outsourcing. The airport industry is susceptible to technological developments and changes in the world. For this reason, it must keep up with all changes in general. In doing so, it tries to do as much as possible without significantly impacting costs. At the same time, it is important for each unit in the organization to focus on its own business and to work safely. The areas where airports go to outsourcing can be listed as ground handling, catering, cleaning, fuel, cargo facilities, and maintenance of airport facilities. In May 2004, when the growing importance of private equity and market-driven decisions began, a project was launched to outsource handling, which includes all tasks related to the embarkation, disembarkation, transit, and land transportation of passengers; this was entrusted jointly by the State to ONDA and Royal Air Maroc, following a directive of the European Commission relating to the management of the European sky ("Open-Sky" to which Morocco is associated), recommending the introduction of competition in the handling activity.

The Moroccan airport's authority currently manages a network of 25 airports, including 19 international ones. The Mohammed V airport is one of the six main airports with air traffic of more than 1 million passengers and managed 41% of total traffic in 2019, followed by the airports of Marrakech (26%), Agadir (8%), Fez (6%), Tangier (5%), and Rabat (4%) [17]. Due to the rapid improvement of technology and the complexity of installations, the airport industry has been re-engineering its perspectives toward operation management and seeking to cut costs to improve efficiencies. Considering good maintenance management and outsourcing maintenance activities has become a method to realize the full benefit [18]. However, the airport industry is a highly regulated sector in which decisions about the process of outsourcing maintenance must be evaluated not only based on cost, according to airport maintenance managers, but the primary purpose of a maintenance service is also to ensure passengers and users by providing safety, security, quality, and maintaining optimal performance for all airport installations.

Since 2013, ONDA has started outsourcing maintenance EDS equipment along with other airport face-upkeep tasks. Financial considerations solely drive this wave of outsourcing. The absence of an internal ability to deliver these services was an additional element supporting this phenomenon. Then, outsourcing expanded to include all forms of terminal equipment. A new maintenance policy based on the outsourcing of the maintenance of airport equipment and facilities by launching complete maintenance contracts (labor and spare parts) of the Service Level Objective type has been adopted within the framework of the ONDA strategic plan, and more specifically, the rationalization of expenses and the improvement of maintenance performance (corrective and preventive maintenance) [19]. This new strategy of outsourcing the maintenance of airport equipment has been adopted to overcome the following:

- The intricate nature of the airport's machinery and the technological installations
- The human resources available and the competencies they possess
- The importance of certain pieces of machinery and cutting-edge technology in connection with airport operations
- The challenges and restrictions associated with the availability of replacement components
- Regularity, safety, and security issues

Despite the substantial investment, there are still not enough maintenance services available. Because of the rapid advancement of knowledge and the opening of the aviation markets [20], technology, globalization, and changes in the microenvironment [21], as well as intense competition [22,23], there is an increasing need for high-quality ground services. Each country is working extremely hard to meet the growing demand for airport facilities in terms of human and material resources. Providing excellent customer service that satisfies people's wants and expectations while monitoring shifts in passenger behavior is one of the managers' top priorities [24].

29.3   THE METHODOLOGY OF SCIENTIFIC ANALYSIS

We adopted the classic five-stage literature review process established by [25] to identify the development of outsourcing research, as shown in Figure 29.1. This method allows an objective synthesis of current data. Articles were identified by searching the four academic publication indexing databases: Science Direct, Scopus, Emerald Insight, and IEEE Xplore. A list of keywords, including outsourcing, Airports, Maintenance, Morocco, and a mixture of AND/OR operators, filtered out relevant articles. Finally, the information collected from the database was reviewed and organized into several parts for this article.

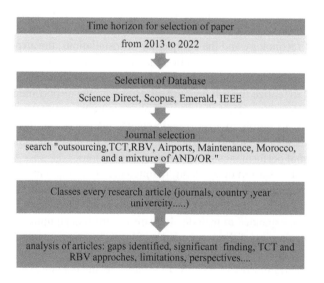

FIGURE 29.1   A literature review process.

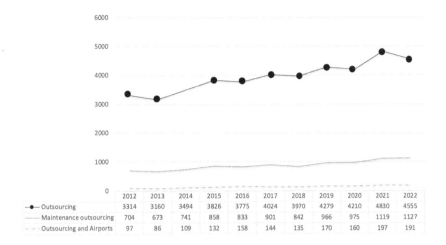

	2012	2013	2014	2015	2016	2017	2018	2019	2020	2021	2022
Outsourcing	3314	3160	3494	3826	3775	4024	3970	4279	4210	4830	4555
Maintenance outsourcing	704	673	741	858	833	901	842	966	975	1119	1127
Outsourcing and Airports	97	86	109	132	158	144	135	170	160	197	191

FIGURE 29.2 The number of publications on Outsourcing, Maintenance outsourcing, and outsourcing in Airports.

29.4 THEORETICAL TOOLS

Figure 29.2 shows that while the number of studies on outsourcing exists and is steadily increasing massively in the literature, research on maintenance outsourcing in conjunction with the airport industry is still in its infancy. For example, in 2012, there were 3314 papers on outsourcing. Still, only 704 papers were published in the context of maintenance outsourcing and even fewer in the context of outsourcing to the airport sector, with only 97 papers. This figure also illustrates the parallel evolution of research in these areas over the years. For the year 2021, there are 4830 articles on outsourcing. These articles deal with different methods for making this decision. We can see that no specific conceptual framework addresses the topic of outsourcing decisions.

To gain a better understanding of this phenomenon, however, the Transaction Costs (TCT) and Resource-Based View (RBV) methodologies, which are both tried-and-true methods, are utilized. It is essential to have a solid understanding of these two theoretical frameworks (RBV and TCT), which serve as the focal point of our research, to comprehend the phenomenon of outsourcing maintenance labor in the airport industry. Analyzing its traditional tools prompts the following queries: What tasks can be contracted out? Why do businesses choose to do this? How does outsourcing maintenance work for airport facilities? Public corporations outsource their support tasks to make the most of their resources, gain access to knowledge that is not readily available, and adapt to their environment.

29.4.1 THE THEORY OF TRANSACTION COSTS (TCT)

A general theoretical approach applied to the issue of outsourcing is the Transaction Cost Theory, essentially developed by Williamson [26], for firms' boundary decisions. Transaction cost theory emphasizes using the market if it reduces transaction costs compared to doing the activity internally [27]. Once the core and non-core activities are identified, TCT can help determine which elements of the non-core competencies are desirable to outsource. The idea behind this theory is that you have a good argument for the company's principal concern in outsourcing maintenance: minimizing production and transaction costs.

Savings is the primary concern of economic organizations. Williamson [28] explains that the firm prevails over the market when it reduces transaction costs. Simply put, when applied to the problem of outsourcing, the hierarchical organization is considered an alternative solution when it reduces information costs (costs of finding a partner). However, it reveals other organizational costs of supervision or control costs related to bureaucracy and supervision of long-term contracts (the costs of negotiating and concluding the agreement and the costs resulting from incompleteness or default of the contract) in addition to the opportunity costs represented by the immobilization of certain assets to ensure compliance with the terms of the agreement (the bond).

29.4.2 THE RESOURCE-BASED VIEW (RBV)

The TCT lost its relevance as cost was the only decision criterion. The resource-based theory of RBV was born. It considers the firm as a set of assets and resources that, if used distinctly, can create a competitive advantage. It determines the factors that underlie the competitive advantage of firms [29]. Four criteria determine whether a resource is the basis of the core business and whether it constitutes a competitive advantage for the firm: the value, the scarcity, the inimitability, and the non-substitutability of the resource.

The concept of a company's core business and the causes that lead to outsourcing are indeed topics covered by resource theory. Through outsourcing, the theory's logic provides a theoretical framework that enables the corporation to refocus on its "core business." This approach involves transferring a company's non-strategic tasks to an outside service provider to refocus its internal resources on strategic competencies. Outsourcing is gaining access to assets and competencies not present within. The strategic competencies of the firm should not be affected by outsourcing because the key competencies are supposed to be activities that offer a competitive advantage in the long term. They must therefore be imperatively protected and rigorously controlled [30]. It is also essential to focus on the resources and skills forming the "core business," which the company must keep in-house, and the other activities to be outsourced. Therefore, the kind of outsourcing is determined by the level of performance. It is recommended to outsource peripheral activities without a long-term competitive advantage and to protect resources with superior skills and capabilities to those in the market [31]. According to these theories, as they are applied to the airport industry, where maintenance is crucial and valuable because it imposes some control over the service, poor maintenance can damage safety, security, and quality. The importance and irreplaceability of RBV must therefore be considered when deciding whether to outsource the upkeep of airport infrastructure; in other words, the outsourcing organization must concurrently grow, refresh, and adapt its capabilities [32].

29.5 DISCUSSION

According to several research findings, transaction costs might be considered an indicator of the choice to outsource. More specifically, one should consider outsourcing a task whenever a minimal cost is associated with the transaction. These expenses are determined by the particular asset, the transaction frequency (whether infrequent or not), and the degree of uncertainty involved (future supplier behavior, environment). Williamson's argument in [8] strongly emphasizes asset specificity (geographic specificity, physical specificity, human specificity, dedicated assets, site specificity, and temporal specificity). Therefore, outsourcing is less attractive when there is a higher level of support specificity, and the opposite is true when there is a lower level of support specificity.

According to the TCT, we may draw the following conclusions based on these factors: the more specialist the equipment is, the less desirable it is to outsource airport equipment maintenance, and vice versa.

The maintenance service value chain needs to be controlled more as there is more uncertainty. It is advised to insource if there is uncertainty over a provider's performance since the impact of the

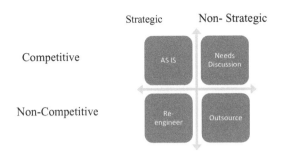

FIGURE 29.3 Matrix decision source.

Contribution to Operational Performance

FIGURE 29.4 Matrix decision performance source.

provider's underperformance can be highly destructive to the outsourcing business [12]. However, it is the opposite regarding technological uncertainty [33].

However, to reap the benefits of the TCT, it is necessary to ensure the following things: airport maintenance managers are aware of the internal costs (production and transactions) and external management costs; the development of the specificity of their assets (airport facilities) (technical nature and benefits); the internal production costs and the total acquisition cost; and airport equipment maintenance department needs in terms of nature, depth, and breadth of knowledge.

It is possible to employ various configurations for RBV when applied to airport equipment maintenance, provided that the scope is well specified. This is important since the incorrect option might negatively affect quality, safety, and cost. If the company has too many different lines of business to engage in, and because of this, it wants to concentrate on its primary one, the direction of the firm is taken into account while adapting decision support matrices [34] (Figures 29.3 and 29.4).

29.6 LIMITATIONS

Outsourcing involves handing over direct control of a support function or process to a third party. As such, it carries certain risks. The outsourcing project should not be considered haphazardly, as several risks can arise if the project is not well thought out:

- Timeframes for service delivery.
- Confidentiality and security.
- Lack of flexibility: the contract may be too rigid to accommodate change.
- Management difficulties: changes in the outsourced company may cause friction.
- Instability: problems with providers (bankruptcy).

29.7 CONCLUSION

This evaluation of the relevant literature contributes to the existing body of information regarding outsourcing maintenance. It covers a significant body of literature on the evolution of the business environment in the airport sector, which has recently undergone considerable change and is becoming more complex and unpredictable, forcing organizations to reassess their various resource optimization and management strategies. The airport sector has recently undergone considerable change and is becoming more complex and unpredictable. We have strongly emphasized a set of tools that, when combined, form a decision-making framework. This framework will assist in comprehending the phenomenon of outsourcing maintenance at airports. It will be utilized by airport managers to specify the tasks that will be outsourced for their respective businesses.

Because it considers asset specificity, frequency, and uncertainty as the most relevant determinants of the transaction, the transactional perspective developed in this chapter has proven to be very useful in evaluating the maintenance outsourcing process. This is because it addresses the economic aspect of the decision to outsource maintenance, which is one of the primary considerations in making the decision. However, because of the complexity of the airport sector, it might be challenging to make clear commercial decisions. To accomplish this goal, we devised a strategy considering resource and transaction cost theories. In the future, we will discuss the organizational characteristics that have the potential to influence the efficiency of an outsourcing operation. These characteristics include the size of the company, the level of expertise it possesses, and the organization's significance in maintaining its assets.

REFERENCES

1. Sood, V.: Strategic Frameworks for Outsourcing Decisions Part Three | Reuters Events | Supply Chain & Logistics Business Intelligence. https://www.reutersevents.com/supplychain/strategic-frameworks-outsourcing-decisions-part-three.
2. Quinn, J.B., Hilmer, F.G.: Strategic Outsourcing – ProQuest. https://www.proquest.com/openview/b40046da33ff1c364aaa3a81ba23237a/1?pq-origsite=gscholar&cbl=26142.
3. Holcomb, T.R., Hitt, M.A.: Toward a model of strategic outsourcing. *Journal of Operations Management*, 25, 464–481 (2007). https://doi.org/10.1016/j.jom.2006.05.003.
4. McFarlan, F.W., and Nolan, R.L.: How to Manage an IT Outsourcing Alliance. https://sloanreview.mit.edu/article/how-to-manage-an-it-outsourcing-alliance/.
5. Huber, R.L.: How Continental Bank Outsourced Its "Crown Jewels". https://hbr.org/1993/01/how-continental-bank-outsourced-its-crown-jewels (1993).
6. Chase, R.B., Aquilano, N.J.: *Operations Management for Competitive Advantage*. Irwin Professional Pub, Boston, MA (2004).
7. Barthelemy, J.: Stratégies d'externalisation, 3ème édition. Préparer, décider et mettre en oeuvre l'externalisation d'activités stratégiques - broché. Achat Livre | fnac. https://livre.fnac.com/a1928196/Jerome-Barthelemy-Strategies-d-externalisation-3eme-edition.
8. Williamson, O.E.: *The Mechanisms of Governance*. Oxford University Press, Oxford, New York (1996).
9. Graham, R.: Outsourcing and keeping control: The key legal issues. *Property Management* 11, 141–145 (1993). https://doi.org/10.1108/02637479310026703.
10. Muller, T.: De l'externalisation à la fragmentation des entreprises – Baromètre Outsourcing Europe, Cabinet of Ernst & Young. https://www.bibliobaseonline.com/notice.php?NUMERO=97730&OLD=97545%7C92737.
11. Embleton, P.R., Wright, P.C.: A practical guide to successful outsourcing. *Empowerment in Organizations*, 6, 94–106 (1998). https://doi.org/10.1108/14634449810210832.
12. Barthélémy, J.: Stratégies d'externalisation, 3ème edition. Librairie Eyrolles. (2007).
13. Simões, J.M., Gomes, C.F., Yasin, M.M.: Changing role of maintenance in business organisations: Measurement versus strategic orientation. *International Journal of Production Research*, 54, 3329–3346 (2016). https://doi.org/10.1080/00207543.2015.1106611.
14. Deshmukh, S.G., Sharma, A., Yadava, G.S.: A literature review and future perspectives on maintenance optimization. *Journal of Quality in Maintenance Engineering*, 17, 5–25 (2011). https://doi.org/10.1108/13552511111116222.

15. Cegos: The Economic Slowdown: What has L&D learned from the economic slowdown?, Retrieved from: https://static.cegos.com/wp-content/uploads/2012/03/Cegos-What-has-LD-learned-from-the-economic-slowdown-21.4.11.pdf (2011).

16. Chater, Y., Talbi, A.: Developpement d'une méthodoligie d'externalisation de la maintenance (2011).

17. ONDA, Financial Report 2019, Retrieved from: https://www.onda.ma/content/download/11758/107594/version/1/fichier/RAPPORT+FINANCIER+2019.pdf (2019).

18. Salonen, A., Deleryd, M.: Cost of poor maintenance: A concept for maintenance performance improvement. *Journal of Quality in Maintenance Engineering*, 17, 63–73 (2011).

19. ONDA, Comprehensive maintenance contract for the public address systems installed at the Kingdom's various airports, https://www.onda.ma/Media/AO-Consultation/AOO-103-16 (2016).

20. Augustyniak, W., López-Torres, L., Kalinowski, S.: Performance of Polish regional airports after accessing the European Union: Does liberalisation impact on airports' efficiency? *Journal of Air Transport Management*, 43, 11–19 (2015). https://doi.org/10.1016/j.jairtraman.2015.01.001.

21. Itani, N., O'Connell, J.F., Maison, K.: The impact of emigrants' homeland relations on air travel demand in a security volatile market: A case study on Lebanon *Journal of Transport Geography*, 30, 170–179 (2013). https://proxy.univh2c.ma:2073/science/article/pii/S0966692312002128.

22. Forsyth, P., Gillen, D., Muller, J., Niemeir, H.-M.: *Airport Competition: The European Experience*. Routledge, London. https://www.taylorfrancis.com/books/edit/10.4324/9781315566481/airport-competition-peter-forsyth-david-gillen-jurgen-muller-hans-martin-niemeier (2010).

23. Forsyth, P.: Airport competition: A perspective and synthesis. In P. Forsyth, D. Gillen, J. Muller, & H-M. Niemeier (Eds.), *Airport Competition: The European Experience* (1st ed., pp. 427–436). Ashgate Publishing Limited, Routledge, London. https://research.monash.edu/en/publications/airport-competition-a-perspective-and-synthesis (2010).

24. Castillo-Manzano, J.I., López-Valpuesta, L., Sánchez-Braza, A.: When the mall is in the airport: Measuring the effect of the airport mall on passengers' consumer behavior. *Journal of Air Transport Management*, 72, 32–38 (2018). https://doi.org/10.1016/j.jairtraman.2018.07.003.

25. Akbari, M.: Logistics outsourcing: a structured literature review. *BIJ*, 25, 1548–1580 (2018). https://doi.org/10.1108/BIJ-04-2017-0066.

26. Williamson, O.E.: *Markets and Hierarchies: Analysis and Antitrust Implications: A Study in the Economics of Internal Organization*. Social Science Research Network, Rochester, NY (1975).

27. Hätönen, J., Eriksson, T.: 30+ years of research and practice of outsourcing: Exploring the past and anticipating the future. *Journal of International Management*, 15, 142–155 (2009). https://doi.org/10.1016/j.intman.2008.07.002.

28. Ghertman, M.: Oliver Williamson et la théorie des coûts de transaction. *Revue francaise de gestion*, 142, 43–63 (2003).

29. Mahoney, J.T., Pandian, J.R.: The resource-based view within the conversation of strategic management. *Strategic Management Journal*, 13, 363–380 (1992). https://doi.org/10.1002/smj.4250130505.

30. Conner, K.R., Rumelt, R.P.: Software piracy: An analysis of protection strategies. *Management Science*, 37, 125–139 (1991).

31. Prahalad, C.K., Hamel, G.: *The Core Competence of the Corporation*. Harvard Business Review, Brighton, MA, 17 (1990).

32. Gebauer, H., Paiola, M., Edvardsson, B.: A capability perspective on service business development in small and medium-sized suppliers. *Scandinavian Journal of Management*, 321–339 (2012). https://doi.org/10.1016/j.scaman.2012.07.001.

33. Balakrishnan, S., Wernerfelt, B.: Technical change, competition and vertical integration. *Strategic Management Journal*, 7, 347–359 (1986).

34. Naidu, H.: The Outsource Decision Matrix. Accel8. https://accel8.com/the-outsource-decision-matrix/ (2017).

30 Sustainable Finance and FinTech

Facilitating a Sustainable Future with the Utilization of Socio-Economic Financial Services

Zerina Bihorac, Azra Zaimovic, and Tarik Zaimovic

30.1 INTRODUCTION

The mere concept of sustainable development was first introduced in 1987 in a report by the United Nations known as the Brundtland Report, where sustainable development was defined as development that meets the needs of the present without compromising the ability of future generations to meet their own needs [1]. Sustainable development is a fundamental objective of the European Union. Maintaining economic efficiency is equally as significant as retaining social and environmental awareness. Integration of the three is what the European Union is trying to implement.

Building a financial system that stands on sustainable growth is a very complex process that still has not been fully formed. The financial system has a role in contributing to the sustainable development of the economy, mainly by allocating funding to its most productive use [2].

The 2030 UN Agenda for Sustainable Development, with its 17 Sustainable Development Goals (hereinafter SDGs), aims to end and completely eradicate extreme poverty by 2030 [3], subsequent to the fact that poverty in society is the most significant obstacle that is obstructing the growth of sustainability.

Financial technology (FinTech) refers to the integration of financial services with information technology, and it is an innovation that has made financial intermediation between users faster and more efficient. The term denotes an industry that stands on Internet-based financial intermediation, such as services delivered through mobile phones or computing devices, and as such. FinTech plays an important role in the facilitation of sustainability as it leverages artificial intelligence to enable a green finance transition between clients and enterprises [4–7]. Green practices can generate financial value by engaging Corporate Social Responsibility and Environmental, Social, and Governance (hereinafter ESG) concerns as a central axis when making financial decisions.

Our goal is to explore the growing importance of financial technology tools in achieving sustainable development. The main research methodology is based on qualitative and bibliometric analysis. It relies on analyzing the relevant regulations and recently published scientific papers about FinTech and sustainable development. We found several prominent FinTech tools to help combat some of the most important social and environmental concerns. Blockchain, RegTech, crowdfunding, green P2P lending, and digital banking are the main FinTech tools for sustainable development.

The chapter is organized as follows: Section 30.2 discusses sustainable finance and the financial system; Section 30.3 analyzes challenges in growth through sustainable financing; Section 30.4 presents a bibliometric analysis of sustainable finance and FinTech; Section 30.5 brings prominent

DOI: 10.1201/9781032667478-35

FinTech solutions for sustainable development; and Section 30.6 explores FinTech and barriers to sustainable development, after which Section 30.7 concludes.

30.2   SUSTAINABLE FINANCE AND THE FINANCIAL SYSTEM

Sustainable finance has been a prominent topic in the world of economics and finance in recent decades. Its use is a growing concern, as it is one of the main conditions for achieving the SDGs. Sustainable financing requires considering ESG factors when making investment decisions. Thus, some also refer to sustainable finance as "green finance". However, the two are not synonymous. Sustainable finance is mandatory for public and private green investments that promote low-carbon development to protect biodiversity and consider water management and landscapes [8].

Some repercussions leave a trail on the financial system under unpredictable events, such as climate, weather, or policy changes. The impacts of climate change on the economy can be hard to estimate and very extreme. According to the IMF, they can affect the economy directly or indirectly [9]. Direct effects such as price impairment and underwriting losses disturb the financial system. Indirect effects stagnate the economy's growth, slow the circular economy, and lead to tighter policies and financial conditions. The need for the integration of sustainability and financial technology was denoted in the crisis caused by the COVID-19 pandemic, as the reliance on technology exponentially increased and countries had to adapt to a new reality. Sustainability is key for the recovery after the pandemic [10].

When the world is faced with a global crisis, such as the global financial crisis of 2008, firms with higher social capital tend to have more significant stock returns than firms with lower social capital [11]. The overall trust between firms and stockholders increases when there are negative economic shocks because climate change risks can be very difficult to combat and expose businesses to substantial losses. Firms that succeed in considering ESG become more valuable than brown firms, according to Pástor et al. [12]. Banks are in a good position to redirect economic flows toward the sustainable development of the economy [13]. By redirecting given loans to facilitate the development of countries and businesses based on sustainability criteria, banks will not only be improving the well-being of society; they will also be improving their relationship with clients over the long term. Robins et al. find that transitioning to a sustainable system is possible through the cooperation of private and public organizations and companies that work to integrate environmental and social risks into their business approaches [14]. Financial institutions and multilateral development banks have recently been continuously promoting sustainable and green finance and adopting practices into their activities to drive capital to sustainable projects.

30.3   CHALLENGES TO THE GROWTH OF THE GLOBAL ECONOMY THROUGH SUSTAINABLE FINANCING

Sustainable finance explains how finance (lending and investing) cooperates with environmental and societal issues. The economy is heavily manipulated by finance, focusing primarily on financial return on investments, risk, and profitability. However, into the equation comes sustainable financing. It is a great challenge for investors and economists worldwide to start considering ESG concerns while also having to speculate if they will generate a positive return on investments. However, to reach a state of a climate-neutral economy, this is what needs to be done.

Integrating sustainability into a business strategy can be achieved in multiple ways, such as by providing training to staff about the importance of ESG factors, laying out designated ESG objectives, or establishing membership in networks that support sustainable finance, as well as having a policy statement that aligns with a sustainable economy [15]. Sustainable finance has many challenges in meeting the SDGs. According to Cattaneo, the main obstacles can be seen in inequalities regarding access to finance; primarily, high-income countries hold around 80% of the global assets under their management and 97% of established investment funds for sustainability [16]. Dembele et al. find that capital markets are crucial because their funding helps close the SDG financing gaps and mobilize capital toward sustainable projects [17]. Considering that high-income countries hold most of the world's capital, they also have better access to sustainability finance such as to budgets and funds designated for sustainability when an economy needs to recover after a crisis or a pandemic.

30.4   SUSTAINABLE FINANCE AND FINTECH: A BIBLIOMETRIC ANALYSIS

To evaluate the current trend in publications about the role of financial technology in achieving sustainable development, we retrieved data from the Clarivate Analytics Web of Science core collection platform, the leading database for academic publishing and the most suitable for bibliometrics. We created a string of the most suitable search terms in order to screen papers ("SDG*" and "fin*tech" or "sustain* finance*" and "fin*tech" or "sustain* develop*" and "fin*tech"). The analysis was performed in September 2022 with the help of VOS Viewer software.

In total, 82 publications were found in the WOS database, and all were included in our bibliometric analysis. FinTech's role in achieving SDGs as a new phenomenon was tracked in the WOS database for the first time in 2018, with the following frequency of published papers (2018-1, 2019-8, 2020-20, 2021-30, 2022-23).

Our analysis shows that the leading publishers are MDPI with 30 documents, Elsevier with 12 papers, Emerald Group Publishing, Taylor & Francis, and Wiley with seven documents each. Sustainability (MDPI) is a leading academic journal with 27 articles and 174 citations, followed by European Business Organization Law Review (Springer) with two documents and 47 citations, and *European Journal of Finance* (Traylor & Francis) with two documents and 38 citations. The topic of FinTech and sustainable development, SDGs, or sustainable finance is not connected only to economics and business finance/studies but even more to the environmental sciences, green sustainable science technology, and environmental studies, and also to law, development studies, geography, international relations, management, and regional urban planning. The most active institutions working on FinTech and sustainable development/finance, or SDGs, are University College Dublin, with 7 publications, and Shanghai University of Finance Economics, with 6 publications. Analysis of co-authorship by countries shows that China is leading with 29 documents and 201 citations, followed by England with 13 papers and 120 citations.

We use citation, co-citation, and keywords analysis as the most common bibliometric tools. The distance between items in VOS Viewer maps can explain the relatedness between them. The most prevalent themes in FinTech and sustainable development/finance, or SDG, can be seen in Figure 30.1. A total of 522 keywords were found in 82 papers, with a threshold of at least 4 occurrences presented in Figure 30.1. "FinTech" is the most frequently used keyword with 40 occurrences, followed by "financial inclusion", "technology", "sustainable development", etc. Papers in the same cluster share a common topic and differ from papers in other clusters. Keywords around FinTech are grouped into four clusters: (1) financial inclusion, policy, financial development, consumption, poverty, etc.; (2) sustainable development, financial technology, blockchain, etc.; (3) performance, impact, sustainability, etc.; and (4) SDGs, innovation, governance, etc.

Co-citation analysis depicts the frequency of two papers being cited together by another paper. Since the researched topic is relatively new, our co-citation analysis of references shows that only two articles have been co-cited 10 times or more. There are 11 authors that have been co-cited 10 times or more.

Based on the analysis of the bibliographic coupling of documents, we find that the most influential articles are those from Arner et al. [18] with 46 citations in the WOS database, Zhao et al. [19] with 36 citations, and Demir et al. [20] with 31 citations (Figure 30.2).

Based on the citation analysis, we found 17 authors cited more than 20 times in the WOS database, 67 cited more than 10 times, and 191 cited one time. The most cited authors are Zetzsche D., with 47 citations from two articles, followed by Arner D., Buckley R., and Veidt R., with 46 citations.

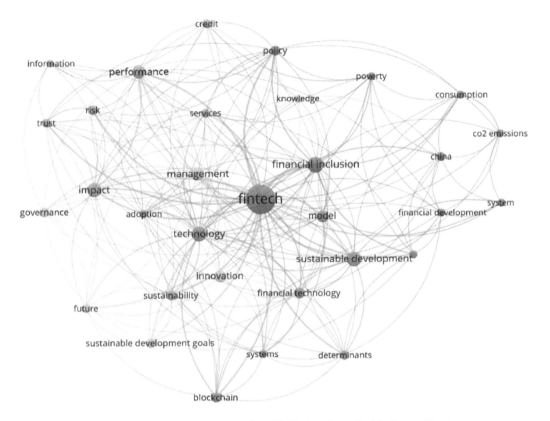

FIGURE 30.1 Keywords co-occurrence network on FinTech and sustainable finance/development.

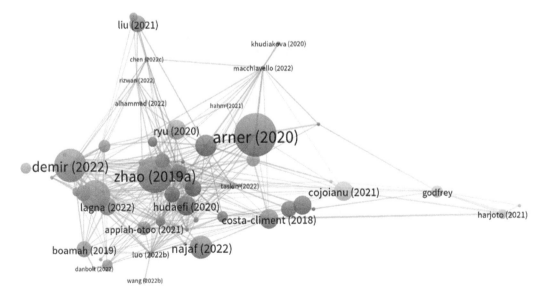

FIGURE 30.2 The most influential articles organized in clusters.

FinTech development has many advantages for customers and financial services providers as market participants, reflected in financial inclusion, lower costs, and enhanced user convenience and experience. Financial markets and regulators benefit from financial inclusion, more efficient and competitive markets, lower costs, greater transparency, risk diversification, and decentralization. However, financial technology raises risks such as data protection, cyber risk, consumer protection, operational risk, money laundering, terrorism financing risk, and market integrity risk.

30.5 PROMINENT FINTECH SOLUTIONS FOR SUSTAINABLE DEVELOPMENT

Acting in accordance with the SDGs and sustainable growth will ensure a stable and prosperous economy, a healthier planet, and, most importantly, an inclusive society where everyone can reach a point of having the same access to resource financing and investment opportunities. The European Commission's Action Plan from 2018 lays out the importance of sustainable financing for the long term. Realization of the 17 SDGs and following the guidelines of the UN's 2030 Agenda for Sustainable Development and the European Green Deal heavily depend on the willingness of investors and companies to consider the impacts of their decisions in the long term [21]. Following sustainable practices and policies that align with the principles of Corporate Social Responsibility and ESG will generate a positive reputation for the company that conforms to them, as it indicates its awareness of the geophysical environment, such as paper waste, water waste, and energy consumption [22].

30.5.1 MITIGATING GREENHOUSE GAS (GHG) EMISSIONS WITH FINTECH

The primary goal of the European Green Deal is to enforce reductions of GHG emissions in the EU by at least 55% by 2030 and, by 2050, to completely remove net emissions of GHGs [23]. Fossil fuels accumulate in the atmosphere, pollute the ocean, increase the greenhouse effect, and, due to increased temperatures, can cause acid rain, contaminating fish populations and wildlife. The emissions essentially change the ecosystem as a whole, and they cause problems for both human and environmental health.

The Paris Agreement was signed in 2015 as the first-ever universal treaty regarding climate change to limit the temperature increase to a maximum of 2°C [24]. Nonetheless, the average planet temperature has increased by 1°C [25]. Any warming above 1.5°C can cause sea rise, extreme weather changes, the loss of biodiversity, and the total extinction of some species. It also causes food scarcity, leading to citizens' worsening health and poverty. Net zero emissions mean reducing these harmful emissions to a total of zero. A very efficient way of reducing these emissions with the help of FinTech is to pay with a card or from your mobile device to prevent unnecessary paper waste. Most businesses today will encourage contactless payment methods as they are environmentally friendlier.

30.5.2 FINTECH AS A SUPPORT INSTRUMENT FOR SOCIAL ENTERPRISES

Stimulating social businesses and small companies by making it possible for them to access finance and capital favors the development of sustainable finance. Social businesses mainly aim to generate social impact instead of profitable ones. Social enterprises have difficulties receiving proper funding from financial institutions, as they usually lack credibility. However, social enterprises are key to economic growth because they help boost the economy, create new jobs, deliver innovative products and services, and promote sustainability [26,27]. FinTech proactively helps social enterprises to utilize their data more efficiently and attract prospective investors. These enterprises can gain greater visibility by showing their propensity to include digital and mobile-based financial intermediation and artificial intelligence.

30.5.3 BLUE ECONOMY AND FINTECH

Considering the world's need for food is at its greatest, biodiversity is compromised, and the ecosystem is degrading. World fish stocks are declining, and they are one of the main components for companies in the seafood sector [26]. Marine resources are from the ocean and are also exposed to climate change. These resources provide food security and the creation of sustainable livelihoods for individuals. According to Laamrich, pollution, climate change, tourism, oil, gas, and unsustainable fishing are some ways these resources are becoming scarce and endangered [28]. Marine resources are especially important for individuals living in coastal communities, and Broom finds that around three-quarters of the world's population live within 50 kilometers of the sea [29].

Financial technology can help with the proper protection of oceans and the sustainable use of marine resources. There should be clear traceability of the seafood supply chain when it is shipped to the supermarket by tracking it with the help of technology. According to the WEF, this can be done with the help of radio frequency identification and GPS trackers, which can track fish supply from a ship to the supermarket and ensure transparency regarding the fish's origin. DNA barcoding can be utilized for the identification of seafood in trade [30].

30.5.4 AGRICULTURAL DIGITAL MARKETPLACE

In the long term, sustainability in the agricultural sector can be realized by several methods, mainly by reducing the use of fertilizers and pesticides and taking advantage of technological solutions. If the financial investments are redirected toward agricultural production, this helps farmers protect their natural resources and biodiversity, which in turn has benefits for all of society. Agricultural production is one of the primary sources of food for rural citizens.

Digital agriculture or e-agriculture is empowering long-term sustainability as it comprises tools that enable gathering data more accurately, which can take into account external factors such as weather changes. According to Klerkx et al., e-agriculture also means a more proper digitalization using augmented reality, sensors, robotics, 3D printing, and artificial intelligence [31]. The use of digitalized agriculture has the potential to achieve the SDGs and improve upon them. Such digitalized advances, proposed by Pauschinger and Klauser, appear in the form of self-driving tractors or drones that detect soil disease and spray crops in vineyards and orchards [32]. The digital marketplace is intertwined with nature-based solutions for financial intermediation. Adaptation of FinTech in the agricultural sector should be via smartphone business transactions, which means buying and paying online. Smartphones are a front-end device in the digital marketplace and are used as a tool for customers and suppliers to manage business remotely [33].

30.6 FINTECH AND BARRIERS TO SUSTAINABILITY

Short-termism is one of the largest barriers to SDG realization [2,26,34]. Short-termism indicates an excessive focus on short-term results while neglecting long-term global sustainability interests. However, sustainable financing is, by nature, a long-term concept. Sustainability occurs over time. SDGs are realized over time. A self-sustainable financial system develops gradually and enables economic growth, but it requires effort and pragmatic solutions. Shareholders' decisions are on a microlevel and are the primary source of short-termism [34]. Their outcomes then influence the macrolevel – financial markets and economy. This is a suboptimal state, as it prevents companies from utilizing their full potential and hinders economic development, growth, and SDG realizations.

TABLE 30.1

Main FinTech Tools for Sustainable Development

FinTech Tool	Sustainable Impact
Blockchain technology	It helps with the transition to clean energy and combating climate change due to efficient data monitoring and platforms for collaboration without intermediaries for cross-border payments.
RegTech	Regulatory compliance software improves the transparency and consistency of regulatory processes through bank monitoring, reporting, and fraud prevention. It is centralized around ESG by enforcing effective governance.
Crowdfunding	Green crowdfunding platforms help raise funds through online platforms for sustainable businesses and enable them to receive proper funding and raise accessible finance.
Green peer-to-peer lending (P2P)	Borrowers can obtain financing from individuals through P2P lending platforms contingent on environmental principles for covering eco-related expenses.
Digital banking	International payment processing and transferring are the main pillars of sustainable digitalization for banks, leading to cost savings, optimized business performance, and customer retention.

According to the G20 Green Finance Study Group, other barriers include information asymmetry and inadequate efforts from the private sectors [35]. Information asymmetry limits the investment of financial resources into green projects, leading to financial activities being jeopardized. As companies don't disclose environmental information, investors' costs increase as they search for green assets, reducing the attractiveness of green projects and investments toward them.

According to Li and Yang, with a lack of information, investors cannot assess environmental risks [36]. The private sector is considered by many to be the engine of economic growth. It has contributed to many technological improvements, such as transportation and energy efficiency; increased life expectancy through innovation in health care; and helped over a billion people living in extreme poverty. The private sector can be the driver of change by limiting the usage of single-use and non-recyclable plastic. This will generate less waste, and consumers will be provided with a sustainable and better option. When private sector efforts are obsolete, implementing sustainable finance becomes exponentially more difficult.

Investors are usually risk-averse. They want safer investments and less uncertainty. FinTech solutions such as blockchain, RegTech, crowdfunding, green P2P lending, and digital banking are at the forefront of addressing the presented concerns (Table 30.1), based on [37,38].

30.7 CONCLUSION

The financial system is most responsible for facilitating a transition toward a socially-inclusive and circular economy. Furthermore, financial technology services can build a resilient economy by improving and automating the financial system. FinTech and sustainable finance both have an important part to play in achieving the UN's 17 SDGs.

FinTech can promote the transition toward renewable energy sources through blockchain technology, RegTech, or artificial intelligence. As indicated, FinTech enables users to be more efficient in a shorter period, and companies are becoming more productive and experiencing increased performance through digitalizing financial services. These companies create competitive advantages for themselves, as they are experiencing faster growth than their traditional financial services peers [5].

Sustainable finance and FinTech are cohesive and consistent, as they have several common aspects since FinTech follows the main principles of ESG, such as facilitating green finance, reducing information asymmetry, promoting efficiency, and valuing nature's assets [38]. A systematic literature review on FinTech and sustainable development would be a natural continuation of our work, which at the same time represents the main limitation of our research. Future research should focus on regulations, which should resolve many risks to which consumers and market participants are increasingly exposed due to technological development, focusing on their protection.

REFERENCES

1. Miralles-Quirós, M. M., & Miralles-Quirós, J. L.: Sustainable finance and the 2030 agenda: Investing to transform the world. *Sustainability*, 13(19), 10505 (2021).
2. Schoenmaker, D.: *Investing for the Common Good: A Sustainable Finance Framework*. Bruegel, Brussels (2017).
3. United Nations. Transforming our world: The 2030 Agenda for Sustainable Development. A/RES/70/1 (2015).
4. Du, M., Zhang, R., Chai, S., Li, Q., Sun, R., & Chu, W.: Can green finance policies stimulate technological innovation and financial performance? Evidence from Chinese listed green enterprises. *Sustainability*, 14(15), 9287 (2022).
5. Moro-Visconti, R., Cruz Rambaud, S., & López Pascual, J.: Sustainability in FinTechs: An explanation through business model scalability and market valuation. *Sustainability*, 12(24), 10316 (2020).
6. Meiling, L., Yahya, F., Waqas, M., Shaohua, Z., Ali, S. A., & Hania, A.: Boosting sustainability in healthcare sector through fintech: Analyzing the moderating role of financial and ICT development. *INQUIRY: The Journal of Health Care Organization, Provision, and Financing*, 58, 00469580211028174 (2021).
7. Hinson, R., Lensink, R., & Mueller, A.: Transforming agribusiness in developing countries: SDGs and the role of FinTech. *Current Opinion in Environmental Sustainability*, 41, 1–9 (2019)
8. Lindenberg, N.: Definition of green finance. Deutsches Institut für Entwicklungspolitik (DIE) Mimeo (2014).
9. International Monetary Fund. Global financial stability report. Lower for Longer. Washington, DC (2019).
10. European Commisssion. Commission puts forward new strategy to make the EU's financial system more sustainable and proposes new European Green Bond Standard. Press release. Brussels (2021).
11. Lins, K. V., Servaes, H., & Tamayo, A.: Social capital, trust, and firm performance: The value of corporate social responsibility during the financial crisis. *The Journal of Finance*, 72(4), 1785–1824 (2017).
12. Pástor, Ľ., Stambaugh, R. F., & Taylor, L. A.: Sustainable investing in equilibrium. *Journal of Financial Economics*, 142(2), 550–571 (2021).
13. United Overseas Bank. How banks can help finance a greener tomorrow. Finance Asia, Haymarket Media Limited https://www.financeasia.com/article/how-banks-can-help-finance-a-greener-tomorrow/470683 (2021).
14. Robins, N., Zadek, S., Agha, M. et al.: Roadmap for a sustainable financial system. The World Bank Group, UN Environment Programme (2017).
15. Coleton, A., Font Brucart, M., Gutierrez, P., Le Tennier, F., & Moor, C.: Sustainable Finance: Market Practices. European Banking Authority Research Paper, (6) (2020). doi: 10.2139/ssrn.3749454.
16. Cattaneo, O.: 7 sustainable finance challenges to fix global inequality. World Economic Forum https://www.weforum.org/agenda/2022/05/sustainable-finance-challenges-global-inequality/ (2022).
17. Dembele, F., Schwarz, R. & Horrocks, P.: *Scaling up Green, Social, Sustainability and Sustainability-linked Bond Issuances in Developing Countries*. OECD Publishing, Paris (2021).
18. Arner, D. W., Buckley, R. P., Zetzsche, D. A., & Veidt, R.: Sustainability, FinTech and financial inclusion. *European Business Organization Law Review*, 21(1), 7–35 (2020).
19. Zhao, Q., Tsai, P. H., & Wang, J. L.: Improving financial service innovation strategies for enhancing China's banking industry competitive advantage during the fintech revolution: A Hybrid MCDM model. *Sustainability*, 11(5), 1419 (2019).
20. Demir, A., Pesqué-Cela, V., Altunbas, Y., & Murinde, V.: Fintech, financial inclusion and income inequality: a quantile regression approach. *The European Journal of Finance*, 28(1), 86–107 (2022).
21. European Commission. Action Plan: Financing Sustainable Growth. 97 final, Brussels (2018).
22. Thayyullathil, R., & Nobanee, H.: Sustainable Financial Management in Europe (2022).
23. European Commission. The European Green Deal. Final report. Brussels (2019).
24. United Nations Framework Convention on Climate Change. Adoption of the Paris Agreement, 21st conference of the Parties, Paris: United Nations (2015).
25. ClientEarth Communications. Fossil fuels and climate change: The facts https://www.clientearth.org/latest/latest-updates/stories/fossil-fuels-and-climate-change-the-facts/ (2022).

26. High-Level Expert Group on Sustainable Finance. Financing a sustainable economy. Final Report (2018).
27. Financier Worldwide. The impact of social entrepreneurship on economic growth. Financier Worldwide Magazine https://www.financierworldwide.com/the-impact-of-social-entrepreneurship-on-economic-growth#.YzciiHbP2M8 (2020).
28. Laamrich, A.: The Sustainable use of Marine Resources. COMHAFAT ATLAFCO (2019).
29. Broom, D.: Only 15% of the world's coastlines remain in their natural state. In *World Economic Forum* (2022).
30. World Economic Forum. Harnessing the Fourth Industrial Revolution for Oceans. Retrieved 09 21 2022, from https://www3.weforum.org/docs/WEF_Harnessing_4IR_Oceans.pdf.
31. Klerkx, L., Jakku, E., & Labarthe, P.: A review of social science on digital agriculture, smart farming and agriculture 4.0: New contributions and a future research agenda. *NJAS-Wageningen Journal of Life Sciences*, 90, 100315 (2019).
32. Pauschinger, D., & Klauser, F. R.: The introduction of digital technologies into agriculture: Space, materiality and the public-private interacting forms of authority and expertise. *Journal of Rural Studies*, 91, 217–227 (2022).
33. Anshari, M., Almunawar, M. N., Masri, M., & Hamdan, M.: Digital marketplace and FinTech to support agriculture sustainability. *Energy Procedia*, 156, 234–238 (2019).
34. Janicka, M., Sajnóg, A., & Sosnowski, T.: Short-termism-the causes and consequences for the sustainable development of the financial markets. In *Innovations and Traditions for Sustainable Development*. Springer, Cham, 485–501 (2021). doi: 10.1007/978-3-030-78825-4_29.
35. G20 Green Finance Study Group. G20 Green Finance Synthesis Report (2016).
36. Li, S., & Yang, B.: Green Investing, Information Asymmetry, and Category Learning (2021).
37. Chen, Y., & Volz, U.: Scaling up sustainable investment through blockchain-based project bonds. ADB-IGF Special Working Paper Series "Fintech to Enable Development, Investment, Financial Inclusion, and Sustainability" (2021).
38. Chueca Vergara, C., & Ferruz Agudo, L.: Fintech and sustainability: Do they affect each other? *Sustainability*, 13(13), 7012 (2021).

31 Information Society Services in Morocco Regulatory Status and Outlook for Development

Khalid Abouelouafa and Hafid Barka

31.1 INTRODUCTION

In recent years, the use of information and communication technologies (ICT) has increased all around the world, including Morocco. The COVID-19 crisis clearly contributed to the increase in ICT usage, mainly for the needs of teleworking, distance learning, socializing, and entertainment. Indeed, eight out of ten people in Morocco have access to the Internet. During 2020, 98% of Internet users in Morocco used social networks; 95.3% used video streaming services; and 86.1% had access to messaging and voice services over the Internet. On the other hand, collaborative services needed for telework and distance learning, e-banking, e-commerce, and e-government services represent a usage rate ranging from about 14% to 30%.[1]

The increased use of information society services (ISS) in daily life is associated with cultural, social, and economic changes. It contributes to economic growth, development, and inclusion. Meanwhile, it raises the issue of new risks and challenges, essentially related to person security and the protection of fundamental rights.

Therefore, this chapter will first assess, the status of the legal and regulatory framework for ISS in Europe under European Commission (EC) directives and regulations. Second, examine the legal and regulatory dispositions in force in Morocco, aiming to promote information society development, secure freedom of expression rights, and prevent illicit or dangerous content that may proliferate via ISS. Third, suggest relevant propositions to deal with the eventual inadequacies of the Moroccan framework and how to evolve it to address information society challenges in Morocco.

31.2 RELATED WORKS

Defining ISS requires a prior definition of information society, which is closely related to the concept of information.

Information is defined by the Merriam-Webster [1] dictionary as "knowledge obtained from investigation, study, or instruction" and as "the attribute inherent in and communicated by one of two or more alternative sequences or arrangements of something … that produce specific effects". Those definitions clearly refer to knowledge and message transmission.

The first definition of information is due to Shannon's information theory [2]. This mathematical theory was originally established to answer the technological question of message or signal transmission over a communication channel. It defines information as a measurable quantity that expresses the uncertainty of the realization of a probable event.

In this sense, the more uncertain the event is, the more information it carries. Similarly, a known event carries no information. We note that Shannon's definition equates the measurement of information with the measurement of knowledge of events. This principle is in line with Machlup's [3] economic theory of information, which also equates the notions of knowledge and information, even though it states that knowledge can sometimes refer to the content while information can specify the process itself.

As for the 'information society', Webster [4] proposes several definitions: (1) Technological definition, based on the technological evolution and convergence of information and communication, which would be at the origin of the information age. Similarly, the invention of the steam engine was at the origin of industrial society. (2) Economic definition, which is in line with the economic approach of Machlup, aiming to measure and quantify the economic value of knowledge and information. (3) Occupational definition, linked to the exercise of activities related to the information sector that dominate agricultural and industrial activities and thus lead to the creation of an information society. (4) Cultural definition: linked to the abundance and increased accessibility and usage of media and information that characterize the information society.

We note that some of these definitions are inducers of the information society, and others are consequences of the information society. Thus, the concept of information society cannot be defined only from a singular point of view. It is a matter for different disciplines: economic, sociological, and technological.

To the author's knowledge, no previous academic work on the legal and regulatory framework of ISS in Morocco has been found so far. Hence the motivation of this study.

31.3   INFORMATION SOCIETY AT ITU LEVEL

The challenges of the information society have aroused the interest of the United Nation's member states since 2003. As a result, the International Telecommunication Union (ITU) organized a world summit dedicated to the information society in Geneva and later in 2005 in Tunis. This summit[2] concluded with a declaration and an action plan for the development of the information society.

The Geneva Declaration affirmed the common vision of member states for the information society. This vision focuses on human dignity, inclusion, and development issues. The aim was to take advantage of the evolution of ICT to fight against extreme poverty and contribute to development and economic growth. Particular attention was paid to the issue of Internet governance, and the Geneva Declaration seeks to be multistakeholder, including states, industry, and international organizations.

Additionally, the declaration emphasized the important role of ICT development, and in particular the development of information society infrastructures, in providing universal access to information and knowledge. The fundamental role of security in ICT services was underlined, particularly regarding the fight against cybercrime, the establishment of an enabling environment favorable to the protection of individuals, intellectual property rights, and state security.

Also, the need to preserve the information society within a framework that guarantees diversity, pluralism, and freedom of expression was mentioned. This framework should ensure the protection of individuals against messages of hatred, xenophobia, pedophilia, or child pornography, as well as trafficking in human beings.

To achieve the above objectives, an action plan was recommended by the summit. It aims to strengthen the role of public authorities in creating an enabling environment for ICT development and to strengthen the role of the private sector in the development of ICT infrastructure and services.

Notwithstanding the fact that the Geneva Declaration gave a concise vision of member states to the question of information society, a declaration is not binding for parties contrary to conventions or international treaties. Hence, the summit declaration was not sufficient to create a harmonized legal framework for the information society.

31.4 EUROPEAN COMMISSION LEGISLATIVE AND REGULATORY FRAMEWORK OF THE INFORMATION SOCIETY

One of the main objectives of the European Union (EU) is to ensure the free movement of goods and services within the EU member states. Therefore, regulations have been introduced to ensure transparency in the establishment of information society technical regulations by member states. Thus, the Transparency Directive [5] established the legal certainty needed for the free movement of services in the internal market.

31.4.1 ISS Definition by EU

The above-mentioned Directive defines ISS as follows: "'service' means any information society service, that is to say, any service normally provided for remuneration, at a distance, by electronic means and at the individual request of a recipient of services."

The remuneration requirement for the service is of concern to us, as a large part of the services of the information society are provided by online platforms, which generally operates in two-sided markets, Feld [6]. The service is provided on one side of the market but remunerated in the other side. For example, the services offered by an Internet search engine, such as Google, are not remunerated by the recipients of the service, but by the advertising publishers who publish on the online interface of the search engine and by websites that can remunerate the platform for the ranking services. However, since the definition sates that the service is normally provided for remuneration can be interpreted as that remuneration is not a necessary condition.

The ISS must be provided at a distance. It does not require the physical presence of the service provider and the recipient of the service simultaneously in the same place. Thus, some services are provided by electronic means. For example, a booking service carried out electronically in the presence of the beneficiary at a travel agency or a service provided by a drink vending machine could not be considered an ISS. Similarly, the definition requires that the service be provided at the individual request of the recipient of the service. This excludes media broadcasting services by any means, including radio broadcasts and the Internet. Yet, on-demand audio-visual services are considered ISS, to the extent that the service is provided at the request of the recipient.

Finally, certain services are automatically excluded from the scope of the Directive, namely those related to notarial and judicial activities. The EC considered not including such activities at this stage.

One of the particularities of ISS is its dematerialized nature, which allows services to be provided to recipients regardless of national boundaries. This characteristic has motivated the adoption of the Transparency Directive with the aim of harmonizing technical regulations at the level of EU member states. Thus, this directive obliges member states to notify the Commission of any legislative, regulatory, or administrative provision and any technical specification or requirement relating to ISS. This notification is made for all technical regulation drafts and for the final version of the draft.

31.4.2 ISS Provider Obligation

While the Transparency Directive aimed to secure a legal and regulatory framework conducive to the development of ISS, the E-Commerce Directive [7] on ISS aims to promote the development of these services by establishing rules to protect ISS providers and service beneficiaries.

The Directive defines in particular:

- The liability of ISS providers for the content they should stock or transmit.
- The information that must be included in commercial communications.
- The rules for concluding online contracts.

ISS cover many activities and are not limited to online contracts but also extend to other activities where there is no price for the service, such as email services and Internet search engines. These services also include hosting, network access, and video streaming. Radio and television broadcasts, as well as near-video on-demand services, are not part of SSI, as they are not provided at the request of an individual recipient as required by the definition in the Transparency Directive.

While the definition of ISS was provided by the Transparency Directive, the E-Commerce Directive introduced new definitions related to the activity of ISS, such as the notion of the establishment of the service provider. This principle requires the effective exercise of economic activity by means of a permanent establishment in the country concerned. Therefore, owning and operating equipment or infrastructure necessary for the provision of the service in a country does not constitute the establishment of the provider. The concept of the provider's establishment is fundamental because it allows the country of origin to make specific obligations to the ISS provider established in this country.

31.4.3 Country of Origin Principle

The E-Commerce Directive adopts the "country of origin" principle. According to this principle, ISS are governed by the law of the country where the service provider is established. By establishment, the Directive means the place where the provider carries out its economic activity through a fixed establishment, not the place where the technological facilities providing the service are established.

Access to ISS provider activity should not be subject to prior authorization by the country of origin. However, the provider is required to notify the recipients of the service and the authorities of the country of origin of a certain amount of information concerning its identity and activity.

31.4.4 Commercial Communications

Commercial communications are one of the ways in which ISS providers are funded. Indeed, many services, such as web search, email, and social networks, are paid for by the commercial communications that they provide to their clients.

Such commercial communications must clearly indicate the person on whose behalf the communication is made and must clearly indicate that it is a commercial communication, a promotional offer, a competition, or a game.

A particular case of commercial communications is that of unsolicited commercial communications from recipients. Member States that allow this type of communication must provide mechanisms to accommodate individuals who do not wish to receive these messages. This is done by means of an 'opt-out' register, which service providers must consult and comply with.

31.4.5 Electronic Contracts

Member States must provide for the legal possibility of concluding electronic contracts. The Directive allows for the legal status of contracts concluded online. However, certain categories of contracts are excluded, such as contracts for the transfer of real estate or contracts relating to family and succession law.

The Directive distinguishes between the notion of "consumer" and the "recipient of the service": the consumer is a natural or legal person who uses the ISS service for purposes that are not part of his or her professional activity. This distinction is intended to provide the consumer with greater legal protection while concluding online contracts.

ISS providers also respect a certain amount of information to be provided to the consumer before the order is placed. Customers should have the possibility of consulting and editing the information entered while making orders.

31.4.6   LIABILITY OF PROVIDERS FOR CONTENT

An intermediary service provider (ISP) is a service provider who provides access to an electronic communication network or transmits information to a recipient of the service. The intermediary provider is not liable for the content of the information transmitted if it does not intervene in the content of the message transmitted or in the selection of the recipient of the content.

The provider is not liable for illegal content belonging to third parties, regardless of the form of the illegal activity: intellectual property rights, defamation, misleading advertising, etc. However, the provider is obliged to comply with any decision issued by a court or administrative authority to prevent or stop any violation.

A distinction is made between the three activities of intermediary providers: (1) simple transport, (2) caching, and (3) and hosting.

- Simple transport: This service consists of transmitting information over a communications network or providing access to a communications network. In this case, the provider is not liable if he does not initiate the transmission, select the recipient of the service, or modify the information transported.
- Cashing storage: This service consists of temporarily storing information to improve the efficiency of its transport to the service's recipient. In this case, the service provider is not responsible for the information stored, provided that the information is not altered by him.
- Hosting: This service consists of the storage of information by the provider at the request of the service recipient. In this case, the service provider is not responsible for the stored content if he has not become aware of the stored illegal content and if he will otherwise remove the illegal content.

Liability exemption does not prevent authorities in the country of origin from requiring the provider to terminate or prevent a breach. To this end, Regulation on Cooperation between National Consumer Protection Authorities [8] provides for several mechanisms to prevent a provider from evading the application of consumer protection laws by establishing itself in another EU country. For example, the national authority has the power to order the removal or suspension of access to an online interface (website or application), to order a hosting provider to remove or withdraw access to an online interface, and to order a registry operator to remove a domain name.

31.5   INFORMATION SOCIETY SERVICES IN MOROCCO

31.5.1   DEFINITION

Up to date, there is no legal or regulatory provision that specifically defines ISS. However, certain ISS are defined by regulatory basis [9,10] as Value-Added Services (VAS). The Telecommunications Act [11] defines VAS as services that add value to the information provided by the customer. This includes formatting, storing, or searching the said information. The list of VAS includes, in particular, electronic messaging services, electronic data interchange, online information services, data access services, including data processing, and Internet service. This definition is like the one given by the European Directive 'Transparency' [12]. It is noted that the list of VAS does not include on-demand audio-visual services, which fall under the scope of the Audio-Visual Communications Act.

31.5.2   REGIME

The provision of VAS is governed by the prior declaration. However, Article 18 of the Telecommunications Act [16] provides that "without prejudice to criminal sanctions, if it appears, following the provision of the service covered by the declaration, that the latter is prejudicial to public safety or order or is contrary to morality, the competent authorities may without delay cancel the said declaration". This provision clearly refers to the liability of the provider regarding the content it may process. Thus, national authorities may order the removal of illegal content from a hosting provider or prohibit access to illegal content from Internet access providers or domain name providers.

The Audio-Visual Communication Act [13] introduced in 2016 the concept of audio-visual service on demand. These are audio-visual services that allow the user to view audio-visual content on demand and at the time chosen by the user. The definition of on-demand audio-visual services excludes media services offered by the electronic press, which are governed by the law on the press and publishing [14], as well as services whose audio-visual content is managed by a private user, such as video streaming platforms, which are not covered by any legal or regulatory framework in Morocco.

31.5.3 ONLINE CONTRACTS

The Consumer Protection Act [15] provides specific provisions regarding the conclusion of a contract online. The scope of the law extends to any supply of goods or services made at a distance or by electronic means to a consumer.

It should be noted that this definition is not limited to services provided at a distance but also extends to services provided by electronic means. However, contracts concluded via vending machines, public telephone booths, and real estate sales transactions are excluded.

The contract concluded at a distance is governed by the provisions of the law on the electronic exchange of legal data [16]. The offer of a distance contract must clearly indicate certain information concerning the identity of the supplier as well as the main characteristics of the product or service subject to the contract.

The law on trust services for digital transactions [17] has made it possible to strengthen the security of contracts concluded at a distance by means of specific provisions on electronic signatures by introducing different levels of digital signature, from the simple signature, which may be based on a reliable identification process managed by the user himself, to the qualified signature, which must be generated by a qualified device and under the control of the national authority in charge.

31.5.4 INTERMEDIARY SERVICE PROVIDER

ISP as defined by European regulations, namely network access providers and data hosting and caching service providers, are not subject to any specific regulation in Morocco, except for network access providers, which are governed by the declaration regime as Internet access providers.

Furthermore, it is noted that the Moroccan regulation does not explicitly address the relevant issue of ISP who provide storage and caching services and does not have any specific procedure for their declaration, even though the Telecommunications Act classifies information processing, including storage and search capability, as VAS.

Intermediary services are of paramount importance in the information society. Indeed, access to the network is a necessary condition to benefit from any service provided at a distance.

While the storage and dissemination of information is at the heart of the information society, given the increased use of intermediary platforms (IPs) in everyday life. Social networks escape any national obligation in Morocco. One of the main functions of social networks, such as Facebook, YouTube, and others, is the storage of information such as text messages, images, audio, or video provided by the user. Such information is stored by the IP and made available to a virtually unlimited number of recipients at the user's request.

Another widely developed category of IPs are those that offer intermediary services between suppliers of goods and services on the one hand and consumers on the other. Examples include tourist accommodation services such as Airbnb and Booking, transport services such as Uber, and online merchandise sales such as Amazon. IPs also escape any obligation in Morocco due to the legal vacuum.

In view of the increased impact and growing use of this type of platform and the associated risk in terms of the illegal content or activity that can be hosted and disseminated on it, preventive and coercive legal and regulatory dispositions are needed to primarily fight against illegal content that may be hosted and disseminated by these IPs.

Similarly, the liability of ISPs for illegal content stored on IPs needs to be clarified to determine to what extent and under what conditions they can benefit from liability exemption. Storage and dissemination functionality amplifies the impact that user-generated content can have to a level unmatched by other ISSs.

31.6 DISCUSSION

Our study shows the existence of a legal and regulatory framework for ISS in Morocco, despite the lack of explicit reference to the term ISS by itself. Some ISS are designated as VAS, others as audio-visual media services. Nevertheless, the list of VAS in Morocco, which, in fact, corresponds to ISS as defined by the EC, does not extend to information services beyond Internet access. This restriction exempts ISPs providing hosting, online storage, caching services, and IPs from any obligation to report their activity. Indeed, the declaration of activity to a national authority is a prerequisite for the application of legal and regulatory provisions related to consumer protection, the protection of users' personal data, and the protection of intellectual property rights and copyrights. Such a legal framework should also secure ISS providers rights regarding their liability for the content they may process, store, or disseminate.

We found that certain audio-visual communication services, such as video streaming services, are also excluded from the scope of the Audio-Visual Communication Act. In fact, the law excludes video content selected and organized under the control of third parties. Taking into consideration the evolution of video streaming platforms, it is necessary to establish a legal and regulatory framework for ISSPs that store and distribute this type of content. Audio-Visual content provided by the online press is also excluded from the scope of the Audio-Visual Communication Act and falls under the Press and Publishing Act. Thus, a person who maintains a news or information website and publishes audio-visual content on that website and on social networks would be exempted from the obligations of a journal online that engages in the same content publishing activity. However, the impact of such activity in terms of audience may be equivalent.

The principles laid down by the Audio-Visual Communication Act, concerning pluralism, diversity, harmful and illegal content, violence, racial discrimination, and terrorism cannot be restricted to the sole audio-visual operators, as the audio-visual landscape is becoming increasingly dominated by IPs.

Removal of illegal content that may be stored by IPs is a concern in terms of feasibility and enforcement. Indeed, IPs established in Morocco must respond to such requests from judicial or administrative authorities. However, IPs established outside Morocco but whose service recipients reside in Morocco have no legal obligation to respond favorably to such requests. Though the Budapest Convention [18], an international treaty ratified by the member states of the Council of Europe and other states such as the United States and Canada, provides a framework for international cooperation in the fight against cybercrime.

ISS are cross-border in nature. It is not necessary for a provider to own relevant infrastructure or equipment in Morocco to provide its services to service recipients. This makes it difficult to apply any regulation to platforms established abroad, unless sanctions prohibiting access to the platform's service from the country in the event of non-compliance should be applied. Therefore, it would be wise to consider supranational legislation by means of international treaties to harmonize the legislative and regulatory framework in the signatory countries and to increase the role of international conventions focusing on cooperation in cybercrime prevention [19].

We deliberately restricted the scope of this study to consumers' and providers' rights and obligations related to ISS. Even though IPs are at the heart of ISS, they need further work regarding market power and competition issues. Also, the comparative study was restricted to EU regulation, even though it could be extended to other regions or countries, such as the United States, for example. Future work could address those aspects.

31.7 CONCLUSION

Examination of the legal and regulatory status of ISS in Morocco and comparisons made with the EU framework allow us to conclude that the Moroccan legal framework for ISS should evolve to explicitly define ISS and to consider media, content, and telecommunication convergence. It should equally take into consideration ISS that are still not covered by Moroccan legislative texts. Also, it would be wise to codify different dispositions related to ISS and emphasize international cooperation and harmonization. Such evolution could be of great benefit to the growing economy of information society while preserving individual fundamental rights and securing stakeholder and investor interests.

NOTES

1 TIC survey in Morocco, 2020.
2 World Summit on the Information Society, Geneva 2003 (wsis2003geneva.org).

REFERENCES

1. Brown, J. (2020). Information. In E. M. Sanchez (Ed.), *Merriam-Webster. Merriam-Webster.* https://www.merriam-webster.com/dictionary/information.
2. Shannon, C. E. (1948). A mathematical theory of communication. *Bell System Technical Journal*, 27(3). https://doi.org/10.1002/j.1538-7305.1948.tb01338.x.
3. Machlup, F. (1981). *Knowledge: Its Creation, Distribution and Economic Significance, Volume I.* Princeton University Press. https://doi.org/10.1515/97814008560085.
4. Webster, F. (1995). *Theories of the Information Society.* London: Routledge (International Library of Sociology). https://doi.org/10.4324/9781315867854.
5. Directive (EU) 2015/1535 of the European parliament and of the council of 9 September 2015 laying down a procedure for the provision of information in the field of technical regulations and of rules on information society services (codified text). ELI: https://data.europa.eu/eli/dir/2015/1535/oj.
6. Feld, H. (2019). The Case for the Digital Platform Act: Market Structure and Regulation of Digital Platforms. Roosevelt Institute, May, 4, 86. The Case for the Digital Platform Act: Market Structure and Regulation of Digital Platforms - Roosevelt Institute. https://www.publicknowledge.org/assets/uploads/documents/Case_for_the_Digital_Platform_Act_Harold_Feld_2019.pdf%0A%0A.
7. Directive 2000/31/EC of the European parliament and of the council of 8 June 2000 on certain legal aspects of information society services, in particular electronic commerce, in the Internal Market ('Directive on electronic commerce'). ELI: https://data.europa.eu/eli/dir/2000/31/oj.
8. Regulation (EU) 2017/2394 of the European parliament and of the council of 12 December 2017 on cooperation between national authorities responsible for the enforcement of consumer protection laws and repealing Regulation (EC) No 2006/2004. ELI: https://data.europa.eu/eli/reg/2017/2394/oj.
9. Decree n°2-97-1024 of 25 February 1998.
10. Order of the Minister of Industry, Trade and New Technologies.
11. n°618-08 of 13 March 2008 supplementing the list of services to be provided by added value.
12. Telecom act n°24/96 on Post and Telecommunications as amended and completed.
13. Directive 2000/31/EC of the European parliament and of the council.
14. Law n° 66-16 modifying and completing law n° 77-03 relating to audiovisual communication.
15. Law n°88-13 on the press and publishing.
16. Law n°31-08 enacting consumer protection measures.
17. Law n°53-05 on the electronic exchange of legal data.
18. Law n° 43-20 on trust services for electronic transactions.
19. Convention on Cybercrime Budapest, 23.XI.2001. European Treaty Series - No. 185.

32 Cartography of Mobile Payment Technologies Used in Morocco

Youness Khourdifi and Abderrahim Hansali

32.1 INTRODUCTION

Over the last two decades, we have observed many technological advances in computing, networking, storage, and processing technologies. These technologies have contributed to the emergence of various types of networks, various user applications, and a vast collection of computing devices that are becoming cheaper and more powerful than ever before.

These new technologies have allowed users to access information anywhere and anytime, from any personal or mobile computing device, and have also given birth to new sectors such as FinTech.

The FinTech field has developed recently, with opportunities to offer online banking services that are secure, high-quality, and easy to use [1].

This innovation in financial technology provides alternative products, business models, and applications that influence financial service provision and industry development while creating a competitive and reputable market culture among a number of different service operators [2,3].

Consequently, financial and technology institutions increased their investments in the innovation of FinTech.

Mobile payment service, or mobile money, is directly associated with FinTech financial institutions and payment services [4]. Therefore, it can be defined as an electronic payment instrument that uses a mobile device to transfer funds between the payer and the payee [5] to benefit from a service or product, and it is increasing people's perception of the way they use mobile devices.

Mobile devices are now considered a necessity for financial activities, which implies that they are more than just for calling, playing, and using the internet.

As a result, face-to-face contact and cash are no longer necessary to conduct transactions and exchange values. Mobile payments have changed how customers and merchants conduct transactions by keeping them easy, fast, and straightforward. In addition, people can make transactions anywhere and anytime, providing commercial banks with comprehensive data on their customers' experiences as the trend toward cashless payments has grown [6].

Every financial institution constantly strives to improve its services by pursuing technological innovation. The mobile payment services that are currently being created in a state of constant improvement serve vital roles in the creation of ecosystems that include regulators, financial institutions, device manufacturers, retailers, and the customer through the use of applications such as Apple Pay [7], Samsung Pay [8], Google Pay [9], Orange Cash, and others.

This research aims to answer the question "What mobile payment technologies are used in Morocco?".

The main objective of this research paper is to browse all the technologies used in mobile payment, including the presentation of the mobile payment service ecosystem in general and especially in Morocco.

In this chapter, we adopted the methodology of documentary research, and we started by searching the research publications located in: Springer Link, Inderscience Publishers, Taylor & Francis, ACM Digital Library, IEEE Xplore Digital Library, Science Direct, and Emerald Insight.

The research aims to concentrate on mobile payments as a subset of financial technology (FinTech) to determine the mobile payment technologies currently used in Morocco.

We use the keyword model to filter the data toward priority in the database search for each search publication to obtain a search that focuses on mobile payment according to the search method. This ensures that we get a search that is relevant to mobile payments.

After the results of the utilized keywords were obtained, the inclusion criteria in the search carried out three separate steps to filter the results in the initial phase of the filtering process (1). After that, the subsequent step is to create the second filter, which examines whether or not the research topic is addressed in the title and abstract of the paper (2). The third and final filter is to read each text collected from a research article (3).

The chapter is organized as follows: the state of the art of mobile payment technologies in Section 32.2, with a description of the mobile payment ecosystem and a detailed explanation of the cartography of mobile payment technologies in general; Section 32.3 presents the mobile payment technologies adopted in Morocco as a case study; a discussion and comparison of our research issue will be presented in Section 32.4; and finally, Section 32.5 presents a conclusion and future research directions.

32.2 STATE OF THE ART OF MOBILE PAYMENT TECHNOLOGIES

Several studies have been conducted on mobile payment and the technologies involved in this field, which has developed into a necessary technology and is increasingly demonstrating a strong presence and a high demand due to its many advantages, including contactless payment, which can be used immediately from any phone and anywhere.

These benefits include speed, security, and simplicity, allowing us to complete our payments for our transactions quickly.

Mobile payments generally have the same properties as all other financial services, including atomicity, the impossibility of non-reputation, security, availability, profitability, integration, convenience, and privacy.

In their analysis of various electronic payment systems, related security concerns, and the potential for mobile payment methods, Zlatko Bezhovski et al. [10] concluded that despite the ease and security offered by mobile electronic payment methods.

It may be argued that mobile payments continue to increase worldwide, surpassing even those done using credit and debit cards.

Other researchers, such as Carlos León [11], studied the mobile payment system of Movii, a FinTech company in Colombia that operates under a non-bank financial license for electronic deposits and payments. Raffaello Balocco et al. [12] proposed a research project that included a census of mPayment applications and an analysis of the most relevant case studies. They began in the first phase with secondary sources that were primarily used. In the second stage, they conducted interviews with senior executives of the companies involved, and for the research results, they discovered that there are 21 mPayment applications identified. Additionally, Francisco Liébana-Cabanillas et al. [13] present a study based on developing a conceptual model to analyze the intention to use mobile payment services in India, which is one of the largest countries.

However, compared to traditional payments, mobile payment problems related to the security of electronic accounts and payment applications present serious challenges [14]. In this effect, we find in the literature several studies that have conducted this component with technical and theoretical studies [15]. In this way, we find Kai Fan et al. [16], who proposed a secure mutual authentication protocol based on the Universal 2nd Factor (U2F) protocol for mobile payment to ensure reliable service. They used an asymmetric cryptographic system to achieve mutual authentication between the server and the client, which can resist fake servers and falsified terminals.

Moreover, to defend against the off-site attack, in particular the possible attackers who could be bad merchants, Liming Fang et al. [17] have proposed a method, and they proposed Secure and Authenticated Payment Protocol (SALP), which is a secure and authenticated payment protocol, using time and location as necessary conditions for payment confirmation. Additionally, they use an identity-based signature to prevent information modification and reduce the overhead of third-party authentication. This allows SALP to be more efficient.

32.2.1 MOBILE PAYMENT ECOSYSTEM

In the mobile payment ecosystem represented in Figure 32.1, banks and financial services organizations supply payers and payees with accounts, payment instruments, and other services.

A single mobile payment transaction necessitates a level of connectivity between distinct ecosystems that is unparalleled.

These exchanges must be quick and safe, necessitating direct, effective, and secret links between the various parties:

- **Financial ecosystem:** Banks that issue payments, payment networks, and acquiring banks.
- **Mobile ecosystem:** Mobile device manufacturers, operating system providers, and network operators.
- **Merchandise ecosystem (enterprise):** Point-of-sale providers and merchants.
- **Marketing Ecosystem:** Includes companies that manage advertising insertion, rewards, proximity/localized offers, and contextual awareness.

32.2.2 MOBILE PAYMENT TECHNOLOGIES

The concept of mobile payment draws inspiration from a wide variety of existing payment systems, including: Unstructured Supplementary Service Data (USSD), Bluetooth Low Energy or Bluetooth Smart (BLE), Short Message Service (SMS), Wireless Application Protocol (WAP), Quick Response Code or two-dimensional bar code or black and white matrix barcode (QRC), and

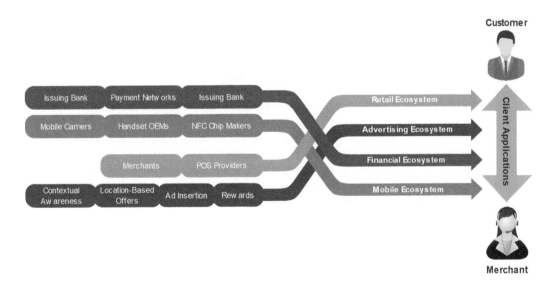

FIGURE 32.1 Mobile payment ecosystems.

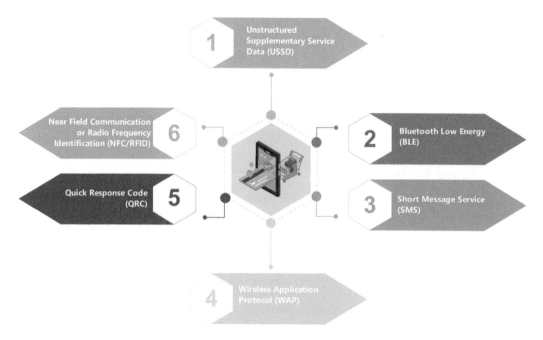

FIGURE 32.2   The technologies used in the mobile payment system.

Near Field Communication or Radio Frequency Identification (NFC/RFID) are the six primary models that are described in the following paragraphs [18].

The six different methods of mobile payment are outlined in Figure 32.2.

Unstructured Supplementary Service Data (USSD) [19]**:** Generally uses the Global System for Mobile Communication/Short Message Service (GSM/SMS) and serves as an interface for customers themselves and between customers and banks. Its advantage is that it is easy to use and compatible with all mobiles. The disadvantage is that the data is not secure enough. The systems usually use a personal identification number (PIN) to authenticate the user at the application level, but as in the previous case, the cryptographic procedures are flawed.

Bluetooth Low Energy (BLE) [20]**:** Uses wireless transmission. Because of its low energy requirements, it has a wide application in the proximity payment sector, hence its limitations. It has the advantage of allowing secure exchanges following the National Institute of Standard and Technology (NIST) recommendations. This organization has described the two modes and security levels for a service between two devices connected via BLE [21–23]:

- For Mode 1, Level 1 doesn't initiate encryption or authentication; Level 2 is only about encryption but not authentication; Level 3 requires both authentication and encryption; and Level 4 requires the use of AES-CMAC elliptic curve keys of 250 bits in length.
- For mode 2, the data signature is considered at two levels depending on whether authentication isn't required at level 1 or level 2 at the beginning of the connection establishment.

In addition, for a service such as mobile payment, where security is needed, NIST recommends Mode 1 and Level 4, where both devices must authenticate each other and exchanges are encrypted using AES-CMAC with p-256 ECC.

Moreover, BLE is compatible with most smartphones. This is an advantage since a mobile payment system with BLE has high availability.

Short Message Service (SMS) [23]: This messaging service can use up to 160 characters. It was developed in the Global System for Mobile Communications (GSM) network. Very simple to use and known throughout the world, it is expensive and can easily become the target of cyber-attacks due to the failure of cryptographic techniques.

Wireless Application Protocol (WAP) [24]: A mobile device can access the Internet through this connection. A very high level of safety is afforded to user data and server authentication procedures by the incorporation of Wireless Transport Layer Security, a protocol that encrypts communications between parties. WAP is highly suggested for use with any online payment methods.

However, its use has been restricted, even though there is significant interest in it.

Quick Response Code (QRC) [25]: It enables reading numeric, alphabetic, or binary data using a smartphone with a camera. Secure QRC was added to the package to tighten encryption measures because it has major security vulnerabilities, such as diverting the user to a hostile site, which is a bogus site capable of collecting financial data. Regarding accessibility, it is akin to BLE because it is available on mid-range phones.

Near Field Communication [26] **or Radio Frequency Identification (NFC/RFID)** [27]: To make contactless payments, NFC is utilized. This kind of radio communication takes advantage of the free ISM band frequency of 13.56 MHz, designated as the Industrial, Scientific, and Medical (ISM) band. It enables the transfer of data and the processing of payments between two devices separated by a distance of between 4 and 10 cm. It is used in two different configurations: one in which a Secure Element (SE) is located within the phone, most notably within a SIM card, and the other in which the SE is embedded in a Cloud Server that is referred to as Host Card Emulation. Both configurations are used to protect sensitive information.

The Host Card Emulation design is susceptible to a relay attack, and with the embedded secure element (SE), the service provider is extremely dependent on the hardware supplier. Deployment can also be expensive, although contactless payment is quick and has been a huge success. In the instance of the replay attack, an adversary can steal a user's banking information as the user walks by and then communicate that information to a third party, who can then use those details to make transactions.

32.3 MOBILE PAYMENT TECHNOLOGIES ADOPTED IN MOROCCO

Mobile payment has proven itself all over the world, not only in developed countries but also in developing countries. In Morocco, some banks and payment institutions have already launched offers related to money transfer via mobile phones, merchant payments, bill payments, and cash withdrawals with innovative and secure solutions that can be used with mobile devices such as mobile phones or tablets.

This movement includes the introduction of what is known as the "M-Wallet", a new means of payment issued either to a payment account held by a payment institution or a bank account held by a banking institution [28].

Instantaneously, safely, and interoperably, the owner of an M-Wallet can conduct the following actions:

- Transfers of funds from person to person by simply entering the recipient's telephone number;
- Payments at accepting merchants displaying the mention "Maroc Pay";
- Payment of water, electricity, and telephone invoices, as well as payment of vignettes, taxes, and telephone credit cards;
- Cash withdrawals.

According to the declarations received from banks and payment institutions, the number of transactions carried out by M-Wallets in 2020 amounted to 1.4 million transactions, for a total amount of MAD 443 million. Almost 51% of the volume of exchanges was carried out by M-Wallets issued by payment institutions, compared with 49% by M-Wallets backed by bank accounts. In terms of value, 35% of the transactions carried out by M-Wallets issued by payment institutions were recorded against 65% of those issued by banks (Figure 32.3).

Regarding the structure of transactions carried out by M-Wallets issued by payment institutions, payments (merchants and invoices) represent, in number, the main part of the transactions carried out, with a portion of 73%, followed by mobile-to-mobile transfers (27%).

Cash withdrawals from ATMs remain almost insignificant, accounting for just 0.4% of total transactions. For M-Wallets connected to bank accounts, payment by M-Wallet remains dominant with 70% of the total, followed by transfer with a 19% proportion and ATM withdrawal (10%) (Figure 32.4).

In contrast, the distribution of transaction value by M-Wallet type is inverted; for payment institutions, person-to-person transfers make up the most significant proportion (63%) of total transactions, followed by payments (35%) and ATM withdrawals (2%).

For M-Wallets issued by banks, transfers account for 54% of all transactions, followed by ATM withdrawals (26%) and payments (20%). (Figure 32.5).

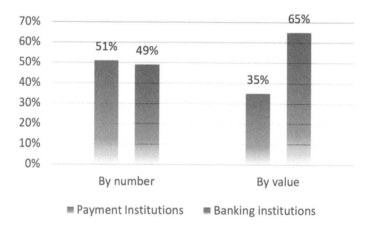

FIGURE 32.3   Distribution of M-Wallets transactions by type of establishment.

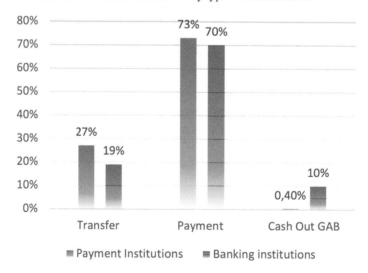

FIGURE 32.4   Distribution of exchanges by M-Wallets by type of transaction (in number).

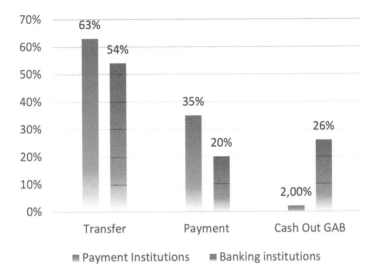

FIGURE 32.5 Allocation of M-Wallets transactions by type of transaction (in value).

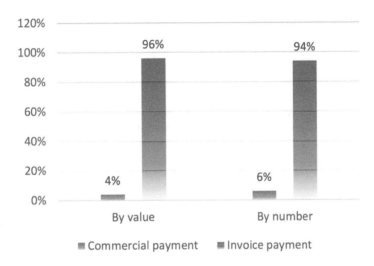

FIGURE 32.6 Distribution of M-Wallet payments.

The payment of bills (water and energy bills, vignettes, taxes, telephone top-ups, etc.) constitutes the vast majority of mobile payments, accounting for 94% in number and 96% in value.

In contrast to the other uses of this new payment method, merchant payments will remain minor in 2020, both for payment institutions and banks (Figure 32.6).

In fact, using mobile payments as a form of payment, especially to settle purchases, is a new idea whose growth is a real challenge. This is especially true because getting merchants to sign up is a crucial and challenging step that requires much work regarding education, awareness, and incentives for acceptance.

32.3.1 MOBILE PAYMENT TECHNOLOGY MODELS ADOPTED IN MOROCCO

The system of accepting payments via mobile devices in Morocco is known as "Maroc Pay." This is a novel method of payment that takes the shape of a mobile application and enables users to make direct payments or transfers and have direct access to their banking or payment accounts. Users can do all of

TABLE 32.1

Payment Institutions Providing Payment Products and Services

Company Name	Group	Product (Application)	Technologies Used	Supported Smartphones
BARID CASH	Barid Bank	Barid Pay	QRC	All smartphones
CASH PLUS SA	Cash Plus	Cash Plus	QRC	All smartphones
CENTRE MONETIQUE INTERBANCAIRE - CMI	CMI	Ibriz	NFC and QRC	NFC-enabled smartphones
DAMANE CASH	Banque Marocaine du Commerce Extérieur Group	Damane Pay	QRC	All smartphones
DIGIFI	Banque Marocaine pour le Commerce et l'Industrie	SmartFlouss	QRC	All smartphones
FAST PAYMENT SA	FAST PAYMENT	Fpay (Rezk-e Business)	SMS and QRC	All smartphones
LANA CASH	Crédit Immobilier et Hôtelier	We Pay	NFC and QRC	NFC-enabled smartphones
MAROC TRAITEMENT DE TRANSACTIONS - M2T	M2T	AmanPay	QRC	All smartphones
MAYMOUNA SERVICES FINANCIERS	Maymouna Services Financiers	Madmoun	QRC	All smartphones
MT CASH	Maroc Télécom	MT Cash	QRC	All smartphones
NAPS SA	NAPS SA	Naps	NFC and QRC	NFC-enabled smartphones
ORANGE MONEY MAROC	Orange	Orange Money	QRC	All smartphones
SOGEPAIEMENT	SGMB	So Pay	QRC	All smartphones
WAFA CASH	Attijariwafa Bank	Jibi	QRC	All smartphones
WANA MONEY	WANA	Inwi Money	QRC	All smartphones
AL FILAHI CASH	Crédit Agricole du Maroc	Filahi Pay	QRC	All smartphones

this without going through their financial institution. This "Maroc Pay" label is your assurance that the payment method you choose complies with the highest requirements of safety, protection of personal data, and the prevention of fraud. Maroc Pay, also known as mobile payment, is subject to the oversight of the regulatory body known as Bank Al Maghrib, just like other forms of payment.

Among the solutions that exist, there are nine linked to banking institutions and 16 payment institutions. In all, there are 25 institutions that offer mobile payment solutions. The tables above summarize the different solutions used in Morocco and the technologies used for payment institutions (Table 32.1) and banking institutions (Table 32.2).

32.4 DISCUSSION AND COMPARISON

From the analysis of the present state of the art and the two tables showing the different mobile payment institutions that offer payment products and services issued either on a payment account held by a payment institution or on a bank account held by a banking institution, we find that among the 16 payment institutions, there are 12 that adopt the QRC and three that adopt the NFC and QRC technology. Finally, only one mobile payment institution uses SMS and QRC technology. For the bank payment institutions, we found only one solution offered by Moroccan Agricultural Credit that allows the customer to use the commercial payment service using QRC technology. The other solutions are limited to money transfers, payment of water, electricity, and telephone bills, as well as payment of vignettes, taxes and telephone top-ups, and cash withdrawals.

TABLE 32.2

Banking Institutions Providing Payment Products and Services

Company Name	Group	Product (Application)	Technologies Used	Supported Smartphones
Attijariwafa Bank	Attijariwafa Bank	Attijari Mobile	WAP	All smartphones
AL Barid Bank	Barid Al Maghrib	Barid Bank Mobile	WAP	All smartphones
Bank of Africa	Banque Marocaine du Commerce Extérieur Group	DabaPay	WAP	All smartphones
Banque Populaire	Banque Populaire BP	Chaabi Pay	WAP	All smartphones
Crédit Immobilier et Hôtelier	Crédit Immobilier et Hôtelier	CIH MOBILE	WAP	All smartphones
Crédit Agricole du Maroc	Crédit Agricole du Maroc	Beztam-E	QRC	All smartphones
Crédit Agricole du Maroc	Crédit Agricole du Maroc	Bank-e	WAP	All smartphones
Casablanca Finance Group	Casablanca Finance Group	CFG Bank	WAP	All smartphones
Société Générale Maroc	Société Générale SA	Société Générale Maroc	WAP	All smartphones

We also found that banking institutions dominate the value of M-Wallets transactions with 65%. However, for M-Wallets transactions by type of transaction, payment institutions have the most significant market share of payments, with 35% and 63% of the transfer.

Finally, for M-Wallets payments of merchants and invoices, we note that invoice payments represent almost all mobile payments, with 96% and 6% of merchant payments, because the merchants who accept mobile payments are very limited.

32.5   CONCLUSION AND FUTURE WORK

This work aims to present various mobile payment methods, which provide a convenient and secure way to pay for goods and services using a mobile device. Several mobile payment technologies are available, each with its unique features and advantages. USSD, BLE, SMS, WAP, QRC, and NFC/RFID are some of the most commonly used mobile payment technologies. USSD is a text-based communication protocol that allows users to access various mobile services. BLE is a wireless communication technology that enables contactless payments. SMS-based mobile payments are widely used in developing countries. WAP is a communication protocol that enables mobile devices to access the internet. QRC codes can be easily generated and are widely supported by mobile devices. NFC/RFID is a wireless communication technology that is secure and widely supported by mobile devices, making it a popular option for mobile payments and many payment applications offered by banking institutions. In the case of Morocco, we will present the mobile payment service ecosystem, which represents a payment service directly connected to FinTech financial institutions and payment services.

The results of our work show that the use of these technologies in Morocco is very limited. With the use of three technologies, which are QRC, SMS, and NFC, this is due to the difference in security and confidence in the technology, and the number of retailers who accept mobile payment is still very limited. Therefore, as a solution, it is necessary to encourage customers to adopt mobile solutions to pay for their transactions.

These challenges responsible for the use of mobile payments in Morocco will be the subject of our future article, which will deal a study detailing these constraints and the solutions proposed to overcome them.

REFERENCES

1. Kang J. Mobile payment in Fintech environment: Trends, security challenges, and services. *Human-centric Comput Inf Sci [Internet]*. 2018;8(1). Available from: https://doi.org/10.1186/s13673-018-0155-4.
2. Thakor A.V. Fintech and banking: What do we know? *J Financ Intermed*. 2020;41:100833.
3. Chandler N., Krajcsák Z. Intrapreneurial fit and misfit: Enterprising behavior, preferred organizational and open innovation culture. *J Open Innov Technol Mark Complex*. 2021;7(1):61.
4. Kang J. Mobile payment in Fintech environment: Trends, security challenges, and services. *Human-centric Comput Inf Sci*. 2018;8(1):1–16.
5. Karthikeyan S.R. Mobile payments: A comparative study between european and non-european markets (Dissertation). Retrieved from: https://kth.diva-portal.org/smash/record.jsf?pid=diva2%3A675943&dswid=-364. 2013.
6. Königstorfer F., Thalmann S. Applications of Artificial Intelligence in commercial banks-A research agenda for behavioral finance. *J Behav Exp Financ*. 2020;27:100352.
7. Liébana-Cabanillas F., García-Maroto I., Muñoz-Leiva F., Ramos-de-Luna I. Mobile payment adoption in the age of digital transformation: The case of Apple Pay. *Sustainability*. 2020;12(13):5443.
8. Mendoza S. Samsung Pay: Tokenized numbers, flaws and issues. *Proc Black Hat USA*. 2016;7:1–11.
9. Poongodi S., Jayanthi P., Ramya M.R. Users perception towards google pay. *PalArch's J Archaeol Egypt/Egyptology*. 2021;18(1):4821–5.
10. Bezovski Z. The future of the mobile payment as electronic payment system. *Eur J Bus Manag*. 2016;8(8):127–32.
11. León C. The adoption of a mobile payment system: The user perspective. *Lat Am J Cent Bank*. 2021;2(4):100042.
12. Balocco R., Ghezzi A., Bonometti G., Renga F. Mobile payment applications: An exploratory analysis of the Italian diffusion process. In: *2008 7th International Conference on Mobile Business*. IEEE; 2008. pp. 153–63.
13. Liébana-Cabanillas F., Japutra A., Molinillo S., Singh N., Sinha N. Assessment of mobile technology use in the emerging market: Analyzing intention to use m-payment services in India. *Telecomm Policy*. 2020;44(9):102009.
14. Ahmed W., Rasool A., Javed A.R., Kumar N., Gadekallu T.R., Jalil Z., et al. Security in next generation mobile payment systems: A comprehensive survey. *IEEE Access*. 2021. https://doi.org/10.1109/ACCESS.2021.3105450.
15. Liu W., Wang X., Peng W. State of the art: Secure mobile payment. *IEEE Access*. 2020;8:13898–914.
16. Fan K., Li H., Jiang W., Xiao C., Yang Y. Secure authentication protocol for mobile payment. *Tsinghua Sci Technol*. 2018;23(5):610–20.
17. Fang L., Li M., Liu Z., Lin C., Ji S., Zhou A., et al. A secure and authenticated mobile payment protocol against off-site attack strategy. *IEEE Trans Dependable Secur Comput*. 2021;19:3564–3578.
18. Lerner T. *Mobile Payment*. Springer, Wiesbaden; 2013.
19. Taskin E. GSM MSC/VLR Unstructured Supplementary Service Data (USSD) Service. (Dissertation). Retrieved from https://urn.kb.se/resolve?urn=urn:nbn:se:kth:diva-136384. 2012.
20. Heydon R., Hunn N. Bluetooth low energy. CSR Present Bluetooth SIG. https://www.bluetooth.org/DocMan/handlers/DownloadDocashx. 2012.
21. Gupta N.K. *Inside Bluetooth Low Energy*. Artech House, Norwood, MA; 2016.
22. Padgette J., Scarfone K., Chen L. Guide to bluetooth security. *NIST Spec Publ*. 2017;800:121.
23. Brown J., Shipman B., Vetter R. SMS: The short message service. *Computer (Long Beach Calif)*. 2007;40(12):106–10.
24. Erlandson C., Ocklind P. WAP–The wireless application protocol. In: *Mobile Networking with WAP*. Springer; 2000. pp. 165–73. doi: 10.1007/978-3-322-86790-2_17.
25. Walsh A. Quick response codes and libraries. *Libr Hi Tech News*. 2009;26:7–9.
26. Want R. Near field communication. *IEEE Pervasive Comput*. 2011;10(3):4–7.
27. Rajaraman V. Radio frequency identification. *Resonance*. 2017;22(6):549–75.
28. Al-Maghrib, B. Rapport annuel sur les infrastructures des marches financiers et les moyens de paiement, leur surveillance et l'inclusion financière – Exercice 2020, ISBN: 978-9920-772-02-0. 2020.

33 A Machine Learning-Based Recommendation System for Smart Mobility Trip Planning in Morocco

El Attar Chaimae, Daoudi Najima, Abourezq Manar,
El Ghali Btihal, Hilal Imane, and Hnida Meriem

33.1 INTRODUCTION

Nowadays, 50% of the world's population is living in cities, and by 2050, this percentage will increase to 70%, according to a new United Nations report [1], highlighting the need for more sustainable urban planning and public service. This population growth is accompanied by an increased number of roads that interconnect them, causing huge traffic problems during the 21st century, such as air pollution, noise, inefficient use of resources, traffic congestion, road accidents, environmental burdens, and mobility difficulties. These problems have negative impacts on users and society such as delays, inconvenience, and economic losses to drivers, in addition to human and financial damages.

To face these challenges and achieve sustainable management of cities, artificial intelligence (AI) can enable smart solutions, creating intelligent transportation planning and engineering, which brings multiple benefits. In fact, planning smart mobility solutions is among the top challenges for large cities around the world and is a major factor in allowing cities to grow.

Our research work is focusing on the question of mobility difficulties to help travelers select the most suitable transport mode and create more free time 'since the average driver spends the equivalent of six weeks driving per year' [2].

The remainder of this chapter is organized as follows: In the second section, we discuss the state of the art of the concept of 'smart mobility' to understand the context of the research work as well as the problem statement. In the third section, we present a literature review of AI techniques used in the field of smart mobility. In the fourth section, we propose a recommendation system for the optimization of single-modal trips, i.e., the use of a single mode of transport for the trip from point A to point B. In the last section, we evaluate the used models to choose the most adequate recommendation system to guarantee better, safer, and fluid mobility.

33.2 BACKGROUND AND PROBLEM STATEMENT

Smart mobility is an important part of developing smart cities. In fact, planning smart mobility allows users to personalize their trip and optimize route times, thereby reducing traffic congestion, road crashes, and injuries and saving energy. We suggest optimizing single-mode trips for Moroccan citizens using recommendations based on AI, which is a crucial part of the successful implementation of smart urban mobility.

DOI: 10.1201/9781032667478-38

The theoretical and conceptual background of our research is essentially smart mobility, which will be addressed in the first subsection, and recommender systems (RS), which will be treated in the second subsection.

33.2.1 SMART MOBILITY

Many different definitions are available today for smart mobility. Sassi and Zombonelli [3] describe smart mobility as the integration of mobility-related data to achieve energy-efficient, secured, personalized, and comfortable mobility solutions. Benevolo et al. [4] state that smart mobility "could be seen as a set of coordinated actions addressed at improving the efficiency, the effectiveness and the environmental sustainability of cities. In other words, smart mobility could consist of a hypothetically infinite number of initiatives often (but not always) characterized by the use of ICT". Smart mobility is an automated, personalized, connected, on-demand, and sustainable mobility [5]. Smart mobility can also be seen as an attempt to integrate all modes and options of transportation, considering user needs by using IT, multiple applications, and intelligent invoicing [6].

Recently, a very similar concept has emerged in the field of intelligent transport named "Mobility as a Service," which "is promising digital packages of personalized mobility that will replace privately owned vehicles and optimize the use and combination of several mobility alternatives" [2]. "The mobility as a Service (MaaS) proposition is often described as a one-stop, travel management platform digitally unifying trip creation, purchase and delivery" [7].

Several research projects and initiatives in many countries have emerged in this field with the objective of making smart mobility more efficient, such as automated vehicles tested in the United Kingdom and France, managing parking in Spain, Germany, the Netherlands, and Italy, chipping tickets or cards in the Netherlands and France, and car sharing and bike sharing in the United Kingdom, Germany, the Netherlands, and France [8].

33.2.2 RECOMMENDER SYSTEMS APPROACHES

The definition of RS has evolved over time as the field has progressed. One of the earliest proposed definitions is by Resnick and Varian [9], who defined them as systems that "assist and augment" the "natural social process" of recommendation by enabling people to "provide recommendations as inputs" and then aggregating and directing the recommendation to "appropriate recipients". Thus, RS help users make choices in a domain where they have some limited information and to sort and evaluate possible alternatives [9,10].

A second definition is that of Burke [11], who defined RS as systems capable of providing targeted recommendations by analyzing users' preferences on items in order to suggest and guide them to choose the products or services that best meet their needs. Indeed, recommendation systems aim at providing users with relevant resources according to their preferences, and therefore they reduce the search time.

More generally, an RS is a system that outputs individualized recommendations or provides personalized guidance to users in order to find the product and/or service that best suits them from a large set of possible options. In order to do so, RS rely on three main filtering approaches [12]:

- Content-based RS uses data about items and about the active user in order to make a recommendation.
- Collaborative RS uses data about users who have similar profiles to the active user and their interactions with items such as ratings and feedback in order to make a recommendation.
- Hybrid RS uses mixed approaches in order to make recommendations, benefiting from the advantages of each method and reducing its disadvantages.

RS are used in many fields, including in the context of smart mobility. These RS are based on different ML algorithms, as presented in the next section.

33.3 RELATED WORKS: ML ALGORITHMS FOR ROUTE RECOMMENDATION

Many machine learning algorithms were used in recommendation systems for smart mobility. In this section, we will present some machine learning algorithms used in the field.

33.3.1 ML TECHNIQUES FOR CONTENT-BASED RECOMMENDATION APPROACH

Random Forest. A random forest (RF) is composed of several individual decision trees. Each tree makes a class prediction, and the class with the most votes becomes the prediction of our model. In fact, a decision tree is viewed as a set of nodes, where each internal (non-leaf) node denotes a test on an attribute, each branch represents the outcome of a test, and each leaf (or terminal) node holds a class label.

Several studies have shown the effectiveness of this algorithm in the recommendation task. For example, RF is used in an Intelligent Transportation System to tackle the issue of traffic congestion, bottlenecks, and incidents [13]. In this work, data is collected from vehicular ad-hoc networks based on the speeds and coordinates of the vehicles, and then it sends traffic alerts. For example, it notifies the driver if any accidents occur on the same route. To achieve this goal, the authors proposed ML algorithms such as RF, Support Vector Machine (SVM), and artificial neural network (ANN) to analyze the behavior of vehicles and detect unusual situations in traffic. The results showed that RF outperformed the ANN and SVM algorithms in this case.

To solve the same issue and in order to recommend the best path to take to a user, an application has been created where the user enters their location details [14]. The application searches for the available routes within the specified locations and gets the nearest Closed-Circuit Television (CCTV) along the user's route. The CCTV videos' frames are used to detect vehicles using the Yolo model. The system uses the previous vehicle count data at that particular CCTV and trains the RF model using the traffic flow data. The RF predicts the traffic using previous data and then stores the predicted values. Then, a graph is created, and the Bellman-Ford algorithm is applied to find out the less congested path from the graph.

Support Vector Machine (SVM). SVM is a non-probabilistic binary linear classifier. Also called maximum margin classifiers, it is a supervised ML method that classifies data by determining the optimal hyperplane that separates observations based on their class labels [15].

The authors of [16] proposed a multi-class classifier to identify transportation modes from the list: driving a car, riding a bicycle, taking a bus, walking, and running. The data collection is based on a smartphone application that stores data from sensors such as GPS, accelerometer, gyroscope, and rotation. For the implementation, the LibSVM library of SVMs was used. As a result, an accuracy of 98.86% and 97.89% were achieved with and without using the GPS data, respectively. The authors proposed a comparison between the SVM classifier, K-Nearest Neighbor (KNN), Decision Tree, Bagging (Bag), and RF methods. The RF and SVM methods were found to give the best performance.

The same issue was addressed in the work of [17]. The authors propose to use alternative data such as TCP throughputs, RSSIs, and cell IDs. Results show that the transportation mode can be predicted with high accuracy using RF, KNN, and SVM.

Artificial Neural Network. ANNs are a non-linear statistical data model used to model a complex relationship between input and output variables for supervised classification. In [18], researchers proposed a new methodological approach for road safety risk modeling, which is a two-stage framework consisting of Data Envelopment Analysis in combination with ANN. In the first phase, the risk level of the studied road segments is calculated by applying Data Envelopment Analysis, and the identification of high-risk segments is done. In the second phase, the ANN is adopted, which seems to be a valuable analytical tool for risk prediction.

An ANN model was also used to develop a vehicle traffic flow prediction model at a signalized road intersection using 434 traffic datasets from seven road intersections [19]. This case study used the intersections connected to the busiest road in South Africa in terms of traffic volume (the N1 Allandale interchange). Sophisticated traffic data equipment, such as inductive loop detectors, video cameras, and geographical positioning equipment, has been used to measure parameters that include the number of vehicles on the road, the speed of each category of vehicles, time, traffic density, and, as a target to analyze, the traffic volume over a ten-day period in 2019. This research's findings include the fact that the ANN predictive approach proposed could be used to predict and analyze traffic flow with a high level of accuracy (99.97%).

33.3.2 ML Techniques for Collaborative Recommendation Approach

K-Nearest Neighbors. This algorithm is used for classification and regression problems. KNN is also non-parametric. Indeed, the KNN algorithm relies on the entire dataset to perform the prediction. In fact, for each new observation that we want to predict, the algorithm will look for the K instances of the dataset that are closest to our observation. Then, it will use the output variables of the neighbors to compute the value of the observation we want to predict.

Several recommendation systems based on collaborative filtering have been inspired by this algorithm in order to make recommendations. The authors in [20] proposed an approach for selecting the most suitable transportation mode based on a GPS sensor. The data collected is about GPS travel data: spatial data (x,y), speed, and time (t). The proposed model achieves an accuracy of 86% when dealing with these labels (e.g., Walk, Bike, Transit, and Private Automobile) and an accuracy of 94% with these labels (e.g., Non-Motorized and Motorized). Another work [21] related to smart mobility using KNN suggests dealing with the issue of traffic congestion, which causes air pollution and increases fuel consumption and travel time. This work identifies traffic congestion patterns in order to classify different road segments based on features like traffic density and average speed of vehicles. Data is collected using sensors installed on roads. The output of the system will help to make a decision, like whether to fly over the road or bypass it, according to the obtained results.

Matrix factorization. Matrix factorization (MF) allows the processing of large amounts of data in a fast and accurate way. It can be perceived as one form of principal component analysis close to singular value decomposition. The principle of MF is to decompose a matrix into several other matrices [22]. The original matrix is the product of these matrices.

The MF approach is also proposed in an urban mobility project [23] to predict the time of arrival of a vehicle using online massively dense GPS trajectories. Data was collected from moving taxis using cameras, mobile devices, and a GPS. As a result, data set of time-based GPS trajectories was constructed. To this end, MF, in combination with the Alternating Least Squares method, was used to estimate the travel time of each road segment. The evaluation of the model is calculated using Mean Absolute Percentage Error, Root Mean Squared Error, and Mean Absolute Error (MAE). For instance, the Mean Absolute Percentage Error is 16.7, the Root Mean Squared Error is 341s, and the MAE is 252s. In another work, the combination of MF with Markov Chains has been suggested by [24] in the field of intelligent traffic management. The proposed approach allows user-specific semantic location prediction, which uses semantic trajectories built upon the movement history of the user to predict his next semantic location. A semantic trajectory is a structured trajectory in which the spatial data are replaced by geo-annotations and semantic annotations such as the aims of stops and moves in the trajectory. Compared to the standard MF and Factorized Markov Chains, this combination outperforms both by approximately 17% and 10% better accuracy, respectively.

33.4 RECOMMENDATION SYSTEM DESIGN AND METHODOLOGY

In this section, we present the design and methodology we adopted in order to implement our recommendation system.

First, we will present our system architecture, as detailed in Figure 33.1. The system considers the data provided by the passenger to provide him with the most appropriate trip. Indeed, in the first step, the passenger specifies his own travel data, including date, time, and place of departure and arrival. A list of trips corresponding to the selected criteria will be selected from a database of all trips. To recommend the best trip, the recommendation system will take this list and the passenger's preferences as input and feed them into the preprocessing module. The result will be the input of the machine learning module. The trained model is used to obtain predictions of trip rates and sort them from the highest scores to the lowest ones.

33.5 DATASET GENERATION AND AUGMENTATION

In this section, we present the process of generation and augmentation of the dataset that will be used to rate the best trips according to users' preferences and consequently enhance their mobility. We used the Python language as it is the most used in the data science field [25].

Finding a suitable dataset to address our problematic issue is the biggest challenge in the smart mobility field. The authors in [26,27] used data generation techniques to overcome the lack of data challenges. To obtain a high-quality data set, we follow a laborious and rigorous process for data generation based on a survey.

Based on the Google Maps API, we started by generating the trip data. So, we generated the geographic data and distances of random trips using the API. Then we used the Faker library, which is a Python library, to add random dates, transport's mean, and departure date to each trip, as these data are variables and then we calculated the trip's duration based on the transport means and, by consequence, calculated the arrival dates. Then, we calculated the price of each trip based on the type of transport, taking into consideration local prices in Morocco.

As the main objective of our study is to rate the generated data depending on the trip's quality, we evaluate the quality of a trip based on many traffic conditions, such as traffic congestion, safety, accidents, and comfort. Each one of these conditions was calculated based on the trip data, such as localization, mean of transport, duration, and the trip period of the week or timing.

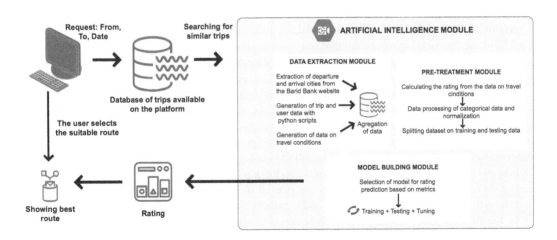

FIGURE 33.1   The architecture of the best route recommendation system for passengers.

Finally, we computed trip scores based on a truth table. So the most-rated trips are the ones that are safe in an infrequent road accident, comfortable, and less congested.

ML algorithms need huge amounts of data to correctly perform their training; thus, we used the Mockaroo platform for user data generation. Based on user preferences collected in the survey, we mapped the generated dataset of 200 trips with the user's categories and came up with 22.000 records in our final dataset.

33.6 IMPLEMENTATION OF THE PREDICTION PROCESS

In this section, we present the implementation of the prediction process. We used both a content-based approach and a collaborative filtering approach in order to compare them and come up with the best solution. We used Python version 3.7 to implement the solution, as it provides large and mature libraries that meet our requirements.

33.6.1 CONTENT-BASED APPROACH

We used RF and ANN as content-based approaches for recommending optimal routes to users based on the history of the routes they have taken before.

Random Forest-Based Model. In order to implement the RF model, we used the "trip dataset" with 21 features (departure city, arrival city, duration, distance, date, price, means of transport, route...) and 688719 data points (samples), divided into 5 different labels (scores from 0 to 4); each label represents a type of trip made by the user. This is a multi-class classification problem, which means that the data are not linearly separable. Indeed, we used 10 estimators in the first step; the dataset was divided into 70% of the training data and 30% of the test data. However, we encountered an overfitting problem. To overcome the problem, we applied cross-validation with five splits to find an accuracy of 97%.

After implementing the RF, we displayed our confusion matrix, which is a performance measure for a machine learning classification problem where the output can be two or more classes. We can conclude that our model predicts class 0 with 90% accuracy, and the probability of predicting it as class 1 is 9.84%. For class 1, or rating equal to 1, our model predicts it with 88% accuracy and 1% accuracy; as for class 0, which is a rating equal to zero, for the other classes, the model predicts them with 100% accuracy.

In order to evaluate the performance of our model, it is necessary to compare it with other models. The second model we used is an ANN.

Artificial Neural Network–Based Model. To predict the rating (score) of a trip, we also used an ANN-based model. Let's consider that our features: trip_ID, user_ID, departure address, arrival address, latitude, longitude, departure date, arrival date, price, distance, duration, number of seats, and means of transport are respectively $X1, X2, X3,..., X13$.

In our project, we have an input layer, represented by our features $(X1, X2,..., X13)$. Each of our features is multiplied by a weight $(W1, W2,..., W234)$. We have 13 features, and in each layer, we have used 18 neurons. The result is 234 assigned weights that represent the importance of the features: Then, the products of the features with the weights will be added and fed by our activation function ReLu to generate the output. This output is considered an input from the hidden layer, with 18 neurons in our case. Indeed, we have used two hidden layers, and in each layer, we have used the activation function ReLu, because it is the most used in the convolutional neural network "CNN" and the artificial neural network "ANN," and it gathers the advantages of Sigmoid and tanh.

In the first implementation of the ANN-based model, we encountered a gradient-exploding problem. Our model was unable to complete processing; the problem occurred when large error gradients accumulated and caused very large updates to the neural network model weights during training. When the magnitudes of the gradients accumulate, an unstable network is likely to occur, which can lead to poor prediction results. The explosion of gradients can lead to problems in the training of ANNs, and the learning cannot be completed. The values of the weights can also become very large and lead to what we call Nan values. Nan values do not represent a number; they represent undefined or unpresentable values, which negatively influence the training of the model.

We cite three techniques we used in our project for solving the Exploding Gradients problem: Redesign the network model, regularize the weights, and add a dropout layer. After several adjustments, we achieved an accuracy of 93%. To be sure of the obtained results, we displayed in Figure 33.2 the confusion matrix, which measures the quality of the model. In this figure, each row corresponds to a real class, and each column corresponds to a predicted class. In the first settings, we found that the model did not predict well the classes 0 and 1, with an accuracy equal to 20% for the class 0 and 50% for the class 1. However, after increasing the volume of the dataset, we could get the desired results with an accuracy equal to 96% for class 0 and 90% for class 1.

33.6.2 Collaborative Filtering Approach

The second approach we propose is collaborative filtering, in order to recommend a trip to a user based on similar users. After a brief benchmarking on collaborative filtering algorithms as shown in Figure 33.3, we used the surprise package, which is a package dedicated to recommendation systems and integrates several algorithms. The first algorithm is K-NN, which is based on similarity. The algorithm calculates the distance between the target trip and all other trips in the database, then ranks the distances and returns the K closest neighboring trips as the results of recommendations of the most similar trips.

The second technique we applied is MF; this method decomposes the trip scores matrix into three matrices using the singular value decomposition: the first contains user data, the second contains feature data, and the third one contains the singular values extracted from the original matrix. The MF algorithm has been used because classical methods rely on explicit evaluations (e.g., ratings) while modern RS should exploit all available interactions, both explicit and implicit, such as tastes and preferences.

We found that KNN Basic, which is based on the score of each user, is the algorithm that provides the best value of MAE, which represents the difference between the actual reel scores and the predicted one.

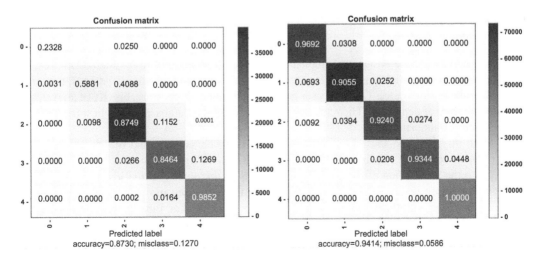

FIGURE 33.2 Visualization of confusion matrices before (left) and after (right) increasing the dataset.

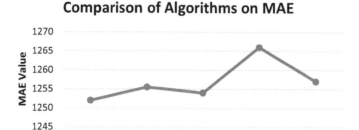

FIGURE 33.3   Visualization of benchmark results.

Table 33.1

The Recommendation Ranked Results for the Query "Rabat to Fes on Wednesday Using the Car as a Mean of Transport"

Id	Price	dep_@time	Weekday dep	arr _time	Weekda_Am	Mean	Nb_place	Read type	Rating
1	68.8	16	Wednesday	19	Wednesday	Car	1	Hightway	4
2	68.8	12	Wednesday	14	Wednesday	Car	1	Hightway	4
3	77	9	Wednesday	11	Wednesday	Car	1	Hightway	4
4	68.79	9	Wednesday	11	Wednesday	Car	1	National road	2
5	68.79	13	Wednesday	15	Wednesday	Car	1	National road	2
6	55.03	18	Wednesday	20	Wednesday	Car	1	National road	2
7	77.05	23	Wednesday	1	Friday	Car	1	National road	1

33.6.3   RECOMMENDATION

After evaluating the content-based and collaborative filtering approaches, we opted for the ANN algorithm, a content-based approach for recommending optimized routes with sorting of the predicted scores from the best scored to the worst scored. Therefore, we developed a test model function to perform the study process. We give our function the departure city, the arrival city, and the non-preprocessed date, and it returns all the routes meeting the query with a score sort as shown in Table 33.1.

33.7   CONCLUSION

This chapter presents the design and implementation of a recommendation system. This recommendation system will allow passengers to choose the most optimal routes by avoiding traffic jams, timetables, and unsafe routes. Thus, the user must fill in his city of departure, his city of arrival, and the date and time of departure. Then, the system searches in the database for the best trips, sorted by their order of importance.

To reach this goal, we started by defining the needs to be satisfied and the objectives of the project, which consist of minimizing the number of accidents and traffic jams on the roads and maximizing the satisfaction of the passengers in terms of the quality of the journeys. After that, we built a literature review of several techniques and methods used in recommendation systems. We then proceeded with the design and implementation of the system. Finally, we proceeded to the evaluation of the models, which showed a high performance of the content-based approach using ANN in recommending the best routes.

In future work, the recommendation system will take into consideration multimodality, which consists of recommending to a passenger more than one means of transport for the same journey.

In this case, the use of graph theory is important to consider for finding the optimal combination of the means of transport. Finally, we project to covert more deep learning algorithms to obtain a better prediction algorithm for trip recommendation in terms of accuracy, cost, and time optimization.

REFERENCES

1. McNabb, D.E.: The population growth barrier. In: McNabb, D. E. (ed.) *Global Pathways to Water Sustainability*. pp. 67–81. Springer International Publishing, Cham (2019).

2. Nikitas, A., Michalakopoulou, K., Njoya, E.T., Karampatzakis, D.: Artificial intelligence, transport and the smart city: Definitions and dimensions of a new mobility era. *Sustainability*. 12, 2789 (2020). https://doi.org/10.3390/su12072789.

3. Sassi, A., Zambonelli, F.: Towards an agent coordination framework for smart mobility services. Presented at the *8th International Workshop on Agents in Traffic and Transportation* (2014).

4. Adeniyi, D.A., Wei, Z., Yongquan, Y.: Automated web usage data mining and recommendation system using K-Nearest Neighbor (KNN) classification method. *Applied Computing and Informatics*. 12, 90–108 (2016). https://doi.org/10.1016/j.aci.2014.10.001.

5. Docherty, I., Marsden, G., Anable, J.: The governance of smart mobility. *Transportation Research Part A: Policy and Practice*. 115, 114–125 (2018). https://doi.org/10.1016/j.tra.2017.09.012.

6. Towards clean and smart mobility - European Environment Agency, https://www.eea.europa.eu/highlights/towards-clean-and-smart-mobility.

7. Wong, Y.Z., Hensher, D.A., Mulley, C.: Mobility as a service (MaaS): Charting a future context. *Transportation Research Part A: Policy and Practice*. 131, 5–19 (2020). https://doi.org/10.1016/j.tra.2019.09.030.

8. Bıyık, C., Abareshi, A., Paz, A., Ruiz, R.A., Battarra, R., Rogers, C.D.F., Lizarraga, C.: Smart mobility adoption: A review of the literature. *Journal of Open Innovation: Technology, Market, and Complexity*. 7, 146 (2021). https://doi.org/10.3390/joitmc7020146.

9. Resnick, P., Varian, H.R.: Recommender systems. *Commun. ACM*. 40, 56–58 (1997). https://doi.org/10.1145/245108.245121.

10. Shardanand, U., Maes, P.: Social information filtering: Algorithms for automating "word of mouth" In: *Proceedings of the SIGCHI Conference on Human Factors in Computing Systems*. pp. 210–217. ACM Press/Addison-Wesley Publishing Co., USA (1995).

11. Burke, R.: Hybrid recommender systems: Survey and experiments. *User Model User-Adap Inter*. 12, 331–370 (2002). https://doi.org/10.1023/A:1021240730564.

12. Rao, K.N.: Application domain and functional classification of recommender systems-a survey. *DESIDOC Journal of Library & Information Technology*. 28, 17–35 (2008). https://doi.org/10.14429/djlit.28.3.174.

13. Dogru, N., Subasi, A.: Traffic accident detection using random forest classifier. In: *2018 15th Learning and Technology Conference (LT)*. pp. 40–45 (2018).

14. Rajeev, G.L., Nancy, R., Megha, S., John, J.M., John, N.E.: Traffic flow prediction using Vrandom forest and bellman ford for best route detection. *International Journal of Engineering Research & Technology*. 9(13), 99–101 (2021).

15. Brandewie, A., Burkholder, R.: RFID based tire classification algorithm using support vector machines. In: *2020 IEEE International Symposium on Antennas and Propagation and North American Radio Science Meeting*. pp. 1359–1360 (2020).

16. Jahangiri, A., Rakha, H.: Developing a support vector machine (SVM) classifier for transportation mode identification by using mobile phone sensor data. Presented at the Transportation Research Board 93rd Annual MeetingTransportation Research Board (2014).

17. Kawakami, W., Kanai, K., Wei, B., Katto, J.: Machine learning based transportation modes recognition using mobile communication quality. In: *2018 IEEE International Conference on Multimedia and Expo (ICME)*. pp. 1–6 (2018).

18. Shah, S.A.R., Brijs, T., Ahmad, N., Pirdavani, A., Shen, Y., Basheer, M.A.: Road safety risk evaluation using GIS-based data envelopment analysis-artificial neural networks approach. *Applied Sciences*. 7, 886 (2017). https://doi.org/10.3390/app7090886.

19. Olayode, I.O., Tartibu, L.K., Okwu, M.O.: Prediction and modeling of traffic flow of human-driven vehicles at a signalized road intersection using artificial neural network model: A South African road transportation system scenario. *Transportation Engineering*. 6, 100095 (2021). https://doi.org/10.1016/j.treng.2021.100095.

20. Nour, A., Casello, J., Hellinga, B.: Developing and optimizing a transportation mode inference model utilizing data from GPS embedded smartphones. Presented at the Transportation Research Board 94th Annual MeetingTransportation Research Board (2015).
21. Mondal, Md.A., Rehena, Z.: Identifying traffic congestion pattern using K-means clustering technique. In: *2019 4th International Conference on Internet of Things: Smart Innovation and Usages (IoT-SIU)*. pp. 1–5 (2019).
22. Bokde, D.K., Girase, S., Mukhopadhyay, D.: Role of matrix factorization model in collaborative filtering algorithm: A survey. *IJAFRC*. 1, (2014). https://doi.org/10.48550/arXiv.1503.07475.
23. Badrestani, E., Bahrak, B., Elahi, A., Faramarzi, A., Golshanrad, P., Monsefi, A.K., Mahini, H., Zirak, A.: Real-time Travel Time Estimation Using Matrix Factorization. arXiv:1912.00455 [cs, eess]. (2019).
24. Karatzoglou, A., Lamp, S.C., Beigl, M.: Matrix factorization on semantic trajectories for predicting future semantic locations. In: *2017 IEEE 13th International Conference on Wireless and Mobile Computing, Networking and Communications (WiMob)*. pp. 1–7 (2017).
25. El Hachimi, C., Belaqziz, S., Khabba, S., Chehbouni, A.: Data Science Toolkit: An all-in-one python library to help researchers and practitioners in implementing data science-related algorithms with less effort. *Software Impacts*. 12, 100240 (2022). ISSN 2665-9638. https://doi.org/10.1016/j.simpa.2022.100240.
26. Demetriou, A., Alfsvåg, H., Rahrovani, S., Haghir Chehreghani, M.: A deep learning framework for generation and analysis of driving scenario trajectories. (2020). https://doi.org/10.1007/s42979-023-01714-3.
27. Shi, H., Li, H., Zhang, D., Cheng, C., Cao, X.: An efficient feature generation approach based on deep learning and feature selection techniques for traffic classification. *Computer Networks*. 132, 81–98 (2018). https://doi.org/10.1016/j.comnet.2018.01.007.

Index

accuracy 9, 28, 29, 32, 33, 34, 35, 36, 37, 38, 43, 48, 49, 50, 51, 52, 68, 69, 73, 74, 75, 76, 77, 78, 79, 80, 81, 82, 85, 86, 90, 91, 97, 98, 99, 100, 101, 102, 105, 106, 107, 108, 109, 110, 111, 112, 113, 114, 115, 118, 119, 123, 129, 133, 146, 284, 309, 314, 315, 316, 317, 318, 319, 320, 321, 381, 382, 384, 385, 387
AdaBoost 28, 29, 30, 33, 34, 35, 314, 316, 318, 320, 323
Africa 1, 22, 36, 46, 50, 51, 65, 78, 80, 81, 121, 138, 139, 142, 163, 165, 213, 217, 224, 229, 233, 234, 235, 236, 237, 239, 240, 262, 274, 294, 301, 377, 382, 387
AI 2, 4, 5, 9, 11, 13, 15, 17, 19, 20, 21, 25, 34, 36, 37, 38, 51, 52, 103, 122, 224, 228, 229, 230, 231, 232, 236, 326, 328, 330, 379; *see also* artificial intelligence
airport sector 347, 350
ANN 36, 37, 38, 41, 43, 47, 50, 51, 53, 108, 315, 322, 323, 381, 382, 384, 385, 386
artificial intelligence 2, 10, 21, 22, 27, 28, 29, 35, 36, 52, 85, 123, 248, 303, 316, 326, 336, 337, 352, 356, 357, 358, 379
audit 5, 13, 283, 284, 288, 289, 291, 292
awareness 14, 153, 163, 164, 165, 204, 205, 221, 263, 274, 352, 356, 371, 375

bank 21, 56, 65, 78, 80, 81, 82, 155, 167, 176, 179, 182, 187, 188, 235, 281, 302, 305, 342, 343, 350, 359, 375, 376, 377
behavior 135, 163, 168, 172, 173, 174, 175, 177, 178, 179, 189, 193, 200, 201, 210, 215, 216, 223, 249, 250, 256
Bfloat-16 84, 103
big data 4, 9, 19, 65, 86, 91, 229, 230, 231, 291, 293, 316, 324, 325, 328, 330, 337
Bitcoin 3, 4, 5, 23, 66, 67, 68, 69, 70, 72, 73, 228, 261, 262, 263, 269, 284, 291
blockchain 2, 3, 5, 6, 10, 11, 21, 22, 23, 111, 119, 121, 122, 230, 259, 261, 263, 265, 269, 270, 283, 284, 285, 286, 292, 293, 295, 297, 300, 301, 302, 305, 306, 307, 308, 309, 310, 311, 312, 328, 340, 342, 352, 358

candlestick 67, 69, 73, 75
candlestick patterns 67, 69, 73, 76
card fraud 27, 28, 29, 34, 35, 84, 85, 87, 90, 91, 92, 94, 96, 102, 103, 105, 106, 107, 108, 109, 110, 111, 112, 114, 119, 120, 121
Casablanca 77, 82, 194, 213, 218, 220, 222, 252, 243, 377
cloud 12, 15, 18, 52, 57, 103, 230, 231, 236, 252, 283, 284, 285, 286, 287, 288, 290, 291, 292, 316, 323, 324, 325, 327, 337, 373
clustering 29, 54, 60, 61, 62, 64, 106, 108, 110, 114, 115, 118, 323
CNN 67, 68, 69, 70, 72, 73, 74, 75, 76, 77, 78, 80, 81, 82, 83, 109, 125, 384
collaborative skills 213, 214, 215, 217, 218, 220

confidentiality 94, 107, 108, 110, 111, 118, 119, 221, 286
consortium 261, 305, 306, 307
COVID-19 206, 238, 239, 240, 241, 245, 246, 247, 248, 249, 308, 324, 325, 326, 327, 328, 330, 331, 335, 339, 353, 361
credit 1, 4, 8, 10, 11, 12, 13, 17, 19, 22, 23, 27, 28, 29, 34, 37, 52, 53, 54, 55, 56, 57, 58, 84, 85, 86, 87, 90, 91, 92, 94, 96, 102, 103, 105, 106, 107, 108, 109, 110, 111, 112, 114, 115, 119, 120, 121, 138, 156, 157, 165, 167, 168, 169, 170, 172, 173, 174, 185, 186, 187, 201, 205, 210, 222, 224, 225, 228, 229, 230, 233, 234, 236, 250, 251, 257, 299, 300, 337, 370, 374
credit card 27, 28, 29, 34, 84, 85, 86, 87, 90, 91, 94, 96, 102, 103, 105, 106, 107, 108, 109, 110, 111, 112, 114, 115, 119, 120, 121, 168, 169, 173, 174, 201, 228, 374
cross-sectional 137, 172
crowdfunding 2, 8, 10, 11, 12, 19, 22, 23, 228, 229, 230, 293, 294, 295, 297, 299, 300, 301, 336, 337, 352, 358
cryptocurrency 6, 67, 68, 69, 73, 228, 229, 230, 232, 233, 262, 263, 284, 298
cryptography 262, 270, 305
CSR 271, 272, 273, 274, 275, 277, 278, 280, 281
CS-SVM 105, 106, 107, 108, 110, 114, 116, 118, 119, 120
customer 1, 3, 4, 5, 7, 9, 10, 11, 12, 13, 15, 16, 17, 18, 20, 23, 27, 28, 29, 30, 34, 35, 37, 84, 85, 105, 106, 107, 108, 110, 113, 114, 118, 129, 163, 173, 193, 201, 202, 206, 207, 208, 209, 214, 215, 217, 220, 221, 222, 224, 229, 231, 236, 263, 266, 276, 239, 294, 295, 297, 298, 308, 338, 339, 346, 356, 357, 358, 365, 369, 372, 376, 377
cyber risk 12, 13, 14, 235, 356
cyberattack 13, 313, 314, 315, 316
cybersecurity 1, 12, 13, 14, 21, 23, 233, 322

data collection 9, 19, 54, 110, 119, 169, 216, 220, 303, 381
data privacy 5, 9, 10, 11, 12, 17, 18, 20, 21, 107, 113, 114, 115, 119
dataset 28, 29, 30, 42, 67, 68, 70, 73, 76, 77, 78, 80, 81, 84, 85, 86, 87, 88, 89, 91, 93, 94, 95, 96, 97, 98, 99, 100, 103, 105, 106, 107, 109, 110, 111, 113, 114, 115, 116, 118, 119, 125, 314, 315, 316, 317, 319, 321, 323, 382, 383, 384, 385
DCGAN 123, 124, 125, 126, 127, 129, 131, 133
decision tree 30, 88, 89, 90, 93, 120, 121, 316, 323, 381
decision-making 5, 9, 10, 12, 16, 17, 37, 145, 169, 215, 263, 264, 266, 267, 268, 269, 270, 343, 350
deduplication 284, 286, 287, 288, 289, 290, 291, 292
Deep learning 78, 80, 81, 82, 83, 103, 109, 120, 121, 123, 124, 125, 133, 316, 328, 387, 388
Development 1, 15, 19, 21, 22, 62, 63, 65, 145, 151, 152, 156, 157, 159, 161, 174, 176, 177, 178, 208, 210, 234, 235, 236, 237, 257, 271, 272, 274, 278, 281, 311, 342, 352, 356, 358, 359, 360, 361
digital assets 5, 15, 229

digital money 2, 3, 4, 5, 9, 228, 262
digital platforms 12, 19, 225
digitization 13, 138, 224, 228

e-commerce 3, 27, 28, 34, 206, 361
economic growth 4, 17, 19, 55, 56, 58, 65, 82, 137, 138,
        139, 140, 141, 142, 143, 144, 146, 147, 150, 156,
        157, 158, 159, 160, 161, 162, 234, 302, 310, 356,
        357, 358, 360, 361, 362
economy 19, 22, 23, 53, 54, 55, 57, 58, 59, 65, 77, 137, 138,
        140, 143, 144, 145, 147, 148, 149, 150, 152, 154,
        156, 157, 158, 159, 160, 161, 162, 164, 170, 182,
        183, 184, 203, 204, 205, 210, 221, 222, 224,
        228, 233, 234, 236, 237, 249, 271, 272, 281,
        293, 305, 352, 353, 354, 356, 357, 358, 360, 368
ecosystem 1, 2, 4, 5, 12, 14, 16, 17, 18, 19, 21, 23, 138, 213,
        214, 215, 216, 217, 218, 219, 221, 222, 223, 224,
        225, 226, 231, 232, 233, 234, 235, 236, 237,
        297, 337, 356, 357, 369, 370, 371, 377
education 12, 16, 22, 163, 165, 166, 167, 168, 169, 171, 173,
        175, 176, 177, 178, 179, 236, 323
Ethereum 5, 111, 228, 262, 263, 267, 294, 296, 297,
        298, 301
Europe 21, 64, 94, 350, 359, 361, 367

federated learning 106, 107, 108, 109, 110, 111, 112, 113,
        114, 116, 119, 122
finance 2, 4, 11, 12, 17, 19, 20, 21, 22, 23, 36, 52, 53, 55, 57,
        65, 111, 119, 138, 145, 147, 156, 157, 158, 162,
        166, 167, 169, 171, 172, 176, 178, 189, 190, 191,
        192, 193, 199, 200, 201, 204, 213, 214, 215, 216,
        220, 224, 225, 228, 229, 233, 234, 237, 250,
        251, 270, 271, 272, 281, 282, 299, 300, 301,
        305, 335, 336, 339, 340, 341, 342, 352, 353, 354,
        355, 356, 358, 359, 378
finance education 166, 167, 171
financial behavior 166, 167, 169, 170, 171, 172, 173, 174,
        175, 176, 177, 178, 179, 180, 221
financial economic 54, 150, 203
financial education 22, 164, 165, 166, 167, 168, 169, 170,
        171, 172, 175, 177, 178
financial inclusion 1, 2, 4, 5, 8, 11, 12, 17, 18, 19, 20, 21,
        23, 137, 138, 139, 140, 141, 142, 159, 189, 205,
        224, 233, 234, 257, 335, 336, 354, 356, 359
financial institutions 2, 3, 4, 7, 9, 11, 12, 13, 15, 16, 18, 19,
        55, 58, 78, 84, 85, 86, 90, 102, 106, 107, 113,
        115, 116, 118, 119, 157, 181, 182, 184, 189, 190,
        200, 201, 203, 204, 215, 220, 221, 224, 231,
        232, 233, 235, 237, 339, 356, 369, 377
financial knowledge 12, 15, 52, 165, 166, 167, 169, 170,
        171, 172, 173, 174, 175, 176, 178, 179
financial literacy 12, 17, 20, 52, 139, 163, 164, 165, 166,
        167, 168, 169, 170, 171, 172, 173, 174, 175, 176,
        177, 178, 179
financial markets 3, 51, 65, 77, 83, 137, 156, 157, 159, 161,
        162, 179, 182, 187, 203, 204, 224, 231, 239, 240,
        357, 360
financial stability 14, 181, 182, 183, 184, 185, 186, 187,
        188, 224, 233, 234, 359
financial system 9, 11, 17, 19, 20, 55, 56, 58, 64, 65, 137,
        157, 159, 181, 182, 183, 184, 185, 186, 187, 203,
        204, 205, 210, 225, 236, 261, 302, 335, 341, 352,
        353, 357, 358, 359

financial technology 2, 3, 23, 228, 237, 250, 293, 352, 353,
        354, 356, 358, 369, 370
Fintech 1, 2, 3, 4, 5, 7, 9, 10, 11, 12, 13, 14, 15, 16, 17, 18,
        19, 20, 21, 22, 23, 25, 163, 211, 213, 214, 215,
        216, 217, 218, 219, 220, 221, 222, 223, 224, 225,
        226, 227, 228, 229, 230, 231, 232, 233, 234,
        235, 236, 237, 250, 251, 257, 293, 300, 301, 333,
        335, 336, 337, 339, 340, 341, 342, 359, 360, 378
Fintech ecosystems 213, 214, 215, 216, 218, 219, 220, 221,
        222, 231, 233, 235
firms 12, 14, 17, 18, 20, 55, 65, 105, 106, 107, 113, 169, 174,
        190, 191, 192, 193, 199, 200, 201, 205, 206, 210,
        213, 214, 215, 220, 221, 222, 231, 294, 336, 343,
        347, 348, 353
Food Industry 203, 205, 210
fraud detection 4, 9, 10, 21, 27, 28, 29, 30, 31, 33, 34, 35,
        84, 85, 86, 87, 89, 90, 91, 92, 93, 94, 95, 96, 97,
        99, 101, 103, 105, 106, 107, 109, 110, 111, 113,
        114, 118, 119, 120, 121, 222, 224
fraudulent transactions 29, 32, 34, 96, 105, 106, 107, 108,
        109, 110, 111, 113, 114, 115, 119

GAN 123, 124, 128, 129
GDP 1, 54, 57, 58, 59, 139, 141, 142, 148, 158, 159, 160,
        161, 168, 183, 184, 274, 302, 303, 304, 305,
        310, 311
GDPR 106, 112, 113, 115, 118, 119
generative 123, 124, 133, 134
government 1, 14, 56, 57, 58, 59, 139, 141, 147, 148, 151,
        152, 153, 154, 215, 217, 219, 220, 221, 225, 231,
        232, 234, 238, 270, 274, 281, 302, 303, 304,
        305, 306, 308, 310, 361
green finance 17, 19, 21, 158, 162, 271, 272, 274, 281, 352,
        353, 358, 359

hash 270, 284, 285, 286, 287, 289, 305, 306, 307

ICT 138, 142, 162, 322, 359, 361, 362, 380
India 7, 60, 65, 104, 120, 157, 163, 164, 165, 166, 171, 176,
        177, 178, 223, 224, 233, 235, 237, 240, 302,
        304, 311, 327, 330, 370, 378
innovation 3, 5, 11, 14, 15, 16, 17, 18, 20, 21, 123, 157, 189,
        199, 203, 204, 205, 206, 210, 213, 214, 215, 217,
        218, 219, 220, 221, 222, 223, 224, 228, 231, 233,
        261, 262, 263, 264, 265, 266, 268, 324, 336,
        338, 339, 352, 354, 358, 359, 369, 378
InsurTech 23, 228, 229, 230, 233, 234
integrity 9, 14, 17, 108, 113, 263, 285, 286, 287, 288, 290,
        293, 294, 306, 356
investment 4, 7, 8, 9, 12, 22, 37, 55, 56, 58, 77, 137, 138,
        141, 143, 144, 145, 146, 147, 148, 149, 150, 151,
        152, 153, 154, 155, 156, 157, 158, 159, 167, 168,
        170, 171, 172, 173, 174, 175, 179, 182, 183, 189,
        204, 210, 221, 222, 224, 225, 228, 229, 233,
        235, 236, 237, 251, 276, 279, 280, 295, 299,
        300, 302, 338, 342, 346, 353, 354, 356, 360
Islamic finance 189, 190, 191, 192, 193, 199, 271, 281, 301

Jordan 143, 144, 147, 148, 149, 150, 151, 152, 153, 154, 155

K-means 106, 107, 108, 110, 111, 114, 115, 116, 118, 119
KNN 315, 316, 381, 382, 385, 386, 387
knowledge sharing 17, 19, 213, 217, 218, 221

ledger 2, 4, 10, 228, 262, 263, 297, 305, 306, 308, 311
lender 251, 252, 253, 254, 255, 256
literacy 12, 17, 20, 22, 52, 139, 163, 164, 165, 166, 167, 168, 169, 170, 171,172, 173, 174, 175, 176, 177, 178, 179
logistic regression 78, 87, 88, 255, 257
LSTM 77, 78, 79, 80, 81, 82, 315

M-Wallet 373, 374, 375
ML 2, 4, 5, 9, 11, 17, 19, 21, 85, 92, 97, 99, 105, 106, 107, 110, 111, 112, 113, 114, 115, 119, 315, 381, 382, 384; *see also* machine learning
machine learning 2, 10, 25, 27, 28, 30, 34, 35, 36, 52, 53, 54, 55, 57, 59, 61, 63, 65, 67, 73, 77, 78, 83, 84, 85, 86, 87, 89, 90, 91, 92, 93, 94, 95, 96, 97, 101, 102, 103, 104, 105, 120, 121, 122, 123, 128, 229, 236, 242, 248, 314, 315, 316, 320, 322, 323, 330, 336, 381, 383, 384
macroprudential policy 181, 185, 187
Malaysia 60, 139, 191, 202, 203, 205, 206, 208, 209, 210, 281
market 18, 34, 36, 37, 39, 41, 43, 45, 47, 48, 49, 50, 51, 53, 177, 238, 249, 307, 308, 311, 337, 359, 363, 368, 387
MAS 283, 284, 286, 288, 290, 290, 291
matrix factorization 382, 388
MLP 314, 315, 316, 318, 319, 320, 323
mobile 1, 2, 3, 6, 8, 10, 11, 18, 21, 23, 27, 122, 138, 204, 206, 224, 225, 228, 229, 230, 232, 233, 234, 235, 295, 331, 336, 337, 339, 352, 356, 369, 370, 371, 372, 373, 374, 375, 376, 377, 378, 382, 387
mobile payment 11, 23, 225, 229, 230, 336, 369, 370, 371, 372, 375, 376, 377, 378
monetary policy 54, 59, 181, 183, 185, 210
money 1, 2, 3, 3, 6, 8, 9, 10, 11, 12, 19, 22, 27, 54, 59, 103, 138, 139, 145, 147, 158, 159, 164, 165, 168, 172, 173, 175, 177, 183, 189, 204, 224, 225, 228, 229, 230, 231, 233, 234, 235, 262, 265, 270, 295, 296, 297, 299, 336, 344, 356, 369, 373, 376
Morocco 47, 50, 60, 62, 189, 190, 193, 194, 199, 200, 201, 217, 273, 274, 275, 276, 277, 278, 280, 295, 345, 346, 361, 362, 366, 367, 368, 369, 370, 373, 375, 376, 377, 379, 383

NLP 78, 82, 112

online banking 3, 201, 204, 369
online payments 28, 137, 163, 204, 228, 229, 262, 373
outsourcing 284, 337, 343, 344, 345, 346, 347, 348, 349, 350, 351

P2P 8, 10, 11, 22, 228, 250, 251, 252, 295, 296, 352, 358
payment service 229, 234, 369, 370, 376, 377
performance 32, 33, 34, 36, 37, 43, 46, 51, 53, 55, 58, 69, 73, 77, 78, 80, 81, 82, 85, 91, 107, 109, 110, 114, 123, 128, 137, 144, 162, 165, 173, 174, 182, 203, 204, 205, 206, 207, 208, 209, 210, 222, 231, 239, 240, 249, 262, 269, 270, 273, 286, 314, 315, 316, 317, 318, 320, 321, 324, 330, 335, 336, 338, 340, 341, 342, 343, 344, 345, 348, 349, 351, 358, 359, 381, 384, 386
PLS 189, 190, 191, 192, 193, 194, 195, 196, 198, 199, 200
prediction 2, 28, 34, 36, 37, 39, 43, 46, 51, 52, 53, 73, 76, 77, 78, 79, 80, 81, 83, 86, 87, 93, 103, 106, 112, 120, 122, 314, 316, 318, 319, 320, 321, 381, 382, 384, 387

privacy 5, 9, 10, 11, 12, 13, 17, 18, 20, 21, 22, 94, 105, 106, 107, 108, 109, 110, 111, 112, 113, 114, 115, 119, 122, 262, 299, 301, 306, 341, 370
products 1, 5, 7, 8, 11, 12, 14, 16, 18, 19, 20, 27, 34, 45, 46, 65, 70, 84, 137, 138, 140, 156, 159, 162, 163, 164, 166, 167, 168, 169, 170, 171, 172, 173, 174, 189, 190, 191, 192, 193, 195, 198, 199, 200, 201, 202, 203, 204, 205, 206, 207, 208, 209, 210, 213, 214, 220, 225, 229, 231, 232, 233, 234, 237, 262, 263, 265, 266, 267, 268, 269, 273, 303, 309, 335, 336, 337, 339, 366, 369, 376, 380, 382, 384

qualitative 128, 164, 170, 216, 221, 222, 226, 304, 352
quantitative 90, 128, 164, 183, 191, 194, 206, 222, 226, 235

random forest 28, 29, 89, 90, 93, 98, 99, 104, 108, 110, 111, 314, 323, 381, 387
real estate 8, 22, 186, 287, 293, 294, 295, 297, 298, 299, 300, 301, 364, 366
Regtech 9, 10, 11, 23, 229, 230
regulations 3, 5, 13,17, 20, 143, 154, 181, 185, 204, 233, 234, 237, 276, 303, 341, 352, 359, 361, 363, 366, 368
resilience 13, 19, 58, 181, 182, 185, 186, 187, 231, 232, 335, 336, 337, 338, 339, 340, 341, 342
risk 4, 6 , 8, 9, 10, 11, 12, 13, 17, 23, 34, 43, 55, 58, 67, 84, 103, 105, 109, 115, 119, 121, 145, 157, 162, 164, 165, 166, 168, 170, 171, 172, 173, 175, 179, 181, 182, 183, 184, 185, 186, 187, 189, 191, 192, 195, 198, 199, 204, 205, 219, 224, 229, 230, 237, 238, 249, 251, 259, 263, 276, 313, 337, 338, 339, 342, 353, 356, 358, 367, 381, 382, 387

security 5, 6, 9, 10, 11, 13, 14, 17, 18, 20, 22, 29, 107, 111, 112, 114, 119, 146, 147, 148, 164, 206, 262, 263, 269, 285, 286, 293, 296, 297, 301, 302, 306, 310, 313, 314, 315, 316, 322, 341, 345, 346, 348, 349, 351, 357, 361, 362, 370, 372, 373, 377, 378
sentiment analysis 28, 82, 238, 248
smart city/smart cities 324, 325, 326, 327, 328, 330, 331, 379, 387
smart contracts 6, 10, 23, 262, 263, 264, 269, 296, 298, 299, 300, 302, 303
SMEs 10, 11, 23, 191, 194, 201, 203, 204, 205, 206, 207, 208, 209, 210, 225, 234, 235, 270, 341
stakeholders 9, 11, 12, 14, 15, 17, 18, 19, 20, 215, 231, 232, 233, 235, 275, 297, 302, 304, 337, 340
startups 3, 4, 8, 224, 231, 232, 233, 234, 336
stock market 2, 36, 37, 38, 46, 48, 50, 51, 52, 53, 66, 67, 68, 73, 76, 77, 78, 82, 83, 121, 139, 141, 157, 159, 161, 170, 172, 178, 222, 238, 239, 240, 245, 246, 248, 249
stock price 37, 52, 53, 77, 78, 79, 80, 82, 83, 239
supervised learning 29, 43, 93, 106, 108, 110, 111, 114, 121, 323
supply chain 2, 5, 210, 261, 262, 263, 264, 265, 266, 267, 268, 269, 270, 305, 309, 335, 336, 337, 338, 339, 340, 341, 342, 357
SupTech 9, 10, 11, 21
sustainability 19, 206, 261, 262, 263, 264, 265, 266, 268, 273, 274, 276, 278, 279, 280, 303, 330, 335, 338, 352, 353, 354, 356, 357, 359, 360, 380

sustainable 17, 18, 19, 23, 53, 149, 157, 158, 222, 224, 264,
        266, 269, 270, 273, 274, 275, 276, 277, 278, 279,
        278, 279, 280, 281, 288, 304, 328, 342, 352,
        353, 354, 355, 356, 357, 358, 359, 360, 379, 380
SVM 36, 37, 38, 44, 48, 50, 51, 53, 67, 70, 72, 73, 76, 89,
        104, 105, 106, 107, 108, 109, 110, 111, 114, 115,
        116, 118, 119, 120, 121, 315, 316, 323, 381, 387

tax 55, 143, 144, 145, 146, 147, 148, 149, 150, 152, 153,
        154, 155, 168, 293, 302, 303, 304, 305, 308,
        309, 310, 311, 312
tax policy 143, 144, 145, 146, 147, 148, 149, 150, 155
trading 4, 5, 10, 23, 44, 58, 66, 67, 69, 73, 76, 78, 173, 174,
        179, 189, 204, 228, 229, 230, 231, 249, 264, 265,
        269, 302, 303, 336, 337
transactions 3, 4, 5, 6, 7, 10, 11, 13, 28, 29, 30, 32, 34, 55,
        78, 84, 85, 86, 88, 93, 94, 95, 96, 97, 103, 105,
        106, 107, 108, 109, 110, 111, 113, 114, 115, 119,
        120, 121, 137, 138, 156, 157, 162, 168, 190, 204,
        205, 225, 228, 229, 251, 262, 263, 266, 267,
        269, 286, 287, 295, 296, 297, 298, 299, 300,
        301, 302, 303, 305, 306, 309, 310, 311, 336, 339,
        340, 341, 343, 347, 348, 349, 350, 351, 357, 366,
        368, 369, 370, 371, 373, 374, 375, 376, 377

transparency 3, 5, 6, 9, 10, 11, 17, 19, 200, 225, 262,
        263, 266, 268, 269, 296, 297, 301, 302,
        303, 305, 306, 308, 310, 311, 340, 356, 357,
        358, 363
trends 2, 4, 9, 14, 21, 51, 52, 67, 78, 83, 133, 174, 175, 179,
        219, 221, 222, 234, 236, 274, 304, 324, 327,
        328, 344
tweets 238, 239, 240, 241, 242, 245, 246,
        247, 248
twitter 238, 240, 241, 248, 249

unsupervised learning 29, 43, 106, 107, 108, 110, 114,
        118, 120

VAT 303, 309, 310, 312
VGGNet 70
Vietnam 139, 156, 158, 159, 161, 162,
        236, 239

wallets 6, 7, 10, 19, 22, 111, 112, 163,
        228, 232

XGBoost 28, 29, 30, 33, 34,
        107, 111